Mathematics and Quantitative Methods
for Business and Economics

Stephen P. Shao

**Chairman, Department of Quantitative Sciences
    in Business and Economics**
Old Dominion University

Published by

M35 **SOUTH-WESTERN PUBLISHING CO.**

CINCINNATI   WEST CHICAGO, ILL.   DALLAS   PELHAM MANOR, N.Y.
PALO ALTO, CALIF.   BRIGHTON, ENGLAND

ISBN: 0-538-13350-3

Library of Congress Catalog Card Number: 74-19541

1 2 3 4 5 6 D 1 0 9 8 7 6

Printed in the United States of America

Preface

The primary purpose of this book is to provide essential mathematical background for students of business administration and economics. It is also intended to introduce the basic topics in quantitative methods. The techniques used in quantitative methods are basically mathematical operations. They provide powerful tools in analyzing problems for business decision making and economic research.

This book is divided into four parts. Part 1 presents basic algebra. It begins with very elementary algebraic concepts, then moves on to cover sufficient material for performing fundamental algebraic operations. Part 2 presents matrix algebra and linear programming. Linear programming is an important topic in the area of quantitative methods (or operations research). Matrix algebra is helpful in understanding the procedure used in solving a linear programming problem by the simplex method. However, if time is not permitted, the two chapters concerning matrices can be omitted without destroying the organization of the part. The three linear programming chapters (8, 9, and 10) have been written as an independent unit.

Part 3 presents mathematics for managerial decisions. Applications of probability, Bayes theorem, expected values, decision rules, and mathematics of finance are emphasized in this part. Part 4 presents calculus and related topics. It covers sequences, series, logarithms, differential calculus, and integral calculus.

The material presented in this book is sufficient for various types of courses. Material that is starred (*) in the table of contents is intended to be optional. The omission of these starred chapters and sections will not interrupt the continuity of the text organization.

The instructor's manual contains details concerning the assignments for these courses and provides detailed solutions to the problems in each chapter. The solutions to the odd-numbered problems are placed at the end of the book. The problems in each exercise are carefully arranged so that either all odd-numbered or all even-numbered problems can be assigned by the instructor without fear of omitting the material that has been illustrated in the examples in the book.

The author is indebted to his colleagues of the School of Business Administration of Old Dominion University for their suggestions made from reading and teaching the manuscript in mimeographed form.

Stephen Pinyee Shao

Contents

Part 3 MATHEMATICS FOR MANAGERIAL DECISIONS

Part 4 CALCULUS AND RELATED TOPICS

Part One Basic Algebra

Chapter
1 Sets

The subject of sets is basic in studying modern mathematics. In recent years, mathematical terms are being expressed more and more in the language of sets. This chapter introduces the concept and the fundamental operations of sets.

1.1 THE CONCEPT OF SETS

The idea of a *set* is generally used in everyday life to represent a group of things having some common property, such as a set of textbooks, dishes, or screwdrivers. Similarly, a set in mathematics is used to represent a well-defined collection of objects, numbers, or symbols.

A member of a set is called an *element* of the set. Each element of a set must be well defined. Thus, we should have no doubt in determining whether or not a given object is an element of the set. For example, let A represent the set of the integers greater than 3 but less than 7. The elements of set A are 4, 5, and 6.

The symbol representing "is an element of" is ϵ. Thus, the expression "4 is an element of set A" may be written symbolically:

$$4 \in A$$

The symbol representing "is not an element of" is $\notin$. Thus, the expression "8 is not an element of set A" may be written symbolically:

$$8 \notin A$$

A set may be *finite* or *infinite*. A set is finite when the number of elements in the set is limited, or the number of elements can be expressed by a positive number. Thus, set A, which has three elements (4, 5, 6) in the above illustration, is a finite set. A set of ten million elements is also a finite set. A set is infinite when the number of elements in the set is unlimited.

EXAMPLE 1 Finite sets:

1. The members of a family (such as 3 members, father, mother, and son).

1

2. The students in an English class (such as 25 students).
3. The integers from 1 to 3, inclusive (or 1, 2, and 3).
4. The odd numbers larger than 5 but smaller than 13 (or 7, 9, and 11).

EXAMPLE 2   Infinite sets:

1. The years in the future (unlimited years).
2. The trials of drawing a white ball from a bag of two white and three black balls; the ball is replaced in the bag after each drawing (unlimited trials).
3. The integers above 3.
4. The numbers between 5 and 8 (or 5.01, 5.001, 5.002, . . .).

A set may contain no elements. A set that has no elements is called an *empty* set or a *null* set, denoted by the symbol ϕ. The role of the empty set in the algebra of sets is similar to that of zero in the operation of the decimal number system.

EXAMPLE 3   Empty sets:

1. The rats in New York City weighing more than five tons.
2. The professors 2,000 years old still teaching.
3. The $3 bills of U.S. currency in circulation during the year 1976.
4. The even integers above 6 but below 8.

1.2 REPRESENTATIONS OF SETS AND ELEMENTS

Letters are frequently used to represent sets and elements. The general practice is to use capital letters for sets and lowercase letters for elements of a set. Thus, we may let letter A represent the set of the heights of the five students in a class and letter x represent the height of a student in the set.

There are two basic methods of specifying the elements of a set. In either case, the elements of a set are written within braces.

(a) The *list* or *roster* method. The names of all elements of a set are listed in braces. This method is convenient when the number of elements of a set is small.

(b) The *descriptive* or *defining* method. A rule or condition by which all elements of the set can be determined is described within braces. This method is convenient when the number of elements of a set is large.

EXAMPLE 4   Let 65, 68, 70, 74, 76 be set A representing the heights in inches of the five students in a class. The set may be specified in two ways:

1. By the list method.

   $A = \{65, 68, 70, 74, 76\}$

2. By the descriptive method.

   $A = \{x | x$ is the height in inches of a student in the class$\}$

The vertical bar (|) or a colon(:) means "such that." Thus, the preceding expression is read "A is the set of those elements x such that x is the height of a student in the class."

EXAMPLE 5   The integers greater than 5 but less than 10,000 are expressed as set B by the descriptive method which follows. The list method is obviously not practical in this case.

$$B = \{x|x \text{ is an integer and } 5 < x < 10{,}000\}$$

The above expression may be read "B is the set of all elements x such that x is an integer and x is greater than 5 but less than 10,000."

1.3   UNIVERSAL SETS AND SUBSETS

A *universal set*, usually denoted by the capital letter U, is the total collection of all elements under consideration in a given problem. The elements of a universal set may form many subsets. A *subset* is defined as follows:

If every element of set A is also an element of set U, then set A is a subset of set U.

The relationship between a universal set and its subsets may clearly be shown by a *Venn diagram*, named in honor of the English logician John Venn (1834–1923).

EXAMPLE 6   Illustration of a Venn diagram:

Let U = the set of all employees in a company,
 A = the set of all employees in the Accounting Department of the company, and
 B = the set of all employees in the Purchasing Department of the company.

The Venn diagram appears as follows:

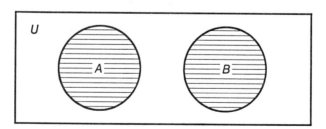

The total area of the rectangle represents the set of all employees in the company, or the universal set. Since every element of set A (an employee in the Accounting Department) is also an element of set U (an employee in the company), set A is a subset of the universal set U, or written symbolically:

$$A \subseteq U$$

The symbol $\subseteq$ represents "is a subset of." Similarly, set B is a subset of U, written:

$$B \subseteq U$$

EXAMPLE 7   Draw a Venn diagram to show the universal set U and the subsets A and B.

$U = \{1, 2, 3, 4, 5, 6, 7, 8\}$
$A = \{1, 2, 3, 4\}$
$B = \{3, 4, 5, 6\}$

The Venn diagram appears as follows:

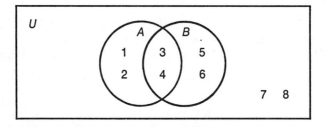

A subset is called a *proper subset* if the number of elements in the subset is *at least one less* than the number of elements in the given set from which the subset is formed. The symbol used to represent "a proper subset of" is $\subset$. Thus, $C \subset D$ represents "C is a proper subset of D." $C \subset D$ can also be written in a reversed order so long as the symbol $\subset$ is also reversed: $D \supset C$. Here every element of set C is also an element of set D, or C is a proper subset of D. In addition, set D contains at least one element not in C. The relationship between proper subset C and set D may be diagrammed as follows:

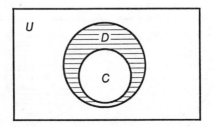

According to the definition of a subset, we may state that a set is always a subset of itself. Thus, U is a subset of U. However, U is not a proper subset of U. Also, the empty set is a subset of every set, but not a proper subset of itself.

The symbols commonly used in set expressions are summarized on page 5. (The last three symbols, $\cup$, $\cap$, and ', are used in the set operations discussed in Section 1.4.)

Symbol	Meaning
ϵ	*is an element of* or *belongs to* (Thus, "object x is an element of set A" is written $x \in A$. ϵ is the Greek letter epsilon.)
$\notin$	*is not an element of* or *does not belong to* (Thus, "object y is not an element of set A" is written $y \notin A$.)
ϕ	*an empty set* or *a null set* (Thus, "A is an empty set" is written $A = \phi$. ϕ is the Greek letter phi, pronounced fī.)
$\subseteq$	*is a subset of* (Thus, "A is a subset of B" is written $A \subseteq B$.)
$\subset$	*is a proper subset of* (Thus, "A is a proper subset of B" is written $A \subset B$, or $B \supset A$.)
$\cup$	*union* (Thus, "the union of sets A and B" is written $A \cup B$.)
$\cap$	*intersection* (Thus, "the intersection of sets A and B" is written $A \cap B$.)
$'$	*the complement of* (Thus, "the complement of set A" is written A'; also may be written $\bar{A}$, $\sim A$, $\tilde{A}$.)

EXERCISE 1–1   Reference: Sections 1.1–1.3

1. Indicate whether each of the following sets is finite, infinite, or empty:
   (a) The students enrolled in this course.
   (b) The points on a line.
   (c) The chickens having four legs.
   (d) The integers above 5 but below 100.
   (e) The letters in the word "development."

2. State whether each of the following collections is a finite set, an infinite set, or an empty set:
   (a) The tables in a restaurant.
   (b) The integers above 30.
   (c) The even integers between 18 and 20.
   (d) The numbers between 6 and 12.
   (e) The numbers satisfying the equation $x + y = 10$.

3. Draw a Venn diagram to show the universal set U and the subsets A and B:
   $$U = \{x \mid x \text{ is an integer and } 3 < x < 14\}$$
   $$A = \{6, 7, 9, 10, 13\}$$
   $$B = \{4, 5, 11\}$$

4. Draw a Venn diagram to show the universal set U and the subsets A and B:
   $$U = \{a, b, c, d, e, f, g, h, i, j, k\}$$
   $$A = \{a, c, e, g, h, i\}$$
   $$B = \{a, b, d, e, f\}$$

5. Express each of the sets given below by both the list method and the descriptive method:
   (a) The first eight letters of the alphabet.
   (b) The integers from 3 to 9, inclusive.
   (c) Partners Carson, Estes, Frenton, and Nixton of Tex Oil Company.

(d) English professors Begg, Griffin, Smith, Meyers, and Taby of King College.

6. Use either the list method or the descriptive method to specify each set given below. Explain your reason for selecting the method in each case.
    (a) The integers above 25.
    (b) The teachers whose names are Bater, Doming, Gibson, Holten, and Smith.

7. Write each of the following expressions by using symbols only:
    (a) Object b is an element of set S.
    (b) Object h is not an element of set S.
    (c) Q is an empty set.
    (d) P is a subset of G.
    (e) H is a proper subset of M.

8. Refer to Problem 4. Indicate whether each statement below is true or false.
    (a) $A \subseteq U$ (b) $B \subseteq U$ (c) $B \subseteq A$ (d) $A \in U$
    (e) $b \in U$ (f) $m \notin U$ (g) $n \notin A$ (h) $g \in A$
    (i) $C \subseteq U$ and $C = \phi$ (j) $D \subseteq U$ and $D = \{l, m, n, o, p\}$

1.4   BASIC OPERATIONS WITH SETS

The basic operations with sets are union, intersection, and complement. In many cases the union operation with sets resembles the addition operation in ordinary algebra and the intersection operation resembles multiplication. The three operations with sets are illustrated below with the aid of Venn diagrams. It is assumed within the illustrations that A and B are two subsets of the universal set U, and x represents an element of U. Operations with sets are frequently referred to as Boolean algebra in honor of George Boole, an English logician (1815–1864).

A. Union

The *union* of sets A and B is denoted by the expression $A \cup B$ (read "A union B," or "A cup B"). $A \cup B$ contains all elements of U that belong to A or to B or to both. The set $A \cup B$ is also called the sum of sets A and B and sometimes is written $A + B$. The sum is indicated by the shaded areas in each of the cases shown by the Venn diagrams on page 7.

Observe in the diagrams that every element of A and every element of B belongs to $A \cup B$. Symbolically, the definition of the union is:

$$A \cup B = \{x \in U \mid x \in A \text{ or } x \in B\}$$

The expression is read: "A union B is the set of those elements x of U such that x is an element of either A or B or both."

EXAMPLE 8   Let all six salespeople in a company be the universal set:

$$U = \{\text{Dent, Estern, Fort, Green, Hanston, Iron}\}$$

Case a: Case b: Case c:

$A \cup B$ $A \cup B = A$ $A \cup B$

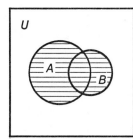   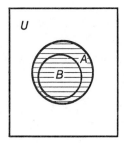   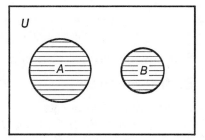

Also, let the three salespeople who come from Virginia be set A:

$$A = \{\text{Dent, Estern, Fort}\}$$

Let the four salespeople who have college educations be set B:

$$B = \{\text{Estern, Fort, Green, Hanston}\}$$

Then:

$$A \cup B = \{\text{Dent, Estern, Fort, Green, Hanston}\}$$

$A \cup B$ represents the set of five salespeople who are either from Virginia or college educated. (See Case a in the above diagram.)

Note that in Example 8 salespeople Estern and Fort are not listed twice in the set $A \cup B$. Note also that the change of the order of the elements listed in the braces does not change the property of the set. Thus, without changing the meaning the sum may be written:

$$A \cup B = \{\text{Estern, Green, Fort, Hanston, Dent}\}$$

In general, two given sets A and B are said to be *equal* if and only if every element of A is an element of B and every element of B is also an element of A. Thus, let the given sets be:

$$A = \{\text{Steve, Dale, Lawrence, Alan}\}$$
$$B = \{\text{Dale, Alan, Steve, Lawrence}\}$$

representing four boys in a family and a club respectively. Then, the two sets are equal, or $A = B$, since sets A and B contain identical elements.

EXAMPLE 9   Let $U = \{a, b, c, d, e, f\}$
$A = \{a, b, c, d\}$
$B = \{a, b, c\}$

Then, $A \cup B = \{a, b, c, d\} = A$.
(See Case b in the above diagram.)

EXAMPLE 10   Let $U = \{x \mid x$ is an integer and $4 < x < 25\}$
$A = \{5, 6, 8, 17, 22\}$
$B = \{7, 10, 15\}$

Then, $A \cup B = \{5, 6, 7, 8, 10, 15, 17, 22\}$
(See Case c in the diagram on page 7.)

Similarly, for any set A we may state that $A \cup \phi = A$.

B. Intersection

The *intersection* of sets A and B is denoted by the expression $A \cap B$ (read "A intersection B," or "A cap B"). $A \cap B$ contains all elements of U that belong to both A and B. The set $A \cap B$ is also called the *product* of sets A and B and sometimes is written $A \cdot B$. The product is indicated by the shaded areas in each of the cases as shown by the following Venn diagrams:

Case a: Case b: Case c:
$A \cap B$ $A \cap B = B$ $A \cap B = \phi$

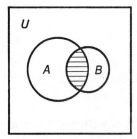 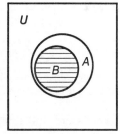 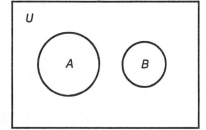

Observe in the diagrams that every element which belongs to A and to B belongs to $A \cap B$. In Case c, set $A \cap B$ is an empty set since there are no elements common to A and B. A and B in Case c are thus called *disjoint* or *mutually exclusive* sets.

Symbolically, the definition of the intersection is:

$$A \cap B = \{x \in U \mid x \in A \text{ and } x \in B\}$$

The expression is read: "A intersection B is the set of those elements x of U such that x is an element of both A and B."

EXAMPLE 11   Refer to Example 8. Find the set $A \cap B$.

$A \cap B = \{$Estern, Fort$\}$

Also, see Case a in the above diagram. $A \cap B$ represents the set of the two salespeople who are from Virginia and college educated.

EXAMPLE 12   Refer to Example 9. Find the set $A \cap B$.

$A \cap B = \{a, b, c\} = B$

Also, see Case b in the above diagram.

EXAMPLE 13   Refer to Example 10. Find the set $A \cap B$.

$A \cap B = \phi$ (the empty set)

Also, see Case c in the preceding diagram. A and B are disjoint since their intersection is the empty set. Notice that each of the elements of set B (7, 10, 15) is not an element of set A.

By definition, for any set A we may state that $A \cap \phi = \phi$.

C. Complement

The *complement* of set A is denoted by the set A' (read A prime; also may be written $\bar{A}$, $\sim A$, or $\tilde{A}$), which contains all elements in U that do not belong to A. Set A' is indicated by the shaded area in the following Venn diagram:

Complement of A ($= A'$)

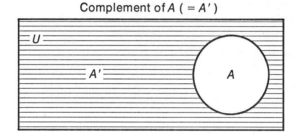

Symbolically, the definition of the complement is:

$$A' = \{x \in U : x \notin A\}$$

The expression is read: "the complement of A is the set of those elements x of U such that x is not an element of A."

EXAMPLE 14   Refer to Example 8. Find set A' and set B'.

$A' = \{$Green, Hanston, Iron$\}$

The set A' represents all salespeople in the company who are not from Virginia.

$B' = \{$Dent, Iron$\}$

The set B' represents all salespeople in the company who are not college educated.

EXAMPLE 15   Refer to Example 9. Find set A'.

$A' = \{e, f\}$

EXERCISE 1–2   Reference: Section 1.4

1. Let the board of directors of SPS Company be the universal set U. The directors are: Anderson, Darr, Bell, Griffin, Kello, and Meyers. Let the

directors Anderson, Darr, and Kello, who are golf players, be set A. Also, let the directors Bell, Darr, Griffin, and Kello, who are officers of the company, be set B. Express the sets representing the groups of directors who are:

(a) either golf players or officers of the company.

(b) golf players and officers.

2. Refer to Problem 1. Express the sets representing the groups of directors who are:

(a) not golf players.

(b) not officers of the company.

3. Let $U = \{1, 2, 3, 4, 5, 6, 7, 8\}$
$$A = \{2, 3, 5, 7\}$$
$$B = \{1, 3, 4, 7, 8\}$$

Draw a Venn diagram to show the universal set U and the subsets A and B. Find sets:

(a) $A \cup B$ (b) $A \cap B$ (c) $A' \cup B$ (d) $A' \cap B'$

4. From the information given in Problem 3, find sets:

(a) $U \cup A$ (b) $B \cap U$ (c) $B' \cap A$ (d) $A' \cup B'$

5. Draw a Venn diagram to show the universal set U and the subsets A and B:
$$U = \{x \mid x \text{ is an integer and } 3 < x < 14\}$$
$$A = \{6, 7, 9, 10, 13\}$$
$$B = \{4, 5, 11\}$$

Find sets: (a) $A \cup B$ (b) $A \cap B$ (c) $A \cup B'$ (d) $A \cap B'$

6. From the information given in Problem 5, find sets:

(a) $A \cup U$ (b) $U \cap B$ (c) $B \cup A'$ (d) $A' \cap B'$

7. Draw a Venn diagram to show the universal set U and the subsets A and B:
$$U = \{a, b, c, d, e, f, g, h, i, j, k\}$$
$$A = \{a, c, e, g, h, i\}$$
$$B = \{a, b, d, e, f\}$$

Find sets: (a) $A \cup B$ (b) $A \cap B$ (c) A' (d) $A \cup A'$

8. From the information given in Problem 7, find sets:

(a) $A \cup U$ (b) $A \cap U$ (c) B' (d) $A \cap A'$

1.5 OTHER OPERATIONS WITH SETS

In this section, we shall cover two additional types of operations: operations of difference (subtraction) and operations with three or more sets.

A. Operations of Difference (or Subtraction)

Let A and B be two subsets of the universal set U. The difference of sets A and B is denoted by the set $A - B$ (read "A minus B"). The difference, or the remainder of subtracting set B from set A, is obtained by taking all elements that belong to B away from A. The set $A - B$ is indicated by the shaded areas in each of the following Venn diagrams:

Case a   Case b   Case c   Case d
                                                    $A - B = A$ $A - B = \phi$

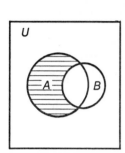   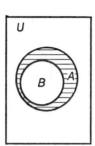   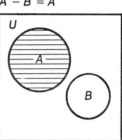   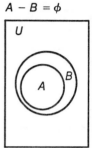

In the diagrams, every element that belongs to $A - B$ also belongs to both A and B', or:

$$A - B = A \cap B'$$

Since the subtraction operation $(A - B)$ may be replaced by the combined operations of intersection and complement $(A \cap B')$, subtraction is generally not regarded as a basic operation in the algebra of sets.

EXAMPLE 16   Let $U = \{1, 2, 3, 4, 5, 6, 7, 8\}$
                $A = \{1, 3, 5, 6\}$
                $B = \{2, 3, 4, 5\}$
         Find $A - B$ and $B - A$

   First, draw a Venn diagram.

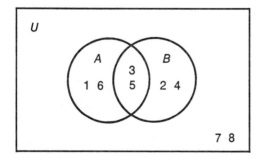

$A - B = \{1, 6\}$, or
$A - B = A \cap B' = \{1, 6\}$,
where $B' = \{1, 6, 7, 8\}$
$B - A = \{2, 4\}$, or
$B - A = B \cap A' = \{2, 4\}$,
where $A' = \{2, 4, 7, 8\}$

B. Operations with Three or More Sets

The methods used in the previous discussion of set operations may be expanded for three or more sets or subsets. Let A, B, and C be three subsets of the universal set U. In the following Venn diagram, we find:

1. Set $A \cap B \cap C'$ 2. Set $A \cap B \cap C$

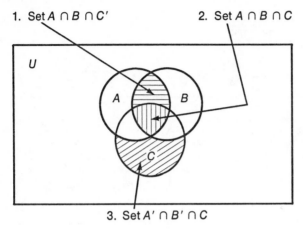

3. Set $A' \cap B' \cap C$

1. The shaded area with horizontal lines represents set $A \cap B \cap C'$.

2. The shaded area with vertical lines represents set $A \cap B \cap C$.

3. The shaded area with diagonal lines represents set $A' \cap B' \cap C$.

In the above three illustrations, only the intersection operations $\cap$ are used in each expression. If there are mixed operations in a given expression, parentheses should be utilized to show the order of operations. If parentheses are not used in the expression, the order of sets in an arrangement is interpreted as the order of performing the operations. Observe that in Example 17 $[(A \cap B) \cup C]$ of part 4 and $[A \cap (B \cup C)]$ of part 5 are different sets.

EXAMPLE 17   Let $U = \{1, 2, 3, 4, 5, 6, 7, 8, 9, 10. 11\}$
$A = \{1, 2, 3, 4\}$
$B = \{2, 4, 5, 6\}$
$C = \{3, 4, 6, 7, 8\}$

Find sets: 1. $A \cap B \cap C'$ 4. $(A \cap B) \cup C$
2. $A \cap B \cap C$ 5. $A \cap (B \cup C)$
3. $A' \cap B' \cap C$ 6. $A \cup (B \cap C)$

First, draw a Venn diagram. The elements of each set are written according to the given sets.

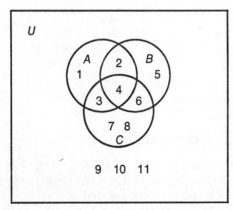

1. $A \cap B \cap C' = \{2\}$, where $C' = \{1, 2, 5, 9, 10, 11\}$

2. $A \cap B \cap C = \{4\}$

3. $A' \cap B' \cap C = \{7, 8\}$, where $A' = \{5, 6, 7, 8, 9, 10, 11\}$, and $B' = \{1, 3, 7, 8, 9, 10, 11\}$

4. $(A \cap B) \cup C = \{2, 3, 4, 6, 7, 8\}$

5. $A \cap (B \cup C) = \{2, 3, 4\}$

6. $A \cup (B \cap C) = \{1, 2, 3, 4, 6\}$

* 1.6 LAWS OF SET OPERATIONS

Operations with sets may be performed directly according to laws of set operations. Venn diagrams are useful in verifying the laws. Many of the laws of union and intersection in set operations are similar to the laws of addition and multiplication in ordinary algebra. As stated previously, the addition sign (+) may replace the cup ($\cup$) for union and the multiplication sign ($\cdot$) may replace the cap ($\cap$) for intersection in set operations. However, we must be cautious in using the addition and multiplication signs in set operations since not all the set operations are similar to the operations of ordinary algebra.

Table 1–1 shows the laws of set operations and their equivalent laws of the operations of ordinary algebra. Only the most important laws of set operations are listed in column 1 of the table. The equivalent expression of each law used in ordinary algebra is listed in column 2. The signs $\cup$, $\cap$, and ϕ (empty set) used in column 1 are replaced by signs $+$, $\cdot$, and 0 (zero) respectively for the expressions in column 2. In column 1, A, B, and C are subsets of the universal set U. However, letters A, B, C, and U in column 2 represent numbers, not sets. The comparison of similarities and dissimilarities between the laws of set operations and the equivalent laws of ordinary algebra is given in column 3. For example, the expression $A \cup A = A$ (or $A + A = A$) for the union operation on sets is dissimilar to the expression $A + A = 2A$ for the addition operation in ordinary algebra.

The applications of the laws of set operations are illustrated in Example 18.

EXAMPLE 18 Let $U = \{1, 2, 3, 4, 5, 6, 7, 8, 9\}$
$A = \{1, 2, 3, 4\}$
$B = \{2, 4, 5, 6\}$
$C = \{3, 4, 6, 7, 8\}$
Find sets:
1. $A \cap (B \cup C)$ 3. $(A \cup B)'$
2. $A \cup (B \cap C)$ 4. $(A \cap B)'$

The elements of the given sets are shown in the following Venn diagram:

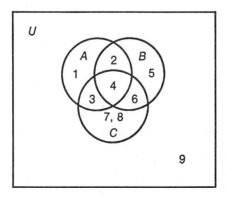

1. Method 1 – By observing the Venn diagram:
$A \cap (B \cup C)$
$= \{1, 2, 3, 4\} \cap$
$\{2, 3, 4, 5, 6, 7, 8\}$
$= \{2, 3, 4\}$

Method 2 – By applying law IV-1 from Table 1–1:

$A \cap (B \cup C)$
$= (A \cap B) \cup (A \cap C)$
$= \{2, 4\} \cup \{3, 4\}$
$= \{2, 3, 4\}$

2. Method 1 — By observing the Venn diagram:
$A \cup (B \cap C) = \{1, 2, 3, 4\} \cup \{4, 6\} = \{1, 2, 3, 4, 6\}$
Method 2 — By applying law IV-2 from Table 1–1:
$A \cup (B \cap C) = (A \cup B) \cap (A \cup C)$
$= \{1, 2, 3, 4, 5, 6\} \cap \{1, 2, 3, 4, 6, 7, 8\}$
$= \{1, 2, 3, 4, 6\}$

3. Method 1 — By observing the Venn diagram:
$(A \cup B)' = $ complement of set $A \cup B = \{7, 8, 9\}$

Table 1–1 Laws of Set Operations and Equivalent Laws of the Operations in Ordinary Algebra

(1) Laws of Set Operations	(2) Laws Equivalent in Ordinary Algebra	(3) Similarity Between (1) and (2)
I. Laws of Union		
*1. $A \cup B = B \cup A$	$A + B = B + A$	yes
**2. $A \cup (B \cup C) =$	$A + (B + C) =$	
$(A \cup B) \cup C$	$(A + B) + C$	yes
3. $A \cup \phi = A$	$A + 0 = A$	yes
4. $A \cup U = U$	$A + U \neq U$, if $A \neq 0$	no
5. $A \cup A = A$	$A + A = 2A$	no
II. Laws of Intersection		
*1. $A \cap B = B \cap A$	$A \cdot B = B \cdot A$	yes
**2. $A \cap (B \cap C) =$	$A \cdot (B \cdot C) =$	yes
$(A \cap B) \cap C$	$(A \cdot B) \cdot C$	
3. $A \cap \phi = \phi$	$A \cdot 0 = 0$	yes
4. $A \cap U = A$	$A \cdot U \neq A$, if $U \neq 1$	no
5. $A \cap A = A$	$A \cdot A = A^2$	no
III. Laws of Complement		
1. $(A')' = A$ $(A')'$ represents the complement of A'.		
2. $U' = \phi$		
3. $\phi' = U$		
IV. Laws of Combined Operations		
#1. $A \cap (B \cup C) =$	$A \cdot (B + C) =$	yes
$(A \cap B) \cup (A \cap C)$	$(A \cdot B) + (A \cdot C)$	
#2. $A \cup (B \cap C) =$	$A + (B \cdot C) \neq$	
$(A \cup B) \cap (A \cup C)$	$(A + B) \cdot (A + C)$	no
3. $A \cup A' = U$		
4. $A \cap A' = \phi$		
##5. $(A \cup B)' = A' \cap B'$		
##6. $(A \cap B)' = A' \cup B'$		
7. $A - B = A \cap B'$		
8. $A - U = \phi$		
9. $U - A = A'$		

  * I-1 and II-1 are called commutative laws.
 ** I-2 and II-2 are called associative laws.
  \# IV-1 and IV-2 are called distributive laws.
 \#\# IV-5 and IV-6 are called De Morgan's laws.

Method 2 – By applying law IV-5 from Table 1-1:
$(A \cup B)' = A' \cap B' = \{5, 6, 7, 8, 9\} \cap \{1, 3, 7, 8, 9\} = \{7, 8, 9\}$

4. Method 1 – By observing the Venn diagram:
$(A \cap B)' = $ complement of set $A \cap B = \{1, 3, 5, 6, 7, 8, 9\}$
Method 2 – By applying law IV-6 from Table 1-1:
$(A \cap B)' = A' \cup B' = \{5, 6, 7, 8, 9\} \cup \{1, 3, 7, 8, 9\} = $
$\{1, 3, 5, 6, 7, 8, 9\}$

The laws of set operations can also be used to simplify or expand set expressions.

EXAMPLE 19 Simplify $A \cup (A \cup B)$.

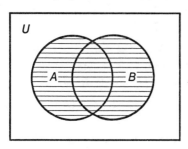

$$A \cup (A \cup B) = (A \cup A) \cup B \quad \text{Law I-2}$$
$$= A \cup B \quad\quad\quad \text{Law I-5}$$

EXAMPLE 20 Expand $(A \cup B) \cap (C \cup D)$. $A, B, C,$ and D are subsets of the universal set U.

Let $P = $ set $(A \cup B)$ for the given expression. Then:

$$(A \cup B) \cap (C \cup D)$$
$$= P \cap (C \cup D)$$
$$= (P \cap C) \cup (P \cap D) \quad\quad\quad\quad\quad\quad\quad \text{Law IV-1}$$
$$= [(A \cup B) \cap C] \cup [(A \cup B) \cap D] \quad\quad \text{Since } P = A \cup B$$
$$= [C \cap (A \cup B)] \cup [D \cap (A \cup B)] \quad\quad\quad \text{Law II-1}$$
$$= [(C \cap A) \cup (C \cap B)] \cup [(D \cap A) \cup (D \cap B)] \quad \text{Law IV-1}$$
$$= (C \cap A) \cup (C \cap B) \cup (D \cap A) \cup (D \cap B)$$

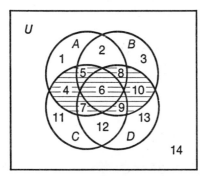

Check: Let:

$U = \{1, 2, 3, 4, 5, 6, 7, 8, 9, 10,$
$\quad\quad 11, 12, 13, 14\}$
$A = \{1, 2, 4, 5, 6, 7, 8\}$
$B = \{2, 3, 5, 6, 8, 9, 10\}$
$C = \{4, 5, 6, 7, 9, 11, 12\}$
$D = \{6, 7, 8, 9, 10, 12, 13\}$

Then, the shaded area in the Venn diagram:

$(A \cup B) \cap (C \cup D)$
$= \{1, 2, 3, 4, 5, 6, 7, 8, 9, 10\} \cap \{4, 5, 6, 7, 8, 9, 10, 11, 12, 13\}$
$= \{4, 5, 6, 7, 8, 9, 10\}$

and:

$$(C \cap A) \cup (C \cap B) \cup (D \cap A) \cup (D \cap B)$$
$$= \{4, 5, 6, 7\} \cup \{5, 6, 9\} \cup \{6, 7, 8\} \cup \{6, 8, 9, 10\}$$
$$= \{4, 5, 6, 7, 8, 9, 10\}$$

Exercise 1–3   Reference: Sections 1.5 and 1.6

1. Draw a Venn diagram to show the universal set U and the subsets A and B:

$$U = \{1, 2, 3, 4, 5, 6, 7, 8\}$$
$$A = \{2, 3, 5, 7\}$$
$$B = \{1, 3, 4, 7, 8\}$$

   Find sets: (a) $A - B$ (b) $B - A$ (c) $A \cap B'$ (d) $B \cap A'$

2. From the information given in Problem 1, find sets:

   (a) $U - A$ (b) $U - B'$ (c) $A' - U$ (d) $B - U$

3. Draw a Venn diagram to show the universal set U and the subsets A and B:

$$U = \{x \mid x \text{ is an integer and } 3 < x < 14\}$$
$$A = \{6, 7, 9, 10, 13\}$$
$$B = \{4, 5, 11\}$$

   Find sets: (a) $A - B$ (b) $A \cap B'$

4. From the information given in Problem 3, find sets:

   (a) $B - A$ (b) $B \cup A'$

5. Let $U = \{1, 2, 3, 4, 5, 6, 7, 8, 9, 10, 11, 12, 13, 14\}$
           $A = \{1, 2, 3, 5, 6\}$
           $B = \{3, 4, 6, 7, 13\}$
           $C = \{5, 6, 7, 8, 9, 10, 13\}$

   List the elements of each of the sets from a Venn diagram:

   (a) $A \cup B$ (e) $A \cap B \cap C$
   (b) $B \cap C$ (f) $(A \cup B) \cap C$
   (c) $A \cap B$ (g) $A' \cap C \cap B'$
   (d) $A - C$ (h) $A \cap B \cap C'$

6. From the information given in Problem 5, list the elements of each of the sets:

   (a) $A \cup C$ (e) $(B \cup C) \cap A$
   (b) $A - B$ (f) $(A \cup C) \cap B$
   (c) $A \cap C \cup B$ (g) $A \cap B' \cap C'$
   (d) $A \cup B \cup C$ (h) $A' \cap B' \cap C'$

7. Let the area inside the circle be set P; inside the square, set Q; and inside the triangle, set R. Shade the area that represents each of the following sets by the type of lines indicated in the parentheses:

(a) $(P \cap Q') \cup R$
(b) $(P - Q) - R$
(c) $(Q - R) \cap P'$
(d) $(R - Q) - P$

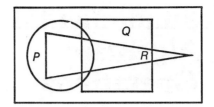

8. Let the area inside the square be set L; inside the circle, set M; and inside the triangle, set N. Shade the area that represents each of the following sets by the type of lines indicated in the parentheses:

(a) $L \cup M \cup N$
(b) $(M \cup N) \cap L$
(c) $(N - M) \cap L'$
(d) $(N \cap M) - L$

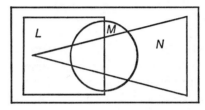

*9. Refer to Problem 5. Use the laws of set operations to expand each of the following sets. Then find the elements of each set:
(a) $C \cap (A \cup B)$ (c) $A \cap (B \cup C)$
(b) $B \cup (A \cap C)$ (d) $C \cup (A \cap B)$

*10. Use the information and instruction given in Problem 9. Find the elements of each set given below:
(a) $(A \cup B)'$ (b) $(A \cap B)'$

*11. Use the laws of set operations to simplify each of the sets given below:
(a) $B \cup (B \cup A)$ (b) $A \cup (A' \cap B)$

*12. Use the laws of set operations to simplify each of the sets given below. (A and B are subsets of the universal set U.)
(a) $A \cap (A \cup B) \cup U$ (b) $A \cup (A \cap B) \cap U$

*13. Expand $(A \cup B) \cap (A' \cup C)$. A, B, and C are subsets of the universal set U. Use the information given in Example 18 to check your answer.

*14. Expand set $(A \cup C) \cap (B \cup D)$. A, B, C, and D are subsets of the universal set U. Use the information given in Example 20 to check your answer.

Chapter

2 Fundamental Algebraic Operations

Algebra deals with numbers and symbols representing the numbers. Section 2.1 introduces the common types of numbers. Because of their importance, real numbers are covered in some detail in Section 2.2. The fundamental algebraic operations of addition, subtraction, multiplication, and division are illustrated in Sections 2.3 through 2.6.

2.1 CLASSIFICATION OF NUMBERS

Numbers are commonly classified as integer, rational, irrational, real, imaginary, and complex.

A. Integers

Integers are the positive and negative whole numbers and zero. Let *I* be the set of integers. Then:

$$I = \{\ldots -3, -2, -1, 0, 1, 2, 3, \ldots\}$$

The positive integers (1, 2, 3, . . .) are also called *natural numbers* or *counting numbers*.

B. Rational Numbers

Rational numbers are numbers that can be expressed as quotients of two integers, such as $-\frac{2}{3}$, $\frac{1}{5}$, and $\frac{2}{7}$, or generally in a fractional form, a/b, where a and b represent integers and b is not 0.

Let Q be the set of irrational numbers. Then:

$$Q = \left\{ \frac{a}{b} \;\middle|\; a \in I, b \in I, b \neq 0 \right\}$$

It is important to remember the condition $b \neq 0$, because any integer divided by zero is undefined. Also, the rational number 6/1 can be conceptually distinguished from the integer 6. However, it is more practical and convenient to identify 6/1 with 6, or consider $6/1 = 6$. Thus, the set of rational numbers includes all integers.

C. Irrational Numbers

Irrational numbers are numbers that have infinite and nonrepeating decimal expansions. The word "irrational" means "not rational," or "a number that cannot be expressed as a quotient of two integers." Examples of irrational numbers are the square root of 2, or $\sqrt{2}$ (the ratio of the length of a diagonal of a square to a side) and π, the Greek letter pi (the ratio of the circumference of a circle to its diameter). $\sqrt{2} = 1.41421 \ldots$ and $\pi = 3.1415927 \ldots$. Let P be the set of irrational numbers. Then:

$$P = \{x|x \text{ has a nonterminating and nonrepeating decimal expansion}\}$$

D. Real Numbers

Real numbers include all rational and irrational numbers.

Let R be the set of real numbers. Then:

$$R = \{x \mid x \in (Q \cup P)\}$$

Or:

Real Numbers
- Rational Numbers
  - Integers
    - Positive whole numbers (1, 2, 3, . . .)
    - Zero (0)
    - Negative whole numbers ($-1, -2, -3, \ldots$)
  - Fractions ($-\frac{2}{3}, \frac{1}{5}, \frac{2}{7}, \ldots$)
- Irrational Numbers ($\sqrt{2}, -\sqrt{3}, \pi, \ldots$)

The set R of real numbers may also be shown geometrically by a number line. The real number corresponding to a point on the line is called the *co-ordinate* of the point. The point representing *0* is called the origin. The numbers represented by the points on the right side of the origin belong to the set R^+ of positive numbers, and the numbers represented by the points on the left side belong to the set R^- of negative real numbers.

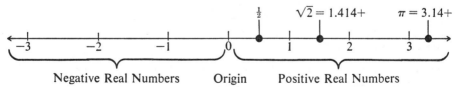

E. Imaginary Numbers

The *imaginary unit,* denoted by the letter i, is the square root of -1, or $i = \sqrt{-1}$. Examples of imaginary numbers are $\sqrt{-4} = 2i$, $\sqrt{-9} = 3i$, and $\sqrt{-16} = 4i$. Let M be the set of imaginary numbers. Then:

$$M = \{x \mid x \text{ is a multiple of } i, \text{ and } i = \sqrt{-1}\}$$

F. Complex Numbers

Complex numbers are formed by imaginary numbers and real numbers and may be represented in the form:

$$a + bi, \text{ where } a \text{ and } b \text{ are real numbers and } i = \sqrt{-1}.$$

Let C be the set of complex numbers. Then:

$$C = \{a + bi \mid a \in R, \, b \in R, \, i = \sqrt{-1}\}$$

2.2 PROPERTIES OF REAL NUMBERS

Let a, b, and c be any elements of the set R of real numbers, written:

$$a, b, c \in R$$

Some properties of real numbers concerning addition and multiplication are stated in the following paragraphs.

A. Closure Property

In adding two real numbers a and b, the sum $a + b$ is a unique element of R. Thus, in $3 + 4 = 7$, the addends 3 and 4 and the sum 7 are real numbers.

In multiplying two real numbers a and b, the product $a \times b$ is a unique element of R. Thus, in $2 \times 5 = 10$, the factors 2 and 5 and the product 10 are real numbers.

B. Commutative Property

In adding two numbers a and b, changing the order of the two numbers does not change the sum. For example:

$$a + b = b + a$$
$$3 + 4 = 4 + 3$$

Similarly, in multiplying two numbers a and b, changing the order of the two numbers does not change the product. For example:

$$a \times b = b \times a$$
$$2 \times 5 = 5 \times 2$$

C. Associative Property

In adding three or more numbers a, b, and c, changing the manner of grouping does not change the sum. For example:

$$(a + b) + c = a + (b + c)$$
$$(3 + 4) + 6 = 3 + (4 + 6)$$

Similarly, in multiplying three or more numbers a, b, c, changing the grouping does not change the product. For example:

$$(a \times b) \times c = a \times (b \times c)$$
$$(2 \times 5) \times 6 = 2 \times (5 \times 6)$$

D. Cancellation Property

If the sum of a and c equals the sum of b and c, then a equals b; that is:

$$\text{if} \quad a + c = b + c$$
$$\text{then} \quad a = b$$

If the product of a and c equals the product of b and c and c is a nonzero real number, then a equals b, or:

$$\text{if} \quad ac = bc, \ (c \neq 0)$$
$$\text{then} \quad a = b$$

E. Distributive Property of Multiplication with Respect to Addition

In multiplying the number a and the sum $(b + c)$, the product of the multiplication is equal to the sum of the products ab and ac.

$$a(b + c) = ab + ac$$
$$2(5 + 6) = (2 \times 5) + (2 \times 6)$$

F. Identity Element

The identity element for addition is zero, since the sum of 0 and any given real number is the given number.

$$a + 0 = a$$
$$3 + 0 = 3$$

The identity element for multiplication is 1, since the product of 1 and any given real number is the given number.

$$a \times 1 = a$$
$$3 \times 1 = 3$$

G. Inverse

Every real number has an *additive inverse;* the sum of the given number and its additive inverse is 0.

$$a + (-a) = 0$$
$$3 + (-3) = 0$$

The expression -3 is called the additive inverse of 3 or the *negative* of 3.

Every nonzero real number has a *multiplicative inverse;* the product of the given number and its multiplicative inverse is 1.

$$a \times \left(\frac{1}{a}\right) = 1$$

$$3 \times \left(\frac{1}{3}\right) = 1$$

The expression $\frac{1}{3}$ is called the multiplicative inverse of 3 or the *reciprocal* of 3.

2.3 SIGNED NUMBERS AND GROUPING

The plus and minus signs may also be used as the indicators of positive and negative numbers. For example, +5 is a positive number and −5 is a negative number. When no sign is written before a number, the number is understood to be positive. Thus, +5 and 5 have the same value. The concept of positive and negative numbers is shown graphically on page 19. When the concept of positive and negative numbers is disregarded, the value of any number is *absolute* and the number is denoted by the sign ||. Thus, |5| indicates the absolute value of 5.

Numbers are often grouped together to represent a single value. The most frequently used symbol of grouping is parentheses ( ). When a subgroup is required within a group, various symbols, such as brackets [ ], braces { }, and the vinculum ‾ are used in addition to parentheses to avoid confusion.

The fundamental operations of signed numbers and the operations indicated by grouping symbols are presented in the following sections.

A. Addition

To add numbers having the same sign, first find the sum of the absolute values of the numbers. Then prefix the common sign to the sum.

EXAMPLE 1 $(+4) + (+7) + (+3.28) = +14.28$

EXAMPLE 2   Add $(−7.31)$, $(−5)$, and $(−3.842)$.

The sum of the absolute values is $|7.31| + |5| + |3.842| = |16.152|$
The common sign is negative. Thus:

$$(−7.31) + (−5) + (−3.842) = −16.152$$

To add numbers having unlike signs, first subtract the smaller absolute from the larger. Then prefix the sign of the larger to the remainder.

EXAMPLE 3   Add $(−5)$ and $(+.28371)$.

$$\begin{array}{r} 5.00000 \\ −0.28371 \\ \hline 4.71629 \end{array}$$

The sign of the larger number, 5, is negative. Thus:

$$(−5) + (+.28371) = −4.71629$$

EXAMPLE 4 $(+38) + (−18.27) = +19.73$

B. Subtraction

To subtract, first change the sign of the subtrahend (the number to be subtracted from the minuend). Then add.

EXAMPLE 5 $(+16) - (+4) = 16 + (-4) = 12$

EXAMPLE 6 $(+35) - (-15) = 35 + (+15) = 50$

EXAMPLE 7 $(-20) - (-18.472) = (-20) + (+18.472) = -1.528$

C. Multiplication

To multiply numbers having the same sign, first multiply the absolute values. Then prefix a positive sign to the product.

EXAMPLE 8 $(+6) \times (+7) = +42$

EXAMPLE 9 $(-3.2) \times (-4) = +12.8$

To multiply numbers having unlike signs, first multiply the absolute values. Then prefix a negative sign.

EXAMPLE 10 $(-5) \times (+4) = -20$

EXAMPLE 11 $(+8.1) \times (-3) = -24.3$

D. Division

To divide numbers having the same sign, first divide the absolute values. Then prefix a positive sign to the quotient.

EXAMPLE 12 $(+24) \div (+4) = +6$

EXAMPLE 13 $(-30) \div (-2) = +15$

To divide numbers having unlike signs, prefix a negative sign to the quotient.

EXAMPLE 14 $(+4.5) \div (-9) = -.5$

E. Grouping Symbols

In carrying out the indicated operations in an expression including several grouping symbols, it is convenient to follow the rules given below to remove the symbols, beginning with the innermost grouping symbol.

If a grouping symbol is preceded by a plus sign, the symbol may be removed without changing the signs of the terms inside the symbol.

EXAMPLE 15 $15 + [6 + (12 + 2)] = 15 + [6 + 12 + 2]$
$$= 15 + [20]$$
$$= 15 + 20 = 35$$
$$\text{or} = 15 + [6 + 14]$$
$$= 15 + 20 = 35$$

If a grouping symbol is preceded by a minus sign, the symbol may be removed only if the signs of the terms inside the symbol are changed; that is, + to − and − to +.

EXAMPLE 16
$$28 - [15 - (3 + 2)] = 28 - [15 - 3 - 2]$$
$$= 28 - [10]$$
$$= 28 - 10 = 18$$
$$\text{or} = 28 - [15 - 5]$$
$$= 28 - 10 = 18$$

EXAMPLE 17
$$6 - \{3 + 2\,[9 - (3 + 4) + 5] - 15\}$$
$$= 6 - \{3 + 2\,[9 - 3 - 4 + 5] - 15\}$$
$$= 6 - \{3 + 2\,[7] - 15\}$$
$$= 6 - 3 - 14 + 15$$
$$= 4$$

The grouping symbols used in Example 17 are arranged in the traditional order: parentheses innermost, then brackets, then braces.

EXERCISE 2–1   Reference: Section 2.3

A. *Perform the indicated operations:*

1. $(-2) + (-6)$
2. $(-7) + (-12.78)$
3. $(-14.234) + (-6.75)$
4. $(+3.25) + (+4.781)$
5. $(-3.2) + (-8.31) + (-13.256)$
6. $(+5) + (+6.34) + (7.218)$
7. $(+4) + (-5.647) + (3.184)$
8. $(+12.832) + (-20.45) + (-5.496)$
9. $(+3.12) - (-4.32)$
10. $(+14.375) - (+18.39)$
11. $(-7.81) - (-10.356)$
12. $(-10.42) - (+3.572)$
13. $(+5.746) - (+3.28) - (+7.314)$
14. $(-4.17) - (-18.942) - (+3.152)$
15. $(+7.564) - (-2.38) - (+4.638)$
16. $(-12.876) - (+8.157) - (-6.325)$

17. $(+2.1) \times (+3.6)$
18. $(+11) \times (-30)$
19. $(-7.2) \times (+4.1)$
20. $(-3.5) \times (-8)$
21. $(+6) \times (+8) \times (-12)$
22. $(-4) \times (-3) \times (+7)$
23. $(-3.2) \times (+5) \times (-6.1)$
24. $(+2.4) \times (-7.3) \times (+9)$
25. $(-56) \div (+8)$
26. $(-108) \div (-12)$
27. $(+36.92) \div (+5.2)$
28. $(+12.04) \div (-2.8)$
29. $(+168) \div (+7) \div (-6)$
30. $(-105) \div (-5) \div (+7)$
31. $(-37.2) \div (+6) \div (-3.1)$
32. $(+64.5) \div (-4.3) \div (+3)$

B. *Remove the symbols of grouping and find the value of each of the following:*

33. $23 + [4 + (7 + 2)]$ 35. $12 - [6 + (3.21 + 4.76)]$
34. $18 + [3 + (2 + 5)]$ 36. $26 - [5 - (7.1 - 3.25)]$
37. $7 + 2\,\{8 - [4.162 + (1.78 - 2.31)] + 2.4\}$
38. $10 + \{5 - 3[7.21 + (4.25 - 6.413)] - 8.23\}$
39. $15 - \{20 - [3.27 - 3(4.26 + 8.54) + 1.5\}$
40. $32 - \{16 - 2[4.31 - (2.67 - 3.45)] - 7.4\}$

2.4 OPERATIONS OF ALGEBRAIC EXPRESSIONS

The basic operations of algebraic expressions are the same as those in arithmetic—addition, subtraction, multiplication, and division. However, before the operations can be discussed, some algebraic terminologies should be fully understood.

An *algebraic expression,* also simply called an *expression,* is any symbol or combination of numbers and symbols that represents a single quantity.

EXAMPLE 18 $(2a^2 + 5b^2 - 18)$ and $7x^2y$ are expressions; the letters used as symbols (a, b, x, and y) represent numbers.

If an expression consists of several parts that are connected by plus and minus signs, each of the parts, together with the sign preceding it, is called a *term.* When there is no sign preceding a term, the sign is understood to be positive. In the first expression of Example 18, $2a^2$ or $+2a^2$ is the first term, $+5b^2$ is the second term, and -18 is the third term. The second expression, $7x^2y$ or $+7x^2y$, has only one term.

An expression consisting of one term is called a *monomial;* an expression consisting of more than one term is called a *multinomial* or a *polynomial.* An expression consisting of two terms is called a *binomial,* whereas one of three terms is a *trinomial.*

EXAMPLE 19   The expression $3ax^2$ is a monomial, the expression $(2a + 6b)$ is a binomial, and the expression $(5x + 6y^2 - 7)$ is a trinomial.

When two or more numbers are multiplied, each number or the product of any of the numbers is called a *factor.* Any factor of a term is also called the *coefficient* of the remaining part. If a factor is an explicit number, it is called the *numerical coefficient* of the term; other factors are called *literal coefficients.*

EXAMPLE 20   In the term $7xy$, each number or symbol (7, x, and y) or the product of any of them ($7x$, $7y$, and xy) is a factor. The number 7 is the numerical coefficient of xy, the symbol x is the literal coefficient of $7y$, and the symbol y is the literal coefficient of $7x$.

The product of equal factors is called a *power* of one of the factors. Thus, 4×4 is the second power of 4 and is written 4^2, and $a \times a \times a$ is the third power of a and is written a^3. In the expression a^3, the symbol a is called the *base,* and the small number 3, which indicates the number of equal factors, is the *exponent.* The expression of an indicated power of a given symbol or number is also called an *exponential term,* such as 4^2 and a^3.

The exponent for the first power of a base is 1, which is understood and is usually not written. For example, a^1 is written as a for simplicity. A second power is also called a *square;* 4^2 is called 4 squared. A third power is a *cube;* a^3 is called a cubed.

The exponent of a symbol representing an unknown in a term is also called the *degree* in the unknown. For example, -5 is the numerical coefficient in the term $-5x^3$, and the exponent 3 indicates that the term is of the third degree in x. The degree of a term referring to certain unknowns is the sum of the exponents of those unknowns. Thus, the expression $4x^3y^2z$ is of the third degree in x, the second degree in y, the first degree in z, the fifth degree in x and y, the fourth degree in x and z, the third degree in y and z, and the sixth degree in x, y, and z, since the sum of the exponents is $3 + 2 + 1 = 6$.

The four basic operations of algebraic expressions are presented in the following sections.

A. Addition

Like terms (terms with the same literal factors and the same exponents of the factors) can be combined or added. The addition of *unlike terms* is simply indicated by a plus (+) sign.

EXAMPLE 21   Add $2xy$, $5xy$, and $3ab$.
$2xy$ and $5xy$ are like terms. $2xy + 5xy = 7xy$.
But $7xy$ and $3ab$ are unlike terms.
Thus, $2xy + 5xy + 3ab = 7xy + 3ab$.

EXAMPLE 22   Add $(4x + 5y - 6)$, $(2x - 3y + 7)$, and $(3x + 2y - 5)$.

$$\begin{array}{r} 4x + 5y - 6 \\ 2x - 3y + 7 \\ \underline{(+)\ 3x + 2y - 5} \\ 9x + 4y - 4 \end{array}$$

EXAMPLE 23   Add $4x^2y$, $8x^2y$, $3xy$, $2x^2$, and $5xy$.
$4x^2y + 8x^2y + 3xy + 5xy + 2x^2 = 12x^2y + 8xy + 2x^2$

B. Subtraction

A term may be subtracted from a like term. When one polynomial is subtracted from another, simply change the sign (+ to − and − to +) of each term in the subtrahend and then add.

EXAMPLE 24   Subtract $7abc$ from $15abc$.

$$15abc - 7abc = 8abc$$

EXAMPLE 25   Subtract $(-3a + 5b - 7)$ from $(6a - 8b + 12)$.

$$\begin{array}{l} 6a -\ \ 8b + 12 \quad \text{(Minuend)} \\ \underline{(-)\ -3a +\ \ 5b -\ \ 7} \quad \text{(Subtrahend)} \end{array}$$

$$\begin{array}{l} 6a -\ \ 8b + 12 \quad \text{(Minuend with the same sign)} \\ \underline{(+)\ \ \ 3a -\ \ 5b +\ \ 7} \quad \text{(Subtrahend with the signs} \\ 9a - 13b + 19 \quad\ \text{changed)} \end{array}$$

Or:

$$(6a - 8b + 12) - (-3a + 5b - 7) = 6a - 8b + 12 + 3a - 5b + 7$$
$$= 9a - 13b + 19$$

EXAMPLE 26 Subtract $(4x^3y - 3xyz)$ from $(7x^3y + 5yz)$.

$$(7x^3y + 5yz) - (4x^3y - 3xyz) = 3x^3y + 5yz + 3xyz$$

C. Multiplication

When multiplication involves exponential terms, the following laws and definitions may be applied to simplify computation.

Law 1 — Bases are the same.

$$a^m \cdot a^n = a^{m+n}$$

EXAMPLE 27 $2^2 \cdot 2^3 = 2^{2+3} = 2^5$
Or: $2^2 \cdot 2^3 = (2 \cdot 2)(2 \cdot 2 \cdot 2) = 2^5$

EXAMPLE 28 $x^3 \cdot x^4 = x^{3+4} = x^7$
Or: $x^3 \cdot x^4 = (x \cdot x \cdot x)(x \cdot x \cdot x \cdot x) = x^7$

Law 1 gives the following two definitions:

I. $a^0 = 1$ $(a \neq 0)$
The sign $\neq$ represents "not equal to."

Illustration:

$$a^m \cdot a^0 = a^{m+0} = a^m$$
$$a^0 = \frac{a^m}{a^m} = 1$$

Thus: $5^0 = 1$ and $37^0 = 1$

II. $a^{-m} = \frac{1}{a^m}$

Illustration:

$$a^m \cdot a^{-m} = a^{m+(-m)} = a^0 = 1$$
$$a^{-m} = \frac{1}{a^m}$$

Thus:

$$3^{-2} = \frac{1}{3^2} \quad \text{and} \quad 10^{-4} = \frac{1}{10^4}$$

Law 2 — Bases are different but exponents are the same.

$$a^m \cdot b^m = (ab)^m$$

EXAMPLE 29 $3^2 \cdot 5^2 = (3 \cdot 5)^2 = 15^2$
Or: $3^2 \cdot 5^2 = 3 \cdot 3 \cdot 5 \cdot 5 = (3 \cdot 5)(3 \cdot 5) = (3 \cdot 5)^2 = 15^2$

EXAMPLE 30 $x^3 \cdot y^3 = (x \cdot y)^3$

Or: $x^3 \cdot y^3 = (x \cdot x \cdot x)(y \cdot y \cdot y) = (xy)(xy)(xy) = (xy)^3$

Law 3 — Base is an exponential term.

$$(a^m)^n = a^{mn}$$

Or:

$$(a^{1/m})^n = a^{n/m}$$

EXAMPLE 31 $(4^3)^2 = 4^{3 \cdot 2} = 4^6$

Or: $(4^3)^2 = (4 \cdot 4 \cdot 4)(4 \cdot 4 \cdot 4) = 4^6$

EXAMPLE 32 $(x^2)^3 = x^{2 \cdot 3} = x^6$

Or: $(x^2)^3 = (x \cdot x)(x \cdot x)(x \cdot x) = x^6$

EXAMPLE 33 $(6^{1/2})^4 = 6^{(1/2)4} = 6^{4/2} = 6^2$

Or: $(6^{1/2})^4 = (6^{1/2})(6^{1/2})(6^{1/2})(6^{1/2}) = 6^{1/2+1/2+1/2+1/2} = 6^{4/2} = 6^2$

EXAMPLE 34 $(x^{1/4})^3 = x^{(1/4)3} = x^{3/4}$

Or: $(x^{1/4})^3 = (x^{1/4})(x^{1/4})(x^{1/4}) = x^{1/4+1/4+1/4} = x^{3/4}$

In multiplying a monomial by another monomial, the product is equal to the product of their numerical coefficients multiplied by the product of their literal factors.

EXAMPLE 35 $4ab^2 \cdot 6abc = 24a^{1+1}b^{2+1}c = 24a^2b^3c$

In multiplying a polynomial by a monomial, multiply each term of the polynomial by the monomial.

EXAMPLE 36 Multiply $(3x + 5y)$ by $2x$.

$$\begin{array}{r} 3x + 5y \\ 2x \\ \hline 6x^2 + 10xy \end{array}$$

In multiplying a polynomial by another polynomial, multiply each term of the multiplicand by each term of the multiplier.

EXAMPLE 37 Multiply $(2a - 6b)$ by $(3a^2 + 5c)$.

$$\begin{array}{l} 2a - 6b \\ 3a^2 + 5c \\ \hline 6a^3 - 18a^2b \\ + 10ac - 30bc \\ \hline 6a^3 - 18a^2b + 10ac - 30bc \end{array}$$

EXAMPLE 38 Multiply $(3x + 5)$ by $(2x - 4)$.

$$\begin{array}{r} 3x + 5 \\ 2x - 4 \\ \hline 6x^2 + 10x \\ - 12x - 20 \\ \hline 6x^2 - 2x - 20 \end{array}$$

D. Division

When division involves exponential terms the following laws can be applied to simplify computation.

Law 1—Bases are the same.

$$a^m \div a^n = \frac{a^m}{a^n} = a^{m-n}$$

If $m > n$, the exponent $m - n$ is a positive number. If $m < n$, the exponent $m - n$ is a negative number. Thus:

$$\frac{a^m}{a^n} = a^{-(n-m)} = \frac{1}{a^{n-m}}$$

If $m = n$, the exponent $m - n = 0$. Thus:

$$\frac{a^m}{a^n} = a^0 = 1$$

EXAMPLE 39 $3^5 \div 3^3 = \dfrac{3^5}{3^3} = 3^{5-3} = 3^2 = 9$

Or: $3^5 \div 3^3 = \dfrac{3 \cdot 3 \cdot \cancel{3} \cdot \cancel{3} \cdot \cancel{3}}{\cancel{3} \cdot \cancel{3} \cdot \cancel{3}} = 3^2 = 9$

EXAMPLE 40 $2^3 \div 2^7 = \dfrac{2^3}{2^7} = 2^{3-7} = 2^{-4} = \dfrac{1}{2^4} = \dfrac{1}{16}$

EXAMPLE 41 $x^6 \div x^4 = \dfrac{x^6}{x^4} = x^{6-4} = x^2$

EXAMPLE 42 $y^2 \div y^8 = \dfrac{y^2}{y^8} = y^{2-8} = y^{-6} = \dfrac{1}{y^6}$

EXAMPLE 43 $x^2y \div x^2y = \dfrac{x^2y}{x^2y} = 1$

Law 2—Bases are different but exponents are the same.

$$\frac{a^m}{b^m} = \left(\frac{a}{b}\right)^m$$

EXAMPLE 44 $3^2 \div 6^2 = \dfrac{3 \cdot 3}{6 \cdot 6} = \dfrac{3}{6} \cdot \dfrac{3}{6} = \left(\dfrac{3}{6}\right)^2 = \left(\dfrac{1}{2}\right)^2 = \dfrac{1}{4}$, or:

$$\frac{3^2}{6^2} = \left(\frac{3}{6}\right)^2 = \left(\frac{1}{2}\right)^2 = \frac{1}{4}$$

EXAMPLE 45 $\dfrac{x^5}{y^5} = \left(\dfrac{x}{y}\right)^5$

In dividing a monomial by another monomial, the quotient of the division is the product of the quotient of the numerical coefficients and the quotient of the literal coefficients.

EXAMPLE 46 Divide $-36x^5$ by $3x^2$.

$$\frac{-36x^5}{3x^2} = \frac{-36}{3} \cdot \frac{x^5}{x^2} = -12x^3$$

EXAMPLE 47 Divide $18a^7b^2c^6$ by $6a^2b^5x$.

$$\frac{18a^7b^2c^6}{6a^2b^5x} = \frac{18}{6} \cdot \frac{a^7}{a^2} \cdot \frac{b^2}{b^5} \cdot \frac{c^6}{1} \cdot \frac{1}{x} = 3a^5b^{-3}c^6x^{-1} = \frac{3a^5c^6}{b^3x}$$

In dividing a polynomial by a monomial, divide each term of the polynomial by the monomial. The quotient of the division is the sum of the partial quotients.

EXAMPLE 48 Divide $27a^5b^3 - 18a^3b^2 + 2ab^4$ by $3a^2b^3$.

$$\frac{27a^5b^3}{3a^2b^3} = 9a^3; \quad \frac{-18a^3b^2}{3a^2b^3} = \frac{-6a}{b}; \quad \frac{2ab^4}{3a^2b^3} = \frac{2b}{3a}$$

Thus: $\dfrac{27a^5b^3 - 18a^3b^2 + 2ab^4}{3a^2b^3} = 9a^3 - \dfrac{6a}{b} + \dfrac{2b}{3a}$

The procedure of dividing a polynomial by another polynomial is different from the above and is illustrated in Example 49.

EXAMPLE 49 Divide $(8x^3 + 25 - 6x^2)$ by $(-3x + 2x^2 + 7)$

$$
\begin{array}{r}
4x + \ 3 \quad \text{(Quotient)} \\
\text{(Divisor)} \quad 2x^2 - 3x + 7 \overline{\smash{)}8x^3 - \ 6x^2 \qquad + 25} \quad \text{(Dividend)} \\
\underline{8x^3 - 12x^2 + 28x \qquad\quad} \\
+ \ 6x^2 - 28x + 25 \\
\underline{6x^2 - \ 9x + 21} \\
- 19x + \ 4 \quad \text{(Remainder)}
\end{array}
$$

Thus, the solution equation is:

$$(8x^3 - 6x^2 + 25) \div (2x^2 - 3x + 7) = (4x + 3) + \frac{-19x + 4}{2x^2 - 3x + 7}$$

Observe the above arrangement of division:

1. The terms of both the dividend and the divisor are first arranged in descending powers of the same letter x.

   Dividend: $8x^3 - 6x^2 + 25$ (from $8x^3 + 25 - 6x^2$)
   Divisor: $2x^2 - 3x + 7$ (from $-3x + 2x^2 + 7$)

   Note: If a term of a polynomial is missing, its coefficient is considered to be zero, such as $0x$ in the dividend.

2. The quotient in each step of division is obtained by dividing the first term of the dividend (or remainder) by the first term of the divisor.

$$4x = 8x^3 \div 2x^2 \text{ and } +3 = 6x^2 \div 2x^2$$

3. The quotient in each step is then multiplied by the divisor, or:
$$8x^3 - 12x^2 + 28x = 4x(2x^2 - 3x + 7)$$
$$6x^2 - 9x + 21 = 3(2x^2 - 3x + 7)$$
The first product is subtracted from the dividend to obtain the first remainder, and the second product is subtracted from the first remainder to obtain the second remainder. When a remainder is neither zero nor a lower power than the divisor, continue the division by the procedure used above. Here the second remainder $(-19x + 4)$ is the final remainder since it is a lower power (first-degree) than the divisor (second-degree polynomial).

The result of the division may be checked by either one of the following two methods:

METHOD 1   Let $x = 2$ (or any number except 0 and 1).
Substitute the x value in the solution equation:

$$\text{Left side} = (8 \cdot 2^3 - 6 \cdot 2^2 + 25) \div (2 \cdot 2^2 - 3 \cdot 2 + 7)$$
$$= (64 - 24 + 25) \div (8 - 6 + 7)$$
$$= 65 \div 9 = 7\frac{2}{9}$$

$$\text{Right side} = (4 \cdot 2 + 3) + \frac{-19 \cdot 2 + 4}{2 \cdot 2^2 - 3 \cdot 2 + 7}$$
$$= 11 + \frac{-34}{9}$$
$$= 7\frac{2}{9}$$

METHOD 2   Use the relationship:
Quotient $\times$ Divisor $+$ Remainder $=$ Dividend

$$(4x + 3)(2x^2 - 3x + 7) + (-19x + 4)$$
$$= 4x(2x^2 - 3x + 7) + 3(2x^2 - 3x + 7) - 19x + 4$$
$$= 8x^3 - 12x^2 + 28x + 6x^2 - 9x + 21 - 19x + 4$$
$$= 8x^3 - 6x^2 + 25 = \text{the given dividend}$$

EXERCISE 2-2   Reference: Section 2.4

A. *Add:*

1. $25x + 14x$
2. $15x + (-8x)$
3. $(-15ab) + (-23ab)$
4. $(-7cd) + (-10cd)$

5. $20g + 12g + (-5g)$

6. $16h + (-5h) + 8h$

7. $(-4x) + (-5y) + (-16x) + 31y$
8. $6a + (-3a) + (-8b) + (-12b)$
9. $(25xy - 18ab) + (7ab - 12xy)$
10. $(2a - 5b + 7c)$
    $+ (3a + 6b - 15c)$
11. $(3a^2b + 4cd^2 - 7abc)$
    $+ (4a^2b + 7cd^2 - 3abc)$
12. $(8x + 2x^2y + 6abx^2)$
    $+ (6x + 3x^2y + 5abx^2)$

B. *Subtract:*

13. $20xy - 7xy$

14. $16x - (-9x)$

15. $(-45ab) - (-12ab)$

16. $(-6cd) - (-12cd)$

17. $18e - 15e - (-6e)$

18. $24f - (-8f) - 5f$

19. $(-3x) - (-6y) - (-5x) - 10y$

20. $4a - (-5a) - (-7b)$
    $\quad - (-14b)$

21. $(16fg + 51hi)$
    $\quad - (4fg - 7hi)$

22. $(6xy + 5x + 7)$
    $\quad - (4xy - 3x - 16)$

23. $(4xy^2 - 7x^2y^3 - 10xy)$
    $\quad - (8xy^2 + 2x^2y^3 - 3xy)$

24. $(16ax^2 - 3abx - 9a^2bx^3)$
    $\quad - (10ax^2 - 6abx + 5a^2bx^3)$

C. *Multiply:*

25. $3^2 \cdot 3^4$

26. $6^3 \cdot 6^0$

27. $a^3 \cdot a^2$

28. $b^4 \cdot b^{-3}$

29. $c^5 \cdot c^{-2}$

30. $d^{-3} \cdot d^{-4}$

31. $7^2 \cdot 8^2$

32. $a^3 \cdot b^3$

33. $(5^2)^3$

34. $(a^3)^4$

35. $(6^{2/5})^3$

36. $(x^{1/2}y^{1/4})^8$

37. $(-4x^2yz^3)(5xy)$

38. $(-6abc)(-3ab^2c)$

39. $(-3x)(4x - 5y + 6)$

40. $(4a + 7)(5a - 2)$

41. $(3x^2 - 4)(2x + 5)$

42. $(4ab^2 - 3a + 6)(2a - 8)$

D. *Divide:*

43. $5^3 \div 5^2$

44. $6^2 \div 6^5$

45. $4^3 \div 4^3$

46. $x^5 \div x^2$

47. $x^2y^3 \div x^5y^2$

48. $ab^2c \div a^2b^4c$

49. $4^2 \div 8^2$

50. $x^3 \div y^3$

51. $(-45xyz^3) \div (5xzw^2)$

52. $(24ab^2c) \div (-3a^3b^5y)$

53. $(8ab^2 - 15a^3bc^5 + 20a^2b^3)$
    $\quad \div (5a^2b)$

54. $(-6xy^2z + 20x^2y^3z^2 - 18x^3y^4z^3)$
    $\quad \div (2x^2y^3z)$

55. $(12x^2 - 7x - 10) \div (3x + 2)$

56. $(6a^2 + 14a - 43) \div (3a - 5)$

57. $(10x^3 + 35 - 15x^2)$
    $\quad \div (5x^2 + 7 - 5x)$

58. $(12x^3 + 4x + x^2)$
    $\quad \div (4x^2 - 5x - 2)$

2.5 FACTORING

When each term in an expression contains the same factor, called the *common factor*, the expression may be written as the product of the common factor and another factor. The other factor is obtained by dividing the given expression by the common factor. The process of finding the factors in an expression is called *factoring*. The most common types of factoring are presented in this section.

A. Monomial Factor

Let the given expression be $ax + ay$. Each term of the expression contains the common monomial factor a. Dividing the expression by a, the quotient is:

$$\frac{ax + ay}{a} = x + y$$

Thus, in general, a monomial factor can be written in the following form:

$$ax + ay = a(x + y)$$

The left side of the equation is called the *expanded form* and the right side is called the *factored form*.

EXAMPLE 50 Factor $3x + 3y$.

$$3x + 3y = 3(x + y)$$

EXAMPLE 51 Factor $8x - 8y$.

$$8x - 8y = 8(x - y)$$

EXAMPLE 52 Factor $10ax + 15ay - 20az$.

$$10ax + 15ay - 20az = 5a(2x + 3y - 4z)$$

B. Binomial Factor

In general, a binomial factor can be written in the following form:

$$a(x + y) + b(x + y) = (a + b)\,(x + y)$$

Each term of the expanded expression (on the left side) contains the common binomial factor $(x + y)$. Factor $(a + b)$ on the right side is obtained by dividing the expanded expression by the common factor.

EXAMPLE 53 Factor $4ax - ay + 12bx - 3by$.

$$4ax - ay + 12bx - 3by = a(4x - y) + 3b(4x - y)$$
$$= (a + 3b)\,(4x - y)$$

C. Difference of Two Squares

In general, the factored form of the difference of two squares is as follows:

$$x^2 - y^2 = (x + y)\,(x - y)$$

This equation can be verified by multiplying the factors on the right side:

$$
\begin{array}{r}
x + y \\
(\times)\ \underline{x - y} \\
x^2 + xy \\
\underline{-xy - y^2} \\
x^2 \qquad - y^2
\end{array}
$$

EXAMPLE 54 Factor $16x^2 - y^2$.

$$16x^2 - y^2 = (4x)^2 - y^2 = (4x + y)(4x - y)$$

EXAMPLE 55 Factor $9a^2 - 25b^2$.

$$9a^2 - 25b^2 = (3a)^2 - (5b)^2 = (3a + 5b)(3a - 5b)$$

D. Trinomials — Perfect Squares

In general, the expanded terms and the factored form of perfect squares can be written:

$$x^2 + 2xy + y^2 = (x + y)^2$$
And:
$$x^2 - 2xy + y^2 = (x - y)^2$$

These two equations can also be verified by multiplying the factors on the right sides.

EXAMPLE 56 Factor $4x^2 + 12xy + 9y^2$.

$$4x^2 + 12xy + 9y^2 = (2x)^2 + 2(2x)(3y) + (3y)^2$$
$$= (2x + 3y)^2$$

EXAMPLE 57 Factor $25a^2 - 10ab + b^2$.

$$25a^2 - 10ab + b^2 = (5a)^2 - 2(5a)b + b^2$$
$$= (5a - b)^2$$

E. Trinomials — General Case

The expanded terms and their factors can be written:

$$acx^2 + (ad + bc)x + bd = (ax + b)(cx + d)$$
Since:

$$
\begin{array}{r}
ax + b \\
(\times) \quad cx + d \\
\hline
acx^2 + bcx \quad\quad\quad\quad \\
+ adx \quad\quad + bd \\
\hline
acx^2 + (ad + bc)x + bd
\end{array}
$$

The letters a, b, c, and d represent constants. The constants in the factored form may be determined by the trial-and-error method when the numerical coefficients of a given trinomial are arranged in columnar form as follows:

Observe the arrangement:

Step $1 = ac$
Step $2 = bd$
Step $3 = ad + bc$

EXAMPLE 58   Factor $10x^2 + 11x + 3$.

STEP 1: $ac = 10$. The factors of 10 are 2 and 5 or 1 and 10.

STEP 2: $bd = 3$. The factors of 3 are 1 and 3.

STEP 3: $ad + bc = 11$. Place the two pairs of numbers $(a, c; b, d)$ in appropriate positions by the trial-and-error method. The results are:

$$a = 2 \quad b = 1 \qquad a = 2, b = 1$$
$$c = 5, d = 3$$
$$(1) \quad (3) \quad (2) \qquad \text{Or: } ad + bc = (2)(3) + (1)(5) = 11$$
$$c = 5 \quad d = 3$$

Thus:

$$10x^2 + 11x + 3 = (2x + 1)(5x + 3)$$

EXAMPLE 59   Factor $12x^2 + 7x - 10$.

STEP 1: $ac = 12$. The factors of 12 are 1 and 12, or 2 and 6, or 3 and 4.

STEP 2: $bd = -10$. The factors of (-10) are 1 and -10, or -1 and 10, or 2 and -5, or -2 and 5.

STEP 3: $ad + bc = 7$. By the trial-and-error method, the results are:

$$a = 3 \quad b = -2 \qquad a = 3, b = -2$$
$$c = 4, d = 5$$
$$(1) \quad (3) \quad (2) \qquad \text{Or: } ad + bc = (3)(5) + (-2)(4)$$
$$c = 4 \quad d = 5 \qquad \qquad = 15 - 8 = 7$$

Thus: $12x^2 + 7x - 10 = (3x - 2)(4x + 5)$

The following procedure is an alternative way to factor a trinomial:

STEP 1   Multiply the coefficient of x^2 by the term not having x.

STEP 2   Find two numbers whose algebraic sum is the coefficient of x and whose product equals the product obtained in Step 1.

STEP 3   Use the two numbers found to replace the coefficient of x in the given trinomial and factor the new expression in the usual manner.

EXAMPLE 60   Factor $10x^2 + 11x + 3$. (Same as Example 58.)

STEP 1:   The coefficient of x^2 is 10 and the term not having x is $+3$. $10(3) = 30$.

STEP 2:   The two numbers are 5 and 6 by the trial-and-error method, since $5 + 6 = 11$ and $(5)(6) = 30$.

STEP 3:   Rewrite the given trinomial and factor:

$$10x^2 + (5 + 6)x + 3 = 10x^2 + 5x + 6x + 3$$
$$= 5x(2x + 1) + 3(2x + 1)$$
$$= (5x + 3)(2x + 1)$$

EXAMPLE 61   Factor $12x^2 + 7x - 10$. (Same as Example 59.)

STEP 1: $12(-10) = -120$

STEP 2:   The two numbers are 15 and (-8) since $15 + (-8) = 7$ and $(15)(-8) = -120$.

STEP 3:   Rewrite the given trinomial and factor:

$$12x^2 + (15 - 8)x - 10 = 12x^2 + 15x - 8x - 10$$
$$= 3x(4x + 5) - 2(4x + 5)$$
$$= (3x - 2)(4x + 5)$$

EXERCISE 2–3   Reference: Section 2.5

Factor:

1. $7a + 7b$
2. $9c - 9d$
3. $5x + 10$
4. $12y - 3$
5. $18x + 6y - 12z$
6. $-16ab + 8ac - 4az$
7. $2ax + 6bx - ay - 3by$
8. $6ax - 15bx + 8ay - 20by$
9. $cd + dx + cy + xy$
10. $ef - fx - ey + xy$
11. $4x^2 - y^2$
12. $9a^2 - 16y^2$

13. $25b^2 - 4m^2$
14. $16c^2 - 25d^2$
15. $4x^2 + 4xy + y^2$
16. $9a^2 + 12ab + 4b^2$
17. $25b^2 - 20bm + 4m^2$
18. $16x^2 - 24xy + 9y^2$
19. $8x^2 + 22x + 15$
20. $6x^2 + 5x - 4$
21. $15x^2 + 14x - 8$
22. $8x^2 - 34x + 21$
23. $10x^2 + 29x - 72$
24. $12x^2 + x - 35$

2.6 FRACTIONS

A fraction is a rational number and is also an indicated division. For example, the fraction $\frac{1}{4}$ is a rational number and also indicates division $(1 \div 4)$. However, the division is not actually carried out when it represents the number. When the indicated division is carried out, on the other hand, the result is called a *decimal fraction,* such as .25 since $\frac{1}{4} = 1 \div 4 = .25$.

In a common fraction, there are two terms: the *numerator* or *dividend* is written above the line and the *denominator* or *divisor* is written below the line. Thus, in the fraction $\frac{1}{4}$, the numbers 1 and 4 are the terms of the fraction; 1 is the numerator and 4 is the denominator.

Common fractions may be classified into three types according to the values of numerators and denominators:

In a *proper fraction,* the numerator is smaller than the denominator, such as $\frac{1}{4}$ and $\frac{16}{25}$.

In an *improper fraction,* the numerator is equal to or larger than the denominator, such as $\frac{3}{3}$, $\frac{5}{2}$, and $\frac{17}{8}$.

In a *complex fraction,* there are one or more fractions in the numerator, the denominator, or both. The line which separates the numerator from the denominator should be longer than the lines used in the numerator or denominator, such as:

$$\frac{\frac{2}{3}}{15} \qquad \frac{7}{\frac{1}{4}} \qquad \frac{\frac{1}{2}}{\frac{5}{23}}$$

When a number consists of a whole number and a fraction, it is called a *mixed number*, such as $2\frac{3}{5}$, which is the same as $2 + \frac{3}{5}$.

The terms of a fraction can be changed to either higher or lower terms *without changing the value* of the fraction, provided that both the dividend and the divisor are *multiplied or divided* by the same number. In general:

$$\frac{a}{b} = \frac{a \times c}{b \times c} \qquad \frac{a}{b} = \frac{a \div c}{b \div c}$$

EXAMPLE 62 Change $\frac{2}{5}$ to higher terms.

$$\frac{2}{5} = \frac{2 \times 3}{5 \times 3} = \frac{6}{15}; \quad \frac{6}{15} = \frac{6 \times 2}{15 \times 2} = \frac{12}{30}; \quad \frac{12}{30} = \frac{12 \times c}{30 \times c}; \text{ and so on.}$$

There is an unlimited number of higher terms of $\frac{2}{5}$.

EXAMPLE 63 Change $\frac{18}{24}$ to lower terms.

$$\frac{18}{24} = \frac{18 \div 3}{24 \div 3} = \frac{6}{8}; \quad \frac{6}{8} = \frac{6 \div 2}{8 \div 2} = \frac{3}{4}.$$

In Example 63, 3 is the common divisor of 18 and 24; 2 is the common divisor of 6 and 8; but there is no common divisor for the terms of the fraction $\frac{3}{4}$. When the numerator and the denominator have no common divisor, the fraction has been changed to its *lowest terms*, or its *simplest form*.

Further, if the numerator and the denominator of a fraction can be factored and divided by a common factor, the fraction can be changed to a simpler form.

EXAMPLE 64 Simplify: $\dfrac{x^2 - 16}{x^2 + 6x + 8}$

$$\frac{x^2 - 16}{x^2 + 6x + 8} = \frac{(x + 4)(x - 4)}{(x + 4)(x + 2)} = \frac{x - 4}{x + 2}$$

A mixed number may be expressed as an improper fraction having equal value.

EXAMPLE 65 Change the mixed number $2\frac{3}{5}$ to an improper fraction.

$$2\frac{3}{5} = 2 + \frac{3}{5} = \frac{2}{1} + \frac{3}{5} = \frac{2 \times 5}{1 \times 5} + \frac{3}{5} = \frac{10}{5} + \frac{3}{5} = \frac{13}{5}$$

$$\text{Or: } 2\frac{3}{5} = \frac{(2 \times 5) + 3}{5} = \frac{13}{5}$$

Conversely, an improper fraction may be expressed as a mixed number.

EXAMPLE 66 Change the improper fraction $\frac{19}{8}$ to a mixed number.

$\frac{19}{8} = 19 \div 8 = 2$ with a remainder of 3, which may be further divided by 8, written: $\frac{19}{8} = 2\frac{3}{8}$.

The four fundamental operations involving fractions—addition, subtraction, multiplication, and division—are discussed in the following sections.

A. Addition

When fractions are added, they should be changed to fractions having a *lowest common denominator* (l.c.d.). The sum of the numerators of all new fractions and the common denominator are the two terms of the answer. In adding mixed numbers, it is not necessary to change the numbers to improper fractions. It is simpler to add the whole numbers and fractions separately. The fractional part of an answer should always be expressed in its lowest terms.

EXAMPLE 67 Add $\frac{2}{3}$ and $\frac{4}{5}$.

When there is no common divisor of the denominators of the given fractions, the product of the two denominators is the l.c.d.

$3 \times 5 = 15$, which is the l.c.d. of the two fractions.

$$\frac{2}{3} = \frac{2 \times 5}{3 \times 5} = \frac{10}{15} \qquad \frac{4}{5} = \frac{4 \times 3}{5 \times 3} = \frac{12}{15}$$

$$\frac{2}{3} + \frac{4}{5} = \frac{10}{15} + \frac{12}{15} = \frac{10 + 12}{15} = \frac{22}{15} = 1\frac{7}{15}$$

EXAMPLE 68 $\dfrac{3}{a} + \dfrac{4}{b}$

The l.c.d. is ab.

$$\frac{3}{a} + \frac{4}{b} = \frac{3b}{ab} + \frac{4a}{ab} = \frac{3b + 4a}{ab}$$

EXAMPLE 69 $\frac{1}{6} + \frac{3}{10} + 5\frac{7}{20}$

When there is a common divisor or divisors of the denominators of given fractions, the l.c.d. may be found by the following division. (The product of the common divisors and the final quotients is the l.c.d.)

$$\begin{array}{l} 2)\overline{6,\ 10,\ 20} \text{ (denominators)} \\ 5)\overline{3,\ \ 5,\ 10} \\ \quad 3,\ \ 1,\ \ 2 \end{array}$$ The l.c.d. is $2 \times 5 \times 3 \times 1 \times 2 = 60$.

Notice that in each step of the division. at least two of the denominators are divided by their common divisor. The first divisor, 2, is the common divisor to all denominators. However. the second divisor. 5. is the common divisor to 5 and 10 only; the number 3 is not divisible by 5 and remains unchanged.

Now the three given fractions are changed to fractions having the common denominator 60.

$$\frac{1}{6} = \frac{1 \times 10}{6 \times 10} = \frac{10}{60} \qquad \frac{3}{10} = \frac{3 \times 6}{10 \times 6} = \frac{18}{60} \qquad 5\frac{7}{20} = 5\frac{7 \times 3}{20 \times 3} = 5\frac{21}{60}$$

Thus:
$$\frac{1}{6} + \frac{3}{10} + 5\frac{7}{20} = \frac{10}{60} + \frac{18}{60} + 5\frac{21}{60} = 5\frac{10 + 18 + 21}{60} = 5\frac{49}{60}$$

EXAMPLE 70

Add and simplify: $\dfrac{4x}{x^2 - y^2} + \dfrac{5}{x + y}$

The l.c.d. is $x^2 - y^2 = (x + y)(x - y)$.

$$\frac{4x}{x^2 - y^2} + \frac{5}{x + y} = \frac{4x}{(x + y)(x - y)} + \frac{5(x - y)}{(x + y)(x - y)}$$

$$= \frac{4x + 5(x - y)}{(x + y)(x - y)} = \frac{9x - 5y}{(x + y)(x - y)}$$

EXAMPLE 71

Add and simplify: $\dfrac{3}{x + 1} + \dfrac{10x + 1}{2x^2 + 7x + 5}$

The l.c.d. is $2x^2 + 7x + 5 = (x + 1)(2x + 5)$

$$\frac{3}{x + 1} + \frac{10x + 1}{2x^2 + 7x + 5} = \frac{3(2x + 5) + (10x + 1)}{(x + 1)(2x + 5)}$$

$$= \frac{6x + 15 + 10x + 1}{(x + 1)(2x + 5)}$$

$$= \frac{16\cancel{(x + 1)}}{\cancel{(x + 1)}(2x + 5)}$$

$$= \frac{16}{2x + 5}$$

B. Subtraction

In subtraction, fractions should also be changed to fractions with a lowest common denominator. The difference between the numerators of the two new fractions and the common denominator are the two terms of the answer.

EXAMPLE 72

Subtract $\dfrac{3x}{5}$ from $\dfrac{7x}{8}$.

The l.c.d. is $5(8) = 40$.

$$\frac{7x}{8} - \frac{3x}{5} = \frac{35x - 24x}{40} = \frac{11x}{40}$$

Mixed numbers are not required to be changed to improper fractions in subtraction unless the fractional part of the minuend (the first number) is smaller than the fractional part of the subtrahend (the second number).

EXAMPLE 73   Subtract $2\frac{1}{6}$ from $6\frac{4}{5}$.

$$6\frac{4}{5} - 2\frac{1}{6} = 6\frac{24}{30} - 2\frac{5}{30} = (6-2) + \left(\frac{24-5}{30}\right) = 4\frac{19}{30}$$

If the mixed numbers are changed to improper fractions, the subtraction becomes more complicated since larger numbers are involved in the numerators.

$$6\frac{4}{5} - 2\frac{1}{6} = \frac{34}{5} - \frac{13}{6} = \frac{204}{30} - \frac{65}{30} = \frac{139}{30} = 4\frac{19}{30}$$

EXAMPLE 74   Subtract and simplify: $\dfrac{2x}{x-y} - \dfrac{2x}{x+y}$

The l.c.d. is $(x - y)(x + y)$.

$$\frac{2x}{x-y} - \frac{2x}{x+y} = \frac{2x(x+y)}{(x-y)(x+y)} - \frac{2x(x-y)}{(x-y)(x+y)}$$

$$= \frac{(2x^2 + 2xy) - (2x^2 - 2xy)}{(x-y)(x+y)}$$

$$= \frac{4xy}{(x-y)(x+y)}$$

C. Multiplication

In multiplication, the given fractions are not changed to have a common denominator. The product of the numerators and the product of the denominators are the two terms of the answer. In multiplication involving mixed numbers, a simple method is to change each mixed number to an improper fraction before multiplying.

EXAMPLE 75   Multiply $\dfrac{6x}{7y}$ by $\dfrac{5y}{12x}$.

$$\frac{6x}{7y} \cdot \frac{5y}{12x} = \frac{30xy}{84xy} = \frac{5}{14}$$

EXAMPLE 76   Multiply $5\frac{4}{9} \times \frac{2}{7}$.

$$5\frac{4}{9} \times \frac{2}{7} = \frac{49}{9} \times \frac{2}{7} = \frac{98}{63} = \frac{98 \div 7}{63 \div 7} = \frac{14}{9} = 1\frac{5}{9}$$

The fractions may be simplified before multiplication as follows:

$$5\frac{4}{9} \times \frac{2}{7} = \frac{\overset{7}{\cancel{49}}}{9} \times \frac{2}{\underset{1}{\cancel{7}}} = \frac{14}{9} = 1\frac{5}{9}$$

EXAMPLE 77 Multiply and simplify: $\dfrac{5y}{x-y} \cdot \dfrac{x^2 - y^2}{7x^2}$

$$\frac{5y}{x-y} \cdot \frac{x^2 - y^2}{7x^2} = \frac{5y}{\cancel{x-y}} \cdot \frac{\cancel{(x-y)}(x+y)}{7x^2} = \frac{5y(x+y)}{7x^2}$$

D. Division

The following three methods are generally used in division of fractions. Method 1 is relatively popular although it is not superior in every case.

METHOD 1 Multiply the dividend by the reciprocal of the divisor. The reciprocal of a fraction is the fraction inverted; that is, the numerator becomes the denominator and the denominator becomes the numerator.

EXAMPLE 78 Divide $\dfrac{8}{15}$ by $\dfrac{7}{10}$.

$$\frac{8}{15} \div \frac{7}{10} = \frac{8}{\underset{3}{\cancel{15}}} \times \frac{\overset{2}{\cancel{10}}}{7} = \frac{16}{21}$$

The division in Example 78 may be written as a complex fraction, then simplified to the proper fraction:

$$\frac{\dfrac{8}{15}}{\dfrac{7}{10}} = \frac{8}{15} \div \frac{7}{10} = \frac{16}{21}$$

EXAMPLE 79 Divide $\dfrac{x+2}{x-y}$ by $\dfrac{3x-7}{5(x-y)}$.

$$\frac{x+2}{x-y} \div \frac{3x-7}{5(x-y)} = \frac{x+2}{\cancel{x-y}} \cdot \frac{5\cancel{(x-y)}}{3x-7} = \frac{5(x+2)}{3x-7}$$

METHOD 2 After changing both the dividend and the divisor to fractions having the lowest common denominator, cancel the common denominators and divide.

Example 78 is computed by Method 2 as follows:

$$\frac{8}{15} \div \frac{7}{10} = \frac{16}{\cancel{30}} \div \frac{21}{\cancel{30}} = 16 \div 21 = \frac{16}{21}$$

Example 79 is computed by Method 2 as follows:

$$\frac{x+2}{x-y} \div \frac{3x-7}{5(x-y)} = \frac{5(x+2)}{5(x-y)} \div \frac{3x-7}{5(x-y)} = 5(x+2) \div (3x-7)$$

$$= \frac{5(x+2)}{3x-7}$$

METHOD 3 Divide after multiplying both the dividend and the divisor by their lowest common denominator.

Example 78 is computed by Method 3 as follows:

$$\tfrac{8}{15} \div \tfrac{7}{10} = (\tfrac{8}{15} \times 30) \div (\tfrac{7}{10} \times 30) = 16 \div 21 = \tfrac{16}{21}$$

Example 79 is computed by Method 3 as follows:

$$\frac{x+2}{x-y} \div \frac{3x-7}{5(x-y)} = \frac{x+2}{x-y} \cdot 5(x-y) \div \left[\frac{3x-7}{5(x-y)} \cdot 5(x-y) \right]$$

$$= 5(x+2) \div (3x-7) = \frac{5(x+2)}{3x-7}$$

EXERCISE 2–4 Reference: Section 2.6

Perform the following indicated operations and simplify all fractions in answers to lowest terms.

A. *Addition*

1. $\dfrac{3}{4} + \dfrac{5}{8}$

2. $\dfrac{7}{9} + \dfrac{4}{5}$

3. $\dfrac{5a}{7} + \dfrac{3a}{10}$

4. $\dfrac{2}{3b} + \dfrac{7}{4a}$

5. $2\dfrac{5}{7}x + \dfrac{8x}{3} + 4\dfrac{1}{6}x$

6. $\dfrac{3}{24}y + 3\dfrac{5}{8}y + 6\dfrac{2}{3}y$

7. $\dfrac{5x}{x^2 - y^2} + \dfrac{8}{x-y}$

8. $\dfrac{4}{2x-1} + \dfrac{3x+6}{2x^2+3x-2}$

B. *Subtraction*

9. $\dfrac{5}{6} - \dfrac{7}{9}$

10. $6\dfrac{4}{7} - 2\dfrac{7}{8}$

11. $\dfrac{207x}{4} - 23\dfrac{14}{15}x$

12. $50\dfrac{2}{5}y - 16\dfrac{3}{8}y$

13. $\dfrac{2x}{7} - \dfrac{x}{3}$

14. $\dfrac{4}{5x} - \dfrac{7}{8x}$

15. $1 - \dfrac{a}{a+b}$

16. $\dfrac{a}{a-b} - \dfrac{3ab}{a^2-b^2}$

17. $\dfrac{-5}{x+1} - \dfrac{-20x+15}{3x^2-x-4}$

18. $\dfrac{4}{2x+1} - \dfrac{2x-6}{2x^2-5x-3}$

C. *Multiplication*

19. $\dfrac{4}{7} \times \dfrac{12}{23}$

20. $3\frac{5}{6} \times 4\frac{3}{7}$

21. $\dfrac{5x^3y^4}{9y} \times \dfrac{4x^2}{5xy^2}$

22. $\dfrac{3a^2b^3}{4a^3c} \times \dfrac{2a^4c^3}{7bc^5}$

23. $\dfrac{2x^2}{3y} \times \dfrac{8y}{9x} \times \dfrac{15y}{16xy}$

24. $2\frac{7}{8}a^2 \times \dfrac{4ab}{b^2} \times \dfrac{3a^2}{7b}$

25. $\dfrac{8}{3x-15} \times \dfrac{x-5}{4}$

26. $\dfrac{5x-3}{7x^2-14} \times \dfrac{x^2-2}{10x-6}$

27. $\dfrac{4x+10}{2x^2-x-15} \times \dfrac{x-3}{2x+4}$

28. $\dfrac{3x-2}{2x+6} \times \dfrac{2x-8}{3x^2-14x+8}$

D. *Division*

29. $\dfrac{5x^2y^3}{6x^3} \div \dfrac{3x^4}{8x^2y}$

30. $\dfrac{5a^2b}{12a^3b^2} \div \dfrac{3a^5b^4}{7a^3b}$

31. $7\frac{4}{5} \div \dfrac{6x}{85}$

32. $\dfrac{7x}{8} \div 5\frac{2}{7}$

33. $\dfrac{x+1}{x-2y} \div \dfrac{6x-4}{2x-4y}$

34. $\dfrac{3x-2}{2x^2+x-3} \div \dfrac{9x-6}{2x-2}$

35. $\dfrac{x+\dfrac{3}{y^2}}{\dfrac{3+x}{y^2}}$

36. $1 - \dfrac{1}{5-\dfrac{1}{4-\dfrac{2}{3}}}$

Chapter

3 Linear Equations

The fundamental algebraic operations presented in Chapter 2 are required in the process of solving various types of equations. The procedures for solving linear equations presented in this chapter, in turn, are indispensable tools in solving problems in subsequent chapters.

3.1 EQUATIONS

An *equation* is a statement indicating that two algebraic expressions are equal. The two expressions are called the *sides* or *members* of the equation and are connected by an equal sign. There are two types of equations: identical and conditional.

When the two sides of an equation are equal for any value that may be substituted for the letter or letters included, the equation is called an *identical equation*, or simply an *identity*. Thus, $3x + x = 4x$ is an identity since the two sides are equal when x represents any value. For example, when $x = 1$, the left side is $3(1) + 1 = 4$ and the right side is also 4, or $4(1) = 4$.

When the two sides of an equation are equal for only certain values of the letters included, the equation is called a *conditional equation*, or simply an *equation*. Thus, $3x + 1 = 7$ is a conditional equation since the two sides are equal only when x represents 2, or $3(2) + 1 = 7$. The value 2, which satisfies the equation, is called the *solution* or the *root*. The letter (or letters) whose value is required is called the *unknown*.

The *degree of a term* referring to certain unknowns in the term is the sum of the exponents of those unknowns, as stated in Section 2.4. On the other hand, the *degree of an equation* in certain unknowns is the degree of the highest term in the unknowns among all terms in the equation. Thus, $3x + 1 = 7$ is a first degree equation in x, since x representing the unknown has the exponent 1, or $x = x^1$. An equation of the first degree is also called a *linear equation* and can be represented by a straight line on a graph. The equation $6x^2y^3z + 8x^4 - 10yz^2 = 0$ is of the fourth degree in x, the third degree in y, and the second degree in z. In determining the degree of an equation in more than one unknown, such as in x and z, it will be helpful to rewrite each term with the same

44

number of unknowns, or in the form $6x^2y^3z^1 + 8x^4y^0z^0 - 10x^0y^1z^2 = 0$. Now, it is easily seen that this equation is of the fourth degree in x and z, since $4 + 0 = 4$ in the second term. Equations of the second, third, and fourth degrees are also called *quadratic*, *cubic*, and *quartic* (or *biquadratic*) equations, respectively.

A. Linear Equations in One Unknown

In solving a linear equation in one unknown, we must perform the same operation with the same quantity on both sides of the equation. We may add the same quantity to both sides, subtract the same quantity from both sides, and multiply or divide both sides by the same quantity, other than zero. Finally, we isolate the unknown x on one side to obtain the desired solution on the other side of the equation. The procedure for finding the solution of an equation in one unknown is outlined as follows:

(1) Add or subtract the same quantity to or from both sides so that the resulting equation will have the term with the unknown on one side and all other terms on the other side.

(2) Divide both sides of the new equation by the coefficient of the unknown to obtain the solution.

EXAMPLE 1   Solve $5x + 4 = 8x - 17$.

STEP 1:   Subtract $8x$ from both sides to remove $8x$ from the right:
$5x + 4 - 8x = 8x - 17 - 8x$

Collect the like terms on both sides:
$-3x + 4 = -17$

Subtract 4 from both sides to remove 4 from the left:
$-3x + 4 - 4 = -17 - 4$

Collect the like terms on both sides:
$-3x = -21$

STEP 2:   Divide both sides by the coefficient of the unknown (-3):
$$\frac{-3x}{-3} = \frac{-21}{-3}$$
$$x = 7$$

Check: $5(7) + 4 = 8(7) - 17$
$35 + 4 = 56 - 17$
$39 = 39$

Steps 1 and 2 may be simplified by moving the terms from one side of the equation to the other after changing their signs (from $+$ to $-$, $-$ to $+$, $\times$ to $\div$, and $\div$ to $\times$.) In the moving process, all the terms containing the unknown are moved to the left side and all other terms are moved to the right side until the unknown remains alone on the left side. Thus, Example 1 may be simplified as follows:

$$5x + 4 = 8x - 17$$
$$5x - 8x = -17 - 4$$
$$-3x = -21$$
$$x = \frac{-21}{-3}$$
$$x = 7$$

EXAMPLE 2 Solve: $7(x - 4) = x + 26$

$$7x - 28 = x + 26$$
$$7x - x = 26 + 28$$
$$6x = 54$$
$$x = \frac{54}{6} = 9$$

Check: $7(9 - 4) = 9 + 26$
$$35 = 35$$

EXAMPLE 3 Solve: $3ax + b = 2c$ for x.

$$3ax = 2c - b$$
$$x = \frac{2c - b}{3a}$$

Check: $3a\left(\frac{2c - b}{3a}\right) + b = 2c$
$$2c - b + b = 2c$$
$$2c = 2c$$

B. Solving Statement Problems

A *statement* or *word problem* may conveniently be solved by the use of algebraic equations. The steps of solving a statement problem by a linear equation in one unknown are:

1. Represent one of the unknown quantities by the letter x and express other unknown quantities, if any, in terms of the same letter x.
2. Translate the quantities based on the statement of the given problem into algebraic expressions and set up an equation.
3. Solve the equation for x and find the other unknowns from the solution.
4. Check the solution based on the information given in the problem.

EXAMPLE 4 Salesperson A and Salesperson B together sold $842 worth of merchandise. If A sold $278 more than B, how much did each sell?

STEP 1: Let $x =$ one desired unknown amount, or A's sales.
Then, $x - \$278 =$ B's sales.

STEP 2: Set up an equation based on the statement that A and B together sold $842.

$$x + (x - 278) = 842$$

STEP 3: Solve the equation.

$x + x - 278 = 842$

$2x = 842 + 278 = 1.120$

$x = \dfrac{1.120}{2} = \$560$ (A's sales)

Find the other unknown.

$x - 278 = 560 - 278 = \$282$ (B's sales)

STEP 4: Check: $560 + 282 = \$842$ (A and B total sales)

$560 - 282 = \$278$ (A sold more than B)

EXERCISE 3–1 Reference: Section 3.1

A. *Solve each equation for x:*

1. $4x + 7 = 19$
2. $6x - 5 = 7$
3. $8 + 3x = 2$
4. $12 - 7x = 47$
5. $3x + 8 = 2x + 5$
6. $5x - 6 = 3x + 8$
7. $10 + 7x = 46 - 2x$
8. $2x + 15 = 8x + 69$

9. $7x + 5 = 14x + 3$
10. $9 + 8x = 16x + 6$
11. $3a + 2x = 4$
12. $5cx + 8 = 3$
13. $6b + 7x = 12b - 5$
14. $-2x + 3y = 14x - 7y$
15. $4(2a - x) = 3(5x + 2)$
16. $7(3c + 2x) = 2(6x - 5c)$

B. *Solve each statement problem. Let x = one unknown.*

17. Peter and Nancy together have $181.00. If Nancy has $131.50 less than Peter, how much does Peter have?
18. Jack has 29 quarters and dimes totaling $4.40. How many dimes does Jack have?
19. If a number is added to eight times the number, the sum is 63. What is the number?
20. If 6 is subtracted from four times the number, the remainder is 42. What is the number?
21. Steve was twice as old as Alan ten years ago. The sum of their ages today is 32 years. How old is Steve now?
22. Dale is six times as old as his sister. The difference between their ages is 15 years. What is Dale's age?
23. A grocery store has two grades of candy, one selling for $1.20 a pound and the other for $1.80 a pound. How many pounds of each would it take to make 30 pounds of a mixture that can be sold for $1.40 a pound?
24. A man bought 100 shares of a certain stock. He sold 30 shares at twice their cost and the remaining shares at 5 cents on each dollar he paid for the stock. What was the cost per share if his net loss was $365?
25. A car which averaged 45 miles per hour left a place 40 minutes before another car which averaged 55 miles per hour. If both cars took the

same route, how much time was needed for the second car to catch up with the first and how far had they traveled?

26. Two trains left a station at the same time for East City and West City respectively. The eastbound train traveled at a speed 30 miles per hour faster than the westbound. At the end of 5 hours they were 750 miles apart. What was the average speed of each train?

27. Mary has twice as many dimes as quarters and three times as many half-dollars as dimes. The sum of her money is $34.50. How many coins of each kind does she have?

28. During one month's time in the ABC Company, a salesgirl worked $\frac{1}{3}$ of her time in the shoe department, $\frac{2}{9}$ of her time in the music department, and the remaining 78 hours in the toy department. Altogether, how many hours did she work?

29. Larry, John, and Peter made $10,600 net profit from their partnership at the end of the year. Larry's share of the profit was $\frac{3}{5}$ as much as John's and John's share was $\frac{7}{10}$ as much as Peter's. How much did each receive?

30. A man sold $\frac{7}{10}$ of his land in January and 38 acres in February. He then had left 4 acres less than $\frac{1}{3}$ of his land. How many acres were in his land?

3.2 SYSTEMS OF LINEAR EQUATIONS

A system of equations is a set of two or more equations. A linear equation in one unknown has only one solution. A linear equation in two unknowns, however, has an unlimited number of solutions. For example, $x + y = 5$ is satisfied by unlimited pairs of numbers such as $x = 1$, $y = 4$; $x = 2$, $y = 3$; and so on. If x is equal to any numerical value, there is a solution for y in the equation.

In general, if a system has two linear equations in two unknowns, there is only one solution for each unknown that satisfies both equations. The two equations are called *independent simultaneous equations,* or simply, *independent equations*. If the two equations in a system can be reduced to the same form, they are called *dependent equations*. If we multiply both sides of equation $x + y = 3$ by 2, we have equation $2x + 2y = 6$. Thus, $x + y = 3$ and $2x + 2y = 6$ are dependent equations. A system of dependent equations also has an unlimited number of solutions. If there is no common solution for two linear equations in two unknowns, they are called *inconsistent equations*. For example, $x + y = 2$ and $x + y = 3$ in a system are inconsistent equations.

In solving two independent linear equations simultaneously, first eliminate one of the two unknowns from the two equations. Second, solve the resulting equation in one unknown. Third, find the answer to the eliminated unknown by substituting the solution found in the second step in either one of the given equations or their derived equivalents. Two methods of elimination are illustrated in Example 5. Only one method is used in subsequent examples.

EXAMPLE 5   Solve: $\begin{cases} 2x - y = 11 & (1) \\ 3x + y = \ 9 & (2) \end{cases}$

METHOD A   Elimination by Addition or Subtraction

    I. Eliminate y by addition:

Add (1) and (2).
$$\begin{array}{r} 2x - y = 11 \\ 3x + y = \ 9 \\ \hline 5x \quad\ = 20 \end{array}$$
$$x = \frac{20}{5} = 4$$

Substitute $x = 4$ in (1).
$$\begin{aligned} 2(4) - y &= 11, \\ -y &= 11 - 8 \\ y &= -3 \end{aligned}$$

    II. Eliminate x by subtraction:

Multiply (1) by 3. $6x - 3y = 33$ (3)
Multiply (2) by 2. $6x + 2y = 18$ (4)

Subtract: $(3) - (4)$.
$$\begin{aligned} -5y &= 15, \\ y = \frac{15}{(-5)} &= -3 \end{aligned}$$

Note that 6, the coefficients of x in (3) and (4), is the lowest common multiple of 2 and 3, which are the coefficients of x in (1) and (2) respectively.

Substitute $y = -3$ in (1).
$$\begin{aligned} 2x - (-3) &= 11 \\ 2x &= 11 - 3 \\ x = \frac{8}{2} &= 4 \end{aligned}$$

Check: Substitute $x = 4$ and $y = -3$ in (1) and (2).
In (1). $2(4) - (-3) = 11, 8 + 3 = 11, 11 = 11$
In (2). $3(4) + (-3) = 9, 12 - 3 = 9, 9 = 9$

METHOD B   Elimination by Substitution.

Solve (2) for y in terms of x. $y = 9 - 3x$ (5)

Substitute (5) in (1). (Note: Do not substitute in (2) since (5) is derived from (2).)

$$\begin{aligned} 2x - (9 - 3x) &= 11 \\ 2x + 3x &= 11 + 9 \\ 5x &= 20 \\ x = \frac{20}{5} &= 4 \end{aligned}$$

Substitute $x = 4$ in (1).

$$\begin{aligned} 2(4) - y &= 11 \\ -y &= 11 - 8 \\ y &= -3 \end{aligned}$$

Check: See Method A.

EXAMPLE 6   A radio is sold for \$68. The gross profit is 70% of the cost. Assume that the selling price is the sum of the cost and the gross profit. Find the cost and the gross profit.

                 The four steps listed in Section 3.1B are also applicable in solving statement problems involving two unknowns.

STEP 1   Let $x =$ the cost of the radio, and $y =$ the gross profit.

STEP 2 $x + y = 68$ (1)   (Cost + gross profit = price)

                      $y = 70\%(x)$ (2)   (Gross profit = 70% of cost)

STEP 3   Substitute (2) in (1).

$$x + .7x = 68$$
$$1.7x = 68$$
$$x = \frac{68}{1.7} = 40$$

                 Cost $(x) = \$40$

                 Gross profit $(y) = 70\%(40) = \$28$

STEP 4   Check: $40 + 28 = \$68$ (price)

                              $28 \div 40 = .7 = 70\%$ (gross profit)

EXAMPLE 7   The cost of two coats and three hats is \$310. The cost of one coat and four hats is \$205. What is the unit cost of each item?

STEP 1   Let $x =$ unit cost of coat, and

                      $y =$ unit cost of hat.

STEP 2 $2x + 3y = \$310$ (1)

                 $1x + 4y = \$205$ (2)

STEP 3   Multiply (2) by 2.

                 $2x + 8y = \$410$ (3)

                 Subtract (1) from (3)

$$
\begin{array}{rll}
2x + 8y = 410 & (3) \\
2x + 3y = 310 & (1) \\
\hline
5y = 100 & \\
y = \$20 & \text{(one hat)}
\end{array}
$$

                 Substitute $y = \$20$ in (2).

$$1x + 4(20) = 205$$
$$x = 205 - 80 = \$125 \qquad \text{(one coat)}$$

STEP 4   Check:

                 In (1). $2(125) + 3(20) = 310, \; 310 = 310.$

                 In (2). $1(125) + 4(20) = 205, \; 205 = 205.$

When there are three independent linear equations in three unknowns, a solution can be found which will satisfy the three equations. The methods of solving a system of three linear equations in three unknowns are the extension of the methods used in solving two equations in two unknowns. Similarly, the methods may be extended to n number of independent equations in n number of unknowns.

EXAMPLE 8   Solve: $\begin{cases} x + y + z = 5 & (1) \\ 2x - y + 3z = 20 & (2) \\ 3x + 2y - 5z = -15 & (3) \end{cases}$

Eliminate y from (1) and (2).

$(1) + (2)$ $3x + 4z = 25$ (4)

Eliminate y from (2) and (3).

$(2) \times 2$ $4x - 2y + 6z = 40$ (5)
$(3) + (5)$ $7x + z = 25$ (6)

Eliminate z from the new system of equations (4) and (6).

$(6) \times 4$ $28x + 4z = 100$ (7)
$(7) - (4)$ $25x = 75$

$$x = \frac{75}{25} = 3$$

Substitute $x = 3$ in (6).

$$7(3) + z = 25$$
$$z = 25 - 21 = 4$$

Substitute $x = 3$ and $z = 4$ in (1).

$$(3) + y + (4) = 5$$
$$y = 5 - 7 = -2$$

Check: Substitute the solution $x = 3$, $y = -2$, and $z = 4$ in (2).

$$2(3) - (-2) + 3(4) = 20$$
$$6 + 2 + 12 = 20$$
$$20 = 20$$

EXERCISE 3–2   Reference: Section 3.2

A. *Solve for x and y (elimination by addition or subtraction):*

1. $x + y = 5$
    $3x - y = 7$

2. $x + 2y = 1$
    $-x - 3y = -3$

3. $x - 4y = 6$
    $x + 3y = -1$

4. $2x + y = 14$
    $6x + y = 34$

5. $3x - 5y = 12$
    $x + 4y = -13$

6. $2x - y = 7$
    $7x + 3y = 5$

7. $x + 3y = -3$
    $5x - 7y = 29$

8. $4x + y = 10$
    $-2x - 5y = -32$

9. $2x + 3y = 17$
    $5x - 2y = -5$

10. $7x + 3y = -11$
    $6x + 5y = -7$

B. *Solve for x and y (elimination by substitution):*

11. $x - y = 2$
    $2x + 3y = 9$

12. $x + y = 4$
    $5x - 4y = -7$

13. $2x + 3y = 5$
    $x - 2y = 6$

14. $3x + y = 1$
    $2x - 5y = 12$

15. $4x + y = 11$
    $7x + 2y = 20$

16. $-x + 2y = 5$
    $4x + 3y = 2$

C. *Solve for x, y, and z:*

17. $2x + y - z = 1$
    $x + 2y + z = 8$
    $3x - y - z = -2$

18. $x - y + z = 4$
    $4x + y - 2z = 5$
    $2x + y + 3z = 6$

19. $3x + 2y - z = 1$
    $2x - y + 5z = 5$
    $6x + 3y - 2z = -1$

20. $4x - 3y + 2z = -19$
    $3x + 2y - 7z = 24$
    $2x - 3y + z = -11$

D. *Solve each statement problem. Let x = one unknown and y = another unknown.*

21. The selling price of 2 pounds of beef and 7 cans of coffee is $18.81, and the selling price of 1 pound of beef and 5 cans of coffee is $12.33. Find the price of beef and the price of coffee.

22. The cost of 3 shirts and 5 brushes is $46.50 and the cost of 2 shirts and 1 brush is $22.60. What is the unit cost of the shirt and the brush?

23. The sum of two numbers is 235. The sum of the larger number times $\frac{1}{3}$ and the smaller number times $\frac{1}{7}$ is 55. What are the numbers?

24. Two workers made a total of $527 from a job. One worker made $45 more than the other worker. How much did each make?

25. A theater sold 242 tickets totaling $1,096. The tickets were sold to adults for $5.40 and to children for $2.00. How many tickets of each kind were sold?

26. A company mailed 100 letters and paid a total of $11.20 for postage. Some of the letters cost 10¢ each and others cost 13¢ each. Find the number of letters at each rate.

27. Peter and Nancy together have $181.00. If Nancy has $131.50 less than Peter, how much does each have?

28. Jack has 29 quarters and dimes totaling $4.40. Find the number of quarters and the number of dimes.

29. Steve was twice as old as Alan ten years ago. The sum of their ages today is 32 years. How old is Steve now? Alan?

30. Dale is six times as old as his sister. The difference between their ages is 15 years. What is Dale's age? His sister's age?

31. A grocery store has two grades of candy, one selling for $1.20 a pound and the other for $1.80 a pound. How many pounds of each would it take to make 30 pounds of a mixture that can be sold for $1.40 a pound?

32. Two trains left a station at the same time for East City and West City respectively. The eastbound train traveled at a speed 30 miles per hour faster than the westbound train. At the end of 5 hours they were 750 miles apart. What was the average speed of each train?

3.3 FRACTIONAL EQUATIONS

In solving an equation involving fractions, first multiply both sides by the lowest common denominator. The result is an equation that contains no fractions. This process is called *clearing* an equation of fractions. Next, solve the new equation for the unknown as discussed in Section 3.1.

When a fractional equation is cleared, multiplying by a denominator other than the lowest common denominator may introduce solutions which are not solutions of the original equation. Those additional solutions are called *extraneous roots*. Also, if we multiply both sides of a fractional equation by the same expression containing the unknown, or raise both sides to the same integral power, the resulting equation may have more solutions than the original equation. The extraneous roots are discarded in solving statement problems. However, if we divide both sides of a fractional equation by an expression containing the unknown, the new equation may have fewer roots than the original equation. Furthermore, division or multiplication by zero must be excluded.

EXAMPLE 9 Solve $\dfrac{x}{x-8} = 3.$

Clear the equation.

$$\frac{x}{x-8}(x-8) = 3(x-8)$$
$$x = 3(x-8)$$

Solve the new equation.

$$x = 3x - 24$$
$$x - 3x = -24$$
$$-2x = -24$$
$$x = 12$$

Check: $\dfrac{12}{12-8} = \dfrac{12}{4} = 3$

EXAMPLE 10 Solve $\dfrac{x-1}{x+3} = \dfrac{x+2}{x+10}$

Clear the equation. Here the lowest common denominator of the two fractions is $(x+3)(x+10)$.

$$\frac{x-1}{x+3}(x+3)(x+10) = \frac{x+2}{x+10}(x+3)(x+10)$$
$$(x-1)(x+10) = (x+2)(x+3)$$

Solve the new equation.

$$x^2 - x + 10x - 10 = x^2 + 2x + 3x + 6$$
$$x^2 + 9x - 10 = x^2 + 5x + 6$$
$$x^2 + 9x - x^2 - 5x = 6 + 10$$
$$4x = 16$$
$$x = 4$$

Check: $\dfrac{x-1}{x+3} = \dfrac{4-1}{4+3} = \dfrac{3}{7}$

$\dfrac{x+2}{x+10} = \dfrac{4+2}{4+10} = \dfrac{6}{14} = \dfrac{3}{7}$

3.4 RATIO, PROPORTION, AND PERCENT

Numerous types of business problems, such as discounts, commissions, taxes, interests, annuities, and insurance premiums, involve the use of percentages. Percentage problems can conveniently be solved by applying the operations used for solving a linear equation. A percentage is a special type of ratio and is closely associated with proportion. We therefore begin with a discussion of ratio and proportion.

A. Ratio

Numbers may be compared in terms of actual or relative values. For example, in one class, there are 45 men and 15 women. Instead of stating actual numbers, it is more convenient to state relative numbers. Therefore, the number of men compared with the number of women is 3 to 1, since:

$$\frac{45 \text{ men}}{15 \text{ women}} = \frac{3 \text{ men}}{1 \text{ woman}}$$

A ratio is a way of expressing the relative values of different things. Thus, the comparison above may be stated as "the ratio of the number of men to the number of women in the class is 3 to 1," which may be written 3:1, or $\frac{3}{1}$.

In general, when a number, called the *first term*, is divided by another number, called the *second term*, the quotient is the ratio of the first term to the second term.

$$\begin{pmatrix} \text{Ratio of first} \\ \text{term to} \\ \text{second term} \end{pmatrix} = \begin{pmatrix} \text{First} \\ \text{term} \end{pmatrix} : \begin{pmatrix} \text{Second} \\ \text{term} \end{pmatrix} = \frac{\text{First term}}{\text{Second term}}$$

Thus, the ratio of a to b is written $a{:}b$, or a/b, or $a \div b$; the ratio of b to a is written $b{:}a$, or b/a, or $b \div a$; and the ratio of the number of women to the number of men in the class is 1 to 3, which is written 1:3, or $\frac{1}{3}$.

Since a ratio may be expressed as a fraction, the rules applying to fractions likewise apply to ratios. For example, the fraction $\frac{2}{8}$ may be reduced to its lowest terms, $\frac{1}{4}$. Thus, the ratio of 2 to 8 is equal to the ratio of 1 to 4, or $2{:}8 = 1{:}4$.

B. Proportion

A proportion is a statement of the equality of two ratios. For example, $a{:}b = c{:}d$, or $a/b = c/d$, is a proportion. It is read "a is to b as c is to d," or "the ratio of a to b is equal to the ratio of c to d." The letters a, b, c, and d are called the *terms* of the proportion.

Since proportions are equations, the rules and operations of equations apply to proportions. Thus, if both sides of the proportion:

$$\frac{a}{b} = \frac{c}{d}$$

are multiplied by the common denominator bd, then:

$$\frac{a}{\not{b}} \cdot \not{b}d = \frac{c}{\not{d}} \cdot b\not{d}$$

$$ad = bc$$

The result indicates that by *cross multiplication* of the terms in the proportion, the two products are equal.

In the proportion $a:b = c:d$, a and d are called the *extremes*, and b and c are called the *means*. It follows that the product of the extremes (ad) equals the product of the means (bc). When any three of the four terms in a proportion are given, the other unknown term can always be determined by the equation $ad = bc$.

EXAMPLE 11 Solve for x: $\dfrac{x}{10} = \dfrac{3}{20}$

Using cross multiplication of the terms, the given proportion becomes:

$$20x = (10)(3)$$
$$x = \tfrac{30}{20} = \tfrac{3}{2} = 1.5$$

Check: $\dfrac{1.5}{10} = \dfrac{3}{20}$

$$\frac{1.5 \times 2}{10 \times 2} = \frac{3}{20}$$

$$\frac{3}{20} = \frac{3}{20}$$

EXAMPLE 12 Solve for y. $5 : y = \tfrac{1}{4} : 10$

The product of the extremes equals the product of the means. Thus:

$$(5)(10) = y(\tfrac{1}{4})$$
$$50 = \frac{y}{4}$$
$$y = 50(4) = 200$$

Check: $\dfrac{5}{200} = \dfrac{\frac{1}{4}}{10}$

$$\frac{5}{200} = \frac{\frac{1}{4} \times 20}{10 \times 20}$$

$$\frac{5}{200} = \frac{5}{200}$$

EXAMPLE 13 A food store charges $10.60 for 4 pounds of beef. How much will it charge for 5.6 pounds?

Let x be the price of 5.6 pounds of beef. Then, the problem may be stated in proportional language as follows: $10.60 is to 4 pounds as x is to 5.6 pounds, since the price per pound is constant.

Price per pound   Price per pound

$$\frac{\$10.60}{4 \text{ pounds}} = \frac{x}{5.6 \text{ pounds}}$$

Solve for x: $\dfrac{10.60}{4} = \dfrac{x}{5.6}$

$$4x = (10.60)(5.6) = 59.36$$
$$x = \tfrac{59.36}{4} = \$14.84$$

Check: $\tfrac{10.60}{4} = \$2.65$ per pound
$\tfrac{59.36}{5.6} = \$2.65$ per pound

C. Percent

The percent unit is one hundredth, which can be written in the decimal form .01, in the fractional form $\tfrac{1}{100}$, or by using the symbol %. Thus, a percent is also a type of ratio. For example, 3% may be written as $\tfrac{3}{100}$, which is the ratio of 3 to 100.

The number of hundredths (such as 3%) is also called the *percent rate*, or simply *rate*. The rate must be expressed on a base. A *base* is the number which is regarded as a whole or is equivalent to 100%. The product of the base and the rate is called the *percentage*, written:

Base × Rate = Percentage

The following examples illustrate the applications of this equation.

EXAMPLE 14 What is 3% of $80?

3% is the rate, and $80 is the base from which 3% is taken.

$$80 \times 3\% = 80 \times .03 = \$2.40 \text{ (percentage)}$$

Check: Use the proportion method. Let $x =$ the percentage. $80 is to 100% (base) as x is to 3%, or:

$$\frac{\$80}{100\%} = \frac{\$x}{3\%}, \ x(100\%) = 80(3\%), \ x = \$2.40.$$

EXAMPLE 15 What percent of $620 is $49.60?

Let $x =$ percent rate.

$$\$620 \times x = \$49.60$$
$$x = \frac{49.60}{620} = .08 = 8\%$$

EXAMPLE 16 If 12% of a number is 42, what is the number?

Let $x =$ the number (base).

$$x(12\%) = 42$$
$$x = \frac{42}{12\%} = \frac{42}{.12} = 350$$

EXAMPLE 17 A grocery store priced a box of candy at $8.91. The price was 32% more than the cost. Find the cost.

Let $x =$ cost.

$$x + 32\%(x) = \$8.91$$
$$x(1 + 32\%) = \$8.91$$
$$x = \frac{8.91}{1.32} = \$6.75$$

Check: $6.75(32\%) = \$2.16$
$6.75 + \$2.16 = \$8.91$ (price)

EXAMPLE 18 John had $4,230 at the end of last month after losing 6% of his investment during the month. Find the amount of his investment at the beginning of the month.

Let $x =$ the amount of his investment at the beginning of the month.

$$x - 6\%x = \$4,230$$
$$x(1 - 6\%) = 4,230$$
$$x(94\%) = 4,230$$
$$x = \frac{4,230}{.94} = \$4,500$$

Check: $4,500(6\%) = \$270$ (amount lost)
$4,500 - \$270 = \$4,230$ (amount remaining)

EXERCISE 3–3 Reference: Sections 3.3 and 3.4

A. *Solve for x:*

1. $\dfrac{x}{x - 40} = 9$

2. $\dfrac{x}{10 + 2x} = \dfrac{3}{7}$

3. $\dfrac{x}{8} + \dfrac{3x}{4} = 14$

4. $\dfrac{2x}{5} - \dfrac{5x}{12} = -2$

5. $\dfrac{x + 1}{x - 2} = \dfrac{x + 9}{x + 3}$

6. $\dfrac{x - 4}{x + 5} = \dfrac{x + 4}{x + 22}$

7. $\dfrac{x + a}{x - 1} = \dfrac{x + 2a}{x - 3}$

8. $\dfrac{x + 3b}{x + b} = \dfrac{x - 2}{x - 5}$

9. $\dfrac{x}{20} = \dfrac{4}{15}$

10. $\dfrac{5}{18} = \dfrac{25}{x}$

11. $12 : x = 8 : \frac{1}{3}$

12. $17 : 35 = x : 40$

B. *Statement problems:*

13. If 40 lamps cost $140, what will 25 lamps cost?
14. If 15 cases of milk cost $153.75, what will 8 cases cost?
15. A distance of $1\frac{1}{2}$ inches on a map is used to represent 375 miles. How many miles are represented by $4\frac{1}{2}$ inches?
16. If a car travels 25 miles in 20 minutes, how long will it take to travel 180 miles?
17. If a group of 35 workers can make 630 toys, how many toys can a group of 42 workers make?
18. When a building casts a shadow 66 feet long, the shadow of a boy $4\frac{1}{2}$ feet tall is 6 feet long. How high is the building?
19. What is 18% of $350?
20. What is 75% of $265?
21. What percent of $580 is $145?
22. $59.22 is what percent of $169.20?
23. If 24% of a number is 84, what is the number?
24. $1.43 is 5% of what amount?
25. A retail store sold a watch at 35% more than its cost. What was the selling price if the cost of the watch was $50?
26. Bob and Jack agreed to divide equally the profit from a business after having paid 20% of the profit for the rent of a truck. The profit was $5,000. How much did each person receive?
27. $279 is what percent less than $465?
28. There were 1,200 employees last year and 1,380 employees this year in a company. Find the percent of increase.
29. A store sold an old washing machine for $110, a loss of 12% on the purchase price. What was the purchase price?
30. An adding machine was sold for $312. The gross profit was 30% of the cost. Assume that the selling price is equal to the gross profit plus the cost. What is the cost?

3.5 GRAPHS AND EQUATIONS

This section is divided into two parts: rectangular coordinates and graph of an equation.

A. Rectangular Coordinates

An equation can be represented by a graph. A graph is constructed according to the system of *rectangular coordinates*. This system requires two straight reference lines perpendicular to each other in a plane as shown in Figure 3–1. The horizontal line is the *X-axis* and the vertical line is the *Y-axis*.

The two lines divide the plane into four parts, called *quadrants.* The quadrants are numbered I, II, III, and IV as indicated in the figure.

Figure 3-1 **Rectangular Coordinates**

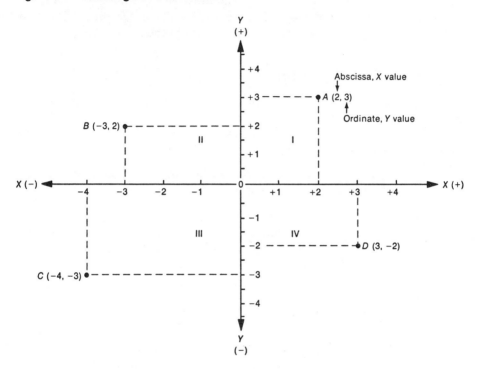

The point of intersection of the two axes, the *origin,* is usually used to represent zero value on both horizontal and vertical scales. The numbers on the scale to the *right* of the origin are conventionally designated as *positive,* whereas those to the *left* of the origin are *negative.* The numbers on the scale *above* the origin are *positive* and those *below* the origin are *negative.*

Two related variables, *dependent* and *independent,* can be described in the plane. The dependent variable is so called because its value depends upon the value of the independent variable. Once the value of the independent variable is assigned, the corresponding value of the dependent variable can be determined from the accompanying equation. The independent variable, also called the X-variable, is usually placed on the X-axis. The dependent variable, also called the Y-variable, is usually placed on the Y-axis.

The steps in placing a point on the plane for a set of corresponding X and Y values are as follows:

1. Draw a line parallel to the Y-axis, starting at the X-axis at a distance equal to the X-value. This X-distance is also called the *abscissa.*
2. Draw a line parallel to the X-axis, starting at the Y-axis at a distance equal to the Y-value, until it intersects the first line. This Y-distance is also called the *ordinate.* The point of intersection is the desired location.

The abscissa and the ordinate are called the *coordinates* of the point. For example, in quadrant I of Figure 3–1 the abscissa of point A is 2 and the ordinate of the point is 3. The values of 2 and 3 are the coordinates of A. The coordinates are usually written in parentheses with the abscissa before the ordinate. Thus, the coordinates of A are written (2, 3). Observe that the co-ordinates of points B, C, and D are also written in the same manner.

B. Graph of an Equation

The procedure of drawing the graph of an equation is illustrated in the following two examples.

EXAMPLE 19 Draw a graph for the equation $2x - y = 11$.

There are an infinite number of values for x and y which will satisfy the given equation. For example, when $x = 6$, $y = 2x - 11 = 2(6) - 11 = 1$.

Three pairs of x and y values are computed here for drawing a straight line on Figure 3–2.

x	$y = 2x - 11$
0	−11
4	−3
6	1

Actually, only two points are required to determine a straight line. However, the third point can be used as a checking point. Observe that the three points form the straight line A. Any point on line A, in turn, will satisfy the given equation. In general, a first-degree equation in one or two unknowns can be represented by a straight line. Therefore, it is frequently referred to as a linear equation.

EXAMPLE 20 Graph the equation $3x + y = 9$.

There are also unlimited values for x and y to satisfy the given equation. For example, when $x = 4$, $y = 9 - 3x = 9 - 3(4) = -3$.

Again, three pairs of x and y values are computed here for drawing a straight line.

x	$y = 9 - 3x$
0	9
4	−3
6	−9

Line B is drawn in Figure 3–2 based on the three points representing the computed values.

Lines A and B in Figure 3–2 intersect at the point (4, −3). The values $x = 4$ and $y = -3$ satisfy both equations given in Examples 19 and 20. It is obvious that the graphic method can be used to solve a system of two independent linear equations in two unknowns. Thus, the graphic solution of the system of the two equations $2x - y = 11$ (Example 19) and $3x + y = 9$ (Example 20) is $x = 4$ and $y = -3$. This graphic solution is the same as the solution obtained by the elimination methods presented in Example 5, page 49.

Figure 3-2 **Graphs of Two Linear Equations** **(Examples 19 and 20)**

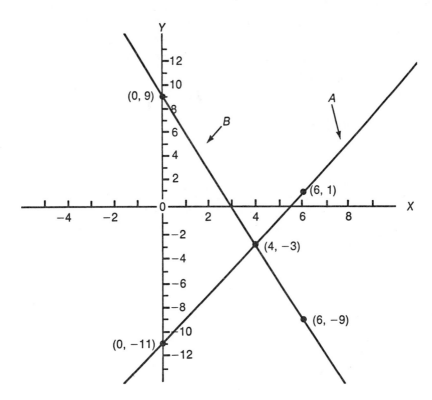

EXERCISE 3–4   Reference: Section 3.5

Solve for x and y graphically:

1. $x + y = 5$
   $3x - y = 7$

2. $x + 2y = 1$
   $-x - 3y = -3$

3. $x - 4y = 6$
   $x + 3y = -1$

4. $2x + y = 14$
   $6x + y = 34$

5. $3x - 5y = 12$
   $x + 4y = -13$

6. $2x - y = 7$
   $7x + 3y = 5$

7. $x + 3y = -3$
   $5x - 7y = 29$

8. $4x + y = 10$
   $-2x - 5y = -32$

9. $2x + 3y = 17$
   $5x - 2y = -5$

10. $7x + 3y = -11$
    $6x + 5y = -7$

11. $x - y = 2$
    $2x + 3y = 9$

12. $x + y = 4$
    $5x - 4y = -7$

3.6 FUNCTIONS

In the equation $y = 2x - 11$, a relationship between variables y and x is expressed such that if x values are assigned, the corresponding values of y can be determined. Thus, if we assign:

$$x = 0, \quad \text{then} \quad y = 2(0) - 11 = -11$$
$$x = 1, \qquad\qquad y = 2(1) - 11 = -9$$
$$x = 2, \qquad\qquad y = 2(2) - 11 = -7, \text{ and so on.}$$

If two variables x and y are so related that, for a value assigned to x, the value of y is determined, then y is said to be a function of x.

A. Functional Notation

A notation to represent "y is a function of x" is:

$$y = f(x)$$

which is read, "y equals f of x." In this notation, since the value of y depends upon the value assigned to x, we usually call x the *independent variable* and y the *dependent variable*. Observe that the expression $f(x)$ does not mean f times x and that y is determined *after* a value has been assigned to x. For the above illustration, now we may write:

$$y = f(x) = 2x - 11$$

EXAMPLE 21 If $f(x) = 3x + 5$, find: (a) $f(0)$, (b) $f(2)$, and (c) $f(-2)$.

We simply replace x by the assigned value in parentheses wherever x occurs in $f(x)$ in each problem.

(a) $x = 0$ $\quad$ $f(0) = 3(0) + 5 = 5$
(b) $x = 2$ $\quad$ $f(2) = 3(2) + 5 = 11$
(c) $x = -2$ $\quad$ $f(-2) = 3(-2) + 5 = -1$

EXAMPLE 22 If $f(x) = 4 + 7x$, find $f(x + 2)$.

Replace x by the assigned value $(x + 2)$ in the given equation.

$$f(x + 2) = 4 + 7(x + 2) = 4 + 7x + 14 = 18 + 7x$$

EXAMPLE 23 If $y = f(x) = 2x - 11$, find $f(y)$ in terms of x.

Replace x by the assigned value $y = 2x - 11$ in the given equation.

$$f(y) = 2y - 11 = 2(2x - 11) - 11$$
$$= 4x - 22 - 11$$
$$= 4x - 33$$

The choice of letter f as the name for the word *function* is arbitrary. We could use other letters, such as g, G, or F, and write the function of x as $g(x)$, $G(x)$, or $F(x)$. In fact, it is desirable to use different letters if we wish to distinguish different functional relationships occurring in the same problem.

The functional notation for two or more independent variables is the extension of the notation for one independent variable, such as in the expression:

$$t = 3x + 5y - 1$$

where t depends upon both x and y for its value. Thus, the expression indicates that t is a function of x and y, or written:

$$t = f(x,y) = 3x + 5y - 1$$

Additional illustrations are:

$$F(r,s) = 2r - 4s + 3$$
$$g(x,y,z) = x + 3y - 2z + 6$$

B. Zero of a Function

If a function is equal to zero, the solution value of the independent variable is called the *zero of the function.*

EXAMPLE 24   If $f(x) = 2x - 11$, find the zero of the function.

First, let the function be equal to zero, or:

$$f(x) = 2x - 11 = 0$$

Next, solve the equation for x.

$$2x - 11 = 0$$
$$2x = 11$$
$$x = 5\tfrac{1}{2}$$

Thus, $5\tfrac{1}{2}$ is the zero of the function $(2x - 11)$.

EXAMPLE 25   If $f(x) = 9 - 3x$, find the zero of the function.

$$f(x) = 9 - 3x = 0$$
$$-3x = -9$$
$$x = 3$$

Thus, 3 is the zero of the function $(9 - 3x)$.

The concept of the zero of a function can be used to solve equations graphically. The steps of the graphical method are:

1. Collect all terms on the left side and let the right side be zero.
2. Let $f(x)$ represent the collected left side and graph the equation $y = f(x)$ on a chart.
3. Find those values of x for which $y = 0$ on the graph.

The graphical method is very useful in solving equations of higher degree, such as the quadratic equations presented in Section 4.4D. The following two examples illustrate this method for solving linear equations in one unknown.

EXAMPLE 26 Solve for x from the equation $5x + 8 = 3x + 19$.

First, collect: $5x + 8 - 3x - 19 = 0$
$$2x - 11 = 0$$

Second, graph: $y = f(x) = 2x - 11$
See Line A in Figure 3–2.

Third, find x on line A:

When $y = f(x) = 0$, line A intersects the X-axis at $x = 5\frac{1}{2}$.

This answer, $x = 5\frac{1}{2}$, is the same as the solution for x in Example 24.

EXAMPLE 27 Solve for x from the equation $2x - 7 = 5x - 16$.

Collect: $2x - 7 - 5x + 16 = 0$
$$-3x + 9 = 0$$

Graph: $y = f(x) = 9 - 3x$
See Line B in Figure 3–2.

Find x on line B:

When $y = f(x) = 0$, line B intersects the X-axis at $x = 3$.

This answer, $x = 3$, is the same as the solution for x in Example 25.

EXERCISE 3–5   Reference: Section 3.6

A. *Find the values of the given functions:*

1. If $f(x) = 10x$, find $f(2)$
2. If $f(x) = -25x$, find $f(5)$
3. If $f(x) = 7 + 3x$, find $f(4)$
4. If $f(x) = 5x - 17$, find $f(-2)$
5. If $f(x) = \dfrac{1}{2x + 3}$, find $f(-1)$
6. If $f(x) = \dfrac{3}{7x - 5}$, find (6)
7. If $g(x) = 5x + 14$, find $g(-3)$
8. If $F(y) = 6 + 7y$, find $F(3)$
9. If $y = f(x) = 4x + 5$, find $f(y)$
10. If $z = G(x) = 9 - 2x$, find $G(z)$
11. If $f(x) = 2x - 3$, find $f(x + 5) - f(x)$
12. If $f(x) = \dfrac{3}{10 - x}$, find $f(3x + 4)$

B. *Solve each equation for x graphically:*

13. $4x + 7 = 19$
14. $6x - 5 = 7$
15. $8 + 3x = 2$
16. $12 - 7x = 47$
17. $3x + 8 = 2x + 5$
18. $5x - 6 = 3x + 8$
19. $10 + 7x = 46 - 2x$
20. $2x + 15 = 8x + 69$
21. $7x + 5 = 14x + 3$
22. $9 + 8x = 16x + 6$

Chapter

4 Radicals and Quadratic Equations

This chapter introduces the basic algebraic operations involving numbers expressed in radicals. It also discusses the methods of finding square roots and solving equations with radicals. Those operations and methods are useful in solving the quadratic equations presented in the last section of this chapter.

4.1 RADICALS

A *radical* is another form of expressing an exponential number. As mentioned in Chapter 2, the product of equal factors can be expressed in an exponential form, such as $a \cdot a \cdot a \cdot a = a^4$; a is the *base* and 4 is the *power*. Now, let the exponential term equal a given quantity b, or $a^4 = b$. Then, a can also be called the 4th root of b and is written in a radical form $a = \sqrt[4]{b}$.

In the radical $\sqrt[4]{b}$, the number 4 written at the upper left corner of the *radical sign* ($\sqrt{}$) is called the *index* and the quantity represented by the letter b under the sign is called the *radicand*. When the index is 2 in a radical, the index is usually omitted and the radical is understood to be the square root. Thus, $\sqrt[2]{25} = \sqrt{25}$.

When the index is an *even* integer, a *positive* radicand has two numerically equal roots: one is *positive* and the other is *negative*. The positive root is called the *principal root* and is denoted by the symbol $\sqrt{}$. For example, the number 25 has two square roots, $+5$ and -5, since $(+5)(+5) = 25$ and $(-5)(-5) = 25$. The positive square root (or the principal square root) is written $\sqrt{25} = +5$. The radical sign $\sqrt{}$ in this case is restricted to represent only the positive root. Thus, $\sqrt{25}$ is not $= -5$. To indicate a negative root, we must place a minus sign before the radical, or $-\sqrt{25} = -5$.

When the index is an *odd* integer, a *positive* radicand has only a *positive* root and a *negative* radicand has only a *negative* root. For example:

$\sqrt[3]{8} = +2$, not -2, since only $(+2)(+2)(+2) = 8$; and
$\sqrt[3]{-8} = -2$, not $+2$, since only $(-2)(-2)(-2) = -8$.

When the index is an *even* integer, a radical with a *negative* radicand represents an *imaginary number* (introduced in Chapter 2). For example, $\sqrt{-25}$ is an imaginary number since neither $+5$ nor -5 qualifies as a square root of the radicand -25.

A. Changing Radical Form

A number with a fractional exponent may be changed to the form of a radical and vice versa. There are also many ways to change the radical form itself. However, the changes must be made in conformity with the laws of exponents as stated in Chapter 2.

In order to change a number with a fractional exponent to a radical or vice versa, the following definition is needed:

$$a^{n/m} = \left(\sqrt[m]{a} \right)^n = \sqrt[m]{a^n}$$

When $n = 1$, the definition becomes:

$$a^{1/m} = \sqrt[m]{a}$$

The above definition can be obtained as follows:

Let $n = 1$ and $m = 2$. Then, $a^{n/m} = a^{1/2}$. When $a^{1/2}$ is raised to its mth (or 2d) power:

$$a^{1/2} \cdot a^{1/2} = a^{1/2 + 1/2} = a^1 = a$$

Observe that $a^{1/2}$ is one of the two equal factors whose product is a. Thus, $a^{1/2}$ is the *2d* (or square) root of a, written:

$$a^{1/2} = \sqrt[2]{a} = \sqrt{a}$$

Similarly, let $n = 2$ and $m = 3$. Then $a^{n/m} = a^{2/3}$. When $a^{2/3}$ is raised to its mth (or 3d) power:

$$a^{2/3} \cdot a^{2/3} \cdot a^{2/3} = a^{2/3+2/3+2/3} = a^2$$

Observe that $a^{2/3}$ is one of the three equal factors whose product is a^2. Thus, $a^{2/3}$ is the *3d* (or cube) root of a^2, written:

$$a^{2/3} = \sqrt[3]{a^2}$$

Extending the above illustration, we have:

$$a^{n/m} = \sqrt[m]{a^n} \quad \text{and} \quad a^{1/m} = \sqrt[m]{a}$$

Also:

$$a^{n/m} = (a^{1/m})^n = \left(\sqrt[m]{a} \right)^n$$

The application of the above definition is further illustrated by the following two examples:

EXAMPLE 1   Express the exponential forms in their equivalent radical forms.

$$15^{1/3} = \sqrt[3]{15} \qquad\qquad\qquad b^{1/4} = \sqrt[4]{b}$$

$$258^{3/5} = \sqrt[5]{258^3} \quad \text{or} \quad \left(\sqrt[5]{258} \right)^3 \qquad x^{4/7} = \sqrt[7]{x^4} \quad \text{or} \quad \left(\sqrt[7]{x} \right)^4$$

EXAMPLE 2 Express the radicals to their equivalent exponential forms.

$$\sqrt{65} = 65^{1/2} \qquad\qquad \sqrt[5]{x} = x^{1/5}$$

$$\sqrt[8]{146^2} = 146^{2/8} = 146^{1/4} \qquad \sqrt[4]{c^3} = c^{3/4}$$

The following two changes in radical form are frequently needed in radical computations. The changes are illustrated by examples.

1. Placing the coefficient of a radical under the radical sign.

EXAMPLE 3 $\quad a\sqrt{b} = a^{2/2} \cdot b^{1/2} = (a^2 b^1)^{1/2} = \sqrt{a^2 b}$

$\qquad\qquad a\sqrt[3]{b} = a^{3/3} \cdot b^{1/3} = (a^3 b^1)^{1/3} = \sqrt[3]{a^3 b}$

$\qquad\qquad x^3 \sqrt{y} = x^{6/2} \cdot y^{1/2} = (x^6 y^1)^{1/2} = \sqrt{x^6 y}$

Observe the answers in Example 3. We may derive the following two steps to simplify the process of placing the coefficient of a radical under the radical sign:

1. Raise the coefficient to the power equal to the index of the radical.
2. Write the result obtained in (1) as a factor to the original radicand.

Based on these two steps, the illustrations in Example 3 are simplified as follows:

$\quad a\sqrt{b} = \sqrt{a^2 b}$, because the index of the radical is 2.

$\quad a\sqrt[3]{b} = \sqrt[3]{a^3 b}$, because the index of the radical is 3.

$\quad x^3 \cdot \sqrt{y} = \sqrt{(x^3)^2 \cdot y} = \sqrt{x^6 y}$, because the index of the radical is 2.

Example 4 gives additional illustrations of the applications of the two steps.

EXAMPLE 4 $\qquad 5\sqrt{3} = \sqrt{5^2 \cdot 3} = \sqrt{75}$

$\qquad\qquad\qquad 4\sqrt[3]{2} = \sqrt[3]{4^3 \cdot 2} = \sqrt[3]{128}$

$\qquad\qquad 2^3 \cdot \sqrt{5} = \sqrt{(2^3)^2 \cdot 5} = \sqrt{2^6 \cdot 5} = \sqrt{320}$

$\qquad 3a \cdot \sqrt[4]{2b} = \sqrt[4]{(3a)^4 \cdot 2b} = \sqrt[4]{3^4 a^4 \cdot 2b} = \sqrt[4]{162 a^4 b}$

2. Changing a factor of the radicand to a coefficient.

The procedure of changing a factor of a given radicand to a coefficient is the reverse of the procedure used in Examples 3 and 4. This is illustrated in Example 5.

EXAMPLE 5 Remove all possible factors from the given radicands and express the factors as the coefficients of new radicals.

$$\sqrt{75} = \sqrt{25 \cdot 3} = \sqrt{5^2 \cdot 3} = 5\sqrt{3}$$

$$\sqrt[3]{a^3 b} = a\sqrt[3]{b}$$

$$\sqrt{72x^3} = \sqrt{36x^2 \cdot 2x} = \sqrt{(6x)^2 \cdot 2x} = 6x \cdot \sqrt{2x}$$

B. Computing Radicals

The basic operations of computing radicals are addition, subtraction, multiplication, and division. They are illustrated in subsections 1 through 3.

1. ADDITION AND SUBTRACTION OF RADICALS. Radicals having the same index and the same radicand are called *like radicals* and can be added or subtracted. The sum or remainder of the coefficients of the like radicals is the coefficient of the common radical in the answer.

EXAMPLE 6   Combine $5\sqrt{2} + 8\sqrt{2} - 4\sqrt{2}$.

The three radicals are like radicals. Thus:

$$5\sqrt{2} + 8\sqrt{2} - 4\sqrt{2} = (5 + 8 - 4)\sqrt{2} = 9\sqrt{2}$$

EXAMPLE 7   Combine $\sqrt{75} - \sqrt{108} + \sqrt{300}$.

The three radicals are unlike radicals. However, they can be changed to like radicals as follows:

$$\sqrt{75} = \sqrt{5^2 \cdot 3} = 5\sqrt{3} \quad \text{(See Example 5.)}$$
$$\sqrt{108} = \sqrt{6^2 \cdot 3} = 6\sqrt{3}$$
$$\sqrt{300} = \sqrt{10^2 \cdot 3} = 10\sqrt{3}$$
$$\text{Thus: } \sqrt{75} - \sqrt{108} + \sqrt{300} = 5\sqrt{3} - 6\sqrt{3} + 10\sqrt{3}$$
$$= (5 - 6 + 10)\sqrt{3}$$
$$= 9\sqrt{3}$$

EXAMPLE 8   Combine $7\sqrt{3a} + 9\sqrt{3a} - 5\sqrt{2b}$

$$7\sqrt{3a} + 9\sqrt{3a} - 5\sqrt{2b} = 16\sqrt{3a} - 5\sqrt{2b}$$

2. MULTIPLICATION OF RADICALS. When the indexes of the radicals are the same, use the following rule in multiplying the radicals:

$$\sqrt[n]{a} \cdot \sqrt[n]{b} = \sqrt[n]{ab}$$

where a and b are greater than zero.

The rule is obtained by applying the laws of exponents as follows:

$$\sqrt[n]{a} \cdot \sqrt[n]{b} = a^{1/n} \cdot b^{1/n} = (ab)^{1/n} = \sqrt[n]{ab}$$

EXAMPLE 9   Multiply $5\sqrt{6}$ by $3\sqrt{2}$.

$$(5\sqrt{6})(3\sqrt{2}) = (5)(3)(\sqrt{6})(\sqrt{2}) = 15\sqrt{6 \cdot 2}$$
$$= 15\sqrt{2 \cdot 3 \cdot 2} = 15(2)\sqrt{3} = 30\sqrt{3}$$

When the indexes are different, first apply the laws of exponents to change the given radicals to equivalent radicals having the same index. Next, multiply the equivalent radicals.

EXAMPLE 10 Multiply $\sqrt{3}$ by $\sqrt[3]{5}$.

$$\sqrt{3} \cdot \sqrt[3]{5} = 3^{1/2} \cdot 5^{1/3} = 3^{3/6} \cdot 5^{2/6} = (3^3 \cdot 5^2)^{1/6}$$
$$= \sqrt[6]{3^3 \cdot 5^2} = \sqrt[6]{27 \cdot 25} = \sqrt[6]{675}$$

Observe that the fractional exponents $\frac{1}{2}$ and $\frac{1}{3}$ above have been reduced to equivalent fractions having the lowest common denominator (or common index) 6 in the multiplication.

3. DIVISION OF RADICALS. When the indexes of the radicals are the same, use the following rule in dividing the radicals:

$$\frac{\sqrt[n]{a}}{\sqrt[n]{b}} = \sqrt[n]{\frac{a}{b}}$$

where a and b are greater than zero.
The rule is obtained by applying the laws of exponents as follows:

$$\frac{\sqrt[n]{a}}{\sqrt[n]{b}} = \frac{a^{1/n}}{b^{1/n}} = \left(\frac{a}{b}\right)^{1/n} = \sqrt[n]{\frac{a}{b}}$$

EXAMPLE 11 Divide $\sqrt{27}$ by $\sqrt{3}$.

$$\sqrt{27} \div \sqrt{3} = \frac{\sqrt{27}}{\sqrt{3}} = \sqrt{\frac{27}{3}} = \sqrt{9} = 3$$

The quotient obtained from the division of radicals is frequently expressed in fractional form. Fractions involving radicals are considered in simplest form when the denominators include no radicals. The process of removing radicals from a denominator is also known as *rationalizing the denominator*.

EXAMPLE 12 Divide $\sqrt{2}$ by $\sqrt{5}$.

$$\frac{\sqrt{2}}{\sqrt{5}} = \frac{\sqrt{2} \cdot \sqrt{5}}{\sqrt{5} \cdot \sqrt{5}} = \frac{\sqrt{10}}{5}$$

When the indexes are different, first apply the laws of exponents to change the given radicals to equivalent radicals having the same index. Next, divide the equivalent radicals to get the answer.

EXAMPLE 13 Divide $\sqrt{5}$ by $\sqrt[3]{2}$.

$$\frac{\sqrt{5}}{\sqrt[3]{2}} = \frac{5^{1/2}}{2^{1/3}} = \frac{5^{3/6}}{2^{2/6}} = \frac{5^{3/6}}{2^{2/6}} \cdot \frac{2^{4/6}}{2^{4/6}} = \frac{(5^3 \cdot 2^4)^{1/6}}{2^{2/6+4/6}} = \frac{(125 \cdot 16)^{1/6}}{2^{6/6}}$$

$$= \frac{\sqrt[6]{2.000}}{2}$$

Observe that the fractional exponents $\frac{1}{2}$ and $\frac{1}{3}$ above have been reduced to equivalent fractions having the lowest common denominator (or common index) 6 in the division.

EXERCISE 4–1   Reference: Section 4.1

A. *Express the following in radical form:*

1. $a^{1/4}$
2. $b^{3/4}$
3. $c^{2/5}$
4. $d^{4/7}$
5. $43^{1/3}$
6. $58^{2/6}$
7. $(26x)^{3/8}$
8. $(134xy)^{5/3}$

B. *Express the following in exponential form:*

9. $\sqrt[3]{e^2}$
10. $\sqrt[4]{f^3}$
11. $\sqrt[6]{g^5}$
12. $\sqrt{h^4}$
13. $\sqrt{3x}$
14. $\sqrt[3]{7x^2}$
15. $\sqrt[5]{34xy^3}$
16. $\sqrt[4]{257x^3y^2}$

C. *Express the coefficient of each radical under the radical sign:*

17. $5\sqrt{3}$
18. $6\sqrt{10}$
19. $x\sqrt{2y}$
20. $ab\sqrt{3a}$
21. $3^2 \cdot \sqrt{5}$
22. $2^3 \cdot \sqrt[3]{3a}$
23. $a^2b \cdot \sqrt[5]{4ab}$
24. $x^4y^2 \cdot \sqrt[3]{xy^2}$

D. *Compute the following:*

25. $4\sqrt{3} + 7\sqrt{3} - 5\sqrt{3}$
26. $10\sqrt{6} - 2\sqrt{6} + 7\sqrt{6}$
27. $\sqrt{50} + \sqrt{98} - \sqrt{18}$
28. $\sqrt{108} - \sqrt{48} + \sqrt{192}$
29. $6\sqrt{5a} + 4\sqrt{5a} + 3\sqrt{5b}$
30. $8\sqrt{xy} + \sqrt{xyz} - 3\sqrt{xy}$
31. $(3\sqrt{4a})(5\sqrt{2ab})$
32. $(6\sqrt{25})(3\sqrt{14})$
33. $(2\sqrt{3})(\sqrt[3]{4})$
34. $(3\sqrt{5})(2 \cdot \sqrt[3]{2})$
35. $\sqrt{25} \div \sqrt{9}$
36. $\sqrt{48} \div \sqrt{2}$
37. $\sqrt{15} \div \sqrt{7}$
38. $\sqrt{27} \div \sqrt{6}$
39. $\sqrt{3} \div \sqrt[3]{2}$
40. $\sqrt[3]{4} \div \sqrt{2}$

4.2 FINDING SQUARE ROOTS

Square roots are the most frequently used radicals in analyzing business and economic problems. The method of finding the square root of a value is illustrated in the following examples:

EXAMPLE 14 Find $\sqrt{61.4656}$.

The steps of finding the square root of the given value 61.4656 are:

1. Separate the given number into groups of two digits, starting at the decimal point and moving in both directions, or $\overline{61}.\overline{46}\,\overline{56}$. Each group will give one digit in the final answer. The first digit of the answer (7) is the largest integral square root of the leftmost group (61). Write the square of the digit ($7^2 = 49$) under the first group and subtract as in long division:

$$\begin{array}{r} 7. \\ \sqrt{61.\ \overline{46}\ \overline{56}} \\ 7^2 = \underline{49}\ (-) \\ \hline 12 \end{array}$$

2. Bring down the next group of two digits (46) to the right side of the result of the subtraction ($12 = 61 - 49$) to make the first remainder (12 46). Multiply the first digit of the answer 7 by 20 to get the partial divisor (or $7 \times 20 = 140$). Divide the first remainder by the partial divisor to estimate the next digit of the answer (8, since $1246 \div 140 = 8+$). Add this digit to the partial divisor to obtain the complete divisor ($140 + 8 = 148$). Multiply this digit by the complete divisor and subtract again as in long division, or:

$$\begin{array}{r} 7.\quad 8 \\ \sqrt{61.\ \overline{46}\ \overline{56}} \\ 7^2 = \underline{49} \\ \hline 12\quad 46 \end{array}$$

$7 \times 20 = 140$

$\underline{+8}$

$148 \times 8 = \underline{11\quad 84}(-)$

$\overline{\quad\quad 62}$

Estimating the second digit of the answer:

$1246 \div 140 = 8+$

3. Repeat step 2 in each successive step for each group of two digits in the given number. However, the *accumulated digits* in the answer are multiplied by 20 for each partial divisor, and *only the digit* used as the answer in each step is added to the partial divisor in obtaining the complete divisor. The complete arrangement of finding the square root is shown in the following steps:

STEP 1:

$$\begin{array}{r} 7.\quad 8\quad 4\quad \text{(Answer)} \\ \sqrt{61.\ \overline{46}\ \overline{56}} \\ 7^2 = \underline{49} \\ \hline 12\quad 46 \end{array}$$

Estimating quotient digit

$1.246 \div 140 = 8+$

STEP 2: $7 \times 20 = 140$

$\underline{+8}$

$148 \times 8 = \underline{11\quad 84}$

$\quad\quad\quad 62\quad 56$ $6,256 \div 1,560 = 4+$

STEP 3: $78 \times 20 = 1,560$

$\underline{+4}$

$1,564 \times 4 = \quad\underline{62\quad 56}$

Check: $7.84^2 = 61.4656$

The following example gives additional illustrations of the above three steps.

EXAMPLE 15 Find $\sqrt{614.656}$ to two decimal places.

This problem requires three decimal places before rounding to two places in the final answer. Thus, three groups ($\overline{65}\ \overline{60}\ \overline{00}$) after the decimal point in the given number are supplied for the process of finding the square root.

$$
\begin{array}{r}
\phantom{\sqrt{6}}\ \ 2\ \ \ \ 4.\ \ \ 7\ \ \ 9\ \ \ 2 \\
\sqrt{6}\ \ \overline{14.}\ \ \overline{65}\ \ \overline{60}\ \ \overline{00}
\end{array}
$$

		(Answer before rounding)
$2^2 = 4$		
$2 \times 20 = 40$ $2\ \ \overline{|14}$		$214 \div 40 = 5+$
$\underline{+4}$		(The actual quotient
$44 \times 4 = 1\ \ \overline{|76}$		is 4, since 5 is too
$24 \times 20 = 480$ $38\ \ \overline{|65}$		high.)
$\underline{+7}$		$3,865 \div 480 = 7+$
$487 \times 7 =34\ \ |09$		
$247 \times 20 = 4,940$ $4\ \ \overline{|56\ \ 60}$		$45,660 \div 4,940 =$
$\underline{+9}$		$9+$
$4,949 \times 9 =4\ \ |45\ \ 41$		
$2,479 \times 20 = 49,580$ $11\ \ \overline{|19\ \ 00}$		$111,900 \div 49,580 =$
$\underline{+2}$		$2+$
$49,582 \times 2 =|9\ \ 91\ \ 64$		
$0.\ \ 01\ \ 27\ \ 36$		(Final remainder)

Answer: $\sqrt{614.656} = 24.792$, rounded to 24.80.

Check: $24.80^2 = 615.0400$, which is slightly above the given value.
              $24.79^2 = 614.5441$, which is slightly below the given value.

Also, the square of the answer before rounding plus the final remainder is equal to the given value:

$$
\begin{array}{r}
24.792^2 + 0.012736 = 614.643264 \\
\underline{+0.012736} \\
614.656000
\end{array}
$$

4.3 EQUATIONS WITH RADICALS

The usual method of solving an equation with radicals is first to remove the radicals. We may remove a radical by raising the radical to the power equal to the index of the radical as illustrated in the following examples. Next, solve the derived equation with no radicals in the usual manner. Since the operation of raising both sides to powers may introduce extraneous roots, it is necessary to test the values obtained for the unknown by substitution in the original equation.

EXAMPLE 16   Solve the equation $\sqrt{x-3} = 2$.

Square both sides of the equation:

$$
\begin{aligned}
(\sqrt{x-3})^2 &= 2^2 \\
x - 3 &= 4 \\
x &= 4 + 3 = 7
\end{aligned}
$$

Test: Substitute $x = 7$ in the given equation:

$$
\sqrt{x-3} = \sqrt{7-3} = \sqrt{4} = 2
$$

Thus, $x = 7$ is the correct root.

EXAMPLE 17 Solve the equation $\sqrt{x-2} = x - 4$.

Square both sides of the equation,

$$(\sqrt{x-2})^2 = (x-4)^2$$
$$x - 2 = x^2 - 8x + 16$$
$$x^2 - 9x + 18 = 0$$
$$(x - 3)(x - 6) = 0$$

When $x - 3 = 0$, $x = 3$; when $x - 6 = 0$, $x = 6$.

Test: Substitute $x = 3$ in the given equation:

$$\sqrt{3-2} = 3 - 4$$
$$1 \neq -1$$

Thus, we reject the extraneous root $x = 3$.

Substitute $x = 6$ in the given equation:

$$\sqrt{6-2} = 6 - 4$$
$$2 = 2$$

Thus, $x = 3$ is the correct root.

EXAMPLE 18 Solve the equation $\sqrt{x+2} - \sqrt{x-3} = 1$.

Transpose: $\qquad\qquad\qquad \sqrt{x+2} = 1 + \sqrt{x-3}$

Square both sides: $\qquad (\sqrt{x+2})^2 = (1 + \sqrt{x-3})^2$

Combine and simplify: $\qquad x + 2 = 1 + 2\sqrt{x-3} + (x-3)$

$$2\sqrt{x-3} = 4$$
$$\sqrt{x-3} = 4/2 = 2$$

Square both sides: $\qquad (\sqrt{x-3})^2 = 2^2$
$$x - 3 = 4$$
$$x = 4 + 3 = 7$$

Test: Substitute $x = 7$ in the given equation:

$$\sqrt{7+2} - \sqrt{7-3} = 3 - 2 = 1$$

Thus, $x = 7$ is the correct root.

EXAMPLE 19 Solve the equation $\sqrt[3]{x-1} = 2$.

Cube both sides of the equation,

$$(\sqrt[3]{x-1})^3 = 2^3$$
$$x - 1 = 8$$
$$x = 9$$

Test: Substitute $x = 9$ in the given equation:

$$\sqrt[3]{9-1} = \sqrt[3]{8} = 2$$

Thus, $x = 9$ is the correct root.

EXERCISE 4-2   Reference: Sections 4.2 and 4.3

A. *Find the square root of the following:*

 1. $\sqrt{625}$ 5. $\sqrt{552.25}$
 2. $\sqrt{2,304}$ 6. $\sqrt{332.6976}$
 3. $\sqrt{17,424}$ 7. $\sqrt{203.376}$ (to two decimal places)
 4. $\sqrt{71,289}$ 8. $\sqrt{1,379.825}$ (to two decimal places)

B. *Solve each equation for x:*

 9. $\sqrt{x+4} = 3$ 14. $\sqrt{x+6} - \sqrt{x-9} = 3$
10. $\sqrt{6+x} = 7$ 15. $\sqrt{2x-3} - \sqrt{x-2} = 1$
11. $\sqrt{x-6} = x - 8$ 16. $\sqrt{3x-5} + \sqrt{x-2} = 3$
12. $\sqrt{x+13} = x - 7$ 17. $\sqrt{3-x} + \sqrt{x+2} = \sqrt{2x+5}$
13. $\sqrt{x+1} + \sqrt{x-4} = 5$ 18. $\sqrt{x-2} + \sqrt{2x-5} - \sqrt{x+1} = 0$

4.4 QUADRATIC EQUATIONS

A quadratic equation in one unknown contains the unknown with the highest degree of 2. The standard form of the quadratic equation in the unknown x is written:

$$ax^2 + bx + c = 0$$

The left side of the equation is arranged in order of descending powers of the unknown x. Letters a, b, and c represent constants. The letter a may have any value other than zero, and the letters b and c may have any values including zero. If $a = 0$, the equation is reduced to the form $bx + c = 0$, which is linear, not quadratic.

A quadratic equation in one unknown has two solutions or roots. There are four common methods that may be used to solve a quadratic equation.

A. Solution by Factoring

To solve a quadratic equation by factoring, first write the problem in the standard form of the quadratic equation in the unknown. Second, factor the left side of the quadratic equation. Third, apply the principle that if a product equals zero, one or more factors of the product equals zero.

EXAMPLE 20 Solve $3x^2 + 2x = 8$.

Transpose: $3x^2 + 2x - 8 = 0$

Factor: $(x + 2)(3x - 4) = 0$

Equate each factor to zero and solve for x:

$x + 2 = 0$ $3x - 4 = 0$
      $x = -2$ $x = \frac{4}{3}$

Check: Substitute each root in the original equation.

$x = -2$: $3(-2)^2 + 2(-2) = 12 - 4 = 8$
$x = \frac{4}{3}$: $3(\frac{4}{3})^2 + 2(\frac{4}{3}) = \frac{16}{3} + \frac{8}{3} = \frac{24}{3} = 8$

EXAMPLE 21   Solve $25x^2 = 100$.

Transpose: $25x^2 - 100 = 0$

Factor: $(5x)^2 - 10^2 = 0$
$(5x + 10)(5x - 10) = 0$

Equate each factor to zero and solve for x:

$5x + 10 = 0$ $\qquad$ $5x - 10 = 0$

$x = \dfrac{-10}{5}$ $\qquad$ $x = \dfrac{10}{5}$

$x = -2$ $\qquad\qquad$ $x = 2$

Check: Substitute each root in the original equation.

$x = -2$: $25(-2)^2 = 25(4) = 100$

$x = 2$: $25(2)^2 = 25(4) = 100$

B. Solution by Completing the Squares

When the left side of a quadratic equation in the standard form cannot be factored by inspection, the roots can be found by a process called *completing the square*. In fact, this process can apply to all quadratic equations whether or not a solution can be found by the factoring method illustrated above. The details of the process are illustrated in Example 22.

EXAMPLE 22   Solve for x: $ax^2 + bx + c = 0$.

Subtract c from both sides of the given equation:

$$ax^2 + bx = -c$$

Divide each side by a. $\qquad$ $x^2 + \dfrac{b}{a}x = -\dfrac{c}{a}$

Add the squares of one half the coefficient of x to both sides.

$$x^2 + \frac{b}{a}x + \left(\frac{b}{2a}\right)^2 = -\frac{c}{a} + \left(\frac{b}{2a}\right)^2$$

The left side of the above equation now becomes a perfect square, or:

$$x^2 + \frac{b}{a}x + \left(\frac{b}{2a}\right)^2 = \left(x + \frac{b}{2a}\right)^2$$

The right side can be written in a fractional form, or:

$$-\frac{c}{a} + \left(\frac{b}{2a}\right)^2 = \frac{b^2}{4a^2} - \frac{c}{a} = \frac{b^2 - 4ac}{4a^2}$$

Thus:

$$\left(x + \frac{b}{2a}\right)^2 = \frac{b^2 - 4ac}{4a^2}$$

Extract the square root of each side:

$$x + \frac{b}{2a} = \pm \frac{\sqrt{b^2 - 4ac}}{2a}$$

Solve for x:

$$x = \frac{-b \pm \sqrt{b^2 - 4ac}}{2a}$$

C. Solution by the Quadratic Formula

When the quadratic equation is written in the standard form,

$$ax^2 + bx + c = 0$$

the two roots of x can be written:

$$x = \frac{-b \pm \sqrt{b^2 - 4ac}}{2a}$$

as illustrated in Example 22.

This x expression is called the *quadratic formula* and can be used to find the roots of any quadratic equation. In the formula, a equals the coefficient of the unknown square (x^2), b equals the coefficient of the unknown (x), and c is a constant.

EXAMPLE 23   Solve $3x^2 + 2x - 8 = 0$.

Here $a = 3$, $b = 2$, $c = -8$.

Substitute the values in the quadratic formula:

$$x = \frac{-(2) \pm \sqrt{(2)^2 - 4(3)(-8)}}{2(3)} = \frac{-2 \pm \sqrt{4 + 96}}{6}$$

$$= \frac{-2 \pm \sqrt{100}}{6} = \frac{-2 \pm 10}{6}$$

$$x = \frac{-2 + 10}{6} = \frac{4}{3}$$

$$x = \frac{-2 - 10}{6} = -2$$

This solution is the same as that of Example 20.

EXAMPLE 24   Solve $25x^2 - 100 = 0$.

Here $a = 25$, $b = 0$, $c = -100$.

$$x = \frac{-(0) \pm \sqrt{(0)^2 - 4(25)(-100)}}{2(25)} = \frac{\pm \sqrt{10.000}}{50}$$

$$= \frac{\pm 100}{50} = \pm 2$$

When $x = \dfrac{+100}{50}$, $x = 2$

When $x = \dfrac{-100}{50}$, $x = -2$

This solution is the same as that of Example 21.

EXAMPLE 25 Solve $x^2 - 4x + 4 = 0$.

Here $a = 1$, $b = -4$, $c = 4$.

$$x = \frac{-(-4) \pm \sqrt{(-4)^2 - 4(1)\,(4)}}{2(1)} = \frac{4 \pm \sqrt{16 - 16}}{2}$$

$$= \frac{4 \pm \sqrt{0}}{2} = 2$$

Observe that when the value of $(b^2 - 4ac)$ is zero, the two roots $(2 + 0$ and $2 - 0)$ are equal (2).

EXAMPLE 26 Solve $x^2 + 2x + 5 = 0$.

Here $a = 1$, $b = 2$, $c = 5$.

$$x = \frac{-(2) \pm \sqrt{(2)^2 - 4(1)(5)}}{2(1)} = \frac{-2 \pm \sqrt{4 - 20}}{2}$$

$$= \frac{-2 \pm \sqrt{-16}}{2} = \frac{-2 \pm 4i}{2}$$

When $x = \dfrac{-2 + 4i}{2}$, $x = -1 + 2i$.

When $x = \dfrac{-2 - 4i}{2}$, $x = -1 - 2i$.

Observe that when the value of $(b^2 - 4ac)$ is negative (-16), the two roots are imaginary.

EXAMPLE 27 Solve $x - 2x^2 = -3$.

First rewrite the given quadratic equation in the standard form:

$$-2x^2 + x + 3 = 0$$

Here $a = -2$, $b = 1$, $c = 3$.

$$x = \frac{-(1) \pm \sqrt{(1)^2 - 4(-2)(3)}}{2(-2)} = \frac{-1 \pm \sqrt{1 + 24}}{-4}$$

$$= \frac{-1 \pm 5}{-4}$$

When $x = \dfrac{-1 + 5}{-4}$, $x = -1$

When $x = \dfrac{-1-5}{-4}$, $x = \dfrac{3}{2} = 1.5$.

D. Solution by Graphs

The graph of any quadratic function $f(x) = ax^2 + bx + c$ is a parabola. In finding the two roots of a quadratic function written in the standard form, we must examine the connection between the parabola and the X-axis. Three possible connections may arise. Based on the concept of the zeros of a function presented in Chapter 3, we may determine the roots for the different connections as follows:

Case 1. If the parabola crosses the X-axis in two distinct points, the roots of $f(x) = 0$ are real numbers and are unequal. (See the graphs in Examples 23, 24, and 27.)

Case 2. If the parabola only touches, does not cross, the X-axis, the roots of $f(x) = 0$ are real numbers and are equal. (See the graph in Example 25.)

Case 3. If the parabola does not touch or cross the X-axis, the roots of $f(x) = 0$ include imaginary numbers. (See the graph in Example 26.)

Also, it has been proven in analytic geometry that:

The parabola opens upward if a is positive. (See the graphs in Examples 23, 24, 25, and 26.)

The parabola opens downward if a is negative. (See the graph in Example 27.)

The parabola has its vertex at the point where $x = -b/2a$. (See individual graphs that follow.)

Graphic solution for Example 23

Graph: $y = f(x) = 3x^2 + 2x - 8$.

x	-3	-2	-1	0	1	2	3
y	13	0	-7	-8	-3	8	25

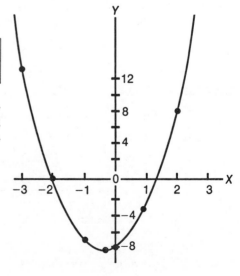

When $y = f(x) = 0$, the curve intersects the X-axis at two points: $x = -2$ and between 1 and 2, or $1\frac{1}{3}$ approximately by observation.

Vertex: $x = -\dfrac{b}{2a} = -\dfrac{2}{2(3)} = -\dfrac{1}{3}$

$y = 3\left(-\dfrac{1}{3}\right)^2 + 2\left(-\dfrac{1}{3}\right) - 8$

$= -8\dfrac{1}{3}$

Graphic solution for Example 24

Graph: $y = f(x) = 25x^2 - 100$.

x	-3	-2	-1	0	1	2	3
y	125	0	-75	-100	-75	0	125

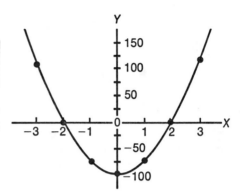

When $y = f(x) = 0$, the curve intersects the X-axis at two points: $x = 2$ and $x = -2$.

Vertex: $x = -\dfrac{b}{2a} = -\dfrac{0}{2(25)} = 0$

$$y = 25(0)^2 - 100 = -100$$

Graphic solution for Example 25

Graph: $y = f(x) = x^2 - 4x + 4$.

x	5	4	3	2	1	0	-1
y	9	4	1	0	1	4	9

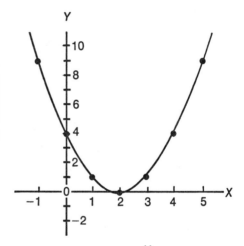

When $y = f(x) = 0$, the curve touches the X-axis at $x = 2$.

Vertex: $x = -\dfrac{-4}{2(1)} = 2$

$$y = (2)^2 - 4(2) + 4 = 0$$

Graphic solution for Example 26

Graph: $y = f(x) = x^2 + 2x + 5$.

x	-4	-3	-2	-1	0	1	2
y	13	8	5	4	5	8	13

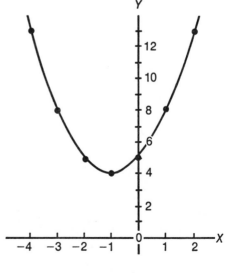

When $y = f(x) = 0$, the curve does not cross or touch the X-axis. Thus, x is imaginary.

Vertex: $x = -\dfrac{2}{2(1)} = -1$

$$y = (-1)^2 + 2(-1) + 5 = 4$$

Graphic solution for Example 27

Graph: $y = f(x) = -2x^2 + x + 3$.

x	-3	-2	-1	0	1	2	3
y	-18	-7	0	3	2	-3	-12

When $y = f(x) = 0$, the curve intersects the X-axis at two points: $x = -1$
and between 1 and 2, or $1\frac{1}{2}$ approximately by observation.

Vertex: $x = -\dfrac{1}{2(-2)} = \dfrac{1}{4}$

$$y = -2\left(\frac{1}{4}\right)^2 + \frac{1}{4} + 3 = 3\frac{1}{8}$$

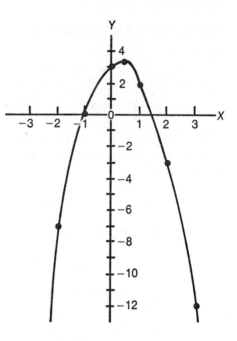

EXERCISE 4–3   Reference: Section 4.4

A. *Solve the following equations by factoring:*

1. $3x^2 = 48$
2. $4x^2 = 36$
3. $5x^2 - 20 = 0$
4. $2x^2 - 50 = 0$
5. $2x^2 + 7x = -6$

6. $3x^2 + 11x = 4$
7. $2x^2 + 5x = 3$
8. $12x^2 + 10x = -2$
9. $-3x^2 - x + 10 = 0$
10. $-6x^2 + 7x + 3 = 0$

B. *Solve the following equations by the quadratic formula and graphs:*

11. $2x^2 + x = 6$
12. $3x^2 - x = 4$
13. $4x^2 + 8x + 3 = 0$
14. $2x^2 + 11x + 12 = 0$
15. $x^2 - 6x + 9 = 0$

16. $x^2 + 4x + 4 = 0$
17. $x^2 - 2x + 10 = 0$
18. $x^2 - 4x + 29 = 0$
19. $-2x^2 + 3x + 2 = 0$
20. $-3x^2 + 8x + 3 = 0$

5 Equations of Degree Higher Than Two

This chapter introduces basic methods for solving polynomial equations of degree higher than two. The general form of the equation is:

$$a_0x^n + a_1x^{n-1} + a_2x^{n-2} + \cdots + a_{n-1}x + a_n = 0$$

where n is a positive integer and the left side is a polynomial of the nth degree in the unknown x. The coefficients $a_0 \, (\neq 0), a_1, a_2, \ldots, a_n$, may be any real or imaginary numbers. However, here we shall restrict our discussions to equations with only real coefficients.

In solving a polynomial equation, it is desirable to know the method of synthetic division, the remainder theorem, and the factor theorem. Those topics are presented in Sections 5.1, 5.2, and 5.3 respectively. The basic methods for solving the equations are introduced in Section 5.4.

5.1 SYNTHETIC DIVISION

The *synthetic division* method can be used to simplify the procedure for dividing a polynomial $f(x)$ by a binomial of the form $(x - r)$. This method omits writing the x values and the coefficients that are merely repetitions in the process of division.

EXAMPLE 1   Divide $(2x^3 + 5x^2 - 4x - 3)$ by $(x - 2)$.

Here $f(x) = 2x^3 + 5x^2 - 4x - 3$
and $x - r = x - 2$
$\quad\quad r = 2$

The ordinary process by long division for this problem is:

$$
\begin{array}{r}
2x^2 + 9x + 14 \text{ (quotient)} \\
x - 2 \overline{)2x^3 + 5x^2 - 4x - 3} \text{ (dividend)} \\
\underline{2x^3 - 4x^2} \\
9x^2 - 4x \\
\underline{9x^2 - 18x} \\
14x - 3 \\
\underline{14x - 28} \\
25 \text{ (remainder)}
\end{array}
$$

(divisor) (dividend)

First we simplify the long division by omitting the x's and the repeated terms. The repeated terms are: 2 under 2 in the x^3 column, 9 under 9 in the x^2 column, -4 under -4 and 14 under 14 in the x column, and -3 under -3 in the constant column. The simplified form of division is:

$$
\begin{array}{r}
2 \quad +9 \quad +14 \\
1-2\overline{)2 \quad +5 \quad -4 \quad -3} \\
\underline{-4} \\
9 \\
\underline{-18} \\
14 \\
\underline{-28} \\
25
\end{array}
$$

In a more compact form:

$$
\begin{array}{r}
2 \quad +9 \quad +14 \\
1-2\overline{)2 \quad +5 \quad -4 \quad -3} \\
\underline{-4 \quad -18 \quad -28} \\
9 \quad 14 \quad 25
\end{array}
$$

Next, we omit 1 in the divisor since the coefficient of x is always 1. We also omit the quotient line but repeat the coefficient 2 under 2 in the remainder line as follows:

$$
\begin{array}{c|rrrr}
-2 & 2 & +5 & -4 & -3 \\
\text{(divisor)} & & -4 & -18 & -28 \\
\hline
& 2 & 9 & 14, & 25
\end{array}
$$

(dividend)
SUBTRACTION
(remainder)

(quotient)

Now the quotient is located in the remainder line.

Finally, for convenience, we replace -2 by 2 in the divisor (2 is the actual value of r) and change the signs of the numbers in the second line (this change allows us to add rather than subtract in the process of division). The final arrangement is as follows:

(dividend) $2 \quad +5 \quad -4 \quad -3 \;\lfloor 2$ (divisor r)
  $\underline{+4 \quad +18 \quad +28}$ ADDITION
 $2 \quad +9 \quad +14, \quad +25$ (remainder)

(quotient)

The process of synthetic division presented in the final arrangement in Example 1 is summarized in the following steps:

1. Write the coefficients of $f(x)$ in order in the first line and write the value of r at the right. Make sure that the powers of x are arranged in descending order and that zeros are supplied for missing powers. In Example 1, the numbers in the first line are:

$$
2 \quad +5 \quad -4 \quad -3 \;\lfloor 2
$$

2. Bring down the left number 2 to the third line. Multiply this number by 2 ($=r$) and write the product 4 ($=2 \times 2$) in the second line under 5. Add 5 and 4 and place the sum 9 in the third line. The numbers in the three lines now are:

$$2 \quad +5 \quad -4 \quad -3 \ \underline{|2}$$
$$\underline{ +4 }$$
$$2 \quad +9$$

3. Continue the multiplication and addition operations as presented in step 2. That is, $9 \times 2 = 18$, $-4 + 18 = 14$, $14 \times 2 = 28$, and $-3 + 28 = 25$.

$$2 \quad +5 \quad \ -4 \quad \ -3 \ \underline{|2}$$
$$\underline{ +4 \quad +18 \quad +28}$$
$$2 \quad +9 \quad +14 \quad +25$$

The last sum, 25, in the third line, is the remainder. The other sums are the respective coefficients of the powers of x arranged in descending order in the quotient. The first term of the quotient is one degree less than that of the dividend.

Steps 1, 2, and 3 are used in illustrating the following example:

EXAMPLE 2   Divide $(3x^4 - x^3 - 21x^2 - 11x + 6)$ by $(x + 1)$.

Here $f(x) = 3x^4 - x^3 - 21x^2 - 11x + 6$, which is already arranged in descending order of the powers of x, and:

$$x - r = x + 1$$
$$r = -1$$

$$3 \quad -1 \quad -21 \quad -11 \quad \ 6 \ \underline{|-1}$$
$$\underline{ -3 \quad \ \ 4 \quad \ \ 17 \quad -6}$$
$$3 \quad -4 \quad -17 \quad \ \ 6 \quad \ 0$$

The quotient is $(3x^3 - 4x^2 - 17x + 6)$, and the remainder is 0.

5.2. REMAINDER THEOREM

The *remainder theorem* is based on the definition of division:

$$\text{Dividend} = \text{Divisor} \times \text{Quotient} + \text{Remainder}$$

According to the above definition and the information obtained from Example 1, we have:

$$f(x) = 2x^3 + 5x^2 - 4x - 3 = (x - 2) \times (2x^2 + 9x + 14) + 25$$

If we let the divisor $x - 2 = 0$, or $x = 2$, then:

$$f(x) = f(2) = (2 - 2) \times (2(2)^2 + 9(2) + 14) + 25 = 0 + 25$$
$$= 25, \text{ the remainder.}$$

In general, if:

$$f(x) = \text{the dividend which is a polynomial in } x,$$
$$x - r = \text{the divisor,}$$
$$q(x) = \text{the quotient, and}$$
$$R = \text{the remainder,}$$

then, we can write the definition:

$$f(x) = (x - r)q(x) + \mathrm{R}$$

Further, if $(x - r) = 0$, or $x = r$, then:

$$f(r) = 0 + R, \text{ or } f(r) = R$$

This fact gives us the remainder theorem, which states that *if a polynomial f(x) is divided by (x − r) until a constant remainder (R) is obtained, this remainder is equal to f(r).*

EXAMPLE 3   Find the remainder based on the remainder theorem when $(2x^3 + 5x^2 - 4x - 3)$ is divided by $(x - 2)$.

$f(x) = 2x^3 + 5x^2 - 4x - 3$
$x - r = x - 2, r = 2$
$f(r) = f(2) = 2(2)^3 + 5(2)^2 - 4(2) - 3 = 25$

The remainder is 25. (Also see Example 1.)

EXAMPLE 4   Find the remainder based on the remainder theorem when $(2x^3 + 5x^2 - 4x - 3)$ is divided by $(x + 3)$.

$f(x) = 2x^3 + 5x^2 - 4x - 3$ (Same as the dividend in Example 3)
$x - r = x + 3, r = -3$
$f(r) = f(-3) = 2(-3)^3 + 5(-3)^2 - 4(-3) - 3 = 0$

Thus, there is no remainder, or $R = 0$.

EXAMPLE 5   Find the remainder based on the remainder theorem when $(3x^4 - x^3 - 21x^2 - 11x + 6)$ is divided by $(x - 1)$.

$f(x) = 3x^4 - x^3 - 21x^2 - 11x + 6$ (Same as the dividend in Example 2)
$x - r = x - 1, r = 1.$
$f(r) = f(1) = 3(1)^4 - (1)^3 - 21(1)^2 - 11(1) + 6 = -24$ (remainder)

5.3   FACTOR THEOREM

Again, based on the definition of division, the result of Example 4 can be written:

$$2x^3 + 5x^2 - 4x - 3 = (x + 3) \times \text{Quotient} + 0$$

Thus, $(x + 3)$ is a factor of $(2x^3 + 5x^2 - 4x - 3)$, or symbolically:

$$f(x) = (x - r)q(x) + 0$$
$$f(x) = (x - r)q(x)$$

This fact gives us the *factor theorem*, which states that *if f(r) = R = 0, then (x − r) is a factor of f(x).* Further, since $(x - r) = 0$, or $x = r$,

$$f(x) = 0q(x) = 0$$

The factor theorem can be stated in a different manner: *If r is a root of the equation f(x) = 0, then (x − r) is a factor of f(x).* This statement is useful in solving a polynomial equation.

EXAMPLE 6 Determine whether or not $(x + 1)$ is a factor of the polynomial $(3x^4 - x^3 - 21x^2 - 11x + 6)$.

$$f(x) = 3x^4 - x^3 - 21x^2 - 11x + 6$$
$$x - r = x + 1, r = -1$$
$$f(r) = f(-1) = 3(-1)^4 - (-1)^3 - 21(-1)^2 - 11(-1) + 6 = 0$$

There is no remainder, or $R = 0$. Thus, $(x + 1)$ is a factor of the polynomial $f(x)$. (Also see the solution of Example 2.)

EXAMPLE 7 Determine whether $(3x - 1)$ is a factor of the given polynomial in Example 6.

Here $x - r = 3x - 1 = 3\left(x - \dfrac{1}{3}\right)$.

Since $x - r = 0$, or $x = r$, we can let:

$$3\left(x - \dfrac{1}{3}\right) = 0$$

When $x - \dfrac{1}{3} = 0$, $x = r = \dfrac{1}{3}$

$$f(r) = f\left(\dfrac{1}{3}\right) = 3\left(\dfrac{1}{3}\right)^4 - \left(\dfrac{1}{3}\right)^3 - 21\left(\dfrac{1}{3}\right)^2 - 11\left(\dfrac{1}{3}\right) + 6 = 0$$

Thus, $(3x - 1)$ is also a factor of the polynomial $f(x)$.

Note that the polynomials given in Examples 5, 6, and 7 are the same. The expression $(x - 1)$, or the divisor given in Example 5 is not a factor of the polynomial $f(x)$, since $f(1) = -24 \neq 0$.

EXERCISE 5–1   Reference: Sections 5.1–5.3

A. *By synthetic division, find the quotient and the remainder in each division:*

1. $(2x^2 - 3x + 5) \div (x - 6)$
2. $(4x + 7x^2 - 20) \div (x + 2)$
3. $(-3x^2 + 4x^3 + 7) \div (x + 1)$
4. $(x^3 + 2x - 7) \div (x - 5)$
5. $(5x^3 + 14x^2 - 3x + 9) \div (x + 4)$
6. $(3x^3 - 2x^2 + 4x - 24) \div (x - 2)$
7. $(2x^4 - 5x^2 + 3x - 108) \div (x + 3)$
8. $(6x^4 + 2x^3 + 3x^2 - 89x - 5)$
   $\div (x - 3)$
9. $(x^5 - 14x^3 - 7x + 10) \div (x - 4)$
10. $(3x^5 + 9x^4 - 10x^2 - 18) \div (x + 5)$
11. Find the value of k so that $(2x^4 + x^3 + 2x^2 - 2kx - 1)$ is divisible by $(x - 1)$.
12. Find the value of k so that $(x^5 + 3x^4 + 4x^3 + kx^2 - 7x + 6)$ is divisible by $(x + 2)$.

B. *By the remainder theorem, find the remainder in each division:*

13. Divide $(2x^2 - 3x + 5)$ by: (a) $(x - 6)$ and (b) $(x + 6)$
14. Divide $(7x^2 + 4x - 20)$ by: (a) $(x + 2)$ and (b) $(x - 2)$
15. Divide $(4x^3 - 3x^2 + 7)$ by: (a) $(x + 1)$ and (b) $(x - 1)$
16. Divide $(x^3 + 2x - 7)$ by: (a) $(x - 5)$ and (b) $(x + 5)$
17. Divide $(5x^3 + 14x^2 - 3x + 9)$ by: (a) $(x + 4)$ and (b) $(x - 4)$
18. Divide $(2x^4 - 5x^2 + 3x - 108)$ by: (a) $(x + 3)$ and (b) $(x - 3)$

C. *By use of the factor theorem, answer the following questions:*

19. Refer to Problem 1. Determine whether $(x - 6)$ is a factor of the given dividend.
20. Refer to Problem 2. Determine whether $(x + 2)$ is a factor of the given dividend.
21. Refer to Problem 3. Determine whether $(x + 1)$ is a factor of the given dividend.
22. Refer to Problem 4. Determine whether $(x - 5)$ is a factor of the given dividend.
23. Determine whether $(x + 2)$ is a factor of $(3x^4 - 7x^2 + 8x - 4)$.
24. Determine whether $(x - 1)$ is a factor of $(2x^5 + 3x^4 - 4x^3 + 2x - 3)$.

5.4 SOLVING EQUATIONS

The following sections describe the method of solving equations of degree higher than two.

A. The Roots of An Equation

The following two theorems are useful in solving a polynomial equation:

1. *Every polynomial of degree n in x, f(x), can be factored into n linear factors.*

Thus, a polynomial of degree 3 in x can be factored into 3 linear factors as illustrated in Example 8.

EXAMPLE 8 Factor the polynomial $f(x) = 2x^3 + 5x^2 - 4x - 3$.

From Example 4, we know that $(x + 3)$ is a factor of the given $f(x)$. Thus:

$$f(x) = (x + 3)Q(x)$$

where $Q(x)$ is a polynomial of degree 2 $(= 3 - 1)$ and is the quotient in the synthetic division:

$$
\begin{array}{rrrr|r}
2 & 5 & -4 & -3 & \underline{-3} \\
  & -6 & 3 & 3 & \\
\hline
2 & -1 & -1 & 0 &
\end{array}
$$

Now: $f(x) = (x + 3)(2x^2 - x - 1)$

The two linear factors of $Q(x) = (2x^2 - x - 1)$ can be obtained by the trial-and-error method (Chapter 2) or by synthetic division:

$$
\begin{array}{rrr|r}
2 & -1 & -1 & \underline{1} \\
  & 2 & 1 & \\
\hline
2 & 1 & 0 &
\end{array}
$$

Or: $Q(x) = 2x^2 - x - 1 = (x - 1)(2x + 1)$

The final form is:

$$f(x) = 2x^3 + 5x^2 - 4x - 3 = (x + 3)(x - 1)(2x + 1)$$

We have factored the given polynomial into 3 linear factors.

2. *Every polynomial equation of degree n in x, f(x) = 0, has exactly n roots or solutions.*

Thus, a polynomial equation of degree 3 in x has exactly 3 roots.

The roots of an equation need not all be distinct. If a number is equal to two of the roots, the number is a *double root;* if equal to three of the roots, it is a *triple root.* A number which is equal to two or more of the roots is also said to be a *multiple root.* For example, a polynomial equation of degree 3 in x may have the three roots: 1, 2, and 2. Here 2 is a double or multiple root.

EXAMPLE 9   Solve the equation $f(x) = 2x^3 + 5x^2 - 4x - 3 = 0$.

From the solution of Example 8,

$$f(x) = (x + 3)(x - 1)(2x + 1) = 0$$

When $x + 3 = 0$, $x = -3$.
When $x - 1 = 0$, $x = 1$.
When $2x + 1 = 0$, $x = -\frac{1}{2}$.

Thus, the given equation of degree 3 has 3 roots.

The solution of Example 9 may also be shown graphically.

Graph: $y = f(x) = 2x^3 + 5x^2 - 4x - 3$.

x	4	3	2	1	$\frac{1}{2}$	0	$-\frac{1}{2}$	-1	-2	-3	-4
y	189	84	25	0	$-3\frac{1}{2}$	-3	0	4	9	0	-35

When $y = f(x) = 0$, the curve intersects the X-axis at three points:

$x = 1$, $x = -\frac{1}{2}$, and $x = -3$.

The three x values are the graphical solution.

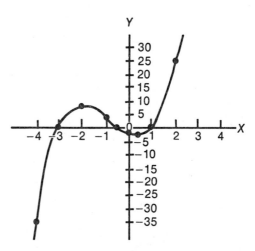

EXAMPLE 10   Solve the equation $f(x) = 3x^4 - x^3 - 21x^2 - 11x + 6 = 0$.

From the solutions of Examples 6 and 7, we know that $(x + 1)$ and $(x - \frac{1}{3})$ are two factors of the given polynomial, or:

$$f(x) = 3x^4 - x^3 - 21x^2 - 11x + 6 = (x + 1)(x - \tfrac{1}{3})Q(x)$$

The value of $Q(x)$ and the two linear factors of $Q(x)$ can be obtained in the following manner:

From the solution of Example 2, we have:

$$f(x) = (x + 1)(3x^3 - 4x^2 - 17x + 6)$$

Since $(x - \tfrac{1}{3})$ is another factor, we may divide the depressed equation $(3x^3 - 4x^2 - 17x + 6)$ by $(x - \tfrac{1}{3})$ to obtain the value of $Q(x)$.

$$\begin{array}{rrrr|l} 3 & -4 & -17 & 6 & \tfrac{1}{3} \\ & 1 & -1 & -6 & \\ \hline 3 & -3 & -18 & 0 & \end{array}$$

Thus:

$$Q(x) = 3x^2 - 3x - 18$$

and:

$$f(x) = (x + 1)(x - \tfrac{1}{3})(3x^2 - 3x - 18) = 0$$

When $x + 1 = 0$, $x = -1$.
When $x - \tfrac{1}{3} = 0$, $x = \tfrac{1}{3}$.

When $(3x^2 - 3x - 18) = 0$, the two roots can be obtained by:

1. Factoring

$$3x^2 - 3x - 18 = (3x + 6)(x - 3) = 0$$

When $3x + 6 = 0$, $x = -\tfrac{6}{3} = -2$
When $x - 3 = 0$, $x = 3$

2. The quadratic formula

$$x = \frac{-b \pm \sqrt{b^2 - 4ac}}{2a}$$

$$x = \frac{-(-3) \pm \sqrt{(-3)^2 - 4(3)(-18)}}{2(3)}$$

$$= \frac{3 \pm \sqrt{9 + 216}}{6} = \frac{3 \pm \sqrt{225}}{6}$$

$$= \frac{3 \pm 15}{6}$$

When $x = \dfrac{3 + 15}{6}$, $x = \dfrac{18}{6} = 3$.

When $x = \dfrac{3 - 15}{6}$, $x = \dfrac{-12}{6} = -2$.

Thus, the given equation of degree 4 has 4 roots: -1, $\tfrac{1}{3}$, 3, and -2.

The graphical solution of Example 10 is shown as follows:

Graph $y = f(x) = 3x^4 - x^3 - 21x^2 - 11x + 6$.

x	-3	-2	-1.5	-1	$-.5$	0	.5	1	1.5	2	3	4
y	120	0	-6.2	0	6.6	6	-4.7	-24	-45.9	-60	0	330

When $y = f(x) = 0$, the curve intersects the X-axis at four points:

$x = -2$, $x = -1$, $x = \frac{1}{3}$, and $x = 3$.

The four x values are the graphical solution.

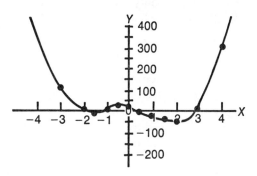

Note that the values of the two roots of a polynomial equation obtained by the quadratic formula,

$$x = \frac{-b \pm \sqrt{b^2 - 4ac}}{2a}$$

depend on the value of the radical $\sqrt{b^2 - 4ac}$. The value of the radical may be rational (such as $\sqrt{(-3)^2 - 4(3)(-18)} = \sqrt{225} = 15$ in Example 10), irrational (such as $\sqrt{2}$), or imaginary (such as $\sqrt{-4} = 2\sqrt{-1} = 2i$). Also, the only difference between the two roots is the "$\pm$ signs" before the radical. Thus, if the number $(A + Bi)$, where A and B are rational but i is an imaginary number since $i = \sqrt{-1}$, is a root of the equation $f(x) = 0$, then the number $(A - Bi)$ is also a root, such as:

$$x = \frac{6 \pm \sqrt{-4}}{2} = 3 \pm 1i$$

The two roots are $(3 + i)$ and $(3 - i)$.

Similarly, if the number $(A + \sqrt{B})$, where A and B are rational but $\sqrt{B}$ is irrational, is a root of the equation $f(x) = 0$, then the number $(A - \sqrt{B})$ is also a root, such as:

$$x = \frac{6 \pm 2\sqrt{2}}{2} = 3 \pm \sqrt{2}$$

The two roots are $(3 + \sqrt{2})$ and $(3 - \sqrt{2})$.

B. Rational Roots

If an equation with integral coefficients has any *rational roots*, they can be determined by the methods introduced below.

An equation $f(x) = 0$ may be formed when its roots are given. We simply equate the product of the corresponding factors $(x - \text{1st root})$ $(x - \text{2nd root})$ $(x - \text{3rd root}) \ldots (x - \text{nth root})$ to zero. If the three roots are 1, 2, and 2, the polynomial equation is:

$$f(x) = (x - 1)(x - 2)(x - 2) = 0$$
$$\text{or: } f(x) = x^3 - 5x^2 + 8x - 4 = 0$$

Observe the constant 4 in the polynomial of the above equation. The constant is determined only by the given roots, 1, 2, and 2, or disregard the signs $(1)(2)(2) = 4$. This fact gives the following theorem:

If the coefficient of the highest power of x is 1, then any integral roots are factors of the constant term of the equation $f(x) = 0$.

In the above equation, the coefficient of the highest power of x, or x^3, is 1. The possible factors of the constant 4 are 1, 2, and 4. We observe that the given roots 1, 2, and 2, are among the group of factors. Also, the theorem indicates nothing in regard to the signs involved.

Further, if the coefficient of the highest power of x is not 1, then this coefficient can be factored from every term of $f(x)$. This new factored equation, along with the theorem stated above, now gives us a more inclusive theorem:

If a rational number b/c, which is in its lowest terms, is a root of a polynomial equation

$$a_0 x^n + a_1 x^{n-1} + \ldots + a_n = 0$$

with integral coefficients, then b is a factor of a_n *and c is a factor of* a_0.

The coefficient of x^3 in the equation given in Example 9,

$$f(x) = 2x^3 + 5x^2 - 4x - 3 = 0$$

is 2. Factor 2 from every term, we have the new equation and the coefficient of x^3 becomes 1, or:

$$f(x) = 2\left(x^3 + \frac{5}{2}x^2 - \frac{4}{2}x - \frac{3}{2}\right) = 0$$

The integral factors of 3 ($= a_n$) are 1 and 3, and the integral factors of 2 ($= a_0$) are 1 and 2. The roots (b/c) of the equation therefore must be found from the group of all possible positive and negative quotients: $\pm\frac{1}{1} = \pm 1$, $\pm\frac{3}{1} = \pm 3$, $\pm\frac{1}{2}$, and $\pm\frac{3}{2}$. We can verify the statement from the solution of Example 9. The actual roots of the equation are $-\frac{1}{2}$, -3, and 1, and the product of the roots is $\left|\frac{3}{2}\right| = (-\frac{1}{2})(-3)(1)$.

The coefficient of x^4 in the equation given in Example 10,

$$f(x) = 3x^4 - x^3 - 21x^2 - 11x + 6 = 0$$

is 3. Factor 3 from every term; we have the new equation and the coefficient of x^4 becomes 1, or:

$$f(x) = 3\left(x^4 - \frac{1}{3}x^3 - 7x^2 - \frac{11}{3}x + \frac{6}{3}\right) = 0$$

The integral factor of 6 ($= a_n$) are 1, 2, 3, and 6, and the integral factors of 3 ($= a_0$) are 1 and 3. The roots (b/c) of the equation must be found from the group of all possible quotients: ± 1, ± 2, ± 3, ± 6, $\pm\frac{1}{3}$, $\pm\frac{2}{3}$. From the solution of

Example 10, the actual roots of the equation are $-1, -2, \frac{1}{3}$, and 3. The product of the roots is $|2| = (-1)(-2)(\frac{1}{3})(3)$.

We now may test all possible roots in each step of synthetic division as illustrated in Examples 8 to 10 to find the actual roots of an equation.

C. Irrational Roots by Linear Interpolation

The method of *linear interpolation* can be used to find the approximate values of the irrational roots of a polynomial equation $f(x) = 0$. This method is illustrated in Example 11.

EXAMPLE 11   Find the real roots (rational or irrational numbers) of the equation:

$$f(x) = x^3 - 2x^2 + 3x - 5 = 0$$

by the linear interpolation method to two decimal places.

If the given equation has any rational roots, we should first find the corresponding linear rational factors and the depressed expression having no rational roots. The possible factors of the constant term -5 are ± 1 and ± 5.

Substitute the factors in the given function $f(x)$ individually (or use the synthetic division to find the remainder):

$$f(+1) = -3, f(-1) = -11, f(+5) = 85, f(-5) = -195.$$

Based on the remainder and factor theorems, we find that none of the factors (± 1 and ± 5) satisfies the given equation $f(x) = 0$. Thus, the equation has no rational roots.

Next, use the graphic method to isolate each irrational root:

Graph $y = f(x) = x^3 - 2x^2 + 3x - 5.$

x	5	4	3	2	1.5	1	0	-1	-2	-3	-4	-5
y	85	39	13	1	-1.625	-3	-5	-11	-27	-59	-113	-195

From the graph we observe that there is only one real root since the curve intersects X-axis at only one point:

$x = 1.8$ approximately (between $x = 1.5$ and $x = 2$)

Since the degree of the given equation in x is 3, the remaining two roots are imaginary.

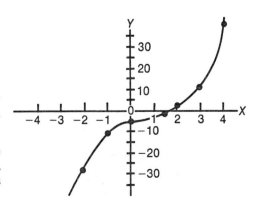

In determining the actual location of this real root. we compute $f(x)$ for successive tenths near the approximated value 1.8 until a change in sign is obtained.

Since $f(1.8) = 1.8^3 - 2(1.8^2) + 3(1.8) - 5 = -0.248,$
and $f(1.9) = 1.9^3 - 2(1.9^2) + 3(1.9) - 5 = +0.339,$

or by synthetic division:

$$
\begin{array}{rrrr|l}
1 & -2 & 3 & -5 & \underline{1.8} \\
  & 1.8 & -0.36 & 4.752 & \\
\hline
1 & -0.2 & 2.64 & -0.248 & \text{(remainder)}
\end{array}
$$

$$
\begin{array}{rrrr|l}
1 & -2 & 3 & -5 & \underline{1.9} \\
  & 1.9 & -0.19 & 5.339 & \\
\hline
1 & -0.1 & 2.81 & +0.339 & \text{(remainder)}
\end{array}
$$

the given equation $f(x) = 0$ must have a root between $x = 1.8$ and $x = 1.9$.
Now we may use the linear interpolation method to determine the digit in the hundredths place:

x	$f(x)$	
		Line
1.9	+0.339	(1)
?	0	(2)
1.8	−0.248	(3)

$\dfrac{(2)-(3)}{(1)-(3)} \quad \dfrac{?-1.8}{0.1} = \dfrac{0.248}{0.587} \quad \begin{array}{l}(4)\\(5)\end{array}$

$? = 1.8 + (0.1)\left(\dfrac{0.248}{0.587}\right)$
$= 1.8 + (0.1)(0.42)$
$= 1.8 + 0.042$
$= 1.842,$ or round to
        1.84 (2 decimal places)

When the linear interpolation method is used. it is assumed that the differences between the x values and the differences between corresponding $f(x)$ values are proportional.

$$
\text{Check:}\quad
\begin{array}{rrrr|l}
1 & -2 & 3 & -5 & \underline{1.84} \\
  & 1.84 & -0.2944 & 4.9783 & \\
\hline
1 & -0.16 & 2.7056 & -0.0217 &
\end{array}
$$

Or. $f(1.84) = -0.0217$, which is close to zero. Thus: $x = 1.84$ (approximately).

EXERCISE 5–2   Reference: Section 5.4

A. *Form equations with integral coefficients from the given roots:*

1. $x = 1, 2,$ and -3 3. $x = 1, -2, 3,$ and -1
2. $x = -1, 2,$ and -4 4. $x = \pm\sqrt{3},$ and 2.

B. *Solve each of the following equations:*

5. $x^3 - 2x^2 - 5x + 6 = 0$ 9. $2x^3 + 5x^2 - 2x - 5 = 0$
6. $x^3 + 4x^2 - 7x - 10 = 0$ 10. $2x^3 + x^2 - 7x - 6 = 0$
7. $x^3 - x^2 - 2x + 2 = 0$ 11. $3x^4 - 2x^3 - 11x^2 + 2x + 8 = 0$
8. $x^3 - 3x^2 - 3x + 1 = 0$ 12. $2x^4 - x^3 - 11x^2 + 4x + 12 = 0$

C. *Find the real roots (rational or irrational numbers) of the following equations by the linear interpolation method to two decimal places.*

13. $x^3 - 2x^2 + x - 3 = 0$ 14. $x^3 + x^2 + 2x + 4 = 0$

Part Two Matrix and Linear Programming

Chapter

6 Vectors and Matrices

This chapter will introduce the basic operations of vectors and matrices, which are very useful tools in analyzing many types of business and economic problems. Related topics and methods of solving systems of linear equations by the use of matrix algebra will be presented in Chapter 7.

6.1 COMMON TYPES OF VECTORS

A *vector* is an ordered set of numbers. The numbers, usually enclosed in brackets or bold-faced parentheses, are also called the *components* or *elements* of the vector. We shall use brackets and lowercase letters to represent the vectors in this chapter.

Common types of vectors are described in the following paragraphs.

A. Row Vector

A *row vector* is a vector in which the numbers in the brackets are written horizontally.

a and b are *two-component row vectors*.

$$a = [2, 3]$$
$$b = [5, 1]$$

B. Column Vector

A *column vector* is a vector in which the numbers in the brackets are written vertically.

c and d are *three-component column vectors*.

$$c = \begin{bmatrix} 4 \\ 3 \\ -1 \end{bmatrix}$$

$$d = \begin{bmatrix} 7 \\ -2 \\ 5 \end{bmatrix}$$

C. Unit Vector

A *unit vector* has only one component equal to 1 and all other components equal to 0.

$$\begin{bmatrix} 1 \\ 0 \end{bmatrix}, [1, 0], \begin{bmatrix} 1 \\ 0 \\ 0 \end{bmatrix}, [0, 1, 0]$$

D. Zero Vector

A *zero vector* has all components equal to 0.

$$\begin{bmatrix} 0 \\ 0 \end{bmatrix}, [0, 0], \begin{bmatrix} 0 \\ 0 \\ 0 \end{bmatrix}, [0, 0, 0]$$

E. Transpose of a Vector

A row vector may be *transposed* into its corresponding column vector, and vice versa. The superscript t is usually used to represent the transpose of a vector. The order of the elements in the given vector is unchanged in the transposed vector.

$$\text{Row vector } a = [2, 3].$$

$$\text{The transpose of vector } a = a^t = [2, 3]^t = \begin{bmatrix} 2 \\ 3 \end{bmatrix}$$

Thus, the transpose of a row vector is a column vector.

$$\text{Column vector } c = \begin{bmatrix} 4 \\ 3 \\ -1 \end{bmatrix}$$

$$\text{The transpose of vector } c = c^t = \begin{bmatrix} 4 \\ 3 \\ -1 \end{bmatrix}^t = [4, 3, -1]$$

Thus, the transpose of a column vector is a row vector.

F. Equal Vectors

Two vectors are *equal* if and only if the two vectors are arranged in the same form (both in rows or both in columns) and their corresponding components are equal.

Equal vectors:

$$[2, 3] = [1 + 1, 1 + 2]; \qquad \begin{bmatrix} 1 \\ 4 \end{bmatrix} = \begin{bmatrix} 1 \\ 4 \end{bmatrix}$$

$$\begin{bmatrix} 7 \\ -2 \\ 5 \end{bmatrix} = \begin{bmatrix} 3 + 4 \\ -2 \\ 5 \end{bmatrix}; \qquad \begin{bmatrix} 0 \\ 7 \end{bmatrix} = \begin{bmatrix} 0 \\ 1 + 6 \end{bmatrix}$$

Unequal vectors:

$$[2, 3] \neq \begin{bmatrix} 2 \\ 3 \end{bmatrix}; \qquad \begin{bmatrix} 4 \\ 3 \\ -1 \end{bmatrix} \neq [4, 3, -1]$$

$$\begin{bmatrix} 5 \\ 2 \\ 1 \end{bmatrix} \neq \begin{bmatrix} 5 \\ 3 \end{bmatrix}; \qquad [1, 0, 7] \neq [0, 1, 7]$$

6.2 BASIC OPERATIONS OF VECTORS

The basic operations of vectors, addition, subtraction, and multiplication, are presented in the sections which follow.

A. Addition

Vectors with the same number of components and the same arrangement (rows and columns) can be added. The sum is also a vector with the same arrangement as the given vectors. The components of the sum are obtained by adding the corresponding components of the given vectors.

EXAMPLE 1 (a) Add the two-component row vectors a and b.

$$a + b = [2, 3] + [5, 1] = [2 + 5, 3 + 1] = [7, 4]$$

This addition may be applied to a practical problem as follows:

Vector a = factory a produced 2 units of product X and 3 units of product Y.
Vector b = factory b produced 5 units of product X and 1 unit of product Y.
Vector $(a + b)$ = factories a and b together produced 7 units of product X and 4 units of product Y.

(b) Add the three-component column vectors c and d.

$$c + d = \begin{bmatrix} 4 \\ 3 \\ -1 \end{bmatrix} + \begin{bmatrix} 7 \\ -2 \\ 5 \end{bmatrix} = \begin{bmatrix} 4 + 7 \\ 3 + (-2) \\ (-1) + 5 \end{bmatrix} = \begin{bmatrix} 11 \\ 1 \\ 4 \end{bmatrix}$$

This addition may be applied to a practical problem as follows:

Vector c = this year's profit (+) or loss (−) in millions of dollars in Departments #1, #2, and #3 of a company: $4 million profit in #1, $3 million profit in #2, and $1 million loss in #3.
Vector d = last year's profit or loss: $7 million profit in #1, $2 million loss in #2, and $5 million profit in #3.
Vector $(c + d)$ = the sum of the two years: $11 million profit in #1, $1 million profit in #2, and $4 million in #3.

The two applications in Example 1 can also be tabulated according to the vector arrangements presented as follows:

(a) Production by Products and Factories			(b) Profit and Loss by Years and Departments			
Factory (vector)	Product		Depart-ment	This Year (Vec-tor c)	Last Year (Vec-tor d)	Two Years (Vec-tor $c + d$)
	X units	Y units				
a	2	3	#1	4	7	11
b	5	1	#2	3	−2	1
			#3	−1	5	4
a and b ($a+b$)	7	4				

B. Subtraction

Vectors with the same number of components and the same arrangement (rows and columns) can be subtracted. The remainder of the subtraction is also a vector with the same arrangement as the given vectors. The components of the remainder are obtained by subtracting the corresponding components of the given vectors.

EXAMPLE 2   (a) Subtract vector b from vector a.

$$a - b = [2, 3] - [5, 1] = [2 - 5, 3 - 1] = [-3, 2]$$

(b) Subtract vector d from vector c.

$$c - d = \begin{bmatrix} 4 \\ 3 \\ -1 \end{bmatrix} - \begin{bmatrix} 7 \\ -2 \\ 5 \end{bmatrix} = \begin{bmatrix} 4 - 7 \\ 3 - (-2) \\ (-1) - 5 \end{bmatrix} = \begin{bmatrix} -3 \\ 5 \\ -6 \end{bmatrix}$$

Additional Example

$$\begin{bmatrix} 6 \\ 2 \end{bmatrix} - \begin{bmatrix} 5 \\ -3 \end{bmatrix} + \begin{bmatrix} -7 \\ 4 \end{bmatrix} = \begin{bmatrix} 6 - 5 + (-7) \\ 2 - (-3) + 4 \end{bmatrix} = \begin{bmatrix} -6 \\ 9 \end{bmatrix}$$

Vectors of different arrangements, such as row vector $[2, 3]$ and column vector $\begin{bmatrix} 6 \\ 2 \end{bmatrix}$, cannot be added or subtracted. Also, vectors with different numbers of components, such as the two-component row vector $[2, 3]$ and the three-component row vector $[1, 0, 7]$, cannot be added or subtracted.

C. Multiplication

A vector may be multiplied by a scalar (a real number) or another vector. The rules for the different types of multiplications are presented individually below.

1. SCALAR MULTIPLICATION.   The scalar, such as 2, $\frac{1}{2}$, −3, and $\sqrt{4}$, must be multiplied by each component of the given vector to obtain the corresponding components of the product. Thus, the product is also a vector with the same number of components and the same arrangement as the given vector.

EXAMPLE 3   (a) Multiply 2 by vector e, where $e = \begin{bmatrix} 5 \\ -3 \\ 6 \end{bmatrix}$.

$$2 \cdot e = 2 \cdot \begin{bmatrix} 5 \\ -3 \\ 6 \end{bmatrix} = \begin{bmatrix} 2 \cdot 5 \\ 2 \cdot (-3) \\ 2 \cdot 6 \end{bmatrix} = \begin{bmatrix} 10 \\ -6 \\ 12 \end{bmatrix}$$

Observe the above multiplication. The product of scalar 2 and vector e may be viewed as the result of the addition:

$$e + e = \begin{bmatrix} 5 \\ -3 \\ 6 \end{bmatrix} + \begin{bmatrix} 5 \\ -3 \\ 6 \end{bmatrix} = \begin{bmatrix} 10 \\ -6 \\ 12 \end{bmatrix}, \text{ or } 2 \cdot e = e + e.$$

This fact explains the rule of scalar multiplication.

(b) Multiply $\frac{1}{2}$ by vector f, where $f = [4, 7]$.

$\frac{1}{2} \cdot f = \frac{1}{2} \cdot [4, 7] = [(\frac{1}{2} \cdot 4), (\frac{1}{2} \cdot 7)] = [2, 3\frac{1}{2}]$

(c) Multiply 0 by vector f.

$0 \cdot f = 0 \cdot [4, 7] = [0, 0]$ (a zero vector)

2. MULTIPLICATION OF TWO VECTORS.   Two vectors with the same number of components can be multiplied if the first on the left (multiplicand) is a row vector and the second on the right (multiplier) is a column vector. Each component of the row vector is multiplied by the corresponding component of the column vector to obtain the partial product. The sum of all partial products, called the *inner product* of the multiplication, is a single component or a number.

EXAMPLE 4   (a) Multiply the two-component row vector g by the two-component column vector h.

$$g \cdot h = [4, 5] \begin{bmatrix} 2 \\ 3 \end{bmatrix} = (4 \times 2) + (5 \times 3) = 8 + 15 = 23$$

The above multiplication may be applied to a practical problem as follows:

Vector g = unit prices: \$4 for product X and \$5 for product Y.
Vector h = units sold: 2 units of product X and 3 units of product Y.
Vector $g \cdot h$ = total amount sold of products X and Y: \$23.

In the multiplication, the first component in the row vector (unit price for product X) is multiplied by the first component in the column vector (units of product X sold), and the second component in the row vector (unit price for product Y) is multiplied by the second component in the column vector (units of product Y sold.)

(b) Multiply the three-component row vector i by the three-component column vector j.

$$i \cdot j = [1, -2, 7] \begin{bmatrix} -3 \\ 5 \\ 8 \end{bmatrix} = (1 \times (-3)) + ((-2) \times 5) + (7 \times 8)$$

$$= (-3) + (-10) + 56 = 43$$

Note:

1. If the number of components in a row vector is not the same as the number of components in a column vector, such as vectors $[2, 4, 1]$ and $\begin{bmatrix} 3 \\ 5 \end{bmatrix}$, the two vectors cannot be multiplied.

2. In multiplication involving matrices, it is permissible to multiply a column vector on the left (multiplicand) by a row vector on the right (multiplier). However, the product is a *matrix* and is also called the *outer product*. (See Example 20 in Section 6.6.)

6.3 THE GEOMETRY OF VECTORS

Vectors and vector operations may be expressed geometrically on a graph. We shall first graph vectors and then graph vector operations.

A. Graphic Representation of a Vector

A straight line drawn on a graph to represent a vector usually begins at an *initial point* and ends at a *terminal point*. The direction of the line is indicated by an arrow. Row vectors and column vectors are not treated differently when the vectors are expressed geometrically on a graph. When the straight line is used to represent a given two-component row vector $[x, y]$ or column vector $\begin{bmatrix} x \\ y \end{bmatrix}$, the line is drawn in a two-dimensional space based on the system of rectangular coordinates. The initial point is at the point of origin ($x = 0, y = 0$) and the terminal point is at the location specified by the components x and y of the given vector. The length of the vector is determined by use of the *Pythagorean theorem*, or:

$$\text{length of a vector} = \sqrt{x^2 + y^2}$$

EXAMPLE 5   Graph and compute the length of each vector:

vector $m = [3, 4]$ or its transposed column vector, $m^t = \begin{bmatrix} 3 \\ 4 \end{bmatrix}$

vector $n = \begin{bmatrix} -2 \\ 5 \end{bmatrix}$ or its transposed row vector, $n^t = [-2, 5]$

The graphs are shown in Figure 6–1. The terminal point of vector m is at A ($x = 3, y = 4$) and that of vector n is at B ($x = -2, y = 5$) in the figure.

Figure 6-1 **Graphical Presentation of Two-Component Vectors *m* and *n* (Example 5)**

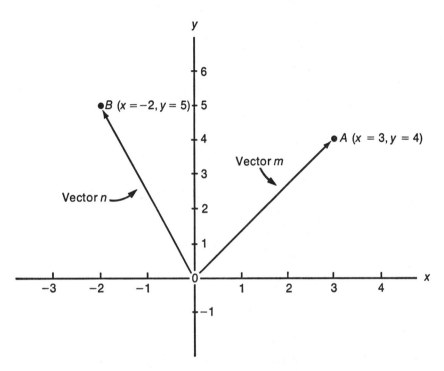

The length of vector $m = \sqrt{x^2 + y^2} = \sqrt{3^2 + 4^2}$

$$= \sqrt{9 + 16} = \sqrt{25} = 5$$

The length of vector $n = \sqrt{(-2)^2 + 5^2} = \sqrt{29} = 5.39$

A vector with three components can be represented graphically in a three-dimensional space. The basic method of plotting a vector on a three-dimensional chart is analogous to that on a two-dimensional chart. However, the third reference line z, which is perpendicular to both x and y lines, is added to the figure.

EXAMPLE 6   Graph vector $p = [3, 4, 2]$ (or its transposed column vector,

$$p^t = \begin{bmatrix} 3 \\ 4 \\ 2 \end{bmatrix})$$

The graph is shown in Figure 6–2. The terminal point of vector p is at point F ($x = 3$, $y = 4$, $z = 2$). Point F is obtained as follows:

1. Draw line AB parallel to the z-axis, starting at the x-axis with $A = 3$.
2. Draw line CD parallel to the x-axis, starting at the z-axis with $C = 2$. The two lines intersect each other at point E.
3. Draw line EF parallel to the y-axis, the length of line $EF = 4$ based on the y-scale.

Figure 6-2 **Graphical Presentation of Three-Component**
          Vector p (Example 6)

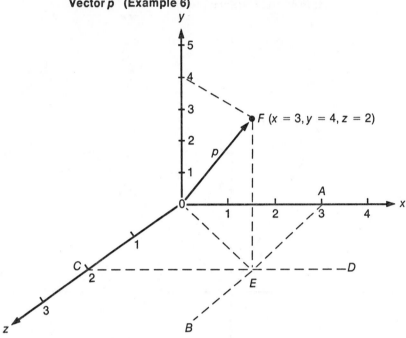

B. Graphic Representation of Vector Operations

The basic operations with vectors can also be shown graphically. For illustration purposes, only the operations of addition and scalar multiplication are presented here.

EXAMPLE 7   Show the following addition of the two-component row vectors a and b graphically.

$a + b = [2, 3] + [5, 1] = [7, 4]$

Vectors a, b, and $a + b$ are shown in Figure 6–3. The lines drawn from the origin ($x = 0$, $y = 0$) to the three terminal points represent the directions and lengths of the three vectors respectively. The line representing vector $a + b$ is the diagonal of the parallelogram formed by lines a and b.

EXAMPLE 8   Show the following scalar multiplications graphically:

(a) $2 \cdot a = 2 \cdot [2, 3] = [4, 6]$
(b) $(-2) \cdot a = (-2) \cdot [2, 3] = [-4, -6]$

Scalar multiplications (a) and (b) are shown in Figure 6–4, page 102. Observe that vectors a, $2a$, and $-2a$ are collinear. The direction of

Figure 6-3   **Graphical Presentation of Two-Component**
Vectors, *a*, *b*, and *a* + *b*   (Example 7)

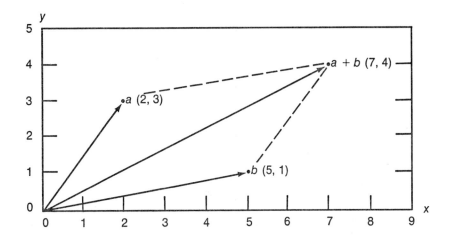

vector $-2a$ is opposite to that of vector $2a$, although their lengths are the same.

6.4 VECTORS AND SYSTEMS OF LINEAR EQUATIONS

Let v_1 and v_2 be two-component vectors, x and y be scalars, and:

$$xv_1 + yv_2 = s$$

Then, the sum s is called a *linear combination* of vectors v_1 and v_2. When the values of v_1, v_2, and s are given, the linear combination may be expressed as a system of linear equations by the rule of scalar multiplication. The values of scalars x and y thus can be found by solving the linear equations simultaneously.

EXAMPLE 9 $v_1 = \begin{bmatrix} 2 \\ 1 \end{bmatrix}$, $v_2 = \begin{bmatrix} 6 \\ 4 \end{bmatrix}$, and $s = \begin{bmatrix} 12 \\ 7 \end{bmatrix}$. Find the scalars x and y.

First, the linear combination s of vectors v_1 and v_2 is written as follows:

$$x \begin{bmatrix} 2 \\ 1 \end{bmatrix} + y \begin{bmatrix} 6 \\ 4 \end{bmatrix} = \begin{bmatrix} 12 \\ 7 \end{bmatrix}$$

Next, derive and solve a system of two equations. By the rule of scalar multiplication, we have:

$$\begin{bmatrix} 2x \\ 1x \end{bmatrix} + \begin{bmatrix} 6y \\ 4y \end{bmatrix} = \begin{bmatrix} 12 \\ 7 \end{bmatrix} \text{ and } \begin{bmatrix} 2x + 6y \\ 1x + 4y \end{bmatrix} = \begin{bmatrix} 12 \\ 7 \end{bmatrix}$$

The two equations are:

$$2x + 6y = 12 \quad (1)$$
$$1x + 4y = 7 \quad (2)$$

Figure 6-4 **Graphical Presentation of Scalar Multiplications**
Vectors *a*, 2*a*, and −2*a* (Example 8)

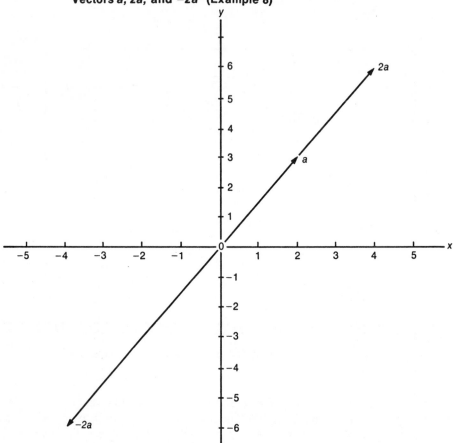

Multiply (2) by 2 and label the resulting equation (2)′.

(2) · 2: $2x + 8y = 14$ (2)′

Subtract (1) from (2)′.

$$(2)' - (1): \quad \begin{array}{r} 2x + 8y = 14 \\ 2x + 6y = 12 \\ \hline 2y = 2 \\ y = 1 \end{array}$$

Substitute $y = 1$ in (2):

$$x + 4(1) = 7$$
$$x = 7 - 4$$
$$x = 3$$

Check: Substitute $x = 3$ and $y = 1$ in the linear combination stated
above as follows:

$$3\begin{bmatrix} 2 \\ 1 \end{bmatrix} + 1\begin{bmatrix} 6 \\ 4 \end{bmatrix} = \begin{bmatrix} 12 \\ 7 \end{bmatrix}$$

$$\begin{bmatrix} 6 \\ 3 \end{bmatrix} + \begin{bmatrix} 6 \\ 4 \end{bmatrix} = \begin{bmatrix} 12 \\ 7 \end{bmatrix}$$

$$\begin{bmatrix} 12 \\ 7 \end{bmatrix} = \begin{bmatrix} 12 \\ 7 \end{bmatrix}$$

In Example 9, there are two linear equations in two unknowns and there is only one solution for each unknown that satisfies both equations ($x = 3$, $y = 1$). The two equations are called *independent equations* and the two given vectors, v_1 and v_2 are called *independent vectors*. Generally, two independent vectors are required to span a two-dimensional space (see vectors m and n in Figure 6–1), three independent vectors for three-dimensional space (such as vector p in Figure 6–2), and n such vectors for n dimensions. These independent vectors, which form the basis of the vector space, are also called *basic vectors*. When the basic vectors are unit vectors, the components of the linear combination vector s are equal to the values of the scalars x and y as shown in Example 10.

EXAMPLE 10 Unit vectors $v_1 = \begin{bmatrix} 1 \\ 0 \end{bmatrix}$ and $v_2 = \begin{bmatrix} 0 \\ 1 \end{bmatrix}$, and scalars $x = 2$ and $y = 5$. Find the linear combination vector s.

SOLUTION $xv_1 + yv_2 = 2 \cdot \begin{bmatrix} 1 \\ 0 \end{bmatrix} + 5 \cdot \begin{bmatrix} 0 \\ 1 \end{bmatrix} = \begin{bmatrix} 2 \\ 5 \end{bmatrix} = s$

If $v_1 = yv_2$ ($y = $ a scalar), the two derived equations from the linear combination equation may be reduced to the same equation. When two equations can be reduced to the same form, they are said to be *dependent*. A linear equation in one unknown has only one solution, but a linear equation in two unknowns has an unlimited number of solutions.

EXAMPLE 11 $v_1 = \begin{bmatrix} 4 \\ 6 \end{bmatrix}$, $v_2 = \begin{bmatrix} 2 \\ 3 \end{bmatrix}$, and $s = \begin{bmatrix} 8 \\ 12 \end{bmatrix}$. Find the scalars x and y.

$$x\begin{bmatrix} 4 \\ 6 \end{bmatrix} + y\begin{bmatrix} 2 \\ 3 \end{bmatrix} = \begin{bmatrix} 8 \\ 12 \end{bmatrix}$$

The two derived equations are:

$4x + 2y = 8$ (1)
$6x + 3y = 12$ (2)

Multiply (1) by 3.

(1) · 3 $12x + 6y = 24$ (1)′

Multiply (2) by 2.

(2) · 2 $12x + 6y = 24$ (2)′

The two equations, (1)' and (2)', are the same equation. Thus, there are unlimited number of x and y values that can satisfy the equation $12x + 6y = 24$, such as $x = 1$ and $y = 2$. (1) and (2) are *dependent equations*.

Observe the given vectors: $v_1 = 2v_2 = 2\begin{bmatrix} 2 \\ 3 \end{bmatrix} = \begin{bmatrix} 4 \\ 6 \end{bmatrix}$.

Vector v_1 is a scalar multiple of vector v_2. v_1 and v_2 are *dependent vectors* and span on only one-dimensional space (see Figure 6–4 for the one-dimension — the straight line.)

If $v_1 = yv_2$ ($y =$ a scalar) and only the left sides of the two derived equations from the linear combination equation can be reduced to the same form — the right sides (the constants) are not the same — the two derived equations are said to be *inconsistent*. There is no common solution for two linear equations in two unknowns if they are inconsistent equations.

EXAMPLE 12 $v_1 = \begin{bmatrix} 4 \\ 6 \end{bmatrix}$, $v_2 = \begin{bmatrix} 2 \\ 3 \end{bmatrix}$, and $s = \begin{bmatrix} 8 \\ 10 \end{bmatrix}$. Find the scalars x and y.

$$x\begin{bmatrix} 4 \\ 6 \end{bmatrix} + y\begin{bmatrix} 2 \\ 3 \end{bmatrix} = \begin{bmatrix} 8 \\ 10 \end{bmatrix}$$

The two derived equations are:

$4x + 2y = 8$ (1)
$6x + 3y = 10$ (2)

Multiply (1) by 3.

(1) · 3 $12x + 6y = 24$ (1)'

Multiply (2) by 2.

(2) · 2 $12x + 6y = 20$ (2)'

Equations (1)' and (2)' are inconsistent. There is no common solution for x and y that satisfies both equations simultaneously.

EXERCISE 6–1   Reference: Sections 6.1–6.4

A. *State the complete name and write the transpose of each of the following vectors:*

   1. $a = [3, 7]$

   2. $b = [-2, 5]$

   3. $c = \begin{bmatrix} 4 \\ 9 \end{bmatrix}$

   4. $d = \begin{bmatrix} 2 \\ 1 \\ 6 \end{bmatrix}$

   5. $e = \begin{bmatrix} 3 \\ -4 \\ 9 \end{bmatrix}$

   6. $f = [1, 5, 7]$

B. *Compute the indicated operations with vectors. Use the vectors given in Problems 1–6.*

7. $a + b$

8. $d + e$

9. $d - e$

10. $a - b$

11. $e - d$

12. $b - a$

13. $4a$

14. $\frac{1}{4}b$

15. $5c$

16. $3f$

17. $6d$

18. $\frac{1}{2}e$

19. ac

20. bc

21. fd

22. fe

C. *Graph the vectors given in Problems 23–30, and compute the length of each vector given only in Problems 23–26.*

23. $g = [2, 3]$

24. $h = \begin{bmatrix} -3 \\ 4 \end{bmatrix}$

25. $i = \begin{bmatrix} 3 \\ -2 \end{bmatrix}$

26. $j = [-4, -1]$

27. $k = \begin{bmatrix} 2 \\ 3 \\ 5 \end{bmatrix}$

28. $l = [3, 1, 2]$

29. $m = [4, 2, 3]$

30. $n = \begin{bmatrix} 5 \\ 4 \\ 4 \end{bmatrix}$

D. *Graph and compute the indicated operations with vectors.*

31. $p + q = [3, 2] + [1, 4]$

32. $r + s = \begin{bmatrix} 2 \\ 5 \end{bmatrix} + \begin{bmatrix} 6 \\ 1 \end{bmatrix}$

33. $2p = 2[3, 2]$

34. $3r = 3\begin{bmatrix} 2 \\ 5 \end{bmatrix}$

E. *Solve for unknowns and determine if the derived systems of equations are independent, dependent, or inconsistent.*

35. $v_1 = \begin{bmatrix} 1 \\ 2 \end{bmatrix}$, $v_2 = \begin{bmatrix} 1 \\ -1 \end{bmatrix}$, and $s = \begin{bmatrix} 8 \\ 7 \end{bmatrix}$. Find scalars x and y.

36. $v_1 = \begin{bmatrix} 3 \\ 1 \end{bmatrix}$, $v_2 = \begin{bmatrix} -1 \\ 4 \end{bmatrix}$, and $s = \begin{bmatrix} 5 \\ 19 \end{bmatrix}$. Find scalars x and y.

37. $v_1 = \begin{bmatrix} 2 \\ 3 \end{bmatrix}$, $v_2 = \begin{bmatrix} 3 \\ -2 \end{bmatrix}$, and $s = \begin{bmatrix} 23 \\ 28 \end{bmatrix}$. Find scalars x and y.

38. $v_1 = \begin{bmatrix} 2 \\ 4 \end{bmatrix}$, $v_2 = \begin{bmatrix} 3 \\ -1 \end{bmatrix}$, and $s = \begin{bmatrix} 5 \\ 17 \end{bmatrix}$. Find scalars x and y.

39. $v_1 = \begin{bmatrix} 3 \\ 1 \end{bmatrix}$, $v_2 = \begin{bmatrix} 6 \\ 2 \end{bmatrix}$, and $s = \begin{bmatrix} 15 \\ 5 \end{bmatrix}$. Find scalars x and y.

40. $v_1 = \begin{bmatrix} 2 \\ 1 \end{bmatrix}$, $v_2 = \begin{bmatrix} 8 \\ 4 \end{bmatrix}$, and $s = \begin{bmatrix} 5 \\ 2 \end{bmatrix}$. Find scalars x and y.

41. $v_1 = \begin{bmatrix} 1 \\ 0 \end{bmatrix}$, $v_2 = \begin{bmatrix} 0 \\ 1 \end{bmatrix}$, and $s = \begin{bmatrix} 5 \\ 7 \end{bmatrix}$. Find scalars x and y.

42. $v_1 = \begin{bmatrix} 0 \\ 1 \end{bmatrix}$, $v_2 = \begin{bmatrix} 1 \\ 0 \end{bmatrix}$, and $s = \begin{bmatrix} -4 \\ 3 \end{bmatrix}$. Find scalars x and y.

43. $v_1 = \begin{bmatrix} 2 \\ 5 \end{bmatrix}$, $v_2 = \begin{bmatrix} 4 \\ 1 \end{bmatrix}$, scalars $x = 3$, and $y = -2$. Find the linear combination vector s.

44. $v_1 = \begin{bmatrix} -3 \\ 1 \end{bmatrix}$, $v_2 = \begin{bmatrix} 2 \\ -5 \end{bmatrix}$, scalars $x = 2$ and $y = 4$. Find the linear combination vector s.

6.5 COMMON TYPES OF MATRICES

A *matrix* is a rectangular array of numbers enclosed in brackets or in bold-faced parentheses. Like the symbols used in vectors, we again shall use brackets. However, instead of using lowercase letters as the representations of vectors, we now use capital letters to represent matrices, such as:

$$A = \begin{bmatrix} 2 & 4 \\ 3 & 1 \end{bmatrix} \qquad B = \begin{bmatrix} 1 & -2 \\ 5 & 8 \end{bmatrix} \quad (A \text{ and } B \text{ are } 2 \times 2 \text{ matrices})$$

$$C = \begin{bmatrix} 1 & -2 \\ 0 & 6 \\ 4 & 1 \end{bmatrix} \qquad D = \begin{bmatrix} 5 & 1 \\ -3 & -4 \\ 0 & 6 \end{bmatrix} \quad (C \text{ and } D \text{ are } 3 \times 2 \text{ matrices})$$

The numbers enclosed in the brackets of a matrix are called *elements*. Thus, the first row elements of matrix A are 2 and 4; the second column elements of matrix C are -2, 6, and 1; and so on.

In specifying a matrix, first designate the number of rows and then the number of columns in the matrix. This type of specification is called the *order* of a matrix. Thus, the order of matrices A and B is 2 by 2, written 2×2, and the order of matrices C and D is 3 by 2, written 3×2.

Common types of matrices are presented as follows:

A. *M* × *N* Matrix

In the *m* × *n matrix*, *m* represents the number of rows and *n* represents the number of columns of the matrix.

> Matrix A is a 2×2 matrix, where $m = 2$ and $n = 2$.
> Matrix C is a 3×2 matrix, where $m = 3$ and $n = 2$.

Thus, a row vector with n components is the same as a *1 × n* matrix, such as [2, 1, 3] which is a *1 × 3* matrix. A column vector with m components is the same as an $m \times 1$ matrix, such as $\begin{bmatrix} 4 \\ 7 \end{bmatrix}$ which is a *2 × 1* matrix.

B. Square Matrix

In a *square matrix*, the number of rows is the same as the number of columns of a matrix, or $m = n$.

A and B are the square matrices of order 2.

C. Identity (or Unit) Matrix

An identity matrix is a square matrix with the value 1 for all elements on its principal diagonal (the line from the upper left corner to the lower right corner) and 0's for all other elements.

$$2 \times 2 \text{ identity matrix } I = \begin{bmatrix} 1 & 0 \\ 0 & 1 \end{bmatrix}$$

$$3 \times 3 \text{ identity matrix } I = \begin{bmatrix} 1 & 0 & 0 \\ 0 & 1 & 0 \\ 0 & 0 & 1 \end{bmatrix}$$

An identity matrix has the same property as 1 in the ordinary algebra involving multiplication. The expression, $A \cdot I = I \cdot A = A$ is true for all square matrices A. (See Example 21.)

D. Zero Matrix

In a zero matrix, every element is equal to 0 and is denoted by the bold-faced symbol **0**.

$$\mathbf{0} = \begin{bmatrix} 0 & 0 \\ 0 & 0 \end{bmatrix} \text{ is a } 2 \times 2 \text{ zero matrix.}$$

$$\mathbf{0} = \begin{bmatrix} 0 & 0 \\ 0 & 0 \\ 0 & 0 \end{bmatrix} \text{ is a } 3 \times 2 \text{ zero matrix.}$$

E. Transpose of a Matrix

The transpose of matrix A is a matrix denoted by A^t. The rows in A^t are the columns of A, and the columns in A^t are the rows of A, correspondingly.

$$A = \begin{bmatrix} 2 & 4 \\ 3 & 1 \end{bmatrix}, \quad A^t = \begin{bmatrix} 2 & 4 \\ 3 & 1 \end{bmatrix}^t = \begin{bmatrix} 2 & 3 \\ 4 & 1 \end{bmatrix}$$

$$C = \begin{bmatrix} 1 & -2 \\ 0 & 6 \\ 4 & 1 \end{bmatrix}, \quad C^t = \begin{bmatrix} 1 & -2 \\ 0 & 6 \\ 4 & 1 \end{bmatrix}^t = \begin{bmatrix} 1 & 0 & 4 \\ -2 & 6 & 1 \end{bmatrix}$$

$$G = \begin{bmatrix} 1 & 2 & 3 \\ 2 & -1 & 0 \end{bmatrix}, \quad G^t = \begin{bmatrix} 1 & 2 & 3 \\ 2 & -1 & 0 \end{bmatrix}^t = \begin{bmatrix} 1 & 2 \\ 2 & -1 \\ 3 & 0 \end{bmatrix}$$

F. Equal Matrices

Two matrices are *equal* if and only if they are of the same order and all their corresponding elements are equal.

Equal matrices:

$$\begin{bmatrix} 2 & 4 \\ 3 & 1 \end{bmatrix} = \begin{bmatrix} 1+1 & 4 \\ 5-2 & 1 \end{bmatrix}, \qquad \begin{bmatrix} 5 & -3 \\ 2 & 0 \\ 1 & 4 \end{bmatrix} = \begin{bmatrix} 5 & 1-4 \\ 2 & 3-3 \\ 1 & 2+2 \end{bmatrix},$$

$$\begin{bmatrix} 6 & 4 & 1 \end{bmatrix} = \begin{bmatrix} 1+5 & 7-3 & 1 \end{bmatrix}$$

Unequal matrices:

$$\begin{bmatrix} 2 & 4 \\ 3 & 1 \end{bmatrix} \neq \begin{bmatrix} 2 & 3 \\ 4 & 1 \end{bmatrix}, \begin{bmatrix} 5 & -3 \\ 2 & 0 \\ 1 & 4 \end{bmatrix} \neq \begin{bmatrix} 5 & 2 & 1 \\ -3 & 0 & 4 \end{bmatrix}, \begin{bmatrix} 6 & 4 & 1 \end{bmatrix} \neq \begin{bmatrix} 6 \\ 4 \\ 1 \end{bmatrix}$$

Thus, $A \neq A^t$.

6.6 BASIC OPERATIONS OF MATRICES

The basic operations with matrices are addition, subtraction, and multiplication. The operations with matrices are basically the same as those with vectors. In fact, when a matrix has only one row or column, it is identical to a row or column vector. Thus, in performing the basic operations with matrices, it is convenient to consider a matrix as a collection of vectors of the same number of components.

A. Addition

We can add the corresponding elements of two matrices with the same order; that is, the two matrices have the same number of rows and columns. The sums obtained are the elements of the new matrix which is the sum of the two given matrices. Thus, the new matrix has the same order as the two original matrices.

EXAMPLE 13　Add the 2×2 matrices A and B.

$$A + B = \begin{bmatrix} 2 & 4 \\ 3 & 1 \end{bmatrix} + \begin{bmatrix} 1 & -2 \\ 5 & 8 \end{bmatrix} = \begin{bmatrix} 2+1 & 4+(-2) \\ 3+5 & 1+8 \end{bmatrix} = \begin{bmatrix} 3 & 2 \\ 8 & 9 \end{bmatrix}$$

Add the 3×2 matrices C and D.

$$C + D = \begin{bmatrix} 1 & -2 \\ 0 & 6 \\ 4 & 1 \end{bmatrix} + \begin{bmatrix} 5 & 1 \\ -3 & -4 \\ 0 & 6 \end{bmatrix} = \begin{bmatrix} 1+5 & (-2)+1 \\ 0+(-3) & 6+(-4) \\ 4+0 & 1+6 \end{bmatrix}$$

$$= \begin{bmatrix} 6 & -1 \\ -3 & 2 \\ 4 & 7 \end{bmatrix}$$

B. Subtraction

We can subtract the corresponding elements of two matrices with the same order. The remainders obtained are the elements of the new matrix which is the result of the subtraction. The new matrix also has the same order as the two original matrices.

EXAMPLE 14 Subtract matrix B from matrix A.

$$A - B = \begin{bmatrix} 2 & 4 \\ 3 & 1 \end{bmatrix} - \begin{bmatrix} 1 & -2 \\ 5 & 8 \end{bmatrix} = \begin{bmatrix} 2-1 & 4-(-2) \\ 3-5 & 1-8 \end{bmatrix} = \begin{bmatrix} 1 & 6 \\ -2 & -7 \end{bmatrix}$$

Subtract matrix D from matrix C.

$$C - D = \begin{bmatrix} 1 & -2 \\ 0 & 6 \\ 4 & 1 \end{bmatrix} + \begin{bmatrix} 5 & 1 \\ -3 & -4 \\ 0 & 6 \end{bmatrix} = \begin{bmatrix} 1-5 & (-2)-1 \\ 0-(-3) & 6-(-4) \\ 4-0 & 1-6 \end{bmatrix}$$

$$= \begin{bmatrix} -4 & -3 \\ 3 & 10 \\ 4 & -5 \end{bmatrix}$$

C. Multiplication

A matrix may be multiplied by a scalar or by another matrix. The methods of the two types of multiplication are presented below.

1. SCALAR MULTIPLICATION. When a matrix is multiplied by a scalar, every element of the matrix must be multiplied by the scalar. The product of the multiplication is a matrix with the same order of the original matrix.

EXAMPLE 15 Multiply 2 by the 3×2 matrix E.

$$2 \cdot E = 2 \begin{bmatrix} 1 & 0 \\ 4 & -6 \\ 3 & 7 \end{bmatrix} = \begin{bmatrix} 2 \times 1 & 2 \times 0 \\ 2 \times 4 & 2 \times (-6) \\ 2 \times 3 & 2 \times 7 \end{bmatrix} = \begin{bmatrix} 2 & 0 \\ 8 & -12 \\ 6 & 14 \end{bmatrix}$$

Multiply $\frac{1}{4}$ by the 2×2 matrix B.

$$\tfrac{1}{4} \cdot B = \tfrac{1}{4} \begin{bmatrix} 1 & -2 \\ 5 & 8 \end{bmatrix} = \begin{bmatrix} \frac{1}{4} & -\frac{1}{2} \\ \frac{5}{4} & 2 \end{bmatrix}$$

Thus, if a given matrix, such as matrix B, is multiplied by the scalar zero, the product is a zero matrix of the same order as the given matrix.

2. MULTIPLICATION OF TWO MATRICES. Two matrices may be multiplied if the number of columns in the first matrix is equal to the number of rows in the second matrix. For example, if the first is a 2×3 matrix and the second is a 3×4 matrix, the two matrices may be multiplied. Each row vector of the first matrix is multiplied by each column vector of the second matrix to obtain the corresponding element of the product matrix of the multiplication. Thus, the product is a 2×4 matrix.

In general, let the first be an $m \times n$ matrix and the second be an $n \times p$ matrix. The element in the ith row and jth column in the product matrix is obtained by multiplying the ith row vector in the first matrix by the jth column vector in the second matrix. The product obtained in this way is an $m \times p$ matrix.

Multiplication of Two Matrices

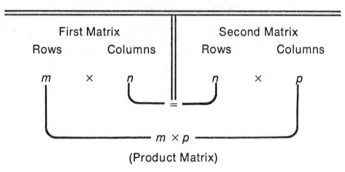

(Product Matrix)

The details of the multiplication of two matrices are presented in the examples below. The order of each matrix in the multiplication is written directly under the matrix for the purpose of checking the answer based on the above diagram.

EXAMPLE 16 Multiply matrix A by matrix F.

$$A \cdot F = \begin{bmatrix} 2 & 4 \\ 3 & 1 \end{bmatrix} \cdot \begin{bmatrix} 5 & 6 \\ 7 & 8 \end{bmatrix} = \begin{bmatrix} [2 \ 4]\begin{bmatrix}5\\7\end{bmatrix} & [2 \ 4]\begin{bmatrix}6\\8\end{bmatrix} \\ [3 \ 1]\begin{bmatrix}5\\7\end{bmatrix} & [3 \ 1]\begin{bmatrix}6\\8\end{bmatrix} \end{bmatrix}$$
$$ {}_{2\times2} \quad {}_{2\times2}$$

$$= \begin{bmatrix} 10 + 28 & 12 + 32 \\ 15 + 7 & 18 + 8 \end{bmatrix} = \begin{bmatrix} 38 & 44 \\ 22 & 26 \end{bmatrix}_{2\times2}$$

The first row vector $[2 \ 4]$ of A is multiplied by the first column vector $\begin{bmatrix}5\\7\end{bmatrix}$ of F to obtain the first row and first column element in the product (38), the second row vector $[3 \ 1]$ of A is multiplied by the first column vector $\begin{bmatrix}5\\7\end{bmatrix}$ of F to obtain the second row and first column element in the product (22), and so on.

Since A and F are 2×2 matrices, the product $A \cdot F$ is also a 2×2 matrix. It can be verified that $A \cdot F \neq F \cdot A$.

EXAMPLE 17 Multiply matrix G by matrix H.

$$G \cdot H = \begin{bmatrix} 1 & 2 & 3 \\ 2 & -1 & 0 \end{bmatrix} \begin{bmatrix} 4 & 1 \\ 1 & 5 \\ 3 & -2 \end{bmatrix} = \begin{bmatrix} \begin{bmatrix} 1 & 2 & 3 \end{bmatrix} \begin{bmatrix} 4 \\ 1 \\ 3 \end{bmatrix} & \begin{bmatrix} 1 & 2 & 3 \end{bmatrix} \begin{bmatrix} 1 \\ 5 \\ -2 \end{bmatrix} \\ \begin{bmatrix} 2 & -1 & 0 \end{bmatrix} \begin{bmatrix} 4 \\ 1 \\ 3 \end{bmatrix} & \begin{bmatrix} 2 & -1 & 0 \end{bmatrix} \begin{bmatrix} 1 \\ 5 \\ -2 \end{bmatrix} \end{bmatrix}$$

$$= \begin{bmatrix} 4 + 2 + 9 & 1 + 10 + (-6) \\ 8 + (-1) + 0 & 2 + (-5) + 0 \end{bmatrix} = \begin{bmatrix} 15 & 5 \\ 7 & -3 \end{bmatrix}_{2 \times 2}$$

The product of a 2×3 matrix (G) and a 3×2 matrix (H) is a 2×2 matrix ($G \cdot H$).

EXAMPLE 18 Multiply matrix G by matrix K.

$$G \cdot K = \begin{bmatrix} 1 & 2 & 3 \\ 2 & -1 & 0 \end{bmatrix}_{2 \times 3} \begin{bmatrix} 5 \\ 4 \\ 6 \end{bmatrix}_{3 \times 1} = \begin{bmatrix} \begin{bmatrix} 1 & 2 & 3 \end{bmatrix} \begin{bmatrix} 5 \\ 4 \\ 6 \end{bmatrix} \\ \begin{bmatrix} 2 & -1 & 0 \end{bmatrix} \begin{bmatrix} 5 \\ 4 \\ 6 \end{bmatrix} \end{bmatrix} = \begin{bmatrix} 5 + 8 + 18 \\ 10 + (-4) + 0 \end{bmatrix} = \begin{bmatrix} 31 \\ 6 \end{bmatrix}_{2 \times 1}$$

The product of a 2×3 matrix (G) and a 3×1 matrix (K, which is also a three-component column vector) is a 2×1 matrix ($G \cdot K$, or a two-component column vector.)

EXAMPLE 19 Multiply matrix L by matrix G.

$$L \cdot G = \begin{bmatrix} 4 & 5 \end{bmatrix}_{1 \times 2} \begin{bmatrix} 1 & 2 & 3 \\ 2 & -1 & 0 \end{bmatrix}_{2 \times 3} = \begin{bmatrix} \begin{bmatrix} 4 & 5 \end{bmatrix} \begin{bmatrix} 1 \\ 2 \end{bmatrix} & \begin{bmatrix} 4 & 5 \end{bmatrix} \begin{bmatrix} 2 \\ -1 \end{bmatrix} & \begin{bmatrix} 4 & 5 \end{bmatrix} \begin{bmatrix} 3 \\ 0 \end{bmatrix} \end{bmatrix}$$

$$= \begin{bmatrix} 4 + 10 & 8 + (-5) & 12 + 0 \end{bmatrix} = \begin{bmatrix} 14 & 3 & 12 \end{bmatrix}_{1 \times 3}$$

The product of a 1×2 matrix (L, or a two-component row vector) and a 2×3 matrix (G) is a 1×3 matrix ($L \cdot G$, or a three-component row vector.) We cannot multiply G (3 columns) by L (1 row.)

EXAMPLE 20 Multiply matrix M by matrix L.

$$M \cdot L = \begin{bmatrix} 2 \\ 3 \end{bmatrix}_{2 \times 1} \cdot \begin{bmatrix} 4 & 5 \end{bmatrix}_{1 \times 2} = \begin{bmatrix} 2 \times 4 & 2 \times 5 \\ 3 \times 4 & 3 \times 5 \end{bmatrix} = \begin{bmatrix} 8 & 10 \\ 12 & 15 \end{bmatrix}_{2 \times 2}$$

The product of a 2×1 matrix (M, or a two-component column vector) and a 1×2 matrix (L, or a two-component row vector) is a 2×2 matrix ($M \cdot L$). However, the product $L \cdot M$ is a single element, or:

$$L \cdot M = [4 \quad 5] \cdot \begin{bmatrix} 2 \\ 3 \end{bmatrix} = (4 \times 2) + (5 \times 3) = 8 + 15 = 23$$

Thus, $M \cdot L \neq L \cdot M$.

EXAMPLE 21   Let the 2×2 matrix $A = \begin{bmatrix} 2 & 4 \\ 3 & 1 \end{bmatrix}$ and the 2×2 identity matrix $I = \begin{bmatrix} 1 & 0 \\ 0 & 1 \end{bmatrix}$. Show that $A \cdot I = I \cdot A = A$.

$$A \cdot I = \begin{bmatrix} 2 & 4 \\ 3 & 1 \end{bmatrix} \cdot \begin{bmatrix} 1 & 0 \\ 0 & 1 \end{bmatrix} = \begin{bmatrix} (2 \times 1) + (4 \times 0) & (2 \times 0) + (4 \times 1) \\ (3 \times 1) + (1 \times 0) & (3 \times 0) + (1 \times 1) \end{bmatrix}$$

$$= \begin{bmatrix} 2 & 4 \\ 3 & 1 \end{bmatrix} = A$$

$$I \cdot A = \begin{bmatrix} 1 & 0 \\ 0 & 1 \end{bmatrix} \cdot \begin{bmatrix} 2 & 4 \\ 3 & 1 \end{bmatrix} = \begin{bmatrix} (1 \times 2) + (0 \times 3) & (1 \times 4) + (0 \times 1) \\ (0 \times 2) + (1 \times 3) & (0 \times 4) + (1 \times 1) \end{bmatrix}$$

$$= \begin{bmatrix} 2 & 4 \\ 3 & 1 \end{bmatrix} = A$$

EXAMPLE 22   Solve for X. $X + \begin{bmatrix} 1 & 3 \\ 4 & -2 \end{bmatrix}^t = \begin{bmatrix} 3 & 9 \\ 4 & 2 \end{bmatrix}$

Solution $X + \begin{bmatrix} 1 & 4 \\ 3 & -2 \end{bmatrix} = \begin{bmatrix} 3 & 9 \\ 4 & 2 \end{bmatrix}$,

$$X = \begin{bmatrix} 3 & 9 \\ 4 & 2 \end{bmatrix} - \begin{bmatrix} 1 & 4 \\ 3 & -2 \end{bmatrix} = \begin{bmatrix} 2 & 5 \\ 1 & 4 \end{bmatrix}$$

EXERCISE 6-2   Reference: Sections 6.5–6.6

A. *State the complete name and write the transpose of each of the following matrices:*

1. $A = \begin{bmatrix} 4 & -8 \\ 3 & 2 \end{bmatrix}$

2. $B = \begin{bmatrix} 1 & 5 \\ 7 & 6 \end{bmatrix}$

3. $C = \begin{bmatrix} 2 & 3 \\ 4 & -6 \\ 1 & 8 \end{bmatrix}$

4. $D = \begin{bmatrix} 7 & -1 \\ 2 & 5 \\ 4 & 9 \end{bmatrix}$

5. $E = \begin{bmatrix} 3 & 2 & 6 \\ 8 & 9 & -1 \end{bmatrix}$

6. $F = \begin{bmatrix} 1 & 4 & -9 \\ 6 & 2 & 3 \end{bmatrix}$

7. $G = \begin{bmatrix} 1 & 4 & 0 \\ -2 & 6 & -5 \\ 3 & -7 & 9 \end{bmatrix}$

8. $0 = \begin{bmatrix} 0 & 0 & 0 \\ 0 & 0 & 0 \\ 0 & 0 & 0 \end{bmatrix}$

B. *Operations with matrices. Use the matrices given in Problems 1–8.*

9. $A + B$

10. $C + D$

11. $C - D$

12. $A - B$

13. $A + (-A)$
14. $C + (-C)$
15. $2A$
16. $3B$
17. $\frac{1}{5}C$
18. $\frac{2}{3}E$
19. $A \cdot B$
20. $B \cdot A$
21. $A \cdot E$
22. $B \cdot F$
23. $A \cdot F$
24. $B \cdot E$
25. $E \cdot C$
26. $F \cdot D$

27. $C \cdot A$
28. $D \cdot B$
29. $F \cdot C$
30. $E \cdot D$
31. $C \cdot E$
32. $D \cdot F$
33. $C \cdot F$
34. $D \cdot E$
35. $G + 0$
36. $G \cdot D$
37. $G \cdot C$
38. $E \cdot G$
39. $F \cdot G$
40. $G \cdot 0$

C. *Show that $A \cdot I = A$, and $I \cdot A = A$.*

41. $A = \begin{bmatrix} 5 & -1 \\ 2 & 3 \end{bmatrix}$, and $I = \begin{bmatrix} 1 & 0 \\ 0 & 1 \end{bmatrix}$.

42. $A = \begin{bmatrix} 4 & 1 \\ 6 & -5 \end{bmatrix}$, and $I = \begin{bmatrix} 1 & 0 \\ 0 & 1 \end{bmatrix}$.

43. $A = \begin{bmatrix} 1 & 3 & 8 \\ 2 & -6 & 7 \\ 5 & 4 & 9 \end{bmatrix}$, and $I = \begin{bmatrix} 1 & 0 & 0 \\ 0 & 1 & 0 \\ 0 & 0 & 1 \end{bmatrix}$.

44. $A = \begin{bmatrix} 4 & -1 & 5 \\ -7 & 2 & 0 \\ 8 & -3 & 6 \end{bmatrix}$, and $I = \begin{bmatrix} 1 & 0 & 0 \\ 0 & 1 & 0 \\ 0 & 0 & 1 \end{bmatrix}$.

D. *Solve matrix equations for X.*

45. $X + \begin{bmatrix} 2 & -1 \\ 3 & 4 \end{bmatrix} = \begin{bmatrix} 7 & -1 \\ 2 & 6 \end{bmatrix}$

46. $X - \begin{bmatrix} -1 & 0 \\ 0 & 0 \end{bmatrix} = \begin{bmatrix} 4 & -1 \\ 2 & 1 \end{bmatrix}$

47. $X + \begin{bmatrix} 5 & 2 \\ -1 & 4 \end{bmatrix} = \begin{bmatrix} 3 & 7 \\ 2 & 5 \end{bmatrix} + \begin{bmatrix} -5 & -2 \\ 1 & 0 \end{bmatrix}$

48. $\begin{bmatrix} 4 & 3 \\ -1 & 0 \end{bmatrix}^t - \begin{bmatrix} 0 & 1 \\ 1 & 0 \end{bmatrix} = \begin{bmatrix} 5 & -2 \\ -1 & 3 \end{bmatrix}^t - X$

49. Refer to Problems 1 and 2. Show that $c(A + B) = cA + cB$. c is a real number.
50. Refer to Problem 3. Show that $(a + b)C = aC + bC$. a and b are real numbers.

Chapter

7 Solving Equations by Matrix Algebra

The objective of this chapter is to introduce methods for solving systems of linear equations with the use of matrix algebra. In order to accomplish this, two related topics, the determinant and the inverse of a square matrix, must also be introduced.

7.1 FINDING THE VALUE OF A DETERMINANT

A determinant is closely associated with a square matrix. It is written in the same form as its associated matrix except that two vertical lines, instead of brackets, are used. For example, if the 2×2 square matrix is:

$$A = \begin{bmatrix} 2 & 4 \\ 3 & 1 \end{bmatrix}, \text{ the determinant of } A, \text{ denoted by } |A|, \text{ is:}$$

$$|A| = \begin{vmatrix} 2 & 4 \\ 3 & 1 \end{vmatrix}$$

The 2×2 determinant is also called a determinant of order 2.

The value of a determinant is a real number. The number can be used to determine whether or not there is an inverse of a square matrix. The inverse and methods of finding it are presented in Section 7.3. In this section, we shall introduce two methods for finding the value of a determinant: by cross multiplication and by using minors.

A. The Cross Multiplication Method

The use of this method is limited only to the determinants of orders 2 and 3; that is, it applies only to finding determinants of 2×2 or 3×3 matrices. The value of a determinant of higher order (above 3) can be found by using the minors presented in Method B.

For convenience, let the letter a represent the elements of the matrix A, with the first subscript indicating the location of the row and the second subscript indicating the location of the column. Thus:

$$2 \times 2 \text{ square matrix } A = \begin{bmatrix} a_{11} & a_{12} \\ a_{21} & a_{22} \end{bmatrix}$$

where:

a_{11} = the element located in the 1st row and the 1st column
a_{21} = the element located in the 2nd row and the 1st column, and so on

The determinant of A is written and evaluated as follows:

$$|A| = \begin{vmatrix} a_{11} & a_{12} \\ a_{21} & a_{22} \end{vmatrix} = +a_{11}a_{22} - a_{12}a_{21}$$

Secondary   Primary
diagonal   diagonal

(from upper right   (from upper left
to lower left)   to lower right

Observe that the elements of the positive product $(+a_{11}a_{22})$ are on the *primary diagonal* of the determinant and the elements of the negative product $(-a_{12}a_{21})$ are on the *secondary diagonal*. Also, each product has only one element from each row and each column. There are no two elements of each product in the same row or the same column.

EXAMPLE 1   Find the value of the determinant of the matrix A if:

(a) $A = \begin{bmatrix} 2 & 4 \\ 3 & 1 \end{bmatrix}$ and (b) $A = \begin{bmatrix} 1 & -2 \\ 4 & 6 \end{bmatrix}$

(a) $|A| = \begin{vmatrix} 2 & 4 \\ 3 & 1 \end{vmatrix} = (2 \times 1) - (4 \times 3) = 2 - 12 = -10$

(b) $|A| = \begin{vmatrix} 1 & -2 \\ 4 & 6 \end{vmatrix} = (1 \times 6) - ((-2) \times 4) = 6 - (-8) = 14$

The 3×3 square matrix is written symbolically.

$$A = \begin{bmatrix} a_{11} & a_{12} & a_{13} \\ a_{21} & a_{22} & a_{23} \\ a_{31} & a_{32} & a_{33} \end{bmatrix}$$

The determinant of A now is written and evaluated as follows:

$$|A| = \begin{vmatrix} a_{11} & a_{12} & a_{13} \\ a_{21} & a_{22} & a_{23} \\ a_{31} & a_{32} & a_{33} \end{vmatrix} = \begin{aligned} &a_{11}a_{22}a_{33} + a_{12}a_{23}a_{31} + a_{13}a_{21}a_{32} \\ &\quad - a_{13}a_{22}a_{31} - a_{12}a_{21}a_{33} - a_{11}a_{23}a_{32} \end{aligned}$$

The method of cross multiplication for the determinant of order 3 is diagrammed in the following manner:

The three positive products are obtained from the primary diagonals:	The three negative products are obtained from the secondary diagonals:
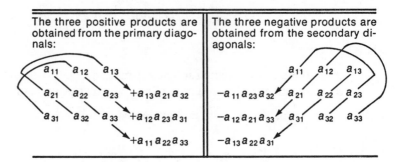	

Observe the diagram. Again, each product is restricted to have only one element from each row and each column. The value of the determinant is the sum of all possible products (+ and −) obtained under the restriction.

EXAMPLE 2   Find the value of the determinant $|A|$.

$$|A| = \begin{vmatrix} 2 & 1 & -5 \\ 4 & 6 & 8 \\ 0 & -7 & 3 \end{vmatrix} = \begin{aligned} &(2 \times 6 \times 3) + (1 \times 8 \times 0) + [(-5) \times 4 \times (-7)] \\ &- [(-5) \times 6 \times 0] - (1 \times 4 \times 3) - [2 \times 8 \times (-7)] \\ &= 36 + 0 + 140 - 0 - 12 + 112 \\ &= 276. \end{aligned}$$

Note that the positive and negative products may be computed in a manner as shown below. The first two columns of the determinant are repeated as the fourth and the fifth columns respectively.

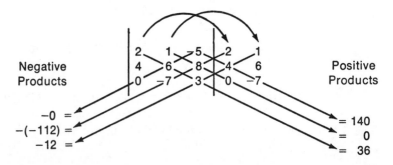

For a determinant of order 3, the three positive products can be obtained from the primary diagonals and the three negative products can be obtained from the secondary diagonals. In general, for a determinant of order 2, 3, or higher, the sign of each product depends on the number of inversions in the second subscripts of the elements of the product when the first subscripts are in natural order (that is, 1, 2, 3, 4, and so on). If the number of inversions is even, the product is positive; if it is odd, the product is negative.

An *inversion* occurs in a sequence of natural numbers when a natural number is preceded by a larger natural number. For example, in the product $a_{13}a_{21}a_{32}$, the first subscripts of the elements are in natural order 1, 2, and 3, and the second subscripts are in the sequence 3, 1, 2. In the sequence 3, 1, 2, there are two inversions: 3 precedes 1, and 3 precedes 2. The even number (two) of inversions gives a positive sign to the product, or $+a_{13}a_{21}a_{32}$.

For the product $a_{13}a_{22}a_{31}$, where the first subscripts are in the natural order 1, 2, 3, the second subscripts are in the sequence 3, 2, 1. This sequence (3, 2, 1) has three inversions: 3 precedes 2, 3 precedes 1, and 2 precedes 1. The odd number (three) of inversions gives a negative sign to the product, or $-a_{13}a_{22}a_{31}$.

These characteristics of determinants of order 2 and order 3 are also true for a determinant of higher order. However, for a determinant of order 4 or higher, the work of listing all possible products, with the restriction of only one element from each row and each column (or no two elements may be taken from the same row or the same column), can not be accomplished simply by using the cross multiplication method. In fact, some of the products formed in higher order determinants contain elements not on the positions of either the primary or the secondary diagonals. For example, in the determinant $|A|$ of order 4, one of the possible products of the determinant is $a_{11}a_{24}a_{32}a_{43}$. This product is positive since the second subscripts are in the sequence 1, 4, 2, 3, which has two inversions: 4 precedes 2, and 4 precedes 3. The four elements, $a_{11}, a_{24}, a_{32}, a_{43}$, are obtained from different rows and columns as shown in the following diagram:

$$
\begin{vmatrix}
\boxed{a_{11}} & \text{—} & \text{—} & \text{—} \\
\text{—} & \text{—} & \text{—} & \boxed{a_{24}} \\
\text{—} & \boxed{a_{32}} & \text{—} & \text{—} \\
\text{—} & \text{—} & \boxed{a_{43}} & \text{—}
\end{vmatrix}
$$

Note that the elements do not reside on the same primary or secondary diagonal.

There are a total of 24 different products which can be formed from a 4×4 determinant. The total number of products from a determinant of order n is *n factorial*, written $n!$. Here $n = 4$ and $n! = 4! = 4 \times 3 \times 2 \times 1 = 24$. If the order of a determinant is 2, $n! = 2! = 2 \times 1 = 2$ products; for a determinant of order 3, $n! = 3! = 3 \times 2 \times 1 = 6$ products. For convenience, the minors method is developed for finding the value of a determinant of any order, but particularly for order 4 or higher.

B. The Minors Method

A minor is specified by an element of a given determinant. In general, the minor of the element a_{ij}, located in the ith row and jth column in a determinant

$|A|$, is denoted by A_{ij}. The minor A_{ij} is obtained by deleting both the ith row and jth column of the determinant $|A|$. Thus, if $|A|$ is a determinant of order 2, then A_{ij} is a determinant of order 1 $(= 2 - 1)$.

EXAMPLE 3   Let $|A|$ be the determinant of order 2, or:

$$|A| = \begin{vmatrix} a_{11} & a_{12} \\ a_{21} & a_{22} \end{vmatrix}$$

then the minor of element a_{11}:

$$A_{11} = \begin{vmatrix} a_{11} & a_{12} \\ a_{21} & a_{22} \end{vmatrix} = |a_{22}| \text{ (order 1)}$$

and so on.

If $|A|$ is a determinant of order 3, then A_{ij} is a determinant of order 2 $(=3-1)$.

EXAMPLE 4   Let $|A|$ be the determinant of order 3, or:

$$|A| = \begin{vmatrix} a_{11} & a_{12} & a_{13} \\ a_{21} & a_{22} & a_{23} \\ a_{31} & a_{32} & a_{33} \end{vmatrix}$$

then the minor of element a_{32}:

$$A_{32} = \begin{vmatrix} a_{11} & a_{12} & a_{13} \\ a_{21} & a_{22} & a_{23} \\ a_{31} & a_{32} & a_{33} \end{vmatrix} = \begin{vmatrix} a_{11} & a_{13} \\ a_{21} & a_{23} \end{vmatrix} \text{ (order 2),}$$

and so on.

EXAMPLE 5 $|A| = \begin{vmatrix} 2 & 4 \\ 3 & 1 \end{vmatrix}$

Find the minors of elements a_{11} and a_{21}.

$$a_{11} = 2, \text{ and } A_{11} = \begin{vmatrix} 2 & 4 \\ 3 & 1 \end{vmatrix} = 1$$

$$a_{21} = 3, \text{ and } A_{21} = \begin{vmatrix} 2 & 4 \\ 3 & 1 \end{vmatrix} = 4$$

EXAMPLE 6 $|A| = \begin{vmatrix} 2 & 1 & -5 \\ 4 & 6 & 8 \\ 0 & -7 & 3 \end{vmatrix}$

Find the minor of element a_{11} and a_{21}.

$$a_{11} = 2, \text{ and } A_{11} = \begin{vmatrix} 2 & 1 & -5 \\ 4 & 6 & 8 \\ 0 & -7 & 3 \end{vmatrix} = \begin{vmatrix} 6 & 8 \\ -7 & 3 \end{vmatrix} = (6 \times 3) - [(-7) \times 8]$$

$$= 18 - (-56) = 74$$

$$a_{21} = 4, \text{ and } A_{21} = \begin{vmatrix} 2 & 1 & -5 \\ 4 & 6 & 8 \\ 0 & -7 & 3 \end{vmatrix} = \begin{vmatrix} 1 & -5 \\ -7 & 3 \end{vmatrix} = (1 \times 3) - [(-5)(-7)]$$

$$= 3 - 35 = -32$$

A minor must be prefixed by a positive or negative sign when it is used for finding the value of a determinant. The sign is positive when the sum of the subscripts $(i + j)$ of the minor A_{ij} is even, and it is negative when the sum $(i + j)$ is odd. A *signed minor* is also called a *cofactor*. Thus, the signed minor (or cofactor) of element a_{11} is $+A_{11}$ since the sum of the subscripts is an even number, $1 + 1 = 2$. The signed minor of element a_{21} is $-A_{21}$ since the sum of the subscripts is an odd number, $2 + 1 = 3$.

The value of an $n \times n$ determinant may be obtained by using signed minors as follows:

1. Multiply each element of any selected column (or any row) by its signed minor.
2. Add the n products. The sum of the products is the value of the determinant.

Thus, if we select the elements of the first column and their signed minors, the value of a 2×2 determinant is:

$$|A| = a_{11} (+A_{11}) + a_{21} (-A_{21}) = a_{11}A_{11} - a_{21}A_{21},$$

and the value of a 3×3 determinant is:

$$|A| = a_{11}A_{11} - a_{21}A_{21} + a_{31}A_{31}$$

EXAMPLE 7   Find the value of the determinant. Use the elements in the first column and their signed minors.

(a) $|A| = \begin{vmatrix} 2 & 4 \\ 3 & 1 \end{vmatrix}$,   (b) $|A| = \begin{vmatrix} 2 & 1 & -5 \\ 4 & 6 & 8 \\ 0 & -7 & 3 \end{vmatrix}$

(a) The elements in the selected first column are 2 and 3. The signed minor of 2 is positive $(+A_{11})$ and that of 3 is negative $(-A_{21})$.

$$|A| = \begin{vmatrix} 2 & 4 \\ 3 & 1 \end{vmatrix} = 2 \begin{vmatrix} 2 & 4 \\ 3 & 1 \end{vmatrix} - 3 \begin{vmatrix} 2 & 4 \\ 3 & 1 \end{vmatrix} = 2|1| - 3|4| = 2 - 12 = -10$$

(b) The elements in the selected first column are 2, 4, and 0. The signed minor of 2 is $+(+A_{11})$, of 4 is $-(-A_{21})$, and of 0 is $+(+A_{31})$.

$$|A| = \begin{vmatrix} 2 & 1 & -5 \\ 4 & 6 & 8 \\ 0 & -7 & 3 \end{vmatrix} = 2 \begin{vmatrix} 2 & 1 & -5 \\ 4 & 6 & 8 \\ 0 & -7 & 3 \end{vmatrix} - 4 \begin{vmatrix} 2 & 1 & -5 \\ 4 & 6 & 8 \\ 0 & -7 & 3 \end{vmatrix}$$

$$+ 0 \begin{vmatrix} 2 & 1 & -5 \\ 4 & 6 & 8 \\ 0 & -7 & 3 \end{vmatrix} = 2 \begin{vmatrix} 6 & 8 \\ -7 & 3 \end{vmatrix} - 4 \begin{vmatrix} 1 & -5 \\ -7 & 3 \end{vmatrix} + 0 \begin{vmatrix} 1 & -5 \\ 6 & 8 \end{vmatrix}$$

$$= 2 [(6 \times 3) - ((-7) \times 8)] - 4 [(1 \times 3) - (-5)(-7)] + 0$$
$$= 2(18 + 56) - 4(3 - 35) + 0$$
$$= 2(74) - 4(-32) = 148 + 128 = 276.$$

In general, the following sign array is a convenient way of determining the sign of a minor A_{ij} at a given position in an $n \times n$ determinant.

$$
\begin{array}{ccccccc}
+ & - & + & - & \cdot & \cdot & \cdot & (-)^{n+1} \\
- & + & - & + & \cdot & \cdot & \cdot & \cdot \\
+ & - & + & - & \cdot & \cdot & \cdot & \cdot \\
- & + & - & + & \cdot & \cdot & \cdot & \cdot \\
\cdot & \cdot & \cdot & \cdot & \cdot & \cdot & \cdot & \cdot \\
\cdot & \cdot & \cdot & \cdot & \cdot & \cdot & \cdot & \cdot \\
\cdot & \cdot & \cdot & \cdot & \cdot & \cdot & \cdot & \cdot \\
(-)^{n+1} & & \cdot & \cdot & \cdot & \cdot & \cdot \\
\end{array}
$$

a. Thus, the signed minors of a 2×2 determinant may be arranged:

$$
\begin{array}{cc}
+A_{11} & -A_{12} \\
\\
-A_{21} & +A_{22}
\end{array}
$$

b. The signed minors of a 3×3 determinant may be arranged:

$$
\begin{array}{ccc}
+A_{11} & -A_{12} & +A_{13} \\
-A_{21} & +A_{22} & -A_{23} \\
+A_{31} & -A_{32} & +A_{33}
\end{array}
$$

EXAMPLE 8   Refer to Example 7. Find the value of each determinant by using the elements in the last row and their signed minors.

(a) The elements in the selected row are 3 and 1. The signs of the minors of the elements are $-$ ($-A_{21}$ of 3) and $+$ ($+A_{22}$ of 1).

$$
|A| = \begin{vmatrix} 2 & 4 \\ 3 & 1 \end{vmatrix} = -3 \begin{vmatrix} 2 & 4 \\ 3 & 1 \end{vmatrix} + 1 \begin{vmatrix} 2 & 4 \\ 3 & 1 \end{vmatrix} = -3|4| + 1|2| = (-12) + 2
$$

$$
= -10
$$

(b) The elements in the selected third row are $0, -7$, and 3. The signs of the minors of the elements are $+$ ($+A_{31}$ of 0), $-$ ($-A_{32}$ of -7), and $+$ ($+A_{33}$ of 3).

$$
|A| = \begin{vmatrix} 2 & 1 & -5 \\ 4 & 6 & 8 \\ 0 & -7 & 3 \end{vmatrix} = 0 \begin{vmatrix} 2 & 1 & -5 \\ 4 & 6 & 8 \\ 0 & -7 & 3 \end{vmatrix} - (-7) \begin{vmatrix} 2 & 1 & -5 \\ 4 & 6 & 8 \\ 0 & -7 & 3 \end{vmatrix}
$$

$$
+ 3 \begin{vmatrix} 2 & 1 & -5 \\ 4 & 6 & 8 \\ 0 & -7 & 3 \end{vmatrix} = 0 \begin{vmatrix} 1 & -5 \\ 6 & 8 \end{vmatrix} - (-7) \begin{vmatrix} 2 & -5 \\ 4 & 8 \end{vmatrix} + 3 \begin{vmatrix} 2 & 1 \\ 4 & 6 \end{vmatrix}
$$

$$
= 0 + 7[(2 \times 8) - (4 \times (-5))] + 3[(2 \times 6) - (4 \times 1)]
$$
$$
= 7(16 + 20) + 3(12 - 4)
$$
$$
= 252 + 24 = 276
$$

The value of a determinant of order higher than 3 may be found in the same manner as in Examples 7 and 8. For instance, the value of a 4×4 determinant may be found by using the elements of the first column in the determinant and their 3×3 signed minors. The value of each 3×3 signed minor may then be found by either the cross multiplication method or by using the 2×2 signed minors.

EXERCISE 7-1   Reference: Section 7.1

A. *Find the value of each determinant given below by (a) the cross multiplication method, and (b) the method of using the signed minors of the first column elements:*

1. $\begin{vmatrix} 2 & 7 \\ 3 & 4 \end{vmatrix}$

7. $\begin{vmatrix} 2 & -5 & 1 \\ 4 & 8 & 6 \\ 9 & 3 & -7 \end{vmatrix}$

2. $\begin{vmatrix} -4 & 8 \\ 3 & -5 \end{vmatrix}$

8. $\begin{vmatrix} 3 & 7 & 9 \\ 0 & -2 & 5 \\ -4 & 1 & 8 \end{vmatrix}$

3. $\begin{vmatrix} 3 & -7 \\ 4 & 5 \end{vmatrix}$

9. $\begin{vmatrix} 1 & 8 & 0 \\ -2 & 4 & -3 \\ 7 & -6 & 5 \end{vmatrix}$

4. $\begin{vmatrix} -2 & 10 \\ -6 & 9 \end{vmatrix}$

10. $\begin{vmatrix} -4 & 6 & 1 \\ 8 & 5 & 0 \\ 3 & 2 & -7 \end{vmatrix}$

5. $\begin{vmatrix} -1 & 6 \\ 12 & 0 \end{vmatrix}$

11. $\begin{vmatrix} 5 & 4 & -2 \\ 6 & 0 & 1 \\ -7 & 3 & -8 \end{vmatrix}$

6. $\begin{vmatrix} 5 & 0 \\ 9 & 14 \end{vmatrix}$

12. $\begin{vmatrix} 1 & -3 & 10 \\ -2 & 5 & -8 \\ 9 & 4 & 7 \end{vmatrix}$

B. *Find the value of each determinant by the signed minors of the elements of the row or column as indicated:*

13. The determinant given in Problem 1. Use the elements in the first row.

14. The determinant given in Problem 2. Use the elements in the first row.

15. The determinant given in Problem 3. Use the elements in the second row.

16. The determinant given in Problem 4. Use the elements in the second row.

17. The determinant given in Problem 7. Use the elements in the second column.

18. The determinant given in Problem 8. Use the elements in the third column.

19. The determinant given in Problem 9. Use the elements in the second row.

20. The determinant given in Problem 10. Use the elements in the third row.

C. *Find the value of each 4 × 4 determinant in the following problems: (Hint: Use the row or column with the most zeros.)*

21. $\begin{vmatrix} 0 & 2 & 0 & 0 \\ -1 & 5 & 7 & 3 \\ 3 & 1 & 2 & 4 \\ 1 & 3 & 0 & 0 \end{vmatrix}$ **23.** $\begin{vmatrix} 3 & 0 & 1 & -1 \\ 4 & 1 & 3 & 2 \\ 0 & 0 & 2 & 0 \\ -1 & 1 & 0 & 1 \end{vmatrix}$

22. $\begin{vmatrix} 1 & -2 & 1 & 0 \\ 0 & 3 & 2 & 1 \\ 2 & 1 & -1 & 0 \\ -1 & 4 & 3 & 2 \end{vmatrix}$ **24.** $\begin{vmatrix} 1 & 2 & 0 & -1 \\ 3 & 1 & 2 & 3 \\ 1 & 0 & 0 & 4 \\ 3 & 2 & 0 & 0 \end{vmatrix}$

7.2 PROPERTIES OF DETERMINANTS AND THEIR USE

Determinants have several properties associated with them which, when properly understood, are useful in simplifying the computation of determinant values.

PROPERTY 1   If every element in any row (or any column) of a determinant is 0, then the determinant is equal to 0.

EXAMPLE 9 $\begin{vmatrix} 0 & 0 \\ 2 & 5 \end{vmatrix} = 0$ $\begin{vmatrix} 1 & 0 & 2 \\ -2 & 0 & 3 \\ 6 & 0 & 4 \end{vmatrix} = 0$

PROPERTY 2   If any two rows (or any two columns) in a determinant have equal corresponding elements, the determinant is equal to 0.

EXAMPLE 10 $\begin{vmatrix} 1 & 4 \\ 1 & 4 \end{vmatrix} = 0$ $\begin{vmatrix} 2 & 1 & 2 \\ 6 & 5 & 6 \\ 9 & 3 & 9 \end{vmatrix} = 0$

PROPERTY 3   If any two rows (or any two columns) of a determinant are interchanged, the determinant is equal to the negative of the resulting determinant.

EXAMPLE 11 $\begin{vmatrix} 1 & 5 \\ 2 & 6 \end{vmatrix} = -\begin{vmatrix} 2 & 6 \\ 1 & 5 \end{vmatrix} = -\begin{vmatrix} 5 & 1 \\ 6 & 2 \end{vmatrix} = -(10-6) = -4$

$\begin{vmatrix} 2 & 1 & -5 \\ 4 & 6 & 8 \\ 0 & -7 & 3 \end{vmatrix} = -\begin{vmatrix} -5 & 1 & 2 \\ 8 & 6 & 4 \\ 3 & -7 & 0 \end{vmatrix} = -(-276) = 276$

PROPERTY 4   If every element in a row (or a column) of a determinant is written as the sum of two terms, the determinant can also be written as the sum of two new determinants. Both new determinants have the same elements as the original determinant except that the first new determinant has the first terms in the row (or column) and the second new determinant has the second terms in the row (or column) correspondingly.

EXAMPLE 12 $\begin{vmatrix} 4 & 5 \\ 8 & 7 \end{vmatrix} = \begin{vmatrix} 4 & 5 \\ 3+5 & 1+6 \end{vmatrix} = \begin{vmatrix} 4 & 5 \\ 3 & 1 \end{vmatrix} + \begin{vmatrix} 4 & 5 \\ 5 & 6 \end{vmatrix} = (-11) + (-1) = -12$, or:

$$\begin{vmatrix} 4 & 5 \\ 8 & 7 \end{vmatrix} = \begin{vmatrix} 4 & 3+2 \\ 8 & 1+6 \end{vmatrix} = \begin{vmatrix} 4 & 3 \\ 8 & 1 \end{vmatrix} + \begin{vmatrix} 4 & 2 \\ 8 & 6 \end{vmatrix} = (-20) + 8 = -12$$

$$\begin{vmatrix} 2 & 1 & -5 \\ 4 & 6 & 8 \\ 0 & -7 & 3 \end{vmatrix} = \begin{vmatrix} 2 & 1 & -5 \\ 1+3 & 2+4 & 3+5 \\ 0 & -7 & 3 \end{vmatrix} = \begin{vmatrix} 2 & 1 & -5 \\ 1 & 2 & 3 \\ 0 & -7 & 3 \end{vmatrix} + \begin{vmatrix} 2 & 1 & -5 \\ 3 & 4 & 5 \\ 0 & -7 & 3 \end{vmatrix}$$

$$= 86 + 190 = 276$$

PROPERTY 5 If every element of one row (or one column) of a determinant is multiplied by a real number k, the determinant is multiplied by k.

EXAMPLE 13 $\begin{vmatrix} 2(3) & 4(3) \\ 1 & 5 \end{vmatrix} = 3 \begin{vmatrix} 2 & 4 \\ 1 & 5 \end{vmatrix} = 3(10 - 4) = 18 \quad (k = 3)$

Conversely: $3 \begin{vmatrix} 2 & 4 \\ 1 & 5 \end{vmatrix} = \begin{vmatrix} 2(3) & 4(3) \\ 1 & 5 \end{vmatrix} = \begin{vmatrix} 6 & 12 \\ 1 & 5 \end{vmatrix} = 30 - 12 = 18,$

Or: $3 \begin{vmatrix} 2 & 4 \\ 1 & 5 \end{vmatrix} = \begin{vmatrix} 2 & 4 \\ 1(3) & 5(3) \end{vmatrix} = \begin{vmatrix} 2 & 4 \\ 3 & 15 \end{vmatrix} = 30 - 12 = 18$

$$\begin{vmatrix} 2 & 1(2) & -5 \\ 4 & 6(2) & 8 \\ 0 & -7(2) & 3 \end{vmatrix} = 2 \begin{vmatrix} 2 & 1 & -5 \\ 4 & 6 & 8 \\ 0 & -7 & 3 \end{vmatrix} = 2(276) = 552 \quad (k = 2)$$

Thus, when a real number is multiplied by a determinant, only the elements in a single row (or column) are multiplied by the number. This process is different from that of the scalar multiplication of a matrix.

PROPERTY 6 If a row (or a column) of a determinant is added (or subtracted) by the product of a real number k and another row (or column), the value of the determinant is unchanged.

EXAMPLE 14 The first row of the following determinant is added by the product of 3 multiplied by the second row correspondingly.

$$\begin{vmatrix} 1 & 2 \\ 5 & 6 \end{vmatrix} = \begin{vmatrix} 1 + 5(3) & 2 + 6(3) \\ 5 & 6 \end{vmatrix} = \begin{vmatrix} 16 & 20 \\ 5 & 6 \end{vmatrix} \quad (k = 3)$$

The first column of the following determinant is added by the product of 5 multiplied by the second column correspondingly.

$$\begin{vmatrix} 2 & 1 & -5 \\ 4 & 6 & 8 \\ 0 & -7 & 3 \end{vmatrix} = \begin{vmatrix} 2 + 1(5) & 1 & -5 \\ 4 + 6(5) & 6 & 8 \\ 0 + (-7)(5) & -7 & 3 \end{vmatrix} = \begin{vmatrix} 7 & 1 & -5 \\ 34 & 6 & 8 \\ -35 & -7 & 3 \end{vmatrix} \quad (k = 5)$$

The following examples are used to illustrate the applications of properties 5 and 6 in finding the values of determinants. Note that the procedure used in Property 6 is also very useful in "row operations" for inversing a square matrix and solving a system of linear equations to be presented in the next two sections.

EXAMPLE 15 Find the value of determinant $|A|$.

$$|A| = \begin{vmatrix} 3 & 4 & 6 \\ 7 & 2 & 14 \\ 8 & 3 & 16 \end{vmatrix} = 2 \begin{vmatrix} 3 & 4 & 3 \\ 7 & 2 & 7 \\ 8 & 3 & 8 \end{vmatrix} = 2(0) = 0$$

The factor 2 in the second determinant is obtained from the third column of the first determinant. (Use property 5.) Since the first and third columns of the second determinant are the same, its value is zero. (Use Property 2.)

EXAMPLE 16 Use Property 5 to find the value of determinant $|A|$.

$$|A| = \begin{vmatrix} 2 & 1 & -5 \\ 4 & 6 & 8 \\ 0 & -7 & 3 \end{vmatrix} = 2 \begin{vmatrix} 2 & 1 & -5 \\ 2 & 3 & 4 \\ 0 & -7 & 3 \end{vmatrix}$$

$$= 2(18 + 0 + 70 - 0 - (-56) - 6)$$

$$= 2(88 + 56 - 6) = 2(138) = 276$$

The factor 2 in the second determinant is obtained from the second row of the first determinant.

There are various methods to apply to Property 6 in finding the value of a determinant. The following two steps are for illustration purposes:

1. Make a single row (or a single column) having all zeros except for one element.
2. Use signed minors to evaluate the determinant.

EXAMPLE 17 Use Property 6 to find the value of determinant $|A|$.

$$|A| = \begin{vmatrix} 2 & 1 & -5 \\ 4 & 6 & 8 \\ 0 & -7 & 3 \end{vmatrix}$$

Since the element located in the first column and the third row of the given determinant is zero, it would be easier to make either (a) the first column or (b) the third row contain two zeros.

To make the first column contain two zeros, change row 2 so that the first element, 4, in the row is equal to zero.

Old row 2	4	6	8	*(1)*
2 × row 1	$2(2) = \underline{4}$	$2(1) = \underline{2}$	$2(-5) = \underline{-10}$	*(2)*
New row 2	0	4	18	*(1)–(2)*

Now use the new row 2 to replace the old row 2 of the given determinant. By applying Property 6 and the signed minors, we have:

$$|A| = \begin{vmatrix} 2 & 1 & -5 \\ 4 & 6 & 8 \\ 0 & -7 & 3 \end{vmatrix} = \begin{vmatrix} 2 & 1 & -5 \\ 0 & 4 & 18 \\ 0 & -7 & 3 \end{vmatrix} = 2 \begin{vmatrix} 4 & 18 \\ -7 & 3 \end{vmatrix} - 0 \begin{vmatrix} 1 & -5 \\ -7 & 3 \end{vmatrix} + 0 \begin{vmatrix} 1 & -5 \\ 4 & 18 \end{vmatrix}$$

$$= 2[12 - (-126)] - 0 + 0 = 2(138) = 276$$

To make the third row contain two zeros, change column 2 to make the third element (-7) in the column equal to zero.

Old Column 2	$\dfrac{7}{3} \times$ Column 3	New Column 2
1	$\dfrac{7}{3} \times (-5) = -\dfrac{35}{3}$	$1 + \left(-\dfrac{35}{3}\right)$ $= -\dfrac{32}{3}$
6	$\dfrac{7}{3} \times 8 \; = \dfrac{56}{3}$	$6 + \dfrac{56}{3} = \dfrac{74}{3}$
-7	$\dfrac{7}{3} \times 3 \; = 7$	$-7 + 7 \; = 0$
(1)	*(2)*	*(1) + (2)*

Now, replace the old column 2 of the given determinant with the new column 2. By applying Property 6 and the signed minors, we have:

$$|A| = \begin{vmatrix} 2 & 1 & -5 \\ 4 & 6 & 8 \\ 0 & -7 & 3 \end{vmatrix} = \begin{vmatrix} 2 & -\frac{32}{3} & -5 \\ 4 & \frac{74}{3} & 8 \\ 0 & 0 & 3 \end{vmatrix} = +0 - 0 + 3 \begin{vmatrix} 2 & -\frac{32}{3} \\ 4 & \frac{74}{3} \end{vmatrix}$$

$$= 3\left[2\left(\tfrac{74}{3}\right) - 4\left(-\tfrac{32}{3}\right)\right] = 3\left[\tfrac{148}{3} + \tfrac{128}{3}\right] = 276$$

EXAMPLE 18   Find the values for x_1 and x_2 from the following equation:

$$\begin{vmatrix} 2 & 1 & -5 \\ 4 & 6 & 8 \\ 0 & -7 & 3 \end{vmatrix} = \begin{vmatrix} 0 & x_1 & x_2 \\ 4 & 6 & 8 \\ 0 & -7 & 3 \end{vmatrix}$$

Change the first row on the left side $(2, 1, -5)$ to the first row on the right side $(0, x_1, x_2)$. Based on Property 6, the change can be written:

Old row 1 (left)	2	1	-5	.......*(1)*
$k \times$ row 2 (left)	$4k$	$6k$	$8k$	.......*(2)*
New row 1 (right) 0		x_1	x_2	.......*(1)–(2)*

From the relationship between the two first elements:

$$2 - 4k = 0$$
$$k = \tfrac{1}{2}$$

Thus:

$$x_1 = 1 - 6k = 1 - 6(\tfrac{1}{2}) = -2$$
$$x_2 = (-5) - 8k = (-5) - 8(\tfrac{1}{2}) = -9$$

EXERCISE 7–2   **Reference: Section 7.2**

A. *Verify each equation by finding the value of each determinant given below. Also, indicate the property number as presented in Section 7.2 if it is applicable to the equation.*

1. $\begin{vmatrix} 0 & 1 \\ 0 & 2 \end{vmatrix} = 0$

2. $\begin{vmatrix} 4 & 3 \\ 4 & 3 \end{vmatrix} = 0$

3. $\begin{vmatrix} 2 & -3 \\ 5 & 1 \end{vmatrix} = - \begin{vmatrix} -3 & 2 \\ 1 & 5 \end{vmatrix}$

4. $\begin{vmatrix} 2 & 4 \\ -1 & 7 \end{vmatrix} = - \begin{vmatrix} -2 & 4 \\ 1 & 7 \end{vmatrix}$

5. $\begin{vmatrix} 5 & 9 \\ 2 & 6 \end{vmatrix} = \begin{vmatrix} 5 & 1 \\ 2 & 7 \end{vmatrix} + \begin{vmatrix} 5 & 8 \\ 2 & -1 \end{vmatrix}$

6. $\begin{vmatrix} 4 & -6 \\ 3 & 2 \end{vmatrix} = \begin{vmatrix} 1 & -2 \\ 3 & 2 \end{vmatrix} + \begin{vmatrix} 3 & -4 \\ 3 & 2 \end{vmatrix}$

7. $\begin{vmatrix} 1 & 4 & 1 \\ -2 & 0 & -2 \\ 3 & 7 & 3 \end{vmatrix} = 0$

8. $\begin{vmatrix} 2 & 0 & 7 \\ 3 & 0 & -2 \\ 5 & 0 & 1 \end{vmatrix} = 0$

9. $\begin{vmatrix} 1 & 2 & 3 \\ 4 & 0 & 7 \\ 2 & 1 & 5 \end{vmatrix} = - \begin{vmatrix} 4 & 0 & 7 \\ 1 & 2 & 3 \\ 2 & 1 & 5 \end{vmatrix}$

10. $\begin{vmatrix} -2 & 5 & -1 \\ 4 & 0 & 3 \\ 5 & 6 & 7 \end{vmatrix} = - \begin{vmatrix} 5 & -2 & -1 \\ 0 & 4 & 3 \\ 6 & 5 & 7 \end{vmatrix}$

11. $2 \begin{vmatrix} 1 & 3 & 5 \\ 2 & 0 & -6 \\ -1 & 4 & 1 \end{vmatrix} = - \begin{vmatrix} 1 & 3 & -10 \\ 2 & 0 & 12 \\ -1 & 4 & -2 \end{vmatrix}$

12. $\begin{vmatrix} 2 & 3 & 0 \\ 4 & 2 & 5 \\ 7 & -1 & 6 \end{vmatrix} = \begin{vmatrix} 2 & -1 & 0 \\ 4 & 3 & 5 \\ 7 & -4 & 6 \end{vmatrix}$

$+ \begin{vmatrix} 2 & 4 & 0 \\ 4 & -1 & 5 \\ 7 & 3 & 6 \end{vmatrix}$

13. $\begin{vmatrix} 1 & 4 & 5 & -2 \\ 3 & 0 & 1 & 0 \\ 1 & 4 & 5 & -2 \\ 2 & 1 & -3 & 1 \end{vmatrix} = 0$

14. $\begin{vmatrix} 5 & 2 & -1 & 6 \\ 1 & -3 & 2 & 4 \\ 0 & 4 & 0 & 1 \\ 2 & -6 & 4 & 8 \end{vmatrix} = 0$

15. $\begin{vmatrix} 2 & 3 & 1 & 4 \\ 0 & 1 & 5 & -1 \\ 1 & -1 & 2 & 3 \\ 0 & 2 & 1 & -2 \end{vmatrix}$

$= \begin{vmatrix} 0 & 5 & -3 & -2 \\ 0 & 1 & 5 & -1 \\ 1 & -1 & 2 & 3 \\ 0 & 2 & 1 & -2 \end{vmatrix}$

16. $\begin{vmatrix} 1 & 0 & 2 & 0 \\ 3 & 5 & 1 & -4 \\ -1 & 3 & 0 & 2 \\ 2 & 1 & 3 & 1 \end{vmatrix} = \begin{vmatrix} 1 & 0 & 2 & 0 \\ 2 & 1 & 3 & -1 \\ -1 & 3 & 0 & 2 \\ 2 & 1 & 3 & 1 \end{vmatrix}$

$+ \begin{vmatrix} 1 & 0 & 2 & 0 \\ 1 & 4 & -2 & -3 \\ -1 & 3 & 0 & 2 \\ 2 & 1 & 3 & 1 \end{vmatrix}$

B. *Find the x or x's in each of the following equations by using Property 6.*

17. $\begin{vmatrix} 4 & 6 \\ 2 & 3 \end{vmatrix} = \begin{vmatrix} 4 & 6 \\ 0 & x \end{vmatrix}$

18. $\begin{vmatrix} 3 & -1 \\ 1 & 2 \end{vmatrix} = \begin{vmatrix} 3 & -1 \\ 4 & x \end{vmatrix}$

19. $\begin{vmatrix} 1 & -3 & 4 \\ 2 & 1 & 3 \\ 0 & 5 & 1 \end{vmatrix} = \begin{vmatrix} 1 & -3 & 4 \\ 0 & x_1 & x_2 \\ 0 & 5 & 1 \end{vmatrix}$

20. $\begin{vmatrix} 2 & -1 & 0 \\ 1 & 3 & 1 \\ 3 & 2 & -1 \end{vmatrix} = \begin{vmatrix} 2 & -1 & 0 \\ 1 & 3 & 1 \\ x_1 & x_2 & 0 \end{vmatrix}$

21. $\begin{vmatrix} -4 & 1 & 2 & 1 \\ 0 & 2 & 1 & 3 \\ 2 & -1 & -2 & 4 \\ 0 & 3 & 2 & 1 \end{vmatrix} = \begin{vmatrix} 0 & x_1 & x_2 & x_3 \\ 0 & 2 & 1 & 3 \\ 2 & -1 & -2 & 4 \\ 0 & 3 & 2 & 1 \end{vmatrix}$

22. $\begin{vmatrix} -1 & 2 & 3 & 1 \\ 1 & 3 & 2 & 4 \\ 0 & 0 & 1 & 1 \\ 3 & -1 & 5 & 2 \end{vmatrix} = \begin{vmatrix} -1 & 2 & x_1 & 1 \\ 1 & 3 & x_2 & 4 \\ 0 & 0 & 0 & 1 \\ 3 & -1 & x_3 & 2 \end{vmatrix}$

C. *Find the value of each of the following determinants. First use Property 6 to convert each determinant to an equal determinant with a single row (or column) having all zeros except one element, and then evaluate. (Hint: See the equivalent determinants computed in problems 19–22 respectively.)*

23. $\begin{vmatrix} 1 & -3 & 4 \\ 2 & 1 & 3 \\ 0 & 5 & 1 \end{vmatrix}$

24. $\begin{vmatrix} 2 & -1 & 0 \\ 1 & 3 & 1 \\ 3 & 2 & -1 \end{vmatrix}$

25. $\begin{vmatrix} -4 & 1 & 2 & 1 \\ 0 & 2 & 1 & 3 \\ 2 & -1 & -2 & 4 \\ 0 & 3 & 2 & 1 \end{vmatrix}$

26. $\begin{vmatrix} -1 & 2 & 3 & 1 \\ 1 & 3 & 2 & 4 \\ 0 & 0 & 1 & 1 \\ 3 & -1 & 5 & 2 \end{vmatrix}$

7.3 THE INVERSE OF A SQUARE MATRIX

Every real number, except 0, has an *inverse* or *reciprocal;* the product of the number and its inverse is always equal to 1. For example, the inverse of the real number 3 is 3^{-1} or $\frac{1}{3}$, and $3 \cdot 3^{-1} = 1$. We may express a similar relationship between a square matrix and its inverse:

Let A = a square matrix, and
A^{-1} = the inverse of A.
Then $A^{-1} \cdot A = A \cdot A^{-1} = I$,
where I = an identity (or unit) matrix of the same order as A and A^{-1}.

The inverse of a square matrix may be found by various methods. We shall introduce three methods in this section: (A) the basic method, (B) the formula method, and (C) the Gaussian method.

A. Basic Method

The *basic method* is based on the expression $A \cdot A^{-1} = I$.

EXAMPLE 19 Find the inverse of the square matrix $A = \begin{bmatrix} 2 & 6 \\ 1 & 4 \end{bmatrix}$.

If A^{-1}, the inverse of A, exists, then we may write $A \cdot A^{-1} = I$ for the given matrix as follows:

$$\begin{bmatrix} 2 & 6 \\ 1 & 4 \end{bmatrix} \cdot \begin{bmatrix} 2 & 6 \\ 1 & 4 \end{bmatrix}^{-1} = \begin{bmatrix} 1 & 0 \\ 0 & 1 \end{bmatrix}$$

Or: $\begin{bmatrix} 2 & 6 \\ 1 & 4 \end{bmatrix} \cdot A^{-1} = \begin{bmatrix} 1 & 0 \\ 0 & 1 \end{bmatrix}$

Since A is a 2×2 square matrix, A^{-1} and I must also be 2×2 square matrices. Let the elements of the matrix A^{-1} be b, with the first and the second subscripts indicating the locations of rows and columns in the matrix respectively, or:

$$\begin{bmatrix} 2 & 6 \\ 1 & 4 \end{bmatrix} \cdot \begin{bmatrix} b_{11} & b_{12} \\ b_{21} & b_{22} \end{bmatrix} = \begin{bmatrix} 1 & 0 \\ 0 & 1 \end{bmatrix}$$

Multiply the left side of the equation.

$$\begin{bmatrix} (2b_{11} + 6b_{21}) & (2b_{12} + 6b_{22}) \\ (1b_{11} + 4b_{21}) & (1b_{12} + 4b_{22}) \end{bmatrix} = \begin{bmatrix} 1 & 0 \\ 0 & 1 \end{bmatrix}$$

Equate corresponding elements of the matrices on both sides. We have two systems of two equations each as follows:

$2b_{11} + 6b_{21} = 1$ ..... (1) $1b_{11} + 4b_{21} = 0$ ..... (2)	$2b_{12} + 6b_{22} = 0$ ..... (3) $1b_{12} + 4b_{22} = 1$ ..... (4)
Solve equations (1) and (2) for b_{11} and b_{21}.	Solve equations (3) and (4) for b_{12} and b_{22}.
(2)×2 $2b_{11} + 8b_{21} = 0$ ... (2)′	(4)×2 $2b_{12} + 8b_{22} = 2$ ... (4)′
(2)′ − (1) $2b_{21} = -1$ $b_{21} = -\frac{1}{2}$	(4)′ − (3) $2b_{22} = 2$ $b_{22} = 1$
Substitute b_{21} value in (2).	Substitute b_{22} value in (4).
$b_{11} + 4\left(-\frac{1}{2}\right) = 0$ $b_{11} = 2$	$b_{12} + 4(1) = 1$ $b_{12} = 1 - 4 = -3$

Thus:

$$A^{-1} = \begin{bmatrix} b_{11} & b_{12} \\ b_{21} & b_{22} \end{bmatrix} = \begin{bmatrix} 2 & -3 \\ -\frac{1}{2} & 1 \end{bmatrix}$$

Check:

$$A \cdot A^{-1} = \begin{bmatrix} 2 & 6 \\ 1 & 4 \end{bmatrix} \cdot \begin{bmatrix} 2 & -3 \\ -\frac{1}{2} & 1 \end{bmatrix} = \begin{bmatrix} 4 + (-3) & (-6) + 6 \\ 2 + (-2) & (-3) + 4 \end{bmatrix} = \begin{bmatrix} 1 & 0 \\ 0 & 1 \end{bmatrix} = I$$

$$A^{-1} A = \begin{bmatrix} 2 & -3 \\ -\frac{1}{2} & 1 \end{bmatrix} \cdot \begin{bmatrix} 2 & 6 \\ 1 & 4 \end{bmatrix} = \begin{bmatrix} 4 + (-3) & 12 + (-12) \\ (-1) + 1 & (-3) + 4 \end{bmatrix} = \begin{bmatrix} 1 & 0 \\ 0 & 1 \end{bmatrix} = I$$

We have shown that $A \cdot A^{-1} = A^{-1} \cdot A = I$. Note that if the two systems of two equations derived by the multiplication were inconsistent, such as:

$$A = \begin{bmatrix} 1 & 1 \\ 3 & 3 \end{bmatrix} \quad \text{then} \quad \begin{cases} 1b_{11} + 1b_{21} = 1 \\ 3b_{11} + 3b_{21} = 0 \end{cases} \quad \text{and} \quad \begin{cases} 1b_{12} + 1b_{22} = 0 \\ 3b_{12} = 3b_{22} = 1 \end{cases}$$

there will be no solutions for the elements b of A^{-1}. Therefore, the square matrix A has no inverse in this case.

For a 3×3 square matrix A, the product of $A \cdot A^{-1}$ according to the basic method will give three sets of three equations each. To solve the three sets of equations is a tedious job. Thus, this method should not be used for the higher order square matrices.

B. The Formula Method

According to the procedure used in the basic method (Example 19), we may derive a simplified or *formula method* in finding the inverse of a square matrix. Let the 2×2 square matrix A be written in a general form with elements being a, or:

$$A = \begin{bmatrix} a_{11} & a_{12} \\ a_{21} & a_{22} \end{bmatrix}$$

If A^{-1} exists, we may write $A \cdot A^{-1} = I$ in the form:

$$\begin{bmatrix} a_{11} & a_{12} \\ a_{21} & a_{22} \end{bmatrix} \cdot \begin{bmatrix} b_{11} & b_{12} \\ b_{21} & b_{22} \end{bmatrix} = \begin{bmatrix} 1 & 0 \\ 0 & 1 \end{bmatrix}$$

After multiplying the matrices on the left side of the above equation, equating corresponding elements of the product matrix with the right side identity matrix, we have the following two systems of two equations each:

$$\begin{cases} a_{11}b_{11} + a_{12}b_{21} = 1 & (1) \\ a_{21}b_{11} + a_{22}b_{21} = 0 & (2) \end{cases} \qquad \begin{cases} a_{11}b_{12} + a_{12}b_{22} = 0 & (3) \\ a_{21}b_{12} + a_{22}b_{22} = 1 & (4) \end{cases}$$

Solve equations (1) and (2) for b_{11} and b_{21}, and solve equations (3) and (4) for b_{12} and b_{22}.*

---

* Solve equations (1) and (2) for b_{11} and b_{21}.

(1) $\times a_{21}$ $\quad a_{21}a_{11}b_{11} + a_{21}a_{12}b_{21} = a_{21}$ $\quad$ (1)′

(2) $\times a_{11}$ $\quad a_{11}a_{21}b_{11} + a_{11}a_{22}b_{21} = 0$ $\quad$ (2)′

(1)′ − (2)′

$$a_{21}a_{12}b_{21} - a_{11}a_{22}b_{21} = a_{21} - 0$$

$$b_{21}(a_{12}a_{21} - a_{11}a_{22}) = a_{21}$$

$$b_{21} = \frac{a_{21}}{a_{12}a_{21} - a_{11}a_{22}}$$

$$= \frac{a_{21}}{-(a_{11}a_{22} - a_{12}a_{21})} = \frac{-a_{21}}{|A|}$$

Substitute b_{21} value in (2):

$$a_{21}b_{11} + a_{22}\left(\frac{a_{21}}{a_{12}a_{21} - a_{11}a_{22}}\right) = 0$$

$$b_{11} = -\frac{a_{22}a_{21}}{a_{12}a_{21} - a_{11}a_{22}} \cdot \frac{1}{a_{21}} = \frac{a_{22}}{|A|}$$

Similarly, we may solve equations (3) and (4) to obtain the values of b_{12} and b_{22}.

$$b_{11} = \frac{a_{22}}{|A|} \qquad b_{12} = \frac{-a_{12}}{|A|}$$

$$b_{21} = \frac{-a_{21}}{|A|} \qquad b_{22} = \frac{a_{11}}{|A|}$$

where
$$|A| = \begin{vmatrix} a_{11} & a_{12} \\ a_{21} & a_{22} \end{vmatrix} = a_{11}a_{22} - a_{12}a_{21} \neq 0$$

The formula for the inverse of the 2×2 square matrix A thus is:

$$A^{-1} = \begin{bmatrix} b_{11} & b_{12} \\ b_{21} & b_{22} \end{bmatrix} = \begin{bmatrix} \frac{a_{22}}{|A|} & \frac{-a_{12}}{|A|} \\ \frac{-a_{21}}{|A|} & \frac{a_{11}}{|A|} \end{bmatrix}$$

Or simply written:

$$A^{-1} = \frac{1}{|A|} \begin{bmatrix} a_{22} & -a_{12} \\ -a_{21} & a_{11} \end{bmatrix}$$

The value of $|A|$ must not be zero since $\frac{1}{0}$ is not defined. Thus, the determinant $|A|$ determines the existence of the inverse of a square matrix A. If, and only if, $|A| \neq 0$, the inverse A^{-1} exists. A square matrix which has an inverse is called an *invertible* or *nonsingular* matrix. A square matrix A for which $|A| = 0$ is called a *singular* matrix.

Example 19 is now solved by the use of the above formula:

$$A = \begin{bmatrix} a_{11} & a_{12} \\ a_{21} & a_{22} \end{bmatrix} = \begin{bmatrix} 2 & 6 \\ 1 & 4 \end{bmatrix} \qquad |A| = \begin{vmatrix} 2 & 6 \\ 1 & 4 \end{vmatrix} = 8 - 6 = 2$$

A has an inverse since $|A| \neq 0$.

$$A^{-1} = \frac{1}{|A|} \begin{bmatrix} a_{22} & -a_{12} \\ -a_{21} & a_{11} \end{bmatrix} = \frac{1}{2} \begin{bmatrix} 4 & -6 \\ -1 & 2 \end{bmatrix} = \begin{bmatrix} 2 & -3 \\ -\frac{1}{2} & 1 \end{bmatrix}$$

Observe the above calculation. In summary, to find the inverse of a 2×2 square matrix A for which $|A| \neq 0$, we use the following steps:

1. Interchange the elements on the principal diagonal (a_{11} and a_{22}).
2. Prefix the other two elements (a_{12} and a_{21}) with a negative sign.
3. Multiply the matrix obtained in steps 1 and 2 by $1/|A|$.

We may also write the inverse of a 2×2 square matrix A in terms of signed minors (or cofactors).

$$A^{-1} = \frac{1}{|A|} \begin{bmatrix} A_{11} & -A_{21} \\ -A_{12} & A_{22} \end{bmatrix} \quad \text{where:} \quad \begin{matrix} A_{11} = a_{22}, & -A_{21} = -a_{12}, \\ -A_{12} = -a_{21}, & A_{22} = a_{11}. \end{matrix}$$

Note that the signed minors in this matrix are in *transposed order*. For example, the subscripts 21 in element a_{21} indicate that the element is in the second row and the first column. However, the subscripts 21 in the signed minor $-A_{21}$ indicate that the minor is in the second column and the first row.

Extending this illustration, we may write a formula for the inverse of a 3 × 3 square matrix in terms of signed minors.

$$A^{-1} = \frac{1}{|A|} \begin{bmatrix} A_{11} & -A_{21} & A_{31} \\ -A_{12} & A_{22} & -A_{32} \\ A_{13} & -A_{23} & A_{33} \end{bmatrix}$$

Again the signed minors in the 3 × 3 matrix are in transposed order. The minors A_{12}, A_{21}, A_{23}, and A_{32} have negative signs since the sum of each pair of subscripts is odd, such as 2 (second column) + 3 (third row) = 5 for the minor A_{23}. The other minors are positive since the sum of an individual pair of subscripts is an even number, such as 1 + 3 = 4 for A_{13}. (Also, refer to page 120 for the sign array.)

We may find the inverse of a 3 × 3 square matrix A for which $A \neq 0$ as follows:

1. Replace each element in A by its signed minor, such as replacing a_{11} by A_{11} and a_{12} by $-A_{12}$.
2. Take the transpose of the matrix obtained in the first step.
3. Multiply the obtained transpose by $\frac{1}{|A|}$.

EXAMPLE 20   Find the inverse of the 3 × 3 square matrix $A = \begin{bmatrix} 2 & 1 & -5 \\ 4 & 6 & 8 \\ 0 & -7 & 3 \end{bmatrix}$

$$|A| = \begin{vmatrix} 2 & 1 & -5 \\ 4 & 6 & 8 \\ 0 & -7 & 3 \end{vmatrix} = 276 \quad \text{(See Example 2.)}$$

Since $|A| \neq 0$, the given matrix A has an inverse.

1. Replace each element in A by its signed minors. For example, the element in the 1st row-1st column (a_{11}) is replaced by its signed minor ($+A_{11}$).

$$+A_{11} = + \begin{vmatrix} 2 & 1 & -5 \\ 4 & 6 & 8 \\ 0 & -7 & 3 \end{vmatrix} = \begin{vmatrix} 6 & 8 \\ -7 & 3 \end{vmatrix}$$

The element in the 2nd row-1st column (a_{21}) is replaced by its signed minor ($-A_{21}$).

$$-A_{21} = - \begin{vmatrix} 2 & 1 & -5 \\ 4 & 6 & 8 \\ 0 & -7 & 3 \end{vmatrix} = - \begin{vmatrix} 1 & -5 \\ -7 & 3 \end{vmatrix}$$

We shall denote the new signed minor matrix (also called *cofactor matrix*) by **B**.

$$
B = \begin{bmatrix}
\begin{vmatrix} 6 & 8 \\ -7 & 3 \end{vmatrix} & -\begin{vmatrix} 4 & 8 \\ 0 & 3 \end{vmatrix} & \begin{vmatrix} 4 & 6 \\ 0 & -7 \end{vmatrix} \\[8pt]
-\begin{vmatrix} 1 & -5 \\ -7 & 3 \end{vmatrix} & \begin{vmatrix} 2 & -5 \\ 0 & 3 \end{vmatrix} & -\begin{vmatrix} 2 & 1 \\ 0 & -7 \end{vmatrix} \\[8pt]
\begin{vmatrix} 1 & -5 \\ 6 & 8 \end{vmatrix} & -\begin{vmatrix} 2 & -5 \\ 4 & 8 \end{vmatrix} & \begin{vmatrix} 2 & 1 \\ 4 & 6 \end{vmatrix}
\end{bmatrix}
$$

$$
= \begin{bmatrix}
18 - (-56) & -(12 - 0) & (-28) - 0 \\
-(3 - 35) & 6 - 0 & -((-14) - 0) \\
8 - (-30) & -(16 - (-20)) & 12 - 4
\end{bmatrix}
$$

$$
= \begin{bmatrix}
74 & -12 & -28 \\
32 & 6 & 14 \\
38 & -36 & 8
\end{bmatrix}
$$

2. Take the transpose of matrix *B* (the transpose of a cofactor matrix is also called *adjoint matrix*), or:

$$
B^t = \begin{bmatrix}
74 & -12 & -28 \\
32 & 6 & 14 \\
38 & -36 & 8
\end{bmatrix}^t = \begin{bmatrix}
74 & 32 & 38 \\
-12 & 6 & -36 \\
-28 & 14 & 8
\end{bmatrix}
$$

3. Multiply $\dfrac{1}{|A|} = \dfrac{1}{276}$ by the transpose, or:

$$
A^{-1} = \frac{1}{|A|} \cdot B^t = \frac{1}{276} \begin{bmatrix}
74 & 32 & 38 \\
-12 & 6 & -36 \\
-28 & 14 & 8
\end{bmatrix} = \frac{1}{138} \begin{bmatrix}
37 & 16 & 19 \\
-6 & 3 & -18 \\
-14 & 7 & 4
\end{bmatrix}, \text{ or:}
$$

$$
= \begin{bmatrix}
\dfrac{37}{138} & \dfrac{16}{138} & \dfrac{19}{138} \\[10pt]
\dfrac{-6}{138} & \dfrac{3}{138} & \dfrac{-18}{138} \\[10pt]
\dfrac{-14}{138} & \dfrac{7}{138} & \dfrac{4}{138}
\end{bmatrix}
$$

Check:

$$
A^{-1} \cdot A = \frac{1}{138} \begin{bmatrix}
37 & 16 & 19 \\
-6 & 3 & -18 \\
-14 & 7 & 4
\end{bmatrix} \cdot \begin{bmatrix}
2 & 1 & -5 \\
4 & 6 & 8 \\
0 & -7 & 3
\end{bmatrix}
$$

$$
= \frac{1}{138} \begin{bmatrix}
138 & 0 & 0 \\
0 & 138 & 0 \\
0 & 0 & 138
\end{bmatrix} = \begin{bmatrix}
1 & 0 & 0 \\
0 & 1 & 0 \\
0 & 0 & 1
\end{bmatrix}
$$

C. The Gaussian Method

When the *Gaussian method* (named in honor of Karl Friedrich Gauss, a German mathematician, 1777–1855) is used to find the inverse of a square matrix, the following steps should be used:

STEP 1   Write a *tableau* formed by the given square matrix A (placed on the left side of a vertical line) and an identity matrix of the same order as A (placed on the right side of the vertical line). The form of the tableau therefore is:

$$[A \mid I]$$

$$\text{If } A = \begin{bmatrix} 2 & 6 \\ 1 & 4 \end{bmatrix} \quad \text{then} \quad [A \mid I] = \begin{bmatrix} 2 & 6 & 1 & 0 \\ 1 & 4 & 0 & 1 \end{bmatrix}$$

This tableau is also called the *augmented matrix.*

STEP 2   Perform *elementary row operations* on both matrices in the tableau until matrix A becomes an identity matrix. The old identity matrix is now replaced by the inverse of matrix A, or A^{-1} which is the solution to our problem. Thus, we are transforming the tableau:

$[A|I]$ into a new tableau, $[I|A^{-1}]$

in finding the inverse of the square matrix A by the Gaussian method.

There are three elementary row operations on matrices. We may use any one of the operations to change a row of a tableau in order to obtain a new row with a desired element.

OPERATION 1 A row in the tableau may be multiplied or divided by a real number k $(k \neq 0)$.

EXAMPLE 21   The first row in the tableau $[A \mid I]$ given in Step 1 is:

2   6   1   0

This row may be converted to a new row by dividing each element by 2 as follows:

$2 \div 2 = 1 \quad 6 \div 2 = 3 \quad 1 \div 2 = \frac{1}{2} \quad 0 \div 2 = 0$

The new first row of the tableau is $1, 3, \frac{1}{2}$, and 0. The given tableau now is changed from:

$$\begin{bmatrix} 2 & 6 & 1 & 0 \\ 1 & 4 & 0 & 1 \end{bmatrix}$$

to:

$$\begin{bmatrix} 1 & 3 & \frac{1}{2} & 0 \\ 1 & 4 & 0 & 1 \end{bmatrix}$$

OPERATION 2 A row may be added or subtracted by the product of a real number k and another row.

EXAMPLE 22   The second row given in Step 1 tableau is:

1   4   0   1

This row may be converted to a new row by multiplying each corresponding element in another row (the first row in this case) by $\frac{1}{2}$ then subtracting each product from each element of the old row.

Old row 2	1	4	0	1 .... (1)
$\frac{1}{2}$ × row 1	$\frac{1}{2}(2) = \underline{1}$,	$\frac{1}{2}(6) = \underline{3}$,	$\frac{1}{2}(1) = \underline{\frac{1}{2}}$,	$\frac{1}{2}(0) = \underline{0}$ .... (2)
New row 2	0	1	$-\frac{1}{2}$	1 .... (1)–(2)

Note that the procedure of this elementary row operation is the same as that of Property 6 of a determinant. (See page 123.)

OPERATION 3 Two rows may be interchanged.

EXAMPLE 23   The two rows in the tableau given in Step 1 may be interchanged; that is, change:

$$\begin{bmatrix} 2 & 6 & | & 1 & 0 \\ 1 & 4 & | & 0 & 1 \end{bmatrix} \quad \text{to} \quad \begin{bmatrix} 1 & 4 & | & 0 & 1 \\ 2 & 6 & | & 1 & 0 \end{bmatrix}$$

The work of transforming the matrix A in $[A|I]$ into the identity I in $[I|A^{-1}]$ may be done systematically by converting the elements of A column by column according to the natural order (first column, second column, and so on). Within each column, we perform two types of conversion:

1. Convert the appropriate element to 1 for obtaining the 1's on the principal diagonal of the identity matrix. Use row Operation 1. Thus, in converting the first column, divide the first row by the first element in the row; in converting the second column, divide the second row by the second element in the row; and so on.
2. Convert all other elements to 0. Use row Operation 2. Each row to be converted is subtracted by the product of the element to be converted to 0 multiplied by the new row obtained from the first conversion.

EXAMPLE 24   Find the inverse of the square matrix $A = \begin{bmatrix} 2 & 6 \\ 1 & 4 \end{bmatrix}$ by the Gaussian method.

STEP 1   Write the initial tableau:

$$[A \mid I] = \begin{bmatrix} 2 & 6 & | & 1 & 0 \\ 1 & 4 & | & 0 & 1 \end{bmatrix} \begin{array}{l} \ldots\ldots (1) \\ \ldots\ldots (2) \end{array}$$

STEP 2   Perform row operations. (Let R denote row.)

1. Convert the elements in column 1 of A from $\begin{smallmatrix} 2 \\ 1 \end{smallmatrix}$ to $\begin{smallmatrix} 1 \\ 0 \end{smallmatrix}$.

(a) Convert the first element 2 in row (1) to 1. Use Operation 1.

$$R(1) \div 2 \qquad \frac{2}{2} = 1 \qquad \frac{6}{2} = 3 \qquad \frac{1}{2} \qquad \frac{0}{2} = 0 \ldots \ldots (1)'$$

(b) Convert the first element 1 in row (2) to 0. Use Operation (2).

$R(2)$	1	4	0	1
$R(1)' \times 1$	1	3	$\frac{1}{2}$	0
Subtracted	0	1	$-\frac{1}{2}$	$1 \ldots \ldots (2)'$

The initial tableau now becomes:

$$\begin{bmatrix} 1 & 3 & | & \frac{1}{2} & 0 \\ 0 & 1 & | & -\frac{1}{2} & 1 \end{bmatrix} \begin{matrix} \ldots \ldots (1)' \\ \ldots \ldots (2)' \end{matrix}$$

2. Convert the elements in column 2 of the new tableau from $\frac{3}{1}$ to $\frac{0}{1}$.

(a) The second element in row (2)' is already 1. Thus, row (2)' should not be changed.

(b) Convert the second element 3 in row (1)' to 0. Use Operation (2):

$R(1)'$	1	3	$\frac{1}{2}$	0
$R(2)' \times 3$	0	3	$-\frac{3}{2}$	3
Subtracted	1	0	$\frac{1}{2}$	$-3 \ldots \ldots (1)''$
			$= 2$	

The final tableau is:

$$[I \mid A^{-1}] = \begin{bmatrix} 1 & 0 & | & 2 & -3 \\ 0 & 1 & | & -\frac{1}{2} & 1 \end{bmatrix} \begin{matrix} \ldots \ldots (1)'' \\ \ldots \ldots (2)' \end{matrix}$$

Thus:

$$A^{-1} = \begin{bmatrix} 2 & -3 \\ -\frac{1}{2} & 1 \end{bmatrix}$$

This answer may be compared with that of Example 19, page 128.

For a 3×3 square matrix, we have to convert the elements in three columns respectively. Example 25 is used to illustrate the operations.

EXAMPLE 25 Find the inverse of the square matrix $A = \begin{bmatrix} 2 & 1 & -5 \\ 4 & 6 & 8 \\ 0 & -7 & 3 \end{bmatrix}$ by the Gaussian method.

STEP 1 Write the initial tableau:

$$[A \mid I] = \begin{bmatrix} 2 & 1 & -5 & | & 1 & 0 & 0 \\ 4 & 6 & 8 & | & 0 & 1 & 0 \\ 0 & -7 & 3 & | & 0 & 0 & 1 \end{bmatrix} \begin{matrix} \ldots (1) \\ \ldots (2) \\ \ldots (3) \end{matrix}$$

STEP 2 Perform row operations.

Convert the elements in column 1 of A from $\begin{matrix} 2 \\ 4 \\ 0 \end{matrix}$ to $\begin{matrix} 1 \\ 0 \\ 0 \end{matrix}$.

(a) Convert the first element 2 in row (1) to 1.

$R(1) \div 2$: $\dfrac{2}{2} = 1$ $\frac{1}{2}$ $-\frac{5}{2}$ $\frac{1}{2}$ 0 $0 \ldots (1)'$

(b) Convert the first element 4 in row (2) to 0.

$R(2)$	4	6	8	0	1	0
$R(1)' \times 4$	4	2	-10	2	0	0
Subtracted	0	4	18	-2	1	$0 \ldots (2)'$

(c) The first element in row (3) is already 0. Thus, row (3) should not be changed.

The derived second tableau is:

$$\begin{bmatrix} 1 & \frac{1}{2} & -\frac{5}{2} & \frac{1}{2} & 0 & 0 \\ 0 & 4 & 18 & -2 & 1 & 0 \\ 0 & -7 & 3 & 0 & 0 & 1 \end{bmatrix} \begin{matrix} \ldots \ldots (1)' \\ \ldots \ldots (2)' \\ \ldots \ldots (3) \end{matrix}$$

2. Convert the elements in column 2 of the second tableau from

$\begin{matrix} \frac{1}{2} \\ 4 \\ -7 \end{matrix}$ to $\begin{matrix} 0 \\ 1. \\ 0 \end{matrix}$

(a) Convert the second element 4 in row (2)' to 1.

$R(2)' \div 4$: 0 1 $\frac{18}{4} = \frac{9}{2}$ $-\frac{2}{4} = -\frac{1}{2}$ $\frac{1}{4}$ $0 \ldots (2)''$

(b) Convert the second element $\frac{1}{2}$ in row (1)' to 0.

$R(1)'$	1	$\frac{1}{2}$	$-\frac{5}{2}$	$\frac{1}{2}$	0	0
$R(2)'' \times \frac{1}{2}$	0	$\frac{1}{2}$	$\frac{9}{4}$	$-\frac{1}{4}$	$\frac{1}{8}$	0
Subtracted	1	0	$-\frac{19}{4}$	$\frac{3}{4}$	$-\frac{1}{8}$	$0 \ldots (1)''$

(c) Convert the second element -7 in row (3) to 0.

$R(3)$	0	-7	3	0	0	1
$R(2)'' \times 7$	0	7	$\frac{63}{2}$	$-\frac{7}{2}$	$\frac{7}{4}$	0
Added	0	0	$\frac{69}{2}$	$-\frac{7}{2}$	$\frac{7}{4}$	$1 \ldots (3)'$

The derived third tableau is:

$$\begin{bmatrix} 1 & 0 & -\frac{19}{4} & \frac{3}{4} & -\frac{1}{8} & 0 \\ 0 & 1 & \frac{9}{2} & -\frac{1}{2} & \frac{1}{4} & 0 \\ 0 & 0 & \frac{69}{2} & -\frac{7}{2} & \frac{7}{4} & 1 \end{bmatrix} \begin{matrix} \ldots \ldots (1)'' \\ \ldots \ldots (2)'' \\ \ldots \ldots (3)' \end{matrix}$$

3. Convert the elements in column 3 of the third tableau from $\begin{matrix} -\frac{19}{4} \\ \frac{9}{2} \\ \frac{69}{2} \end{matrix}$ to $\begin{matrix} 0 \\ 0. \\ 1 \end{matrix}$

(a) Convert the third element $\frac{69}{2}$ in row (3)' to 1.

$R(3) \div \frac{69}{2}$
    (or $\times \frac{2}{69}$): 0 0 1 $-\frac{7}{69}$ $\frac{7}{138}$ $\frac{2}{69} \ldots (3)''$

(b) Convert the third element $-\frac{19}{4}$ in row (1)'' to 0.

$R(1)''$	1	0	$-\frac{19}{4}$	$\frac{3}{4}$	$-\frac{1}{8}$	0
$R(3)'' \times (-\frac{19}{4})$	0	0	$-\frac{19}{4}$	$\frac{133}{276}$	$-\frac{133}{552}$	$-\frac{19}{138}$
Subtracted	1	0	0	$\frac{37}{138}$	$\frac{16}{138}$	$\frac{19}{138}$ .. (1)'''

(c) Convert the third element $\frac{9}{2}$ in row (2)'' to 0.

$R(2)''$	0	1	$\frac{9}{2}$	$-\frac{1}{2}$	$\frac{1}{4}$	0
$R(3)'' \times (\frac{9}{2})$	0	0	$\frac{9}{2}$	$-\frac{63}{138}$	$\frac{63}{276}$	$\frac{18}{138}$
Subtracted	0	1	0	$-\frac{6}{138}$	$\frac{3}{138}$	$-\frac{18}{138}$ ... (2)'''

The final tableau is:

$$[I \mid A^{-1}] = \begin{bmatrix} 1 & 0 & 0 \\ 0 & 1 & 0 \\ 0 & 0 & 1 \end{bmatrix} \begin{array}{ccc} \frac{37}{138} & \frac{16}{138} & \frac{19}{138} \\ -\frac{6}{138} & \frac{3}{138} & -\frac{18}{138} \\ -\frac{7}{69} & \frac{7}{138} & \frac{2}{69} \end{array} \begin{array}{l} \cdots (1)''' \\ \cdots (2)''' \\ \cdots (3)'' \end{array}$$

Thus:

$$A^{-1} = \frac{1}{138} \begin{bmatrix} 37 & 16 & 19 \\ -6 & 3 & -18 \\ -14 & 7 & 4 \end{bmatrix}$$

This answer may be compared with that of Example 20.

The details of performing row operations in Step 2 of Example 25 can be summarized in the following manner:

1. Convert the elements in column 1 of A in the initial tableau. The result is shown in the second tableau. (First compute $R(1)'$ to obtain 1 in column 1.)

$$\begin{array}{l} R(1) \div 2 \\ R(2) - R(1)' \times 4 \\ R(3), \text{ same} \end{array} \begin{bmatrix} 1 & \frac{1}{2} & -\frac{5}{2} & \frac{1}{2} & 0 & 0 \\ 0 & 4 & 18 & -2 & 1 & 0 \\ 0 & -7 & 3 & 0 & 0 & 1 \end{bmatrix} \begin{array}{l} \ldots\ldots(1)' \\ \ldots\ldots(2)' \\ \ldots\ldots(3) \end{array}$$

2. Convert the elements in column 2 of the second tableau. The result is shown in the third tableau. (First compute $R(2)''$ to obtain 1 in column 2.)

$$\begin{array}{l} R(1)' - R(2)'' \times \frac{1}{2} \\ R(2)' \div 4 \\ R(3) - R(2)''(-7) \end{array} \begin{bmatrix} 1 & 0 & -\frac{19}{4} & \frac{3}{4} & -\frac{1}{8} & 0 \\ 0 & 1 & \frac{9}{2} & -\frac{1}{2} & \frac{1}{4} & 0 \\ 0 & 0 & \frac{69}{2} & -\frac{7}{2} & \frac{7}{4} & 1 \end{bmatrix} \begin{array}{l} \ldots\ldots(1)'' \\ \ldots\ldots(2)'' \\ \ldots\ldots(3)' \end{array}$$

3. Convert the elements in column 3 of the third tableau. The result is shown in the final tableau. (First compute $R(3)''$ to obtain 1 in column 3.)

$$\begin{array}{l} R(1)'' - R(3)''(-\frac{19}{4}) \\ R(2)'' - R(3)''(\frac{9}{2}) \\ R(3)' \div \frac{69}{2} \end{array} \begin{bmatrix} 1 & 0 & 0 & \frac{37}{138} & \frac{16}{138} & \frac{19}{138} \\ 0 & 1 & 0 & -\frac{6}{138} & \frac{3}{138} & -\frac{18}{138} \\ 0 & 0 & 1 & -\frac{7}{69} & \frac{7}{138} & \frac{2}{69} \end{bmatrix} \begin{array}{l} \ldots\ldots(1)''' \\ \ldots\ldots(2)''' \\ \ldots\ldots(3)'' \end{array}$$

Note that the first two elementary row operations have been applied consistently in the preceding discussion and examples; that is, operation 1 is used in obtaining 1's and operation 2 is used in obtaining 0's for the transformation of A to I. However, the two operations may also be applied in different manners. For instance, referring to the initial tableau in Example 24:

$$\begin{bmatrix} 2 & 6 & 1 & 0 \\ 1 & 4 & 0 & 1 \end{bmatrix} \begin{array}{l} \ldots\ldots (1) \\ \ldots\ldots (2) \end{array}$$

we may convert the first element 2 in row (1) to 1 by using Operation 2 as follows:

$$
\begin{array}{lcccc}
R(1) & 2 & 6 & 1 & 0 \\
\underline{R(2) \times 1} & \underline{1} & \underline{4} & \underline{0} & \underline{1} \\
\text{Subtracted} & 1 & 2 & 1 & -1 \dots\dots\text{New row } (1)'
\end{array}
$$

The initial tableau then becomes:

$$
\begin{bmatrix} 1 & 2 & | & 1 & -1 \\ 1 & 4 & | & 0 & 1 \end{bmatrix} \begin{array}{l} \dots\dots(1)' \\ \dots\dots(2) \end{array}
$$

Or, if operation 3 is used to interchange rows (1) and (2), the initial tableau is written:

$$
\begin{bmatrix} 1 & 4 & | & 0 & 1 \\ 2 & 6 & | & 1 & 0 \end{bmatrix} \begin{array}{l} \dots\dots(1)' \\ \dots\dots(2)' \end{array}
$$

The same answer for A^{-1} $\left(= \begin{bmatrix} 2 & -3 \\ -\frac{1}{2} & 1 \end{bmatrix} \right)$ may be obtained from the two preceding tableaus.

EXERCISE 7–3   Reference: Section 7.3

A. *Find the inverse of each matrix A if the inverse exists. Use (1) the basic method, (2) the formula method, and (3) the Gaussian method. Check each answer by the relationship $A \cdot A^{-1} = I$.*

1. $A = \begin{bmatrix} 7 & 5 \\ 7 & 5 \end{bmatrix}$ **4.** $A = \begin{bmatrix} 6 & 8 \\ 1 & 3 \end{bmatrix}$

2. $A = \begin{bmatrix} -8 & 20 \\ -2 & 5 \end{bmatrix}$ **5.** $A = \begin{bmatrix} 4 & -5 \\ 2 & 7 \end{bmatrix}$

3. $A = \begin{bmatrix} 6 & -10 \\ 2 & -5 \end{bmatrix}$ **6.** $A = \begin{bmatrix} 8 & -1 \\ 0 & 2 \end{bmatrix}$

B. *Find the inverse of each matrix A if the inverse exists. Use (1) the formula method * and (2) the Gaussian method. Check each answer by the relationship $A^{-1} \cdot A = I$.*

7. $A = \begin{bmatrix} 2 & -5 & 1 \\ 4 & 8 & 6 \\ 9 & 3 & -7 \end{bmatrix}$ **10.** $A = \begin{bmatrix} -4 & 6 & 1 \\ 8 & 5 & 0 \\ 3 & 2 & -7 \end{bmatrix}$

8. $A = \begin{bmatrix} 3 & 7 & 9 \\ 0 & -2 & 5 \\ -4 & 1 & 8 \end{bmatrix}$ **11.** $A = \begin{bmatrix} 5 & 4 & -2 \\ 6 & 0 & 1 \\ -7 & 3 & -8 \end{bmatrix}$

9. $A = \begin{bmatrix} 1 & 8 & 0 \\ -2 & 4 & -3 \\ 7 & -6 & 5 \end{bmatrix}$ **12.** $A = \begin{bmatrix} 1 & -3 & 10 \\ -2 & 5 & -8 \\ 9 & 4 & 7 \end{bmatrix}$

---

* Also see the answers to problems 7–12 in Exercise 7–1.

7.4 SOLVING LINEAR EQUATIONS BY MATRIX ALGEBRA

A system of n linear equations in n unknowns may be solved by the ordinary algebraic methods: the method of elimination by addition or subtraction and the method of elimination by substitution. However, the work of solving a system of three or more linear equations becomes increasingly difficult by the ordinary methods. Matrix algebra offers simplified and systematic methods of solving the equations. A systematic method is usually more convenient for writing a computer program to find the solutions. This section will introduce three methods of solving linear equations by matrix algebra: (1) the inverse of a square matrix, (2) Cramer's rule, and (3) the Gaussian method.

A. Using the Inverse of a Square Matrix

The method of using the *inverse of a square matrix* is illustrated in Example 26 by using only two equations, although it is applicable to a system of more than two linear equations.

EXAMPLE 26   Solve the system of equations.

$$\begin{cases} 2x + 6y = 12 \\ 1x + 4y = 7 \end{cases}$$

The two equations can be written in matrix form as follows:

$$\begin{bmatrix} 2 & 6 \\ 1 & 4 \end{bmatrix} \cdot \begin{bmatrix} x \\ y \end{bmatrix} = \begin{bmatrix} 12 \\ 7 \end{bmatrix}$$

Note that the first matrix on the left side of this equation is formed by the coefficients of unknowns x and y, the second matrix by the unknowns, and the matrix on the right side by the constants. Let A be the coefficients matrix, or:

$$A = \begin{bmatrix} 2 & 6 \\ 1 & 4 \end{bmatrix}$$

Then:

$$A \cdot \begin{bmatrix} x \\ y \end{bmatrix} = \begin{bmatrix} 12 \\ 7 \end{bmatrix}$$

Multiply both sides by A^{-1}, the inverse of the square matrix A.

$$A^{-1} \cdot A \cdot \begin{bmatrix} x \\ y \end{bmatrix} = A^{-1} \cdot \begin{bmatrix} 12 \\ 7 \end{bmatrix}$$

We know that $A^{-1} \cdot A = I$ and $I \cdot \begin{bmatrix} x \\ y \end{bmatrix} = \begin{bmatrix} x \\ y \end{bmatrix}$.

Thus: $\begin{bmatrix} x \\ y \end{bmatrix} = A^{-1} \cdot \begin{bmatrix} 12 \\ 7 \end{bmatrix}$

Now, the problem of solving a system of linear equations becomes a problem of finding the product of the inverse of the coefficients matrix, A^{-1}, and the matrix of constants.

Since $|A| = \begin{vmatrix} 2 & 6 \\ 1 & 4 \end{vmatrix} = 8 - 6 = 2$, A has an inverse.

$$A^{-1} = \frac{1}{2}\begin{bmatrix} 4 & -6 \\ -1 & 2 \end{bmatrix}$$

Substitute the value of A^{-1} in the above derived equation as follows:

$$\begin{bmatrix} x \\ y \end{bmatrix} = A^{-1} \cdot \begin{bmatrix} 12 \\ 7 \end{bmatrix} = \frac{1}{2}\begin{bmatrix} 4 & -6 \\ -1 & 2 \end{bmatrix} \cdot \begin{bmatrix} 12 \\ 7 \end{bmatrix} = \frac{1}{2}\begin{bmatrix} 48 - 42 \\ -12 + 14 \end{bmatrix} = \begin{bmatrix} 3 \\ 1 \end{bmatrix}$$

Or:

$$\begin{bmatrix} x \\ y \end{bmatrix} = \begin{bmatrix} 3 \\ 1 \end{bmatrix}$$

Therefore: $x = 3$
$ y = 1$

Check: Substitute $x = 3$ and $y = 1$ in the given equations:

$$2x + 6y = 2(3) + 6(1) = 12$$
$$1x + 4y = 1(3) + 4(1) = \ 7$$

B. Cramer's Rule

Cramer's rule (named in honor of Gabriel Cramer of Geneva, 1704–1752) uses determinants in solving a system of linear equations.

Let:

$|A|$ = the determinant formed by the coefficients of unknowns in a system of linear equations.

Then, Cramer's rule is stated as follows:

A system of n linear equations in n unknowns has a single solution if and only if the determinant formed by the coefficients of the unknowns is not equal to zero, or $|A| \neq 0$. Each unknown is equal to the product of $\dfrac{1}{|A|}$ and the determinant derived from $|A|$ by replacing the column of coefficients of this unknown with the column of the constants.

The application of this rule is illustrated in Examples 27 and 28.

EXAMPLE 27   Solve the system of equations.

$$\begin{cases} 2x + 6y = 12 \\ 1x + 4y = \ 7 \end{cases}$$

Let $|A|$ = the determinant of the coefficients of the unknowns in the two equations, or:

$$|A| = \begin{vmatrix} 2 & 6 \\ 1 & 4 \end{vmatrix} = 8 - 6 = 2$$

The coefficients 2, 1 of the unknown x in $|A|$ are replaced by the constants 12, 7 in finding the value of x.

x coefficients are
replaced by

$$x = \frac{1}{|A|}\begin{vmatrix} 12 & 6 \\ 7 & 4 \end{vmatrix} = \frac{1}{2}(48 - 42) = 3$$

The coefficients 6, 4 of the unknown y in $|A|$ are replaced by the constants 12, 7 in finding the value of y.

y coefficients
are replaced by

$$y = \frac{1}{|A|}\begin{vmatrix} 2 & 12 \\ 1 & 7 \end{vmatrix} = \frac{1}{2}(14 - 12) = 1$$

EXAMPLE 28 Solve the system of three equations.

$$2x + 1y - 5z = -11$$
$$4x + 6y + 8z = 40$$
$$-7y + 3z = -5$$

Let $|A| =$ the determinant of the coefficients of the unknowns in the three equations.

Since there are three unknowns, we need three equations in order to have a single solution. When one of the three equations includes only two unknowns, we must change it to an equation with three unknowns by adding the third unknown with a zero coefficient. Thus, $-7y + 3z = -5$ is changed to $0x - 7y + 3z = -5$ in forming the determinant $|A|$, or:

$$|A| = \begin{vmatrix} 2 & 1 & -5 \\ 4 & 6 & 8 \\ 0 & -7 & 3 \end{vmatrix} = 276$$

The coefficients 2, 4, 0 of x in $|A|$ are replaced by the constants $-11, 40, -5$ as follows:

x coefficients
are replaced by

$$x = \frac{1}{|A|}\begin{vmatrix} -11 & 1 & -5 \\ 40 & 6 & 8 \\ -5 & -7 & 3 \end{vmatrix} = \frac{1}{276}\left(-11\begin{vmatrix} 6 & 8 \\ -7 & 3 \end{vmatrix} - 40\begin{vmatrix} 1 & -5 \\ -7 & 3 \end{vmatrix} - 5\begin{vmatrix} 1 & -5 \\ 6 & 8 \end{vmatrix}\right)$$

$$= \frac{1}{276}\left(-11(74) - 40(-32) - 5(38)\right) = \frac{1}{276}(276) = 1$$

The coefficients 1, 6, -7 of y in $|A|$ are replaced by the constants:

y coefficients
are replaced by

$$y = \frac{1}{|A|}\begin{vmatrix} 2 & -11 & -5 \\ 4 & 40 & 8 \\ 0 & -5 & 3 \end{vmatrix} = \frac{1}{276}\left(2\begin{vmatrix} 40 & 8 \\ -5 & 3 \end{vmatrix} - 4\begin{vmatrix} -11 & -5 \\ -5 & 3 \end{vmatrix} + 0\right)$$

$$= \frac{1}{276}\left(2(160) - 4(-33 - 25)\right) = \frac{1}{276}(552) = 2$$

The coefficients $-5, 8, 3$ of z in $|A|$ are replaced by the constants:

$$z = \frac{1}{|A|}\begin{vmatrix} 2 & 1 & -11 \\ 4 & 6 & 40 \\ 0 & -7 & -5 \end{vmatrix} = \frac{1}{276}\left(2\begin{vmatrix} 6 & 40 \\ -7 & -5 \end{vmatrix} - 4\begin{vmatrix} 1 & -11 \\ -7 & -5 \end{vmatrix} + 0\right)$$

$$= \frac{1}{276}\left(2(250) - 4(-82)\right) = \frac{1}{276}(828) = 3$$

Check: Substitute $x = 1$, $y = 2$, $z = 3$ in the given equations:

$$2x + 1y - 5z = 2(1) + 2 - 5(3) = -11,$$
$$4x + 6y + 8z = 4(1) + 6(2) + 8(3) = 40,$$
$$-7y + 3z = -7(2) + 3(3) = -5$$

Notes to Cramer's rule:

1. If $|A| = 0$, $\dfrac{1}{|A|} = \dfrac{1}{0}$, which is undefined. Thus, there will be no single solution for a system of linear equations. In other words, the system either has an infinitely large number of solutions, such as for the dependent equations:

$$\begin{cases} x + y = 3 \\ 2x + 2y = 6, \end{cases} \quad |A| = \begin{vmatrix} 1 & 1 \\ 2 & 2 \end{vmatrix} = 0$$

or has no solution, such as for the inconsistent equations:

$$\begin{cases} x + y = 4 \\ x + y = 5, \end{cases} \quad |A| = \begin{vmatrix} 1 & 1 \\ 1 & 1 \end{vmatrix} = 0$$

2. The symbol $|A|$ used in Cramer's rule is sometimes written as Δ (delta).

C. The Gaussian Method

The Gaussian method can be performed two ways in solving a system of linear equations. The two ways are presented individually.

I. *Convert coefficients matrix A to an identity matrix.*

STEP 1 Write a tableau formed by the coefficients matrix A (placed on the left side of a vertical line) and the constants matrix C (placed on the right side of the vertical line). The form of the tableau therefore is:

$$[A|C] = \begin{bmatrix} 2 & 6 & | & 12 \\ 1 & 4 & | & 17 \end{bmatrix}$$

if we use the equations in Example 26.

STEP 2 Perform elementary row operations on both matrices in the tableau until matrix A becomes an identity matrix I. The matrix C is now replaced by the solution matrix X (in the form of a column vector, its first component is the value of x, the second is y, and so on.) Thus, we are transforming the tableau:

$[A|C]$ into a new tableau $[I|X]$

in solving a system of linear equations by the Gaussian method.

EXAMPLE 29   Solve the system of equations. (Same as Example 26.)

$$2x + 6y = 12$$
$$1x + 4y = 7$$

STEP 1   Write the initial tableau:

$$[A|C] = \begin{bmatrix} 2 & 6 & | & 12 \\ 1 & 4 & | & 7 \end{bmatrix} \begin{matrix} \dots (1) \\ \dots (2) \end{matrix}$$

STEP 2   Perform the appropriate row operations.

Convert the elements in column 1 of A from $\begin{smallmatrix} 2 \\ 1 \end{smallmatrix}$ to $\begin{smallmatrix} 1 \\ 0 \end{smallmatrix}$. (First compute $R(1)'$ to obtain 1 in column 1.)

$$\begin{matrix} R(1) \div 2 \\ R(2) - R(1)' \end{matrix} \quad \begin{bmatrix} 1 & 3 & | & 6 \\ 0 & 1 & | & 1 \end{bmatrix} \begin{matrix} \dots (1)' \\ \dots (2)' \end{matrix}$$

Convert the elements in column 2 of the second tableau above from $\begin{smallmatrix} 3 \\ 1 \end{smallmatrix}$ to $\begin{smallmatrix} 0 \\ 1 \end{smallmatrix}$ (First compute $R(2)'$ to obtain 1 in column 2.):

$$\begin{matrix} R(1)' - R(2)' \times 3 \\ R(2)' \text{ (same)} \end{matrix} \quad \begin{bmatrix} 1 & 0 & | & 3 \\ 0 & 1 & | & 1 \end{bmatrix} \begin{matrix} \dots (1)'' \\ \dots (2)' \end{matrix}$$

The final tableau is:

$$[I|X] = \begin{bmatrix} 1 & 0 & | & 3 \\ 0 & 1 & | & 1 \end{bmatrix} \qquad X = \begin{bmatrix} x \\ y \end{bmatrix} = \begin{bmatrix} 3 \\ 1 \end{bmatrix}$$

Thus: $x = 3$, and $y = 1$

EXAMPLE 30   Solve the system of three equations. (Same as Example 28.)

$$2x + 1y - 5z = -11$$
$$4x + 6y + 8z = 40$$
$$-7y + 3z = -5$$

STEP 1   Write the initial tableau.

$$[A|C] = \begin{bmatrix} 2 & 1 & -5 & | & -11 \\ 4 & 6 & 8 & | & 40 \\ 0 & -7 & 3 & | & -5 \end{bmatrix} \begin{matrix} \dots (1) \\ \dots (2) \\ \dots (3) \end{matrix}$$

STEP 2   Perform the appropriate row operations.

Convert the elements in column 1 of A from $\begin{smallmatrix} 2 \\ 4 \\ 0 \end{smallmatrix}$ to $\begin{smallmatrix} 1 \\ 0 \\ 0 \end{smallmatrix}$. (First compute $R(1)'$ to obtain 1 in column 1.)

$$
\begin{array}{l}
R(1) \div 2 \\
R(2) - R(1)' \times 4 \\
R(3), \text{ same}
\end{array}
\quad
\left[
\begin{array}{ccc|c}
1 & \frac{1}{2} & -\frac{5}{2} & -\frac{11}{2} \\
0 & 4 & 18 & 62 \\
0 & -7 & 3 & -5
\end{array}
\right]
\begin{array}{l}
\ldots \ldots (1)' \\
\ldots \ldots (2)' \\
\ldots \ldots (3)
\end{array}
$$

Convert the elements in column 2 of the second tableau from 4 to 1. $\begin{array}{cc} \frac{1}{2} & 0 \\ -7 & 0 \end{array}$

(First compute $R(2)''$ to obtain 1 in column 2.)

$$
\begin{array}{l}
R(1)' - R(2)'' \times \frac{1}{2} \\
R(2)' \div 4 \\
R(3) \ - R(2)''(-7)
\end{array}
\quad
\left[
\begin{array}{ccc|c}
1 & 0 & -\frac{19}{4} & -\frac{53}{4} \\
0 & 1 & \frac{9}{2} & \frac{31}{2} \\
0 & 0 & \frac{69}{2} & \frac{207}{2}
\end{array}
\right]
\begin{array}{l}
\ldots \ldots (1)'' \\
\ldots \ldots (2)'' \\
\ldots \ldots (3)'
\end{array}
$$

Convert the elements in column 3 of the third tableau from $\begin{array}{cc} -\frac{19}{4} & 0 \\ \frac{9}{2} & 0 \\ \frac{69}{2} & 1 \end{array}$ to 0.

(First compute $R(3)''$ to obtain 1 in column 3.)

$$
\begin{array}{l}
R(1)'' - R(3)''(-\frac{19}{4}) \\
R(2)'' - R(3)''(\frac{9}{2}) \\
R(3)' \div \frac{69}{2}
\end{array}
\quad
\left[
\begin{array}{ccc|c}
1 & 0 & 0 & 1 \\
0 & 1 & 0 & 2 \\
0 & 0 & 1 & 3
\end{array}
\right]
\begin{array}{l}
\ldots \ldots (1)''' \\
\ldots \ldots (2)''' \\
\ldots \ldots (3)''
\end{array}
$$

The final tableau is:

$$
X = \begin{bmatrix} x \\ y \\ z \end{bmatrix} = \begin{bmatrix} 1 \\ 2 \\ 3 \end{bmatrix}
$$

Thus, $x = 1$, $y = 2$, and $z = 3$.

II. *Convert coefficients matrix A to an upper triangular matrix. A triangular matrix* is a square matrix that has all zero elements on one side of the principal diagonal. If all zeros are on the lower left side of the principal diagonal, it is an *upper triangular matrix.* Conversely, a square matrix with all zero elements on the upper right side of the principal diagonal is called a *lower triangular matrix.*

STEP 1   Write a tableau formed by the coefficients matrix A and the constants matrix C.

STEP 2   Perform the appropriate elementary row operations to convert the elements in matrix A to 1's on the principal diagonal and to 0's on the lower left side of the diagonal, or in a form of an upper triangular matrix. For a system of three equations, the triangular matrix form will be:

$$
\begin{bmatrix}
1 & a_{12} & a_{13} \\
0 & 1 & a_{23} \\
0 & 0 & 1
\end{bmatrix}
$$

STEP 3   Write the new system of equations based on coefficients obtained from the final tableau — the tableau with an upper triangular matrix. The new equations are now written in *echelon form* and are solved by the method of substitution.

EXAMPLE 31 Solve the system of equations. (Same as Example 29.)

$$2x + 6y = 12$$
$$1x + 4y = 7$$

STEP 1 Write the initial tableau.

$$[A|C] = \begin{bmatrix} 2 & 6 & 12 \\ 1 & 4 & 7 \end{bmatrix} \begin{matrix} \ldots \ldots (1) \\ \ldots \ldots (2) \end{matrix}$$

STEP 2 Perform the appropriate row operations.

Convert the elements in column 1 of A from $\begin{smallmatrix}2\\1\end{smallmatrix}$ to $\begin{smallmatrix}1\\0\end{smallmatrix}$. (First compute $R(1)'$ to obtain 1 in column 1.)

$$\begin{matrix} R(1) \div 2 \\ R(2) - R(1)' \end{matrix} \quad \begin{bmatrix} 1 & 3 & 6 \\ 0 & 1 & 1 \end{bmatrix} \begin{matrix} \ldots \ldots (1)' \\ \ldots \ldots (2)' \end{matrix}$$

Convert the second element in column 2 of the second tableau to 1. The second element in column 2 is already 1. Thus, row (2)' should not be changed. The second tableau is the final tableau.

STEP 3 Write the new system of equations in echelon form based on the co-efficients obtained from the final tableau.

$$1x + 3y = 6 \ldots \ldots (1)'$$
$$1y = 1 \ldots \ldots (2)'$$

Substitute $y = 1$ in (1)'

$$x + 3(1) = 6,$$
$$x = 6 - 3 = 3$$

EXAMPLE 32 Solve the system of three equations given in Example 30.

$$2x + 1y - 5z = -11$$
$$4x + 6y + 8z = 40$$
$$-7y + 3z = -5$$

STEP 1 Write the initial tableau.

$$[A|C] = \begin{bmatrix} 2 & 1 & -5 & -11 \\ 4 & 6 & 8 & 40 \\ 0 & -7 & 3 & -5 \end{bmatrix} \begin{matrix} \ldots \ldots (1) \\ \ldots \ldots (2) \\ \ldots \ldots (3) \end{matrix}$$

STEP 2 Perform the appropriate row operations.

Convert the elements in column 1 of A from $\begin{smallmatrix}2\\4\\0\end{smallmatrix}$ to $\begin{smallmatrix}1\\0\\0\end{smallmatrix}$.

$$\begin{bmatrix} 1 & \frac{1}{2} & -\frac{5}{2} & -\frac{11}{2} \\ 0 & 4 & 18 & 62 \\ 0 & -7 & 3 & -5 \end{bmatrix} \begin{matrix} \ldots \ldots (1)' \\ \ldots \ldots (2)' \\ \ldots \ldots (3) \end{matrix}$$

Convert the second and third elements in column 2 of the second tableau from $\begin{smallmatrix}4\\-7\end{smallmatrix}$ to $\begin{smallmatrix}1\\0\end{smallmatrix}$. (First compute $R(2)''$ to obtain 1 in column 2.)

$R(1)'$ same
$R(2)' \div 4$
$R(3) - R(2)''(-7)$
$$\begin{bmatrix} 1 & \frac{1}{2} & -\frac{5}{2} & -\frac{11}{2} \\ 0 & 1 & \frac{9}{2} & \frac{31}{2} \\ 0 & 0 & \frac{69}{2} & \frac{207}{2} \end{bmatrix} \begin{matrix} \dots (1)' \\ \dots (2)'' \\ \dots (3)' \end{matrix}$$

Convert the third element in column 3 of the third tableau from $\frac{69}{2}$ to 1.

$R(1)'$ same
$R(2)''$ same
$R(3)' \div \frac{69}{2}$
$$\begin{bmatrix} 1 & \frac{1}{2} & -\frac{5}{2} & -\frac{11}{2} \\ 0 & 1 & \frac{9}{2} & \frac{31}{2} \\ 0 & 0 & 1 & 3 \end{bmatrix} \begin{matrix} \dots (1)' \\ \dots (2)'' \\ \dots (3)'' \end{matrix}$$

The fourth tableau is the final tableau since the tableau has all 1's on the principal diagonal.

STEP 3   Write the new system of equations in echelon form based on the co-efficients obtained from the final tableau.

$$1x + \frac{1}{2}y - \frac{5}{2}z = -\frac{11}{2} \dots (1)'$$

$$1y + \frac{9}{2}z = \frac{31}{2} \dots (2)''$$

$$z = 3 \dots (3)''$$

Substitute $z = 3$ in (2)''.

$$y + \frac{9}{2}(3) = \frac{31}{2}$$

$$y = 2$$

Substitute $y = 2$ and $z = 3$ in (1)'.

$$x + \frac{1}{2}(2) - \frac{5}{2}(3) = -\frac{11}{2}$$

$$x = 1$$

EXERCISE 7–4   Reference: Section 7.4

A. *Solve each of the following systems of equations by using (a) the inverse of a square matrix, (b) Cramer's rule, and (c) the Gaussian method:*

1. $3x + y = 11$
   $4x - y = 3$

2. $2x - 3y = 24$
   $x + 7y = -22$

3. $x + 3y = 18$
   $5x - 2y = -29$

4. $2x + y = -6$
   $3x + 4y = 16$

5. $6x - 2y = 46$
   $x + 5y = -3$

6. $4x + 6y = 74$
   $8x - 3y = 13$

B. *Solve each of the following systems of equations by using (a) Cramer's rule, and (b) the Gaussian method:*

7. $2x + y + z = 3$
   $x + 3y + 2z = 1$
   $4x - 5y - 6z = -4$

8. $x + y - z = 6$
   $x - y + 2z = -5$
   $3x + 2y - 4z = 20$

9. $2x + 3y + z = 3$
   $x - 2y - z = 7$
   $3x + 2z = 14$

10. $3x - 2y + 4z = 5$
    $2x + y = -2$
    $x + 2y - 5z = -11$

11. $2x + y - 4z = 12$
    $3x - 2y + 2z = 3$
    $x + y + z = 4$

12. $x + 2y + z = 1$
    $2x - y + 2z = 7$
    $4x + 3y - z = -16$

13. $2x + 3y = 5$
    $3x - 2z = -8$
    $4y - 5z = 7$

14. $3x + 2y - z = 7$
    $2x - y + z = 0$
    $x + y - 2z = 3$

Chapter

8 Linear Programming

Linear programming is a mathematical technique for finding the optimum solution (or best solution) to a given problem from a set of feasible solutions. The technique can be used to solve various complicated business problems, such as maximizing profits and minimizing costs. This chapter will introduce the concept of linear programming. We shall open the discussion with inequalities since they are important to the illustrations of linear programming problems.

8.1 INEQUALITIES

An *inequality* is a statement which indicates that one algebraic expression is greater than ($>$) or less than ($<$) another expression. Thus, the statement, "x is greater than 5," or written symbolically, "$x > 5$," and the statement, "y is less than 7," or written, "$y < 7$," are two inequalities. The three rules for inequality operations are:

1. If the same number is added to or subtracted from both sides of an inequality, the same inequality sign is used for the obtained inequality. Thus let:

    $7 > 2$. Then, $7 + 3 > 2 + 3$, or $10 > 5$. Also, $4 < 9$. Then, $4 - 1 < 9 - 1$, or $3 < 8$.

2. If both sides of an inequality are multiplied or divided by the same positive number, the same inequality sign is used for the obtained inequality. Thus:

    $10 > 6$. Then, $10 \times 2 > 6 \times 2$, or $20 > 12$, and $10 \div 2 > 6 \div 2$, or $5 > 3$.

3. If both sides of an inequality are multiplied or divided by the same negative number, the reversed inequality sign is used for the obtained inequality. Thus:

    $10 > 6$. Then, $10 \times (-2) < 6 \times (-2)$, or $-20 < -12$, and $10 \div (-2) < 6 \div (-2)$, or $-5 < -3$.

The equality sign ($=$) and inequality sign ($>$ or $<$) may be written together. The sign $\geq$ is used to represent "is equal to or greater than" and the sign $\leq$ is used to represent "is equal to or less than." Thus, $x \geq 12$ means that x is greater than or equal to 12, and $y \leq 25$ means that y is less than or equal to 25.

Note that the equal (=) part of a combined sign indicates the limit of the inequality, such as 12 being the lower limit of x in the inequality $x \geq 12$ and 25 being the upper limit of y in the inequality $y \leq 25$.

Inequalities may be shown graphically as in Figure 8–1. The graph is drawn according to the system of rectangular coordinates. Examples 1 and 2 are used to illustrate the graphical method.

EXAMPLE 1　Graph each of the following statements:

(a) $X \geq 0$　　(b) $Y \geq 0$　　(c) $X \leq -2$　　(d) $Y \leq 3$

The four graphs are shown in Figure 8–1. Observe:

(a) The graph of $X = 0$ is the set of points on the Y-axis and the graph of $X > 0$ is the set of points on the right side of the Y-axis. The graph representing $X \geq 0$, or the combination of $X = 0$ and $X > 0$, is shown by the shaded area on graph (a) of the figure. Check: Point K (in the shaded area) has $X = 2$, which meets the restriction stated in $X > 0$.

(b) The graph of $Y = 0$ is the set of points on the X-axis and the graph of $Y > 0$ is the set of points above the X-axis. The graph representing $Y \geq 0$, or the combination of $Y = 0$ and $Y > 0$, is shown by the shaded area on graph (b) in the figure. Check: Point K has $Y = 3$, which meets the restriction stated in $Y > 0$.

(c) The graph of $X = -2$ is the set of points on the vertical line intersecting X-axis at -2 and the graph of $X < -2$ is the set of points on the left side of the vertical line. The graph representing $X \leq -2$ is shown by the shaded area on graph (c). Check: Point K has $X = -3$ in the shaded area.

(d) The graph of $Y = 3$ is the set of points on the horizontal line intersecting Y-axis at 3 and the graph of $Y < 3$ is the set of points below the line. The graph representing $Y \leq 3$ is shown by the shaded area on graph (d) of the figure. Check: Point K has $Y = 2$ in the shaded area.

Figure 8-1　(Example 1)

(a) $X \geqslant 0$

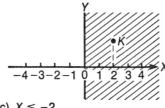

(b) $Y \geqslant 0$

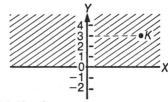

(c) $X \leqslant -2$

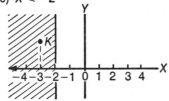

(d) $Y \leqslant 3$

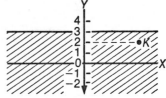

EXAMPLE 2   Graph each of the following statements:

(a) $2X + 6Y \leq 12$
(b) $1X + 4Y \leq 7$
(c) $1X + 4Y \geq 7$

The three graphs are shown in Figure 8–2.

Figure 8-2   (Example 2)

(a) $2X + 6Y \leq 12$ (b) $1X + 4Y \leq 7$
                                         (c) $1X + 4Y \geq 7$

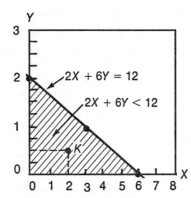   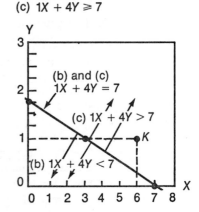

(a) The graph of equation $2X + 6Y = 12$ is first determined by the points representing the following arbitrarily selected three pairs of X and Y values:

X	Y
0	2
6	0
3	1

The three points are on a straight line as shown on graph (a) in the figure. The two points representing $(X = 0, Y = 2)$ on the Y axis and $(X = 6, Y = 0)$ on the X axis are called *terminal points*. The two terminal points will determine the location of the straight line. However, it is always safer to use a third point to check the location of the line.

Next, the graph of the inequality $2X + 6Y < 12$ is determined. The area of the inequality is the set of the points below the straight line as indicated by the shaded area.

Check: Point K in graph (a) has $X = 2$ and $Y = \frac{1}{2}$.
        $2X + 6Y = 2(2) + 6(\frac{1}{2}) = 7$, which is less than 12.

(b) The graph of equation $1X + 4Y = 7$ is determined by the points representing the following three pairs of X and Y values:

X	Y
0	$\frac{7}{4} = 1\frac{3}{4}$
7	0
3	1

The three points are on a straight line as shown on graph (b). The two terminal points represent $(X = 0, Y = \frac{7}{4})$ and $(X = 7, Y = 0)$. The graph of the inequality $1X + 4Y < 7$ is the set of the points below the straight line as indicated by the arrows on the graph.

(c) The graph of equation $1X + 4Y = 7$ is the same as in (b). The graph of the inequality $1X + 4Y > 7$ is the set of points above the straight line as indicated by the arrows on graph (c).

Check: Point K in graph (c) has $X = 6$ and $Y = 1$.
$1X + 4Y = 1(6) + 4(1) = 10$, which is larger than 7.

8.2 LINEAR PROGRAMMING—THE GRAPHIC METHOD FOR MAXIMIZATION PROBLEMS

The basic steps for solving a problem by the linear programming technique are as follows:

1. Derive a group of linear equations and inequalities based on the restraining conditions given by the problem.
2. Solve the group of linear equations and inequalities for an optimum solution based on the objective function.

There are various linear programming methods for solving a group of linear equations and inequalities. We shall first illustrate the graphic method in order to introduce the basic concept of linear programming techniques. This method is first applied to *maximization* problems and then to *minimization* problems.

Note that the examples in this section include only two variables. We usually have greater difficulty in drawing more than two dimensions on a graph. The graphic method thus is not effective in solving a linear programming problem involving three or more variables.

The variables in the following discussion are represented by the letter X with subscripts, such as X_1, X_2, X_3, and so on. Thus, if a linear programming problem involves only two variables, X_1 (instead of X) and X_2 (instead of Y) will be used to represent the two variables. All X variables are of the first power and the equations formed by the X variables thus are linear equations.

The maximization problems illustrated in Examples 3 through 6 concern the subject of "allocation of resources." The objective is to maximize profits which can be generated with the limited resources. The available resources are machine hours and skilled labor hours. Based on the limited resources, the profit is maximized. The optimum solution—the maximized profit—is found by the graphic method.

EXAMPLE 3 A furniture factory is planning to produce tables and chairs. Each table requires 2 hours of machine time and 1 hour of skilled labor. Each chair requires 6 hours of machine and 4 hours of skilled labor. The machine has a maximum availability of 12 hours. The skilled labor has a maximum availability of 7 hours. The profit of each table is $3 and the profit of each chair is $5. Find the best combination of tables and chairs that should be produced in order to maximize the profit.

Let $X_1 =$ the optimum number of tables to be produced and
  $X_2 =$ the optimum number of chairs to be produced.

The given information is summarized in the following table:

Type of Resources	Tables (X_1)	Chairs (X_2)	Maximum Hours Available During the Period
Machine hours	2 hours each	6 hours each	12 hours
Labor hours	1 hour each	4 hours each	7 hours
Profit	$3 each	$5 each	

The objective of our problem, called the *objective function,* is to maximize the profit.

$F = \$3X_1 + \$5X_2$　(*F* denotes profit: X_1 tables at $3 each and X_2 chairs at $5 each.)

The inequalities based on the restraining conditions, called *constraints* (or restraints) are:

Constraint on the machine:

(1) $2X_1 + 6X_2 \leq 12$　(Each table requires 2 hours on the machine and each chair requires 6 hours on the machine. The total number of hours on the machine should be equal to or less than 12 hours.)

Constraint on the skilled labor:

(2) $1X_1 + 4X_2 \leq 7$　(Each table requires 1 hour skilled labor and each chair requires 4 hours skilled labor. The total number of hours should be equal to or less than 7 hours.)

Nonnegative constraints on X_1 and X_2:

(3) $X_1 \geq 0$　(The values of X_1 and X_2 must be positive. We cannot
(4) $X_2 \geq 0$　produce a negative number of tables, X_1, or chairs, X_2.)

The four equations of these constraints are represented respectively by the four straight lines in Figure 8–3: (1) *AB*, (2) *CD*, (3) X_2-axis, and (4) X_1-axis. Lines *AB* and *CD* may be determined by finding the two terminal points for each equation represented. (See Examples 1 and 2.) The four lines form a four-sided polygon, *OCEB* (shaded area) which represents the four inequalities. Any point within the polygon will satisfy the four constraints and thus is a feasible solution. The polygon thus is a *feasible solution area.*

However, our problem is to maximize the objective function, or to find a point in the solution area that will give the highest profit to our products X_1 and X_2. This point must be located on a corner point of the polygon. There are four such corner points on the polygon: *O, C, E,* and *B.* The profits based on the four corner points are computed from the objective function (the profit equation) as follows:

Objective function:　$F = \$3X_1 + \$5X_2$
Point $O - X_1 = 0,\ X_2 = 0$:　$F = 3(0) + 5(0) = \$0$
Point $C - X_1 = 0,\ X_2 = 1\frac{3}{4}$:　$F = 3(0) + 5(1\frac{3}{4}) = \$8\frac{3}{4}$

Point $E - X_1 = 3$, $X_2 = 1$: $F = 3(3) + 5(1) = \$14$
Point $B - X_1 = 6$, $X_2 = 0$: $F = 3(6) + 5(0) = \$18$ (highest)

Figure 8-3   (Example 3)

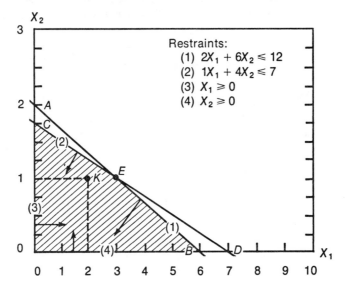

Point B is the optimum solution since it gives the highest profit ($\$18$). It is the best solution from all possible solutions indicated by the points in the shaded area. Thus, the factory should produce 6 tables and no chairs by using the available hours of machine and labor.

Check: The four constraints are satisfied by the optimum solution $X_1 = 6$ and $X_2 = 0$.

(1) $2X_1 + 6X_2 = 2(6) + 6(0) = 12$ (hours available on the machine)
(2) $1X_1 + 4X_2 = 1(6) + 4(0) = 6$ (hours which is less than 7, the available skilled labor hours)
(3) $X_1 = 6$, which is greater than 0.
(4) $X_2 = 0$, which is equal to 0.

Also, select any point in the shaded area, such as point K which has $X_1 = 2$ and $X_2 = 1$. Substitute the X_1 and X_2 values in the objective function: Profit $= F = \$3(2) + \$5(1) = \$11$, which is smaller than the profit based on the optimum solution.

Note that we are interested in theoretical rather than practical value for illustration purposes. The values of X_1 and X_2 must not be negative but can be fractions. For example, point C indicates the production of $1\frac{3}{4}$ chairs. The fractional value, $\frac{3}{4}$, is a portion of the theoretical value. In practice, the factory will not produce a fraction of a chair. However, if the computation unit represents 1,000 chairs, the fractional portion is a practical value since the factory can produce 750 ($= 1,000 \times \frac{3}{4}$) chairs.

EXAMPLE 4   Refer to Example 3. Find the answer if the profit of each table is $3 and the profit of each chair is $10.

The objective function is to maximize the profit: $F = \$3X_1 + \$10X_2$. The profits based on the four extreme points obtained in Example 3 are computed from the objective function which follows:

$$\text{Objective function:} \quad F = \$3X_1 + \$10X_2$$

Point $O - X_1 = 0, X_2 = 0$: $F = 3(0) + 10(0) = \$0$
Point $C - X_1 = 0, X_2 = 1\frac{3}{4}$: $F = 3(0) + 10(1\frac{3}{4}) = \$17\frac{1}{2}$
Point $E - X_1 = 3, X_2 = 1$: $F = 3(3) + 10(1) = \$19$ (highest)
Point $B - X_1 = 6, X_2 = 0$: $F = 3(6) + 10(0) = \$18$

Thus, the factory should produce 3 tables and 1 chair by using the available hours of machine and skilled labor.

Sometimes, a linear programming problem may have more than one optimum solution. Example 5 illustrates this case.

EXAMPLE 5   Again refer to Example 3. Find the answer if the profit for each table is $3 and the profit for each chair is $9.

The objective function now is to maximize the profit: $F = \$3X_1 + \$9X_2$. The profits based on the four extreme points obtained in Example 3 are computed from the objective function which follows:

$$\text{Objective function:} \quad F = \$3X_1 + \$9X_2$$

Point $O - X_1 = 0, X_2 = 0$: $F = 3(0) + 9(0) = \$0$
Point $C - X_1 = 0, X_2 = 1\frac{3}{4}$: $F = 3(0) + 9(1\frac{3}{4}) = \$15\frac{3}{4}$
Point $E - X_1 = 3, X_2 = 1$: $F = 3(3) + 9(1) = \$18$ ⎱ (highest)
Point $B - X_1 = 6, X_2 = 0$: $F = 3(6) + 9(0) = \$18$ ⎰

Thus, the factory may produce either 3 tables and 1 chair (indicated by point E) or 6 tables and no chairs (indicated by point B) to realize the highest profit of $18.

Examples 3, 4, and 5 indicated that the profit equation (the objective function) determined the value of the optimum solution in each case. This fact can also be explained graphically. The profit equation is a linear equation and thus can be represented by a straight line on a graph, when the value of the profit is given. The location of the line changes as the value of the profit (F) changes. However, the lines representing various values of profit based on the same profit equation are parallel. One of the profit lines touches a corner, or sometimes a side of the polygon, at the furthermost distance from the 0 point on the graph. This corner point, or sometimes points, is the optimum solution.

EXAMPLE 6   Refer to Example 3 and Figure 8-3. Plot the objective function (the profit equation) lines on the figure to find the maximized profit.

1. Solve for X_2 from the objective function, $F = 3X_1 + 5X_2$.

$$X_2 = \frac{F - 3X_1}{5} = \frac{F}{5} - \frac{3}{5}X_1$$

The value of the first term $\left(\dfrac{F}{5}\right)$ is called X_2 *intercept* (or the height of the ordinate from the origin (0 point) to the point of intersection of the straight line representing the equation on the X_2-axis.) It equals X_2 when $X_1 = 0$.

The coefficient of X_1 in the second term $\left(-\dfrac{3}{5}\right)$ is the slope of the straight line; it also represents the average amount of change in X_2 variable per unit of X_1 variable.

2. Draw profit lines based on the equation obtained in the first step. We may start a line by letting the profit equal zero, or $F = \$0$. Thus:

$$X_2 = \frac{F}{5} - \frac{3}{5}X_1 = \frac{0}{5} - \frac{3}{5}X_1$$

Or:

$$X_2 = -\frac{3}{5}X_1$$

The straight line of the preceding equation is determined by the points representing the following three pairs of X_1 and X_2 values:

X_1	X_2
0	0
2	$-1\frac{1}{5}$
-2	$1\frac{1}{5}$

The $F = \$0$ line is labeled in Figure 8–4 as Line I.

We then draw more lines (such as Lines II, III, and IV in Figure 8–4) parallel to Line I, moving to upper right direction but within the solution area (polygon $OCEB$). Line II goes through the point $X_1 = 2$ and $X_2 = 0$. Substitute the values in the profit equation:

$F = 3X_1 + 5X_2 = 3(2) + 5(0) = \$6$

Line III goes through the point, $X_1 = 4$ and $X_2 = 0$. Thus:

$F = 3X_1 + 5X_2 = 3(4) + 5(0) = \$12$

Line IV goes through point $B: X_1 = 6$, and $X_2 = 0$. The profit on point B is:

$F = 3X_1 + 5X_2 = 3(6) + 5(0) = \$18$

In fact, any point on Line IV represents $\$18$ profit, such as point Q on the line, $X_1 = 1$, $X_2 = 3$.

$F = 3X_1 + 5X_2 = 3(1) + 5(3) = \$18$

The four lines indicate that as the lines move further away from Line I, the profit increases. Line IV gives the highest profit, $\$18$. However,

Figure 8-4   **(Example 6)**

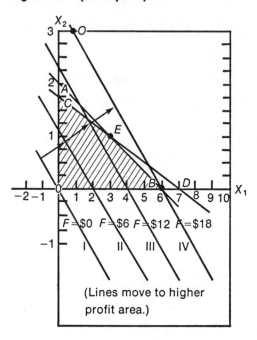

(Lines move to higher profit area.)

only point B (one of the corner points of the solution area) on the line meets the four constraints given in Example 3. Thus, $X_1 = 6$ (tables) and $X_2 = 0$ (chairs), as indicated by point B, are the optimum solution. Point Q does not meet constraints (1) and (2); therefore, it is not a solution point.

Note that when two points give the same maximized profit, the line formed by the two points will coincide with the maximized profit line. For instance, points E and B in Example 5 give the same maximized profit, $18. Any point on line EB thus indicates a combination of the numbers of tables and chairs that may be produced by the factory to obtain the maximized profit of $18 under the conditions given in the example.

8.3 LINEAR PROGRAMMING—THE GRAPHIC METHOD FOR MINIMIZATION PROBLEMS

The minimization problems illustrated in Examples 7 and 8 concern the "blending problem." Restricted materials are blended in order to obtain a new mixture. The objective function in each problem is to minimize the cost of the materials.

EXAMPLE 7   A chemical company plans to make a certain type of mixed powder. The mixture requires the use of material M at a cost of $8 per bag and material N at a cost of $5 per bag. Each bag of material M weighs 4 pounds and each bag of material N weighs 2 pounds. At least 100 pounds of the powder are needed and at least 20 bags of material N must be used in the mixture. How many bags of each material should be used in order to minimize the total cost of the mixed powder? Use the graphic method.

Let X_1 = the number of bags of material M, and
$\quad X_2$ = The number of bags of material N that should be used in the mixture.

The objective function is to minimize the cost.

$F = \$8X_1 + \$5X_2$ (F denotes cost: X_1 bags of material M at \$8 each and X_2 bags of material N at \$5 each.)

Subject to constraints:

(1) $4X_1 + 2X_2 \geq 100$ (pounds — each bag of M weighs 4 pounds, each bag of N weighs 2 pounds, and need at least 100 pounds of mixed powder.)

(2) $\qquad X_2 \geq 20$ (bags — at least 20 bags of N must be used in the mixture.)

(3) $\qquad X_1 \geq 0$ (The values of X_1 and X_2 must be nonnegative.)

(4) $\qquad X_2 \geq 0$

The four equations of constraints 1–4 are represented respectively by four straight lines in Figure 8–5: (1) AB, (2) CD, (3) X_2-axis, and (4) X_1-axis. Lines AB and CD may be determined by finding the terminal points from each equation represented as follows:

Figure 8-5   **(Example 7)**

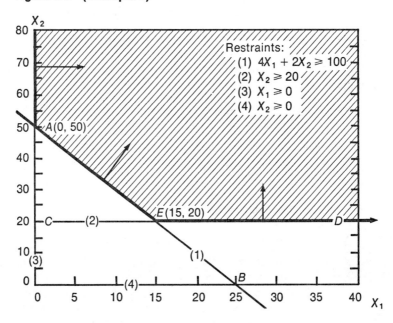

Line *AB*: $4X_1 + 2X_2 = 100$

X_1	X_2
0	50
25	0

Line *CD*: $X_2 = 20$ (constant) $X_1 =$ any value

Constraint (4) is redundant to constraint (2). If the value of X_2 must be greater than or equal to 20, the value must be greater than 0 also. It is not necessary to enter the redundant constraint (4) in the process of finding the solution. The three lines, X_2-axis, *AB*, and *CD*, thus form the feasible solution area (shaded). This area extends indefinitely upper right. There is no finite point in the unbounded area that could make the cost *F* maximum. However, the solution area gives two corner points: *A* and *E*. The costs based on the two extreme points in the solution area are computed from the objective function.

Objective function: $F = \$8X_1 + \$5X_2$
Point $A - X_1 = 0$, $X_2 = 50$: $F = 8(0) + 5(50) = \$250$.
Point $E - X_1 = 15$, $X_2 = 20$: $F = 8(15) + 5(20) = \$220$.

Point *E* is the optimum solution since it gives the lowest cost ($220). The company thus should use 15 bags of material *M* and 20 bags of material *N* for the mixture.

Check: The four constraints are satisfied by the optimum solution, $X_1 = 15$ and $X_2 = 20$.

(1) $4X_1 + 2X_2 = 4(15) + 2(20) = 100$ (pounds of mixture)
(2) and (4) $X_2 = 20$ (bags)
(3) $X_1 = 15$ (bags which are greater than 0)

   Instead of trying every corner point in the solution area to find the minimized cost as shown in Example 7, the optimum solution may also be found by plotting straight lines representing the objective function equation on the graph. Example 8 is used to illustrate the procedure.

EXAMPLE 8   Refer to Example 7 and Figure 8–5. Plot the objective function lines on the figure to find the minimized cost.

1. Solve for X_2 from the objective function $F = 8X_1 + 5X_2$.

$$X_2 = \frac{F - 8X_1}{5} = \frac{F}{5} - \frac{8}{5}X_1$$

The first term $\left(\dfrac{F}{5}\right)$ is the X_2 intercept (which equals X_2 when $X_1 = 0$.)

The coefficient of X_1 in the second term $\left(-\dfrac{8}{5}\right)$ is the slope of the straight lines. Since the slope is constant, the lines representing various values of cost based on the same cost equation are parallel.

2. Draw cost lines based on the X_2 equation obtained in the first step. Since this is a minimization problem, we should start a line by letting the cost equal a higher value, such as cost $F = \$400$. Thus:

$$X_2 = \frac{F}{5} - \frac{8}{5}X_1 = \frac{400}{5} - \frac{8}{5}X_1, \text{ or:}$$

$$X_2 = 80 - \frac{8}{5}X_1$$

X_1	X_2
0	80
20	48

The $F = \$400$ line based on the two pairs of X_1 and X_2 values is labeled in Figure 8–6 as Line I.

Figure 8-6   (Example 8)

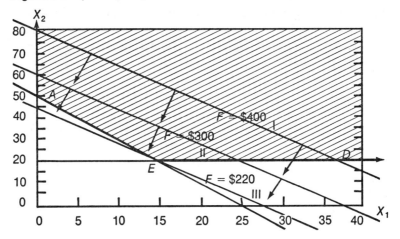

We then draw more lines, such as Lines II and III, parallel to Line I, moving to lower left direction but within the solution area. Line II goes through the point $X_1 = 0$ and $X_2 = 60$. Substitute the values in the cost equation:

$$F = 8X_1 + 5X_2 = 8(0) + 5(60) = \$300$$

Line III goes through point E: $X_1 = 15$ and $X_2 = 20$. The cost on point E is:

$$F = 8X_1 + 5X_2 = 8(15) + 5(20) = \$220$$

The three lines indicate that as the lines move further to the lower left from Line I, the cost decreases each time. Line III gives the lowest cost, $220. Any point on Line III represents $220 cost. However, only point E (one of the two corner points of the solution area) on the line meets the four constraints given in Example 7. Thus, $X_1 = 15$ (bags of material M) and $X_2 = 20$ (bags of material N), as indicated by point E, are the optimum solution.

Examples 7 and 8 used "$\geq$" signs for all constraints. It is also possible to use "$\leq$" and "$=$" signs for some constraints in a minimization problem. Example 9 is used to illustrate the use of different signs for the constraints.

EXAMPLE 9　Refer to Example 7. Assume that the constraints are changed to the following:

(1) $4X_1 + 2X_2 = 100$　(pounds – exactly 100 pounds of mixed powder are required)

(2) $X_1 \leq 12$　(bags of material M – no more than 12 bags of M are allowed.)

(3) $X_2 \geq 20$　(bags of material N – at least 20 bags of N must be used.)

(4) $X_1 \geq 0$ ⎫
(5) $X_2 \geq 0.$ ⎬　(The values of X_1 and X_2 must be nonnegative.)

Use the graphic method to minimize the total cost of the mixture.

The five constraints are plotted in Figure 8–7. The figure shows that line AB is the feasible solution area under the given constraints. The costs based on the two extreme points, A and B, on line AB, are computed from the objective function as follows:

Point $A - X_1 = 0, X_2 = 50$
$\quad F = 8X_1 + 5X_2 = 8(0) + 5(50) = \$250$
Point $B - X_1 = 12, X_2 = 26$
$\quad F = 8(12) + 5(26) = 96 + 130 = \$226$ (smallest cost)

Thus, $X_1 = 12$ (bags of material M) and $X_2 = 26$ (bags of material N), as indicated by point B in the figure, are the optimum solution. The line representing $F = \$226$ (cost) is also plotted in the figure to verify the optimum solution.

In some cases, a linear programming problem may not have a solution. Observe the AB line in Figure 8–7. Had the line fallen outside the area formed by constraints (2), (3), and (4), there would be no feasible solution area. Consider, for example, replacing constraint (1) with: $4X_1 + 2X_2 = 30$ (pounds), represented by line $A'B'$ in the figure. In this case, points A' and B' are located below line CD and do not meet the requirement of constraint (3). Intuitively, we can see that if X_2 must be larger than or equal to 20 (20 bags or $20 \times 2 = 40$ pounds as stated in constraint (3)), X_1 becomes a negative value; or if $X_2 = 20$:

$$4X_1 + 2(20) = 30 \text{ (pounds)}$$
$$X_1 = -2.5 \text{ (bags of 4 pounds each)}$$

Since X_1 variable must not be a negative value in a linear programming problem, there is no solution in this case.

EXERCISE 8–1　　　　　　Reference: Chapter 8

1. Graph each of the following expressions: (Shade areas representing the expressions.)
   (a) $X \geq 3$
   (b) $Y \geq 2$
   (c) $4X + 3Y \leq 12$
   (d) $2X + 3Y \leq -6$

Figure 8-7 **(Example 9)**

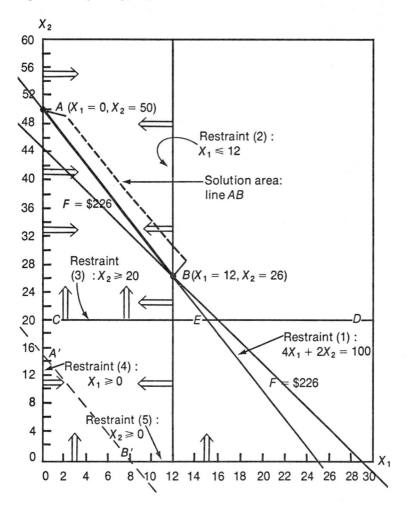

2. Graph each of the following expressions: (Shade areas representing the expressions.)
   (a) $X \leq -1$
   (b) $Y \leq -4$
   (c) $2X + 4Y \geq 8$
   (d) $3X + 5Y \geq -15$

3. Constraints:

$$2X_1 + 3X_2 \leq 12$$
$$4X_1 + 2X_2 \leq 16$$
$$X_1 \geq 0$$
$$X_2 \geq 0$$

Use the graphic method to find the values of X_1 and X_2 which will maximize the objective function F:

(a) $F = 3X_1 + 2X_2$
(b) $F = 2X_1 + 3X_2$
(c) $F = 5X_1 + 2X_2$
(d) $F = 3X_1 + 3X_2$

Also, refer to the objective function in (a) only. Plot the F lines at (1) $X_1 = 0$ and $X_2 = 0$, and (2) the optimum point.

4. Constraints:

$$\tfrac{1}{2}X_1 + \tfrac{1}{4}X_2 \leq 3$$
$$2X_1 + 3X_2 \leq 24$$
$$X_1 \geq 0$$
$$X_2 \geq 0$$

Use the graphic method to find the values of X_1 and X_2 which will maximize the objective function F:

(a) $F = 32X_1 + 12X_2$
(b) $F = 8X_1 + 7X_2$
(c) $F = 4X_1 + 8X_2$
(d) $F = 16X_1 + 8X_2$

Also, refer to the objective function in (a) only. Plot the F lines at (1) $X_1 = 0$ and $X_2 = 0$, and (2) the optimum point.

5. Constraints:

$$2X_1 + 5X_2 \geq 20$$
$$X_1 \geq 5$$
$$X_1 \geq 0$$
$$X_2 \geq 0$$

Use the graphic method to find the values of X_1 and X_2 which will minimize the objective function F:

(a) $F = 4X_1 + 8X_2$
(b) $F = 2X_1 + 7X_2$

Also, refer to the objective function in (a) only. Plot the F lines through (1) $F = 80$, and (2) the optimum point.

6. Constraints:

$$3X_1 + X_2 \geq 12$$
$$X_1 + X_2 = 10$$
$$X_1 + 3X_2 \geq 18$$
$$X_1 \geq 0$$
$$X_2 \geq 0$$

Use the graphic method to find the values of X_1 and X_2 which will minimize the objective function $F = 2X_1 + 5X_2$.

Also, plot the F lines through (1) $F = 60$, and (2) the optimum point.

7. Baxter Company makes doors and windows, among other products. Each door requires 1 hour on machine I and 2 hours on machine II. Each window requires 3 hours on machine I and 1 hour on machine II. The company has a maximum of 9 hours available on machine I and 8 hours available on machine II. Determine the units of doors and windows that should be made in order to maximize profit. Assume that the company can realize a profit of:

   (a) $10 on each door and $20 on each window
   (b) $10 on each door and $35 on each window

   Use the graphic method to find answers.

8. An investment company makes two types of loans: industrial loans at 15% interest rate per year, and residential loans at 10% interest rate per year. The company can lend a maximum of $10 million. For safety, the company has the policy of investing not more than 60% of the total amount in industrial loans and not more than 80% in residential loans. The company wishes to maximize interest income. Use the graphic method to determine the amount that should be invested in each type of loan.

9. A man plans to buy two types of stocks: A and B. He finds that:

   (a) The anticipated dividend rate per year on stock A is 6% and that on stock B is 2%.
   (b) The anticipated increase in market value in one year is $1 per dollar invested in stock A and $2 per dollar invested in stock B.

   He wishes:

   (1) To have at least $300 dividend income each year.
   (2) To have at least $10,000 increase on his investment in one year.

   Use the graphic method to determine the minimum amount that he will spend on each type of stock.

10. Assume that a man is planning to buy certain types of food: meat and fish. Both types can provide nutrient units for his minimum requirements during a given period as follows:

Type of nutrient	Units of nutrient provided in		Minimum units required
	1 pound of meat	1 pound of fish	
A	1	3	9
B	2	2	10
C	4	2	12
Price per pound	$2.50	$1.50	

How many pounds of meat and fish should he buy in order to meet the minimum nutrient requirements at the minimum cost? Use the graphic method.

Chapter
9 Simplex Method for Maximization

The graphic method illustrated in the previous chapter introduced the basic concept of linear programming techniques. It also provided a simple way to solve two-dimensional linear programming problems. However, the graphic method is not practical in solving a problem involving three or more dimensions. The simplex method, also called the *simplex algorithm,* was developed by G. B. Dantzig in 1947 to overcome this weakness. This chapter will first introduce the basic logic of the simplex method and then illustrate the applications of the method to maximization and minimization problems.

⋆9.1 BASIC LOGIC OF THE SIMPLEX METHOD

The simplex method uses matrix algebra operations, as presented in Chapter 7, in solving linear programming problems. The detailed computational procedure of the method is shown in Example 1. The example uses an *algebraic approach* for solving a linear programming problem. The detailed illustration will emphasize the basic logic of the method. In Section 9.2 we will demonstrate the use of simplex tables. These tables greatly simplify the computations. First, however, we should introduce certain terminology in setting up the equations to avoid confusion.

A. Restraints

A *restraint* of a linear programming problem can be classified into two types:

1. **CONSTRAINTS.** A *constraint* is a restraint specified in the problem. Each constraint must be converted into a linear equation in the simplex method.

2. **NONNEGATIVE RESTRAINTS.** A *nonnegative restraint* is a restraint applicable to all types of linear programming problems. It is not necessary to convert a nonnegative restraint into an equation for computational purposes.

164

B. Variables

A *variable* is a set of values and is usually represented by a symbol. Three types of variables are important in setting up the equations:

1. **ORIGINAL VARIABLES.** An *original variable* is used to represent an unknown in a given problem. It is represented by the symbol X.

2. **SLACK OR SURPLUS VARIABLES.** A *slack variable,* or *surplus variable,* is used to convert an inequality into an equation. When the left side is less than ($<$) the right side of an inequality, we add a slack variable; whereas, when the left side is greater than ($>$) the right side, we subtract a surplus variable. A slack or surplus variable is represented by the symbol S.

3. **ARTIFICIAL VARIABLES.** An *artificial variable* is used to obtain an initial solution which will include a logical or nonnegative answer. An artificial variable is represented by the symbol A.

C. Basic Solutions

A *basic solution* is a solution, with nonzero values, to a system of equations. The variables in the solution are called the *basic variables* and the variables not in the solution are called the *nonbasic variables* or *zero variables* (variables with zero values).

ILLUSTRATION 1   A system of two linear equations:

$$2X_1 + 6X_2 + 1S_3 = 12 \qquad (1)$$
$$1X_1 + 4X_2 + 1S_4 = 7 \qquad (2)$$

Let $X_1 = 0$ and $X_2 = 0$.

Substitute the given values in:

(1) $2(0) + 6(0) + 1S_3 = 12$
$$S_3 = 12$$
(2) $1(0) + 4(0) + 1S_4 = 7$
$$S_4 = 7$$

S_3 and S_4 are basic variables with the basic solutions 12 and 7 respectively; X_1 and X_2 are nonbasic variables with zero values.

EXAMPLE 1   The information given in Example 4 of Chapter 8 (page 154) is repeated here:

A furniture factory is planning to produce tables and chairs. Each table requires 2 hours on the machine and 1 hour of skilled labor. Each chair requires 6 hours on the machine and 4 hours of skilled labor. The machine has a maximum of 12 hours available. The skilled labor is limited to 7 hours as the maximum during the period. The profit for each table is $3 and the profit for each chair is $10. Find the best combination of tables and chairs that should be produced in order to maximize the profit.

The preceding information is summarized below:

Maximize the objective function:

$F = \$3X_1 + \$10X_2$ (profit)

Subject to two given restraints:

(1) $2X_1 + 6X_2 \leq 12$ (machine hours available)
(2) $1X_1 + 4X_2 \leq 7$ (skilled labor hours available)
and two nonnegative restraints:

(3) $X_1 \geq 0$ (number of tables to be produced)
(4) $X_2 \geq 0$ (number of chairs to be produced)

Use the algebraic approach to find the values of X_1 and X_2.

Recall that the optimum solution to maximizing the objective function obtained by the graphic method was based on the four corner points (O, C, E, B) of the feasible solution area shown in Figure 8–3. This figure is repeated here in Figure 9–1. Now, instead of finding the corner points by the graphic method, we may find each point by solving the equations derived from the given restraints.

The steps of the simplex method by the algebraic approach in solving the above problem are as follows:

STEP 1   Add one slack variable to each inequality of the given constraints to convert the inequality to an equation.

Add S_3 to restraint (1):

$2X_1 + 6X_2 + 1S_3 = 12$ (1)

Add S_4 to restraint (2):

$1X_1 + 4X_2 + 1S_4 = 7$ (2)

The nonnegative restraints (3) and (4) for X_1 and X_2 respectively are not converted into equations by means of slack variables.

Restraints (3) and (4) are redundant and will not enter the computational procedure directly. A restraint is *redundant* when it is not necessary for a solution. The values of X_1 and X_2 must be positive or zero in restraints (1) and (2). Thus, restraints (3) and (4) are not necessary. (Additional illustration: If there are two restraints, $X_1 \geq 5$ and $X_1 \geq 8$, the restraint $X_1 \geq 5$ is not necessary and is redundant. The value of X_1 must be greater than 5 if it is greater than 8. Therefore we will not convert $X_1 \geq 5$ to an equation for computational purposes.)

The slack variables S_3 and S_4 are also nonnegative. We must remember the nonnegative restraints during the entire process of finding the optimum solution.

There are four variables (X_1, X_2, S_3, S_4) in the two equations converted from the two given constraints. If we assign any values to two of the four variables, the result is to have a system of two equations in two unknowns. The values of the unknowns can be found by solving simultaneously the two equations in the system. For example, the values of slack variables S_3 and S_4 can be computed from the two equations.

ILLUSTRATION 2   Let $X_1 = 2$ and $X_2 = 1$ (point K shown in Figure 9-1). Substitute the given values in:

$$2(2) + 6(1) + 1S_3 = 12$$
$$S_3 = 12 - 4 - 6 = 2 \qquad (1)$$

$$1(2) + 4(1) + 1S_4 = 7$$
$$S_4 = 7 - 2 - 4 = 1 \qquad (2)$$

Figure 9-1   (Example 1)

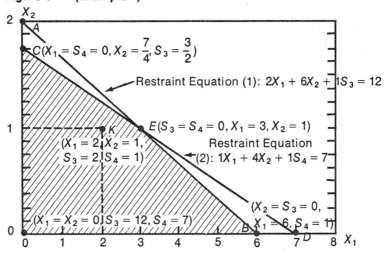

In particular, if we assign zeros to two of the four variables and solve the equations, the solution of the remaining two variables is a basic solution.

ILLUSTRATION 3   Let $X_1 = 0$ and $S_3 = 0$ (Point A in Figure 9-1.) Substitute the given values in:

$$2(0) + 6X_2 + 1(0) = 12 \qquad (1)$$
$$X_2 = 2$$

$$1(0) + 4(2) + 1S_4 = 7 \qquad (2)$$
$$S_4 = -1$$

Here X_2 and S_4 are basic variables with the basic solution 2 and -1 respectively; X_1 and S_3 are nonbasic variables with zero values.

There is a finite number of basic solutions. However, not all basic solutions are feasible, such as $S_4 = -1$ for point A in Illustration 3. A *basic feasible solution* must correspond to an extreme point or a corner point within the feasible solution area. The four corner points are O, C, E, and B. The values of the points can also be obtained by solving Equations (1) and (2). The computations are listed in Table 9–1. (Also see Illustration 1 for point O.)

Table 9–1

Corner Point	Two Zero (Nonbasic) Variables	Equations (1) and (2) After Zero Substitutions		Basic Solution to Equations (Basic Variables)
O	$X_1 = 0$ $X_2 = 0$	$\begin{cases} S_3 = 12 \\ S_4 = 7 \end{cases}$	(1) (2)	$S_3 = 12$ $S_4 = 7$
C	$X_1 = 0$ $S_4 = 0$	$\begin{cases} 6X_2 + S_3 = 12 \\ 4X_2 = 7 \end{cases}$	(1) (2)	$S_3 = \frac{3}{2}$ $X_2 = \frac{7}{4}$
E	$S_3 = 0$ $S_4 = 0$	$\begin{cases} 2X_1 + 6X_2 = 12 \\ 1X_1 + 4X_2 = 7 \end{cases}$	(1) (2)	$X_1 = 3$ $X_2 = 1$
B	$X_2 = 0$ $S_3 = 0$	$\begin{cases} 2X_1 = 12 \\ 1X_1 + 1S_4 = 7 \end{cases}$	(1) (2)	$X_1 = 6$ $S_4 = 1$

The computed values of the slack variables for the four corner points, along with the X_1 and X_2 values, are shown in Figure 9–1. Observe the figure and Table 9–1. We can see that the maximum values of slack variables S_3 and S_4 are 12 and 7 respectively when X_1 and X_2 are both 0 (point O). The minimum values of the slack variables are 0 since they are nonnegatives.

As in the graphic method, the values of the four corner points can be used to test the objective function in finding the optimum solution by the simplex method. The simplex method moves orderly from one feasible corner point to another in such a way that the objective function is being improved if it can be. The orderly movement is presented in detail as follows:

STEP 2   Find the initial solution. (Use Equations (1) and (2).)

Set the initial solution at point O. This solution is reasonable since the least quantities that the furniture factory can produce are no tables ($X_1 = 0$) and no chairs ($X_2 = 0$). This starting point of solving the linear programming problem by the simplex method agrees with that of the graphic method as shown in Figure 8–4 on page 156.

At point O, $X_1 = 0$ and $X_2 = 0$.

The values of S_3 and S_4 can be computed from Equations (1) and (2) expressed in terms of the zero variables X_1 and X_2:

From (1): $S_3 = 12 - 2X_1 - 6X_2$ (3)

Thus: $S_3 = 12 - 2(0) - 6(0) = 12$

From (2): $S_4 = 7 - 1X_1 - 4X_2$ (4)

Thus: $S_4 = 7 - 1(0) - 4(0) = 7$

The initial solution is:

$X_1 = 0,\quad X_2 = 0,\quad S_3 = 12,\quad S_4 = 7$

The objective function is, based on the initial solution:

$F = \$3X_1 + \$10X_2 = 3(0) + 10(0) = \$0$

STEP 3   Find the second solution. (Use Equations (3) and (4).)

In finding the second solution, we hope that the new solution will improve the objective function from $0 profit (in the initial solution) to a higher value profit. The procedure of finding the solution is as follows:

(a) Determine which zero variable (X_1 or X_2) should replace a nonzero variable (S_3 or S_4) in the initial solution to increase the profit.

Observe the F equation in the initial solution. If we increase either X_1 or X_2, we can increase profit; that is, $3 for every unit increase in X_1 or $10 for every unit increase in X_2. Obviously, it is better to increase X_2 units by as many as possible. The result will be a greater profit.

(b) Determine which nonzero variable (S_3 or S_4) should be replaced by X_2.

CASE 1

If S_3 is replaced by X_2, $S_3 = 0$. The value of X_1 is not changed since it is not affected by the replacement, or $X_1 = 0$, as in the initial solution. Solve for X_2 from Equation (3) in Step 2:

$S_3 = 12 - 2X_1 - 6X_2$

$X_2 = \dfrac{12 - 2X_1 - S_3}{6} = 2 - \dfrac{1}{3}X_1 - \dfrac{1}{6}S_3$

Substitute $X_1 = 0$ and $S_3 = 0$.

$X_2 = 2 - \dfrac{1}{3}(0) - \dfrac{1}{6}(0) = 2$

The maximum number of X_2 units that is allowed to replace S_3 is 2.

CASE 2

If S_4 is replaced by X_2, $S_4 = 0$. Also, $X_1 = 0$.

Solve for X_2 from Equation (4) in Step 2:

$$S_4 = 7 - 1X_1 - 4X_2$$

$$X_2 = \frac{7 - 1X_1 - S_4}{4} = \frac{7}{4} - \frac{1}{4}X_1 - \frac{1}{4}S_4$$

Substitute $X_1 = 0$ and $S_4 = 0$:

$$X_2 = \frac{7}{4} - \frac{1}{4}(0) - \frac{1}{4}(0) = \frac{7}{4}$$

The maximum number of X_2 units that is allowed to replace S_4 is $\frac{7}{4}$.

In order to satisfy both restraints stated in Equation (3) (derived from restraint (1)) and in Equation (4) (derived from restraint (2)), we must use a smaller positive number of X_2 for the replacement. Since $\frac{7}{4}$ is smaller than 2, we let $X_2 = \frac{7}{4}$ replace S_4 based on the restraint stated in Equation (4) as analyzed in Case 2.

Note that if $X_2 = 2$ were used for the replacement, Equation (4) becomes:

$$S_4 = 7 - 1X_1 - 4X_2 = 7 - 1(0) - 4(2) = -1$$

The negative value is not allowed for the slack variable S_4.

    (c) Determine the values of the four variables for the second solution.

From the analysis in (b):

$X_1 = 0$ (The value of X_1 is not changed)
$S_4 = 0$ (Since S_4 is replaced by X_2 as indicated in Case 2)

The values of the remaining variables X_2 and S_3 and the objective function F can be expressed in terms of the new zero variables X_1 and S_4 as follows:

$$X_2 = \frac{7}{4} - \frac{1}{4}X_1 - \frac{1}{4}S_4 \text{ (from Case 2)} \tag{5}$$

$$S_3 = 12 - 2X_1 - 6X_2 = 12 - 2X_1 - 6\left(\frac{7}{4} - \frac{1}{4}X_1 - \frac{1}{4}S_4\right) \text{ (from (3))}$$

$$S_3 = \frac{3}{2} - \frac{1}{2}X_1 + \frac{3}{2}S_4 \tag{6}$$

$$F = \$3X_1 + \$10X_2 = 3X_1 + 10\left(\frac{7}{4} - \frac{1}{4}X_1 - \frac{1}{4}S_4\right)$$

$$F = \frac{35}{2} + \frac{1}{2}X_1 - \frac{5}{2}S_4$$

Substitute $X_1 = 0$ and $S_4 = 0$ in Equations (5), (6), and F respectively:

$$X_2 = \frac{7}{4}, \; S_3 = \frac{3}{2}, \; F = \frac{35}{2} = \$17\frac{1}{2}$$

The second solution thus is:

$$X_1 = 0, X_2 = \frac{7}{4}, S_3 = \frac{3}{2}, S_4 = 0, F = \$17\tfrac{1}{2}$$

which is larger than $F = \$0$ in the initial solution. This solution is located at point C in Figure 9-1.

STEP 4 Find the third solution. (Use Equations (5) and (6).)

In finding the third solution, we hope that the new solution will improve the objective function from $17.50 profit (in the second solution) to a higher value profit. The procedure for finding the third solution is basically the same as that illustrated in Step 3.

(a) Determine which zero variable (X_1 or S_4) should replace a nonzero variable (X_2 or S_3) in the second solution to increase the profit.

Observe the F equation in the second solution:

$$F = \frac{35}{2} + \frac{1}{2}X_1 - \frac{5}{2}S_4$$

X_1 has a positive sign: The increase of every unit in X_1 will increase $\$\tfrac{1}{2}$ profit.

S_4 has a negative sign: The increase of every unit in S_4 will decrease $\$2\tfrac{1}{2}$ profit.

Thus, we should increase X_1 units by as many as possible in the replacement.

(b) Determine which nonzero variable (X_2 or S_3) should be replaced by X_1.

CASE 1

If X_2 is replaced by X_1, $X_2 = 0$. The value of S_4 is not changed since it is not affected by the replacement, or $S_4 = 0$, as in the second solution. Solve for X_1 from Equation (5) in Step 3:

$$X_2 = \frac{7}{4} - \frac{1}{4}X_1 - \frac{1}{4}S_4$$

$$X_1 = 4\left(\frac{7}{4} - X_2 - \frac{1}{4}S_4\right) = 7 - 4X_2 - S_4$$

Substitute $X_2 = 0$ and $S_4 = 0$:

$$X_1 = 7 - 4(0) - 0 = 7$$

The maximum number of X_1 units that is allowed to replace X_2 is 7.

CASE 2

If S_3 is replaced by X_1, $S_3 = 0$. Also, $S_4 = 0$. Solve for X_1 from Equation (6) in Step 3:

$$S_3 = \frac{3}{2} - \frac{1}{2}X_1 + \frac{3}{2}S_4$$

$$X_1 = 2\left(\frac{3}{2} - S_3 + \frac{3}{2}S_4\right) = 3 - 2S_3 + 3S_4$$

Substitute $X_3 = 0$ and $S_4 = 0$:

$$X_1 = 3 - 2(0) + 3(0) = 3$$

The maximum number of X_1 units that is allowed to replace S_3 is 3.

In order to satisfy both restraints stated in Equation (5) (derived from restraint (2) and Equations (4) and (2)) and in Equation (6) (derived from restraint (1) and Equations (3) and (1)), we must use the smaller number of X_1 for the replacement. Since 3 is smaller than 7, we let $X_1 = 3$ to replace S_3 based on the restraint stated in Equation (6) as analyzed in Case 2.

(c) Determine the values of the four variables for the third solution.

From the analysis in (b):

$S_3 = 0$ (Since S_3 is replaced by X_1 as indicated in Case 2)
$S_4 = 0$ (The value of S_4 is not changed.)

The values of the remaining variables X_1 and X_2 and the objective function F can be expressed in terms of the zero variables S_3 and S_4 as follows:

$$X_1 = 3 - 2S_3 + 3S_4 \text{ (from Case 2)} \qquad (7)$$

$$X_2 = \frac{7}{4} - \frac{1}{4}X_1 - \frac{1}{4}S_4 = \frac{7}{4} - \frac{1}{4}(3 - 2S_3 + 3S_4) - \frac{1}{4}S_4$$

$$X_2 = 1 + \frac{1}{2}S_3 - S_4 \text{ (from Equation (5))} \qquad (8)$$

$$F = \frac{35}{2} + \frac{1}{2}X_1 - \frac{5}{2}S_4 = \frac{35}{2} + \frac{1}{2}(3 - 2S_3 + 3S_4) - \frac{5}{2}S_4$$

$$F = 19 - 1S_3 - 1S_4 \text{ (From Step 3)}$$

Substitute $S_3 = 0$ and $S_4 = 0$ in Equations (7), (8), and F respectively.

$$X_1 = 3, X_2 = 1, F = \$19$$

The third solution is,

$$X_1 = 3, X_2 = 1, S_3 = 0, S_4 = 0, \text{ and } F = \$19$$

which is larger than $F = \$17.50$ in the second solution. This solution is located at point E in Figure 9–1.

Observe the F equation, $F = 19 - 1S_3 - 1S_4$. Both the coefficients of S_3 and S_4 are negative. An increase in either S_3 or S_4 will force F to decrease. Thus, the third solution is optimal. We now conclude that the optimum solution is to produce:

3 tables ($X_1 = 3$) and 1 chair ($X_2 = 1$)

with the maximized profit of:

$19 (F = \$3X_1 + \$10X_2 = 3(3) + 10(1) = \$19)$

Note that we have examined only corner points O (Step 2), C (Step 3), and E (Step 4) based on the objective function. We do not have to examine every corner point, such as point B, in the above example, for finding the optimum solution by the simplex method.

The equations derived in the steps of Example 1 are summarized in Table 9–2.

Table 9–2

Step Number	Equation Number	Equations Expressed In Zero Variables	Corresponding Corner Point
1. Convert inequalities to equations	(1) (2)	$\begin{cases} 2X_1 + 6X_2 + 1S_3 = 12 \\ 1X_1 + 4X_2 + 1S_4 = 7 \end{cases}$	*(Figure 9–1)*
2. Find the initial solution $(X_1 = 0 \quad X_2 = 0)$	(3) (4)	$\begin{cases} S_3 = 12 - 2X_1 - 6X_2 = 12 \\ S_4 = 7 - X_1 - 4X_2 = 7 \\ F = 3X_1 + 10X_2 = \$0 \end{cases}$	O
3. Find the second solution $(X_1 = 0 \quad S_4 = 0)$	(5) (6)	$\begin{cases} X_2 = \frac{7}{4} - \frac{1}{4}X_1 - \frac{1}{4}S_4 = \frac{7}{4} \\ S_3 = \frac{3}{2} - \frac{1}{2}X_1 + \frac{3}{2}S_4 = \frac{3}{2} \\ F = \frac{35}{2} + \frac{1}{2}X_1 - \frac{5}{2}S_4 = \$\frac{35}{2} \end{cases}$	C
4. Find the third (optimum) solution $(S_3 = 0 \quad S_4 = 0)$	(7) (8)	$\begin{cases} X_1 = 3 - 2S_3 + 3S_4 = 3 \\ X_2 = 1 + \frac{1}{2}S_3 - 1S_4 = 1 \\ F = 19 - 1S_3 - 1S_4 = \$19 \end{cases}$	E

EXERCISE 9–1   Reference: Section 9.1

Use the simplex method as illustrated in Example 1 to find the solution for each of the following problems:

1. Maximize the objective function:

$$F = 3X_1 + 2X_2$$

Subject to restraints:

$$2X_1 + 3X_2 \leq 12$$
$$4X_1 + 2X_2 \leq 16$$
$$X_1, X_2 \geq 0$$

2. Maximize the objective function:

$$F = 32X_1 + 12X_2$$

Subject to restraints:

$$\tfrac{1}{2}X_1 + \tfrac{1}{4}X_2 \leq 3$$
$$2X_1 + 3X_2 \leq 24$$
$$X_1, X_2 \geq 0$$

3. The Baxter Company makes doors and windows. Each door requires 1 hour on machine I and 2 hours on machine II. Each window requires 3 hours on machine I and 1 hour on machine II. The company has a maximum of 9 hours available on machine I and a maximum of 8 hours available on machine II. Determine the units of doors and windows that should be made in order to maximize profit. Assume that the company can realize a profit of $10 on each door and $20 on each window.

4. An investment company makes two types of loans: industrial loans at 15% interest rate per year, and residential loans at 10% interest rate per year. The company can lend a maximum of $10 million. For safety, the company has the policy of investing not more than 60% of the total amount in industrial loans and not more than 80% in residential loans. The company wishes to maximize interest income. Determine the amount that should be invested in each type of loan.

9.2 SIMPLEX TABLES FOR MAXIMIZATION PROBLEMS ($C_j - Z_j$ METHOD)

The presentation of detailed computational steps of the simplex method as illustrated in the previous section can be simplified. Instead of computing the equations of restraints and objective functions in a descriptive manner, we may arrange the entire computation in each step in a *simplex table* with a $C_j - Z_j$ row. The use of the simplex table is illustrated by an example.

EXAMPLE 2 The information given in Example 1 is repeated here:

Maximize the objective function:

$F = \$3X_1 + \$10X_2$ (F = profit)

Subject to two given restraints (constraints):

(1) $2X_1 + 6X_2 \le 12$ (machine hours available)
(2) $1X_1 + 4X_2 \le 7$ (skilled labor hours available)

and two nonnegative restraints:

(3) $X_1 \ge 0$ (number of tables to be produced)
(4) $X_2 \ge 0$ (number of chairs to be produced)

Use the simplex method to find the values of X_1 and X_2.

SOLUTION The steps of the simplex method by using the simplex tables are listed below. The step numbers used in this example correspond to those used in Example 1 of the previous section.

STEP 1 Add one slack variable to each inequality of the given constraints to convert the inequality to an equation.

Add S_3 to constraint (1): $2X_1 + 6X_2 + 1S_3 = 12$
Add S_4 to constraint (2): $1X_1 + 4X_2 + 1S_4 = 7$

Note again that all variables. X_1. X_2. S_3. and S_4. are nonnegative. We must remember the nonnegative restraints throughout the entire process of finding the optimum solution.

Also. arrange the two new equations and the objective function equation to include the same number of variables. This can be done by adding the extra slack variables with zero coefficients to each equation.

Thus. constraints (1) and (2) now become:

$$2X_1 + 6X_2 + 1S_3 + 0S_4 = 12 \quad (1)$$
$$1X_1 + 4X_2 + 0S_3 + 1S_4 = 7 \quad (2)$$

and the objective function becomes:

$$F = \$3X_1 + \$10X_2 + \$0S_3 + \$0S_4$$

STEP 2

Find the initial solution and construct the first simplex table. (Use Equations (1) and (2).) The simplex table is shown in Table 9–3. Details of the finding and the construction are explained as follows:

Table 9-3 **Initial Simplex Table (Example 2)**

C_j	$\$3$	$\$10$	$\$0$	$\$0$	Feasible Solution		V
C_i	X_1	X_2	S_3	S_4	Basic Variable	Value V	X_2
$\$0$	2	6 0	1	0	S_3	12	$12 \div 6 = 2$
$\$0$	1	4 1	0	1	S_4	7	$7 \div 4 = \frac{7}{4}$
Z_j	$\$0$	$\$0$	$\$0$	$\$0$	$F = \$0$		(Smaller
$C_j - Z_j$	$\$3$	$\$10$	$\$0$	$\$0$			Positive)

Largest Positive Value

1. The initial solution must be feasible. The initial feasible solution is obtained from the two restraint equations by letting $X_1 = 0$ and $X_2 = 0$. Then:

   Equation (1) becomes: $2(0) + 6(0) + 1S_3 + 0S_4 = 12$. or
   $\qquad\qquad\qquad S_3 = 12$ (machine hours available)
   Equation (2) becomes: $1(0) + 4(0) + 0S_3 + 1S_4 = 7$. or
   $\qquad\qquad\qquad S_4 = 7$ (skilled labor hours available)

   The initial feasible solution: To have $S_3 = 12$ hours of unused machine time. and $S_4 = 7$ hours of unused skilled labor time. The solution values (12 and 7) are denoted by the symbol V in the simplex table. The two zero variables. X_1 and X_2. are not shown in the solution column.

   The number of variables (slack) in the solution column thus is equal to the number of given restraints. The two restraints require two variables (S_3 and S_4) in the solution column. Note that the variables in the solution column are also called *basic feasible variables*.

2. C_j (the first row heading) = the profit per unit of the variable indicated in the jth column. It is also the coefficient of each variable in the objective function. The products or variables are X_1, X_2, S_3, and S_4, and their unit profits are \$3, \$10, \$0, and \$0 respectively.

    C_i (the first column heading) = the profit per unit of the variable indicated in the ith row. The variables are S_3 and S_4 in the solution column, and their unit profits are \$0 and \$0 respectively.

3. The two numbers under each column headed by the variables X_1, X_2, S_3, and S_4, are the coefficients of each variable found in the two restraint equations. Note that the four numbers under S_3 and S_4 form a 2 × 2 identity matrix $\begin{bmatrix} 1 & 0 \\ 0 & 1 \end{bmatrix}$. S_3 and S_4 are the solution variables. The coefficients of S_3 form a unit vector with 1 on the top (or on S_3 row), and the coefficients of S_4 also form a unit vector with 1 at the bottom (or on S_4 row); that is:

		Feasible Solution	
		Basic	Value
S_3	S_4	Variable	V
1	0	S_3	12
0	1	S_4	7

This fact is true for all variables in a solution to a system of two linear equations in two unknowns. Equations (1) and (2) can be written in a vector form:

$$\begin{bmatrix} 2 \\ 1 \end{bmatrix} X_1 + \begin{bmatrix} 6 \\ 4 \end{bmatrix} X_2 + \begin{bmatrix} 1 \\ 0 \end{bmatrix} S_3 + \begin{bmatrix} 0 \\ 1 \end{bmatrix} S_4 = \begin{bmatrix} 12 \\ 7 \end{bmatrix}$$

If $X_1 = 0$ and $X_2 = 0$, the above two equations can both be reduced into the simple vector form:

$$\begin{bmatrix} 1 \\ 0 \end{bmatrix} S_3 + \begin{bmatrix} 0 \\ 1 \end{bmatrix} S_4 = \begin{bmatrix} 12 \\ 7 \end{bmatrix}$$

and:

$$\begin{bmatrix} S_3 \\ S_4 \end{bmatrix} = \begin{bmatrix} 12 \\ 7 \end{bmatrix}$$

Thus, the final answer is: $S_3 = 12$, and $S_4 = 7$.

4. Z_j (the heading of the 5th row) = the total profit of the variables in the solution column (S_3 and S_4) in terms of the numbers indicated in the jth column. The value of Z_j in each column is the sum of the products obtained by multiplying every C_i by a corresponding row number in the column.

The total profits of variables S_3 and S_4 in terms of the numbers indicated in each (jth) column are:

$$Z \text{ (in } X_1 \text{ column)} = \$0(2) + \$0(1) = \$0$$
$$Z \text{ (in } X_2 \text{ column)} = \$0(6) + \$0(4) = \$0$$

$$Z \text{ (in } S_3 \text{ column)} = \$0(1) + \$0(0) = \$0$$
$$Z \text{ (in } S_4 \text{ column)} = \$0(0) + \$0(1) = \$0$$

The total profit of variables S_3 and S_4 for the solution can be computed in the same manner:

$$F \text{ (in } V \text{ column)} = \$0(12) + \$0(7) = \$0$$

The interpretation of the Z value in each jth column is different from the F value in the V column. For example, the value of Z in the X_2 column ($\$0$) is the total profit of S_3 and S_4 that would be reduced if 1 unit of X_2 is placed in the solution. That is, instead of having the initial solution,

$$X_1 = 0 \text{ and } X_2 = 0 \quad \text{(from which we obtained } S_3 = 12 \text{ and } S_4 = 7)$$

we would have:

$$X_1 = 0 \text{ and } X_2 = 1$$

Substitute $X_1 = 0$ and $X_2 = 1$ in Equation (1):

$$2X_1 + 6X_2 + 1S_3 + 0S_4 = 12$$
$$2(0) + 6(1) + 1S_3 + 0S_4 = 12$$
$$S_3 \qquad\quad = 12 - 6 = 6$$

or 6 units of S_3 are reduced at $\$0$ per unit, or a total of $\$0(6) = \$0$.

Substitute $X_1 = 0$ and $X_2 = 1$ in Equation (2):

$$1X_1 + 4X_2 + 0S_3 + 1S_4 = 7$$
$$1(0) + 4(1) + 0S_3 + 1S_4 = 7$$
$$S_4 = 7 - 4 = 3$$

or 4 units of S_4 are reduced at $\$0$ per unit, or a total of $\$0(4) = \$0$.
Thus, if 1 unit of X_2 is placed into the initial solution, the total profit of variables S_3 and S_4 would be reduced by:

$$\$0(6) + \$0(4) = \$0$$

5. $C_j - Z_j$ (the heading of the last row) = the net profit per unit of the product indicated in the jth column that can be contributed to the total profit if the product is produced or placed into the solution. The C_j value is at the very top of each column.

For example, refer to the X_2 column. If 1 unit in X_2 is placed into solution, the total profit would be increased by $\$10$ $(= C_j)$ but reduced by $\$0$ $(= Z_j)$, or a net profit increase of $\$10$ $(= C_j - Z_j = \$10 - \$0)$. If $\frac{7}{4}$ units in X_2 are placed into solution, the total profit (F) would be increased from $\$0$ to $\$17.50$ $(\$10(\frac{7}{4}) = \$\frac{70}{4} = \$17.50.)$ This result can be verified in the next simplex table (Table 9–4).

STEP 3   Construct the second simplex table to find the second feasible solution. (Use the information obtained from the first simplex table.) See Table 9–4. The details are explained as follows:

1. In the solution column of Table 9–4, X_2 has replaced S_4 in Table 9–3. The replacement is determined in the following manner:

Table 9-4 **Second Simplex Table (Example 2)**

C_j	$3	$10	$0	$0	Feasible Solution		V
C_i	X_1	X_2	S_3	S_4	Basic Variable	Value V	X_1
$0	1/2→1	0	1	−3/2	→ S_3	3/2	$\frac{3}{2} \div \frac{1}{2} = 3$ (Smaller Positive)
$10	1/4→0	1	0	1/4	X_2	7/4	$\frac{7}{4} \div \frac{1}{4} = 7$
Z_j	10/4	10	0	10/4	$F = 70/4$		
$C_j - Z_j$	2/4	0	0	−10/4			

Largest Positive Value

Find out from Table 9–3 which one of the products listed will contribute the highest profit per unit. This is indicated by an arrow in the column which has the largest positive value of $C_j - Z_j$. The highest positive value is $10 and is located in the X_2 column.

Find out from Table 9–3 which one of the variables (S_3 or S_4) is to be replaced by X_2. This is done by dividing each number in the V column by the corresponding row number in the X_2 column as shown in the last column, or:

$$\text{In the } S_3 \text{ row: } \quad V \div X_2 = 12 \div 6 = 2$$
$$\text{In the } S_4 \text{ row: } \quad V \div X_2 = 7 \div 4 = 1\tfrac{3}{4}$$

The smallest quotient obtained by these divisions will allow the maximum number of X_2 units, $1\tfrac{3}{4}$, to enter the solution without making the slack variable S_4 negative. Thus, we replace S_4 in Table 9–3 (also indicated by an arrow) by X_2 in Table 9–4.

Note that if the quotients obtained are negative or undefined (any number divided by 0), they must be eliminated from consideration as smaller quotients.

S_4, which is to be replaced by X_2, is also called the *outgoing variable* and the row with the outgoing variable is also called the *pivot row*. X_2 is now called the *incoming variable* and the column with the incoming variable is called the *pivot column*. The intersection of pivot row and pivot column (shaded) is called the *pivot* and is always circled for easy recognition, such as the pivot ④ as shown in Table 9–3.

2. The solution column in Table 9–4 now has S_3 and X_2 since X_2 replaced S_4 in Table 9–3. The four numbers under the solution variables S_3 and X_2 in Table 9–4 thus must form a 2×2 identity matrix, or:

$$\begin{array}{cc} \underline{S_3} & \underline{X_2} \\ 1 & 0 \\ 0 & 1 \end{array}$$

The coefficient of S_3 must be a unit vector with 1 on the top, and the coefficient of X_2 also must be a unit vector but with 1 at the bottom.

This arrangement of the coefficients of S_3 and X_2 is consistent with the arrangement of the variables in the solution column: S_3 is on the top and X_2 is at the bottom in the table.

Thus, we must transform the coefficients of X_2 from:

$\dfrac{6}{4}$ in Table 9–3

to:

$\dfrac{0}{1}$ in Table 9–4

The work of transforming the coefficients $\dfrac{6}{4}$ in the pivot column to

the unit vector $\dfrac{0}{1}$ can be done systematically by *pivoting*, or performing two types of row operations stated as follows:

OPERATION 1 A row in the tableau may be multiplied or divided by a real number $k \, (k \neq 0)$.

This operation can be used to transform a pivot to 1.

OPERATION 2 A row may be added or subtracted by the product of a real number k and another row.

This operation can be used to transform all numbers in the pivot column other than the pivot to 0's.

We now compute the values of the individual rows of Table 9–4 by row operations.

1. Compute the values in the incoming X_2 row (X_2 is being used to replace S_4) from the pivot row in Table 9–3. Use Operation 1, $k = 4$, the pivot.

X_2 row (in Table 9–4) = S_4 row (in Table 9–3) ÷ 4, or:

$$\dfrac{1}{4} \qquad \dfrac{4}{4} = 1 \qquad \dfrac{0}{4} = 0 \qquad \dfrac{1}{4} \qquad \dfrac{7}{4}$$
(Transform
pivot 4 to 1)

2. Compute the values in the S_3 row (S_3 is kept in the new table — Table 9–4. Note: If there is another variable being kept in the new table, the computational procedure is the same as obtaining S_3 here.) Use Operation 2, $k = 6$, the number in the intersection of the pivot column and S_3 row in Table 9–3.

New S_3 row (in Table 9–4) = Old S_3 row (in Table 9–3) − [6 × X_2 row (obtained from the first step], or:

Old S_3 row	2	6	1	0	12
(−) 6 × X_2 row	$6(\frac{1}{4}) = \frac{3}{2}$	$6(1) = 6$	$6(0) = 0$	$6(\frac{1}{4}) = \frac{3}{2}$	$6(\frac{7}{4}) = 10\frac{1}{2}$
New S_3 row	$\frac{1}{2}$	0	1	$-\frac{3}{2}$	$\frac{3}{2}$

(Transform
6 in pivot
column to 0)

3. Compute the values in the Z row:

$$Z \text{ (in } X_1 \text{ column)} = \$0\left(\frac{1}{2}\right) + \$10\left(\frac{1}{4}\right) = \$\frac{10}{4}, \text{ or } \$2.50$$

$$Z \text{ (in } X_2 \text{ column)} = \$0(0) + \$10(1) = \$10$$
$$Z \text{ (in } S_3 \text{ column)} = \$0(1) + \$10(0) = \$0$$

$$Z \text{ (in } S_4 \text{ column)} = \$0\left(-\frac{3}{2}\right) + \$10\left(\frac{1}{4}\right) = \$\frac{10}{4}, \text{ or } \$2.50$$

$$F \text{ (in } V \text{ column)} = \$0\left(\frac{3}{2}\right) + \$10\left(\frac{7}{4}\right) = \$\frac{70}{4}, \text{ or } \$17.50$$

4. Compute the $C_j - Z_j$ values in each column:

In X_1 column: $\$3 - \$\dfrac{10}{4} = \$\dfrac{2}{4}$, or $\$0.50$

In X_2 column: $\$10 - \$10 = \$0$

In S_3 column: $\$0 - \$0 = \$0$

In S_4 column: $\$0 - \$\dfrac{10}{4} = -\$\dfrac{10}{4}$, or $-\$2.50$

The second simplex table gives the second feasible solution:

$$X_2 = \frac{7}{4} \text{ units, } S_3 = \frac{3}{2} \text{ units}$$

Total profit = \$17.50

Since X_1 and S_4 did not appear in the solution column, the two variables are equal to zero.

The interpretation of the Z value in each of the columns X_1, X_2, S_3, and S_4 is the same as that stated in note 4 in Step 2. For example, the value of Z in X_1 column $\left(\dfrac{10}{4} \text{ or } \$2.50\right)$ is the total profit of S_3 and X_2 that would be reduced if 1 unit of X_1 is placed into the solution. That is, if X_1 is increased by 1 unit in the solution, S_3 is decreased by $\dfrac{1}{2}$ unit at \$0 per unit and X_2 is decreased by $\dfrac{1}{4}$ unit at \$10 per unit. The total profit of variables X_1 and S_3 would be reduced by:

$$Z = \$0\left(\frac{1}{2}\right) + \$10\left(\frac{1}{4}\right) = \frac{10}{4}, \text{ or } \$2.50$$

The profit per unit of X_1 is \$3 $(= C_j)$. Thus, if 1 unit in X_1 is placed into solution, the total profit would be increased by \$3 but at the same time reduced by $\$\dfrac{10}{4}$ $(= Z_j)$, or a net profit increase of $\$\dfrac{2}{4}\left(= C_j - Z_j = \$3 - \$\dfrac{10}{4} = \$\dfrac{2}{4}\right)$. That is:

$$F = \frac{\$70}{4} + \frac{\$2}{4} = \frac{72}{4} = \$18$$

If 3 units in X_1 is placed into the solution, the total profit would be increased to:

$$F = \frac{\$70}{4} + 3\left(\frac{\$2}{4}\right) = \frac{70}{4} = \$19$$

This result can be verified in the third simplex table (Table 9–5).

In general, according to the information given in the $C_j - Z_j$ row of Table 9–4, the total profit may be increased to:

$$F = \frac{70}{4} + \frac{2}{4}X_1 + 0X_2 + 0S_3 - \frac{10}{4}S_4$$

Or:

$$F = \frac{35}{2} + \frac{1}{2}X_1 - \frac{5}{2}S_4$$

This is the F equation presented in Step 3 of Example 1, page 170.

STEP 4   Construct the third simplex table to find the third feasible solution. (Use the information obtained from the second simplex table.) See Table 9–5. The details are explained as follows.

Table 9-5 **Third Simplex Table – Optimal (Example 2)**

C_j	$\$3$	$\$10$	$\$0$	$\$0$	Feasible Solution	
					Basic	Value
C_i	X_1	X_2	S_3	S_4	Variable	V
$\$ 3$	1	0	2	-3	X_1	3
$\$10$	0	1	$-\frac{1}{2}$	1	X_2	1
Z_j	3	10	1	1	$F = 19$	
$C_j - Z_j$	0	0	-1	-1		

All $C_j - Z_j$ values are now negative or zero.

1. In the solution column of Table 9–5, X_1 has replaced S_3 in Table 9–4. The replacement is determined in the following manner:

   Find the highest positive value in the $C_j - Z_j$ row in Table 9–4. The value is $\$\frac{2}{4}$ in the X_1 column and is indicated by an arrow.

   Find out from Table 9–4 which one of the products (S_3 or X_2) is to be replaced by X_1. This is done by dividing each number in the V column by the corresponding row number in the X_1 column.

   In S_3 row: $V \div X_1 = \frac{3}{2} \div \frac{1}{2} = 3$
   In X_2 row: $V \div X_1 = \frac{7}{4} \div \frac{1}{4} = 7$

   The smallest quotient obtained by the above divisions will allow the maximum number of X_1 units, 3, to enter the solution without making the slack variable S_3 negative. Thus, we replace S_3 in Table 9–4 (also indicated by an arrow) by X_1 in Table 9–5.

S_3 now is the outgoing variable and X_1 is the incoming variable. The pivot column and the pivot row are again shaded in Table 9–4. The pivot $\frac{1}{2}$ is also circled.

2. The solution column in Table 9–5 now has the basic variables X_1 (on the top) and X_2 (at the bottom). The four numbers under the solution variables must form a 2×2 identity matrix, or:

$$
\begin{array}{cc}
\underline{X_1} & \underline{X_2} \\
1 & 0 \\
0 & 1
\end{array}
$$

Thus, we must transform the coefficients of X_1 from the pivot column:

$\begin{array}{l}\dfrac{1}{2}\\[4pt]\dfrac{1}{4}\end{array}$ in Table 9–4, to: $\begin{array}{l}1\\0\end{array}$ in Table 9–5

We now compute the values of the individual rows of Table 9–5 by row operations.

1. Compute the values in the incoming X_1 row (X_1 is being used to replace S_3) from the pivot row in Table 9–4. Use Operation 1, $k = \frac{1}{2}$, the pivot.

X_1 row (in Table 9–5) = S_3 row (in Table 9–4) $\div \frac{1}{2}$, or ($\div \frac{1}{2} = \times 2$)

$$2\left(\frac{1}{2}\right) = 1 \qquad 2(0) = 0 \qquad 2(1) = 2 \qquad 2\left(-\frac{3}{2}\right) = -3 \qquad 2\left(\frac{3}{2}\right) = 3$$

(Transform pivot $\frac{1}{2}$ to 1)

2. Compute the values in the X_2 row (X_2 is kept in the new table — Table 9–5.) Use Operation 2, $k = \frac{1}{4}$, the number in the intersection of the pivot column and the X_2 row in Table 9–4.

New X_2 row (in Table 9–5) = Old X_2 row (in Table 9–4) $- [(\frac{1}{4}) \times X_1$ row (in Table 9–5, obtained in the first step)], or:

Old X_2 row	$\frac{1}{4}$	1	0	$\frac{1}{4}$	$\frac{7}{4}$
(–): $\frac{1}{4} \times X_1$ row	$\frac{1}{4}(1) = \frac{1}{4}$	$\frac{1}{4}(0) = 0$	$\frac{1}{4}(2) = \frac{1}{2}$	$\frac{1}{4}(-3) = -\frac{3}{4}$	$\frac{1}{4}(3) = \frac{3}{4}$
New X_2 row	0	1	$-\frac{1}{2}$	$\frac{4}{4}$	$\frac{4}{4}$
				$=1$	$=1$

(Transform $\frac{1}{4}$ in pivot column to 0)

3. Compute the values in the Z_j row:

$$
\begin{array}{lll}
Z \text{ (in } X_1 \text{ column)} = \$3(1) & + \$10(0) & = \quad \$3 \\
Z \text{ (in } X_2 \text{ column)} = \$3(0) & + \$10(1) & = \$10 \\
Z \text{ (in } S_3 \text{ column)} = \$3(2) & + \$10(-\tfrac{1}{2}) & = \quad \$1 \\
Z \text{ (in } S_4 \text{ column)} = \$3(-3) & + \$10(1) & = \quad \$1 \\
F \text{ (in } V \text{ column)} = \$3(3) & + \$10(1) & = \$19
\end{array}
$$

4. Compute the $C_j - Z_j$ values in each column:

In X_1 column: $\$3 - \$3 = \$0$
In X_2 column: $\$10 - \$10 = \$0$
In S_3 column: $\$0 - \$1 = -\$1$
In S_4 column: $\$0 - \$1 = -\$1$

There is no positive value in the $C_j - Z_j$ row. Thus, the final or optimum solution is obtained from the final or third simplex table (Table 9-5):

To produce: $X_1 = 3$ units (tables) and $X_2 = 1$ unit (chair)
Total profit: $F = \$19$

Since S_3 and S_4 did not appear in the solution column, the two variables are equal to zero.

In summary, the answers to X_1 and X_2 of Example 2 obtained by the simplex method are compared with the answers by the graphic method (Example 4, page 154) as follows:

Simplex Table	Units of Product		Total Profit (F — page 154, or F in V column of simplex table)	Point on Figure 8-3 page 153
	X_1 (Tables)	X_2 (Chairs)		
First (Table 9-3)	0	0	$\$0$	O
Second (Table 9-4)	0	$\dfrac{7}{4}$	$\dfrac{\$70}{4} = \$17\frac{1}{2}$	C
Third (Table 9-5)	3	1	$\$19$	E
Fourth (none required)	6	0	$\$18$	B

9.3 A DEGENERATE CASE

In Example 2 there are four variables: two original variables (denoted by X's) and two slack variables (denoted by S's). When any two of the four variables are equated to zero, the two remaining variables can be solved from the system of two equations derived from the two constraints. Thus, the basic solution has two variables. In general, in a linear programming problem of $(n + m)$ variables, including n original variables (X's) and m slack variables (S's), when any n of the $(n + m)$ variables are equated to zero, the m remaining variables can be solved from the system of m equations derived from the m constraints. Thus, the basic solution has m variables. When this property does not hold, that is, the basic feasible solution includes less than m variables because one or more of the basic variables vanish (or equal 0 in the final simplex table), this is a case of *degeneracy*.

Example 3 illustrates a maximization problem involving three original variables and two slack variables in a degenerate case.

EXAMPLE 3   Maximize the objective function:

$$F = \$3X_1 + \$10X_2 + \$2X_3 \quad \text{(profit)}$$

Subject to two restraints (constraints):

(1) $2X_1 + 6X_2 + 2X_3 \leq 60$ (machine hours available)
(2) $1X_1 + 4X_2 + 3X_3 \leq 40$ (labor hours available)

and three nonnegative restraints:

(3) $X_1 \geq 0$ (number of tables to be produced)
(4) $X_2 \geq 0$ (number of chairs to be produced)
(5) $X_3 \geq 0$ (number of cabinets to be produced)

This problem can be interpreted in the same way as the problem in Example 1: assigning available machine and labor hours to three products, X_1, X_2, and X_3. However, some figures are changed in this new problem.

Use the simplex method to find the values of X_1, X_2 and X_3.

SOLUTION   Add slack variables to inequalities of the constraints and arrange the equations to include the same number of variables.
Thus, constraints (1) and (2) now become:

(1) $2X_1 + 6X_2 + 2X_3 + 1S_4 + 0S_5 = 60$
(2) $1X_1 + 4X_2 + 3X_3 + 0S_4 + 1S_5 = 40$

and the objective function becomes:

$$F = \$3X_1 + \$10X_2 + \$2X_3 + \$0S_4 + \$0S_5$$

The first (initial) simplex table based on the above three equations and $X_1 = X_2 = X_3 = 0$ is shown in Table 9-6.

Table 9-6 **Initial Simplex Table (Example 3)**

C_j	$\$3$	$\$10$	$\$2$	$\$0$	$\$0$	Feasible Solution		$\dfrac{V}{X_2}$
C_i	X_1	X_2	X_3	S_4	S_5	Basic Variable	Value V	X_2
$\$0$	2	6	2	1	0	S_4	60	$60 \div 6 = 10$
$\$0$	1	4	3	0	1	S_5	40	$40 \div 4 = 10$
Z_j	$\$0$	$\$0$	$\$0$	$\$0$	$\$0$	$F = \$0$		
$C_j - Z_j$	$\$3$	$\$10$	$\$2$	$\$0$	$\$0$			

Largest Positive Value

The highest positive value of $C_j - Z_j$ is $\$10$ and is located in the X_2 column. Thus, the incoming variable is X_2. Dividing each number in the V column by the corresponding row number in the X_2 column as shown in the last column of Table 9-6, we found two equal quotients, or:

In S_4 row: $V \div X_2 = 60 \div 6 = 10$
In S_5 row: $V \div X_2 = 40 \div 4 = 10$

The equal quotients show the sign of degeneracy. Now we are faced with the problem of selecting the outgoing variable without a definite instruction.

The application of the simplex method to a degenerate problem fortunately requires no special treatment. We can arbitrarily select either S_4 or S_5 to be the outgoing variable for the replacement. If the first selected variable does not lead to the optimum solution (or the simplex tables begin to repeat endlessly), we then use the other variable for the replacement at the sign of degeneracy. Again, the second selected variable may or may not lead to the optimum solution. A degenerate problem may or may not have an optimum solution.

Suppose that we decide to use X_2 to replace S_4 in the second simplex table. The result is shown in Table 9–7.

Table 9–7 Second Simplex Table – Optimal (Example 3)
(Use X_2 to replace S_4)

C_j	$3	$10	$2	$0	$0	Feasible Solution	
C_i	X_1	X_2	X_3	S_4	S_5	Basic Variable	Value V
$10	$\frac{1}{3}$	1	$\frac{1}{3}$	$\frac{1}{6}$	0	X_2	10
$ 0	$-\frac{1}{3}$	0	$\frac{5}{3}$	$-\frac{2}{3}$	1	S_5	0
Z_j	$\frac{10}{3}$	10	$\frac{10}{3}$	$\frac{10}{6}$	0	$F = 100$	
$C_j - Z_j$	$-\frac{1}{3}$	0	$-\frac{4}{3}$	$-\frac{10}{6}$	0		

All $(C_j - Z_j)$ values are now negative or zero.

The coefficients $\begin{smallmatrix}6\\4\end{smallmatrix}$ in the pivot column (X_2) of Table 9–6 are converted to the unit vector $\begin{smallmatrix}1\\0\end{smallmatrix}$ in Table 9–7. The values of the individual rows of Table 9–7 are computed by row operations as follows:

1. Compute the values in the incoming X_2 row. Use Operation 1, $k = 6$, the pivot.

X_2 row (in Table 9–7) $= S_4$ row (in Table 9–6) $\div 6$ or:

$$\frac{2}{6} = \frac{1}{3} \quad \frac{6}{6} = 1 \quad \frac{2}{6} = \frac{1}{3} \quad \frac{1}{6} \quad \frac{0}{6} = 0 \quad \frac{60}{6} = 10$$

2. Compute the values in the S_5 row. Use Operation 2, $k = 4$, the number in the intersection of the pivot column and the S_5 row in Table 9–6.

New S_5 row (in Table 9–7) = Old S_5 row (in Table 9–6) $- [4 \times X_2$ row (obtained from the first step above)], or:

	1	4	3	0	1	40
Old S_5 row						
$(-)$ $4 \times X_2$ row	$4(\tfrac{1}{3})=\tfrac{1}{3}$	$4(1)=4$	$4(\tfrac{1}{3})=\tfrac{1}{3}$	$4(\tfrac{1}{6})=\tfrac{2}{3}$	$4(0)=0$	$4(10)=40$
	$-\tfrac{1}{3}$	0	$\tfrac{5}{3}$	$-\tfrac{2}{3}$	1	0

The values in the Z_j and $C_j - Z_j$ rows are computed in the usual manner. There is no positive value in the $C_j - Z_j$ row. Thus, the optimum solution is obtained.

$X_2 = 10$ (chairs), all other variables $= 0$, and
$F = \$100$ (profit)

Notice that although Example 3 has two constraints, it has only one variable with a nonzero value in the optimum solution ($X_2 = 10$). This fact is true in a degenerate case.

If we had used X_2 to replace S_5 in the second simplex table, we would have to prepare two tables (Tables 9–8 and 9–9) to obtain the same optimum solution.

Table 9-8 **Second Simplex Table (Example 3)**
(Use X_2 to Replace S_5)

C_j	$\$3$	$\$10$	$\$2$	$\$0$	$\$0$	Feasible Solution		$\dfrac{V}{X_1}$
C_i	X_1	X_2	X_3	S_4	S_5	Basic Variable	Value V	
$\$0$	$1/2$	0	$-5/2$	1	$-3/2$	S_4	0	$0 \div \tfrac{1}{2} = 0$
$\$10$	$1/4$	1	$3/4$	0	$1/4$	X_2	10	$10 \div \tfrac{1}{4} = 40$
Z_j	$5/2$	10	$15/2$	0	$5/2$	$F = 100$		
$C_j - Z_j$	$1/2$	0	$-11/2$	0	$-5/2$			

Positive Value

Table 9–9 **Third Simplex Table—Optimal (Example 3)**

C_j	$\$3$	$\$10$	$\$2$	$\$0$	$\$0$	Feasible Solution	
C_i	X_1	X_2	X_3	S_4	S_5	Basic Variable	Value V
$\$ 3$	1	0	-5	2	-3	X_1	0
$\$10$	0	1	2	$-\tfrac{1}{2}$	1	X_2	10
Z_j	3	10	5	1	1	$F = 100$	
$C_j - Z_j$	0	0	-3	-1	-1		

All $C_j - Z_j$ values are now negative or zero.

*9.4 ALTERNATIVE SIMPLEX TABLE METHOD

Simplex tables can be arranged in various ways. The arrangement with the $C_j - Z_j$ row as presented in previous sections is one of the most commonly used methods. Another common arrangement, which is called the *alternative simplex table* in this section, is introduced to show a more apparent agreement with the equations presented in the summary of Example 1 on page 173.

The alternative simplex table is simpler in appearance. It requires no C_i column, Z_j row, and $C_j - Z_j$ row. The coefficients of the objective function F are written in the bottom row instead. However, it does require additional work of performing row operations on the F rows in the process of computing the optimum solution.

EXAMPLE 4   Refer to the information given in Example 2, page 174. The initial simplex table, Table 9–3, is rearranged as shown in Table 9–10. Use the alternative simplex table method to find the optimum solution.

Table 9–10   **Initial Simplex Table (Example 4) (Rearranged from Table 9–3)**

X_1	X_2	S_3	S_4	Basic Variable	V	$\dfrac{V}{X_2}$
2	6	1	0	S_3	12	$12 \div 6 = 2$
1	④	0	1	S_4	7	$7 \div 4 = 1\frac{3}{4}$ (smaller)
\$3	\$10	\$0	\$0		F	

↑
Largest Positive Value

The steps and detailed computations of the simplex method by using the alternative simplex tables are the same as those illustrated in Example 2. The second simplex table is shown in Table 9–11.

Table 9–11   **Second Simplex Table (Example 4) (See Table 9–4 for S_3 and X_2 Rows)**

X_1	X_2	S_3	S_4	Basic Variable	V	$\dfrac{V}{X_1}$
½	0	1	$-\frac{3}{2}$	S_3	$\frac{3}{2}$	$\frac{3}{2} \div \frac{1}{2} = 3$ (smaller positive)
$\frac{1}{4}$	1	0	$\frac{1}{4}$	X_2	$\frac{7}{4}$	$\frac{7}{4} \div \frac{1}{4} = 7$
$\frac{2}{4}$	0	0	$-\frac{10}{4}$		$F - \frac{70}{4}$	

↑
Largest Positive Value

The F row in the second simplex table is obtained by performing the row operation with the pivot ④ in the initial simplex table as follows:

Old F row	3	10	0	0	F
(−): $10 \times (X_2$ row)	$10(\frac{1}{4}) = \frac{10}{4}$	$10(1) = 10$	$10(0) = 0$	$10(\frac{1}{4}) = \frac{10}{4}$	$10(\frac{7}{4}) = \frac{70}{4}$
	$\frac{2}{4}$	0	0	$-\frac{10}{4}$	$F - \frac{70}{4}$

(Transform 10 in pivot column to 0)

The third simplex table is shown in Table 9–12.

Table 9–12   **Third Simplex Table (Example 4)**
(See Table 9–5 for X_1 and X_2 rows)

X_1	X_2	S_3	S_4	Basic Variable	V
1	0	2	−3	X_1	3
0	1	$-\frac{1}{2}$	1	X_2	1
0	0	−1	−1		$F - 19$

All values in the last row (other than $F - 19$) now are negative or zero. Thus, we have the optimum solution.

The F row in the third simplex table is obtained by performing the row operation with the pivot ② in the second simplex table as follows:

Old F row	$\frac{2}{4} = \frac{1}{2}$	0	0	$-\frac{10}{4} = -\frac{5}{2}$	$F - \frac{70}{4}$
(−): $\frac{1}{2} \times (X_1$ row)	$\frac{1}{2}(1) = \frac{1}{2}$	$\frac{1}{2}(0) = 0$	$\frac{1}{2}(2) = 1$	$\frac{1}{2}(-3) = -\frac{3}{2}$	$\frac{1}{2}(3) = \frac{3}{2}$
New F row	0	0	−1	−1	$F - 19$

The new F row as shown in the final simplex table gives the equation:

$$0X_1 + 0X_2 - 1S_3 - 1S_4 = F - 19$$

The table also gives the optimum solution:

$$X_1 = 3, X_2 = 1, S_3 = 0, \text{ and } S_4 = 0$$

Substitute the above values in the F equation:

$$F - 19 = 0$$
$$F = \$19 \text{ (The maximized profit.)}$$

EXERCISE 9–2   Reference: Sections 9.2, 9.3, and 9.4

Use the simplex table method to find the solution for each of the following problems:

1. Maximize the objective function:

$$F = 3X_1 + 2X_2$$

Subject to restraints:

$$2X_1 + 3X_2 \leq 12$$
$$4X_1 + 2X_2 \leq 16$$
$$X_1, X_2 \geq 0$$

2. Maximize the objective function:

$$F = 32X_1 + 12X_2$$

Subject to restraints:

$$\tfrac{1}{2}X_1 + \tfrac{1}{4}X_2 \leq 3$$
$$2X_1 + 3X_2 \leq 24$$
$$X_1, X_2 \geq 0$$

3. Maximize the objective function:

$$F = 3X_1 + 5X_2$$

Subject to restraints:

$$2X_1 + 6X_2 \leq 12$$
$$X_1 + 4X_2 \leq 7$$
$$X_1, X_2 \geq 0$$

4. Maximize the objective function:

$$F = 8X_1 + 5X_2$$

Subject to restraints:

$$4X_1 + 2X_2 \leq 100$$
$$X_2 \leq 20$$
$$X_1, X_2 \geq 0$$

5. Maximize the objective function:

$$F = 2X_1 + 7X_2$$

Subject to restraints:

$$X_1 \leq 5$$
$$X_2 \leq 10$$
$$X_2 \leq 14$$
$$X_1, X_2 \geq 0$$

6. Maximize the objective function:

$$F = 4X_1 + 6X_2 + 10X_3$$

Subject to restraints:

$$X_1 + X_2 \leq 10$$
$$2X_1 + X_3 \leq 9$$
$$X_2 + X_3 \leq 13$$
$$X_1, X_2, X_3 \geq 0$$

7. The Baxter Company makes doors and windows. Each door requires 1 hour on machine I and 2 hours on machine II. Each window requires 3 hours on machine I and 1 hour on machine II. The company has a maximum of 9 hours available on machine I and a maximum of 8 hours available on machine II. Determine the units of doors and windows that should be made in order to maximize profit. Assume that the company can realize a profit of $10 on each door and $20 on each window. Find the solution by the simplex table method.

8. Refer to Problem 7. Assume that the company can realize a profit of $10 on each door and $35 on each window. Find the solution by the simplex table method.

9. An investment company makes two types of loans: industrial loans at 15% interest rate per year, and residential loans at 10% interest rate per year. The company can lend a maximum of $10 million. For safety, the company has the policy of investing not more than 60% of the total amount in industrial loans and not more than 80% in residential loans. The company wishes to maximize interest income. Determine the amount that should be invested in each type of loan. Find the solution by the simplex table method.

10. A calculator company produces two different models of adding machines: X_1 and X_2. The company has a capacity of 60,000 man-hours for the production process. It requires 2 man-hours to produce an X_1 machine and 3 man-hours to produce an X_2 machine. The market for X_1 machines is limited to 15,000 units and that for X_2 machines to 18,000 units. The selling price of each X_1 machine is $100 and its total cost is $90. The selling price of each X_2 machine is $146 and its total cost is $130. Determine the units of X_1 and X_2 that should be produced to maximize profit. Use the simplex table method and the graphic method to find the answer.

10 Simplex Method for Minimization

The methods of using simplex tables to solve minimization problems are illustrated by four examples in this chapter. The first two examples in Section 10.1 illustrate the direct methods. The third example in Section 10.2 illustrates the indirect method which converts a minimization problem to an equivalent maximization problem before using simplex tables to solve the problem. The fourth example in Section 10.3 uses the alternative simplex table method (first introduced in Section 9.4 of Chapter 9) to solve a minimization problem. Section 10.4 gives a summary of the procedures used in the simplex method for solving maximization and minimization problems.

10.1 SIMPLEX TABLES FOR MINIMIZATION PROBLEMS

EXAMPLE 1 The information given in Example 7 of Section 8.3 (page 156) is repeated here:

A chemical company plans to make a certain type of mixed powder. The mixture requires the use of material M at a cost of $8 per bag and material N at a cost of $5 per bag. Each bag of material M weighs 4 pounds and each bag of material N weighs 2 pounds. At least 100 pounds of powder are required and at least 20 bags of material N must be used in the mixture. How many bags of each material should be used in order to minimize the total cost of the mixed powder?

The above information is summarized as follows:

Minimize the objective function:

$F = \$8X_1 + \$5X_2$ ($F = $ cost)

Subject to two given restraints:

(1) $4X_1 + 2X_2 \geq 100$ (pounds of powder required)
(2) $\qquad X_2 \geq \ 20$ (bags of N must be used)

and two nonnegative restraints:

(3) $\qquad X_1 \geq 0$ (number of bags of material M to be used)
(4) $\qquad X_2 \geq 0$ (number of bags of material N to be used)

Use the simplex method to find the values of X_1 and X_2.

SOLUTION   The steps of the simplex method using simplex tables are listed as follows:

STEP 1   Use slack or surplus variables (denoted by S) to convert each inequality of the constraints to an equation.

Constraint (1) has a "greater than" or $>$ sign. Thus, subtract surplus variable S_3 from the left side of the inequality to make:

$$4X_1 + 2X_2 - S_3 = 100 \qquad (1)'$$

Constraint (2) also has a "greater than" sign. Thus, subtract surplus variable S_4 from X_2 to make:

$$X_2 - S_4 = 20 \qquad (2)'$$

STEP 2   Use artificial variables, denoted by A, to obtain a logical initial solution. The initial solution will have all original variables (X's) equal to zero, or $X_1 = 0$ and $X_2 = 0$.

Constraint (1), which has a "greater than" sign, has been converted to Equation (1)'. However, this equation cannot give a logical initial solution. When $X_1 = 0$ and $X_2 = 0$, Equation (1)' becomes:

$$4(0) + 2(0) - S_3 = 100$$
$$S_3 = -100 \text{ pounds}$$

A negative number of pounds of material cannot be used in the mixture. Thus, a positive artificial variable A_5 must be added to the left side of Equation (1)' to make:

$$4X_1 + 2X_2 - S_3 + A_5 = 100 \qquad (1)$$

Now, if we assign $X_1 = X_2 = S_3 = 0$, Equation (1) gives a logical answer, or:

$$4(0) + 2(0) - 0 + A_5 = 100$$
$$A_5 = 100 \text{ pounds}$$

We may use 100 pounds of material represented by the artificial variable A_5.

Likewise, we add artificial variable A_6 to the left side of Equation (2)' to make:

$$X_2 - S_4 + A_6 = 20 \qquad (2)$$

Arrange the two new equations to include the same number of variables. This can be done by adding the extra variables with zero coefficients to each equation.

$$4X_1 + 2X_2 - 1S_3 + 0S_4 + 1A_5 + 0A_6 = 100 \qquad (1)$$
$$0X_1 + 1X_2 + 0S_3 - 1S_4 + 0A_5 + 1A_6 = 20 \qquad (2)$$

Also, write the objective function equation to include the same number of variables.

$$F = \$8X_1 + \$5X_2 + \$0S_3 + \$0S_4 + \$MA_5 + \$MA_6$$

The coefficient of each surplus variable (or slack variable, if any) is 0, whereas the coefficient of each artificial variable is M. M represents a very large number. The objective function is to minimize the value of F (the cost). A large value of the coefficient, such as $M = \$1$ million, will force the artificial variable to be zero. In other words, A_5 and A_6 will be driven out of the final solution to the problem.

STEP 3   Find the initial solution and construct the first simplex table. (Use Equations (1) and (2).) This is done in Table 10-1.

Table 10-1   Initial Simplex Table (Example 1)

C_j	$8	$5	$0	$0	$M	$M	Feasible Solution		$\dfrac{V}{X_1}$
C_i	X_1	X_2	S_3	S_4	A_5	A_6	Basic Variable	Value V	X_1
M	4 1	2	−1	0	1	0	A_5	100	$100 \div 4 = 25$
M	0 0	1	0	−1	0	1	A_6	20	$20 \div 0 =$ Undefined
Z_j	4M	3M	−M	−M	M	M	$F = 120M$		
$C_j - Z_j$	8 − 4M	5 − 3M	M	M	0	0			

Largest Negative Value

1. The initial feasible solution is to use material A_5 (100 bags if we assume that each bag weighs one pound) and A_6 (20 bags). The initial solution is obtained from the two constraint equations as explained in Step 1 by letting $X_1 = X_2 = S_3 = S_4 = 0$. The number of variables in the solution column (A_5 and A_6) thus is equal to the number of constraints. Note that the four numbers under the solution variables, or:

$$\begin{array}{cc} A_5 & A_6 \\ 1 & 0 \\ 0 & 1 \end{array}$$

form a 2×2 identity matrix. The coefficient of A_5 is a unit vector with 1 on the top (or in the $A_5 = 100$ row), and the coefficient of A_6 is also a unit vector with 1 at the bottom (or in the $A_6 = 20$ row).

2. The values in the Z_j and $C_j - Z_j$ rows are computed in the usual manner. The interpretation of the Z value in each material-variable column, however, is different from that in the F column. For example, the value of Z in the X_1 column ($4M + 0M = 4M$) is the total cost of A_5 and A_6 that would be reduced if 1 unit of X_1 is placed into the solution. That is, instead of having the initial solution,

$X_1 = 0$, $X_2 = 0$, $S_3 = 0$, $S_4 = 0$, $A_5 = 100$, and $A_6 = 20$

we would have:

$X_1 = 1$ (increased from 0 to 1 unit)
$X_2 = 0$, $S_3 = 0$, $S_4 = 0$ (no change)

$A_5 = 100 - 4 = 96$ (reduced by 4 — see X_1 column)
$A_6 = 20 - 0 = 20$ (reduced by 0 in X_1 column and A_6 row)

The detailed computations of A_5 and A_6 values are as follows:

Substitute $X_1 = 1$ and $X_2 = S_3 = S_4 = 0$ in Equation (1):

$$4(1) + 2(0) - 1(0) + 0(0) + 1A_5 + 0A_6 = 100$$
$$A_5 = 100 - 4 = 96$$

or 4 units of A_5 are reduced at $\$M$ per unit, or a total of $\$M(4) = \$4M$.
Substitute $X_1 = 1$ and $X_2 = S_3 = S_4 = 0$ in Equation (2):

$$0(1) + 1(0) + 0(0) - 1(0) + 0A_5 + 1A_6 = 20$$
$$A_6 = 20 - 0(1)$$
$$= 20 - 0 = 20$$

or 0 units of A_6 are reduced at $\$M$ per unit, or a total of $\$M(0) = \$0$.
Thus, if 1 unit of X_1 is placed into the initial solution, the total cost of material A_5 and A_6 ($100M + 20M = \$120M$) would be reduced by:

$$4M + 0M = \$4M$$

Each $C_j - Z_j$ value represents the net cost per unit of the material indicated in the jth column that can be increased (or added) to the total cost if the material is used or placed into the solution.

If 1 unit in X_1 is placed into solution, the total cost would be increased by \$8 ($= C_j$) but reduced ($-$) by \$4M ($= Z_j$), or a net cost increase of $\$(8 - 4M)$ ($= C_j - Z_j$). If 25 units in X_1 are placed into the solution, the total cost (F) would be increased (negatively, since M is a very large number) from \$120M (as shown in Table 10–1, F column) to:

$$120M + 25(8 - 4M) = 120M - 100M + 200 = \$(20M + 200)$$

The above result can be verified in the next simplex table (Table 10–2).

Table 10-2 **Second Simplex Table (Example 1)**

C_j	\$8	\$5		\$0	\$0	\$M	\$M	Feasible Solution		V
C_i	X_1	X_2		S_3	S_4	A_5	A_6	Basic Variable	Value V	X_2
\$8	1	1/2	0	−1/4	0	1/4	0	X_1	25	$25 \div \frac{1}{2} = 50$
\$M	0	1	−1	0	−1	0	1	A_6	20	$20 \div 1 = 20$
Z_j	8	$4+M$		−2	−M	2	M	$F = 200 + 20M$		
$C_j - Z_j$	0	$1-M$		2	M	M−2	0			

Only Negative Value

STEP 4   Construct the second simplex table to find the second feasible solution. (Use the information obtained from the first simplex table.) See Table 10-2.

1. In the solution column of Table 10-2, X_1 has replaced A_5 in Table 10-1. The replacement is determined in the following manner:

   Figure from Table 10-1 which one of the materials (variables) will reduce the cost per unit the most. This is indicated by an arrow in the column which has the largest negative value of $C_j - Z_j$. The largest negative value is $8 - 4M$ (since M is a very large number) and is located in the X_1 column. Thus, X_1 is the incoming variable, and the X_1 column is the pivot column.

   Figure from Table 10-1 which one of the materials (A_5 and A_6) is to be replaced by X_1. This is done by dividing each number in the V column by the corresponding row number in the X_1 column in the usual manner as shown in the last column, or:

   In the A_5 row: $V \div X_1 = 100 \div 4 = 25$ (smallest positive quotient)
   In the A_6 row: $V \div X_1 = 20 \div 0 =$ undefined

   We again select the row with the smallest positive quotient for the replacement as done in the maximization problems. Thus, A_5 is the outgoing variable and the A_5 row is the pivot row. The pivot is ④.

2. The solution column in Table 10-2 now has X_1 and A_6 since X_1 replaced A_5 in Table 10-1. The four numbers under the solution variables X_1 and A_6 in Table 10-2 thus must form a 2×2 identity matrix, or:

$$\begin{array}{cc} \underline{X_1} & \underline{A_6} \\ 1 & 0 \\ 0 & 1 \end{array}$$

Thus, we must transform the coefficients of X_1 from the pivot column:

④ in Table 10-1 to: $\dfrac{1}{0}$ in Table 10-2
0

The values of the individual rows of Table 10-2 are computed by the row operations in the same manner as the maximization problems.

Compute the values in the incoming X_1 row from the pivot row in Table 10-1. Use Operation 1, $k = 4$, the pivot.

X_1 row (in Table 10-2) $= A_5$ row (in Table 10-1) $\div 4$ or:

$$\dfrac{4}{4} = 1 \qquad \dfrac{2}{4} = \dfrac{1}{2} \qquad -\dfrac{1}{4} \qquad \dfrac{0}{4} = 0 \qquad \dfrac{1}{4} \qquad \dfrac{0}{4} = 0 \qquad \dfrac{100}{4} = 25$$

(Transform
pivot to 1)

Compute the values in the A_6 row. Since the number in the pivot column (X_1 column) and the A_6 row in Table 10-1 is 0 and the required number in the same location in Table 10-2 is also 0, the values in A_6 are unchanged, or:

New A_6 row (in Table 10-2) = Old A_6 row (in Table 10-1)

Again the values in the Z_j and $C_j - Z_j$ rows are computed in the same manner as the maximization problems.

There is a negative value $(1 - M)$ in the $C_j - Z_j$ row in Table 10-2. Thus, the total cost, $F = 200 + 20M$, can further be reduced.

STEP 5   Construct the third simplex table to find the third feasible solution. (Use the information obtained from the second simplex table.) See Table 10-3.

Table 10-3   **Third Simplex Table—Optimal (Example 1)**

C_j	$8	$5	$0	$0	M	M	Feasible Solution	
C_i	X_1	X_2	S_3	S_4	A_5	A_6	Basic Variable	Value V
$8	1	0	$-\frac{1}{4}$	$\frac{1}{2}$	$\frac{1}{4}$	$-\frac{1}{2}$	X_1	15
$5	0	1	0	-1	0	1	X_2	20
Z_j	8	5	-2	-1	2	1		$F = 220$
$C_j - Z_j$	0	0	2	1	$M - 2$	$M - 1$		

1. In the solution column of Table 10-3, X_2 has replaced A_6 in Table 10-2. The replacement is determined in Table 10-2:

   X_2 is the incoming variable. The X_2 column has the only negative value in the $C_j - Z_j$ row: $1 - M$.

   A_6 is the outgoing variable. A_6 has the smallest positive quotient in the V/X_2 column: 20.

   ① is the pivot.

2. The solution column in Table 10-3 now has X_1 and X_2. We must transform the coefficients of the incoming variable X_2 from the pivot column:

$$\frac{\frac{1}{2}}{①} \text{ in Table 10-2 to: } \frac{0}{1} \text{ in Table 10-3}$$

Since the pivot is already ①, the values in the pivot row are unchanged, or:

X_2 row (in Table 10-3) = A_6 row (in Table 10-2)

Compute the values in the new X_1 row in Table 10-3. Use row Operation 2, $k = \frac{1}{2}$, the number in the intersection of the pivot column and the X_1 row in Table 10-2.

New X_1 row (in Table 10-3) = Old X_1 row (in Table 10-2) − [($\frac{1}{2}$) × X_2 row (in Table 10-3)], or:

$$
\begin{array}{llll}
\text{Old } X_1 \text{ row} & 1 & \tfrac{1}{2} & -\tfrac{1}{4} & 0 \\
(-): \tfrac{1}{2} \times X_2 \text{ row} \quad \tfrac{1}{2}(0)=0 & \tfrac{1}{2}(1)=\tfrac{1}{2} & \tfrac{1}{2}(0)=0 & \tfrac{1}{2}(-1)=-\tfrac{1}{2} \\
\text{New } X_1 \text{ row} & 1 & 0 & -\tfrac{1}{4} & \tfrac{1}{2}
\end{array}
$$

(Transform
$\tfrac{1}{2}$ in pivot
column to 0)

$$
\begin{array}{lll}
 & \tfrac{1}{4} & 0 & 25 \\
\tfrac{1}{2}(0)=0 & \tfrac{1}{2}(1)=\tfrac{1}{2} & \tfrac{1}{2}(20)=10 \\
 & \tfrac{1}{4} & -\tfrac{1}{2} & 15
\end{array}
$$

There is no negative value in the $C_j - Z_j$ row in Table 10–3. Thus, the optimum solution is obtained:

$X_1 = 15$ (bags of material M to be used)
$X_2 = 20$ (bags of material N to be used)

$F = \$8(15) + \$5(20) = \$220$ (the minimized cost)

EXAMPLE 2 Refer to Example 1. Assume that the objective function is not changed, or minimize:

$F \text{ (cost)} = \$8X_1 + \$5X_2$

But the constraints are changed as follows:

(1) $4X_1 + 2X_2 = 100$ (pounds — exactly 100 pounds of mixed powder are required.)

(2) $X_1 \le 12$ (bags of material M — no more than 12 bags of M are allowed.)

(3) $X_2 \ge 20$ (bags of material N — at least 20 bags of N must be used.)

Use the simplex method to find the values of X_1 and X_2 (≥ 0).

SOLUTION (Note: This problem is the same as that of Example 9, Chapter 8. Also see Figure 8–7.)
   The steps of the simplex method by using the simplex tables for this example are basically the same as those in Example 1, only special notes are given as follows:

STEP 1 Use slack or surplus variables to convert each inequality of the constraints to an equation.

   Constraint (1) is an equation. Thus, no change is required.
   Constraint (2) has a "less than" or $<$ sign. Thus, add slack variable S_3 to X_1 to make Equation (2):

$X_1 + S_3 = 12$

   Constraint (3) has a "greater than" or $>$ sign. Thus, subtract surplus variable S_4 from X_2 to make Equation (3)':

$X_2 - S_4 = 20$

STEP 2   Use artificial variables to obtain a logical initial solution. The initial solution will have original variables equal to zero, or $X_1 = 0$ and $X_2 = 0$.

Constraint (1) has an "equal" or = sign. When X_1 and X_2 equal 0, constraint (1) becomes:

$$4(0) + 2(0) = 100$$
$$0 = 100 \text{ (an unlogical solution)}$$

Thus, add artificial variable A_5 to the left side of the constraint to make Equation (1):

$$4X_1 + 2X_2 + A_5 = 100$$

When X_1 and X_2 equal 0, Equation (1) becomes,

$$4(0) + 2(0) + A_5 = 100$$
$$A_5 = 100$$

which indicates that we should use 100 pounds of material represented by the artificial variable A_5.

Constraint (2), which has a "less than" sign, has been converted to Equation (2). The equation can give a logical initial solution because, when $X_1 = 0$, the equation becomes:

$$0 + S_3 = 12, \text{ and } S_3 = 12 \text{ bags}$$

Thus, there is no need to add an artificial variable to Equation (2).

Constraint (3), which has a "greater than" sign, has been converted to Equation (3)'. However, this equation cannot give a logical initial solution. When $X_2 = 0$, Equation (3)' becomes:

$$0 - S_4 = 20, \text{ and } S_4 = -20 \text{ bags}$$

A negative number of bags of material cannot be used in the mixture. Thus, add artificial variable A_6 to the left side of Equation (3)' to make Equation (3):

$$X_2 - S_4 + A_6 = 20$$

If we let $X_2 = 0$ and $S_4 = 0$, Equation (3) becomes:

$$0 - 0 + A_6 = 20$$
$$A_6 = 20 \text{ bags, a logical solution}$$

Now, the three constraints become a system of three equations in 6 unknowns:

(1) $4X_1 + 2X_2 \qquad\qquad + A_5 \qquad = 100$
(2) $X_1 \qquad + S_3 \qquad\qquad = 12$
(3) $\qquad X_2 \qquad - S_4 \qquad + A_6 = 20$

Arrange the three equations and the objective function equation to include the same number of variables. The new system of three equations is:

(1) $4X_1 + 2X_2 + 0S_3 + 0S_4 + 1A_5 + 0A_6 = 100$
(2) $1X_1 + 0X_2 + 1S_3 + 0S_4 + 0A_5 + 0A_6 = 12$
(3) $0X_1 + 1X_2 + 0S_3 - 1S_4 + 0A_5 + 1A_6 = 20$

The objective function becomes:

$$F = \$8X_1 + \$5X_2 + \$0S_3 + \$0S_4 + \$MA_5 + \$MA_6$$

STEP 3 Find the initial solution and construct the first simplex table. (Use Equations (1), (2), and (3).) This is done in Table 10–4.

Table 10-4 **Initial Simplex Table (Example 2)**

C_j	$\$8$	$\$5$	$\$0$	$\$0$	$\$M$	$\$M$	Feasible Solution		V
C_i	X_1	X_2	S_3	S_4	A_5	A_6	Basic Variable	Value V	X_1
$\$M$	4	2	0	0	1	0	A_5	100	$100 \div 4 = 25$
$\$0$	1	0	1	0	0	0	S_3	12	$12 \div 1 = 12$ (Smallest)
$\$M$	0	1	0	-1	0	1	A_6	20	$20 \div 0 =$ Undefined
Z_j	$4M$	$3M$	0	$-M$	M	M	$F = 120M$		
$C_j - Z_j$	$8-4M$	$5-3M$	0	M	0	0			

Largest Negative Value

The initial feasible solution, $A_5 = 100$, $S_3 = 12$, and $A_6 = 20$, is obtained from the three constraint equations as explained in Step 1 by letting X_1, X_2, and the surplus variable with negative coefficient S_4 equal zero. The number of variables in the solution column thus is equal to the number of constraints (3). Note that the nine numbers under the solution variables form a 3×3 identity matrix and the 1's correspond with the row locations of the basic variables in the solution column respectively, or:

			Basic
A_5	S_3	A_6	Variable
1	0	0	A_5
0	1	0 and the 1's located in rows	S_3
0	0	1	A_6

This table also indicates the information to be used in the second simplex table (Table 10–5): The incoming variable is X_1 (with the largest negative value $(8 - 4M)$ in the $C_j - Z_j$ row), the outgoing variable is S_3 (with the smallest positive quotient (12) in the V/X_1 column), and the pivot is ①.

STEP 4 Construct the second simplex table to find the second feasible solution. This is done in Table 10–5.

The solution column in Table 10–5 now has A_5, X_1, and A_6 since X_1 replaced S_3 in Table 10–4. The nine numbers under the solution

Table 10-5   **Second Simplex Table (Example 2)**

C_j	$8	$5	$0	$0	$M	$M	Feasible Solution		$\dfrac{V}{X_2}$
C_i	X_1	X_2	S_3	S_4	A_5	A_6	Basic Variable	Value V	X_2
$M	0	2	−4	0	1	0	A_5	52	$52 \div 2 = 26$
$8	1	0	1	0	0	0	X_1	12	$12 \div 0 =$ Undefined
$M	0	(1)	0	−1	0	1	A_6	20	$20 \div 1 = 20$ (Smallest)
Z_j	8	3M	8 − 4M	−M	M	M	$F = 96 + 72M$		
$C_j − Z_j$	0	5 − 3M	−8 + 4M	M	0	0			

Only Negative Value

variables in Table 10–5 must form a 3×3 identity matrix. Thus, we must transform the coefficients of X_1 from the pivot column:

4		0
1	in Table 10–4 to: 1	in Table 10–5
0		0

Since the second and third numbers in the pivot column (1 and 0) are the same as the required numbers (1 and 0 respectively in the new table), the values in the second row (X_1) and in the third row (A_6) in Table 10–5 are the unchanged values from the respective rows in Table 10–4. Only the values in the new A_5 row need to be computed by row operation (2), $k = 4$ (the number in the intersection of the pivot column and the A_5 row in Table 10–4.)

New A_5 row (in Table 10–5) = Old A_5 row (in Table 10–4) − [$4 \times X_1$ row (in Table 10–5)], or:

Old A_5 row	4	2	0	0	1
(−): $4 \times X_1$ row	$4(1)=4$	$4(0)=0$	$4(1)=4$	$4(0)=0$	$4(0)=0$
New A_5 row	0	2	−4	0	1

	0	100
	$4(0)=0$	$4(12)=48$
	0	52

Table 10–5 also indicates the information to be used in the third simplex table (Table 10–6). The incoming variable is X_2 (with the only negative value $(5 - 3M)$ in the $C_j - Z_j$ row), the outgoing variable is A_6 (with the smallest positive quotient (20) in the V/X_2 column), and the pivot is ①.

STEP 5   Construct the third simplex table to find the third feasible solution. This is done in Table 10–6.

The solution column in Table 10–6 now has A_5, X_1, and X_2 since X_2 replaced A_6 in Table 10–5. The nine numbers under the solution

Table 10-6   Third Simplex Table (Example 2)

C_j	$8	$5	$0	$0	$M	$M	Feasible Solution		V
C_i	X_1	X_2	S_3	S_4	A_5	A_6	Basic Variable	Value V	S_4
$M	0	0	−4	2	1	−1	−2	A_5	12	$12 \div 2 = 6$ (Smallest)
$8	1	0	1	0	0	0	0	X_1	12	$12 \div 0 =$ Undefined
$5	0	1	0	−1	0	0	1	X_2	20	$20 \div (−1) = −20$
Z_j	8	5	$1 − 4M$	$−5 + 2M$	M	$5 − 2M$	$F = 196 + 12M$		
$C_j − Z_j$	0	0	$−1 + 4M$	$5 − 2M$	0	$−5 + 1M$			

Only Negative Value

variables in Table 10–6 must form a 3×3 identity matrix. Thus, we must transform the coefficients of X_2 from the pivot column:

$$
\begin{array}{ccc}
2 & & 0 \\
0 & \text{in Table 10–5 to:} \quad 0 & \text{in Table 10–6} \\
1 & & 1
\end{array}
$$

Since the second and third numbers in the pivot column (0 and 1) are the same as the required numbers (0 and 1 respectively in the new table), the values in the second row (X_1) and in the third row (X_2) in Table 10–6 are the unchanged values from the respective rows in Table 10–5. Only the values in the new A_5 row need to be computed by row operation 2, $k = 2$ (the number in the intersection of the pivot column and the A_5 row in Table 10–5.)

New A_5 row (in Table 10–6) = Old A_5 row (in Table 10–5) − [2 × X_2 row (in Table 10–6)], or:

$$
\begin{array}{ccccc}
\text{Old } A_5 \text{ row} & 0 & 2 & -4 & 0 \\
(-): 2 \times X_2 \text{ row} & 2(0)=0 & 2(1)=2 & 2(0)=0 & 2(-1)=-2 \\
\hline
\text{New } A_5 \text{ row} & 0 & 0 & -4 & 2
\end{array}
$$

$$
\begin{array}{ccc}
 1 & 0 & 52 \\
2(0)=0 & 2(1)=2 & 2(20)=40 \\
\hline
 1 & -2 & 12
\end{array}
$$

Table 10–6 also indicates the information to be used in the fourth simplex table (Table 10–7). The incoming variable is S_4 (with the only negative value ($5 − 2M$) in the $C_j − Z_j$ row), the outgoing variable is A_5 (with the only positive quotient (6) in the V/S_4 column), and the pivot is ②

STEP 6   Construct the fourth simplex table to find the fourth feasible solution. This is done in Table 10–7.

The solution column in Table 10–7 now has S_4, X_1, and X_2 since S_4 replaced A_5 in Table 10–6. The nine numbers under the solution

Table 10-7 **Fourth Simplex Table—Optimal (Example 2)**

C_j	$8	$5	$0	$0	$M	$M	Feasible Solution	
C_i	X_1	X_2	S_3	S_4	A_5	A_6	Basic Variable	Value V
$0	0	0	−2	1	$\frac{1}{2}$	−1	S_4	6
$8	1	0	1	0	0	0	X_1	12
$5	0	1	−2	0	$\frac{1}{2}$	0	X_2	26
Z_j	8	5	−2	0	$\frac{5}{2}$	0	$F = 226$	
$C_j - Z_j$	0	0	2	0	$M - (\frac{5}{2})$	M		

variables in Table 10-7 must form a 3×3 identity matrix. Thus, we must transform the coefficients of S_4 from the pivot column:

$$\begin{matrix} 2 & & 1 \\ 0 & \text{in Table 10-6 to:} & 0 & \text{in Table 10-7} \\ -1 & & 0 \end{matrix}$$

Compute the values in the incoming S_4 row from the pivot row in Table 10-6. Use Operation 1, $k = 2$, the pivot.

S_4 row (in Table 10-7) = A_5 row (in Table 10-6) ÷ 2, or:

$$\frac{0}{2} = 0 \quad \frac{0}{2} = 0 \quad \frac{-4}{2} = -2 \quad \frac{2}{2} = 1 \quad \frac{1}{2} \quad \frac{-2}{2} = -1 \quad \frac{12}{2} = 6$$

Compute the values in the X_1 row. Since the number in the pivot column and the X_1 row in Table 10-6 already is 0, the values in the X_1 row are unchanged, or:

New X_1 row (in Table 10-7) = Old X_1 row (in Table 10-6)

Compute the values in the X_2 row. Use row operation 2, $k = -1$ (the number in the intersection of the pivot column and the X_2 row in Table 10-6.)

New X_2 row (in Table 10-7) = Old X_2 row (in Table 10-6) − [(−1) × S_4 row (in Table 10-7)], or:

Old X_2 row	0	1	0
(−): (−1) × S_4 row	(−1)0 = 0	(−1)0 = 0	(−1)(−2) = 2
New X_2 row	0	1	−2

−1	0	1	20
(−1)1 = −1	(−1)$\frac{1}{2}$ = −$\frac{1}{2}$	(−1)(−1) = 1	(−1)6 = −6
0	$\frac{1}{2}$	0	26

Table 10-7 shows that there is no more negative value in the $C_j - Z_j$ row. Thus, the optimum solution is obtained:

$X_1 = 12$ (bags of material M to be used)
$X_2 = 26$ (bags of material N to be used)
$S_4 = 6$ (bags of material N as surplus from constraint 3)

$F = \$8(12) + \$5(26) = 96 + 130 = \$226$ (the minimized cost)

Check the constraints:

Substitute $X_1 = 12$ and $X_2 = 26$ in constraint (1), $4X_1 + 2X_2 = 4(12) + 2(26) = 100$ (pounds exactly); in constraint (2), $X_1 = 12$ (bags which are exactly equal to the maximum limit 12); and in constraint (3), $X_2 = 26$ (bags which are 6 bags above the minimum limit 20. The excess amount is equal to $S_4 = 6$ in the final solution.)

Thus, all constraints are satisfied by the optimum solution.

10.2 CONVERTING A MINIMIZATION PROBLEM TO A MAXIMIZATION PROBLEM

The previous discussions separated maximization and minimization problems as if they were unrelated. Actually, any minimization problem can be converted into a maximization problem, and vice versa. To convert a minimization problem to a maximization problem, simply multiply the given objective function (F) by -1. The given constraints are unchanged.

$$\text{Let } G = (-1)F = -F.$$

Then, we may change a *minimize F* problem to a *maximize G* problem for solving the minimization problem. This relationship is illustrated in Example 3.

EXAMPLE 3   Use the maximization simplex method to find the values of X_1 and X_2 (≥ 0) for the information given in Example 2.

Minimize the cost F:

$$F = 8X_1 + 5X_2$$

Constraints:

(1) $4X_1 + 2X_2 = 100$
(2) $\qquad X_1 \leq 12$
(3) $\qquad X_2 \geq 20$ (same as Example 2)

SOLUTION   The given objective function is changed to Maximize G:

$$G = -F = -(8X_1 + 5X_2)$$

The steps for "maximize G" by simplex tables are as follows:

STEP 1   Convert the three given constraints to equations by using slack, surplus, and artificial variables.

The method of conversion is the same as that used in Step 1 of Example 2. The resulting new system of three equations is also the same as that of Example 2 (page 198), or:

$$
\begin{aligned}
(1) \quad & 4X_1 + 2X_2 && + A_5 && = 100 \\
(2) \quad & X_1 && + S_3 && = 12 \\
(3) \quad & X_2 && - S_4 && + A_6 = 20
\end{aligned}
$$

However, the new objective function for this maximization problem now is:

$$G = -(8X_1 + 5X_2 + 0S_3 + 0S_4 + MA_5 + MA_6)$$

which includes the same number of variables as in the system of three equations, or written:

$$G = -8X_1 - 5X_2 - 0S_3 - 0S_4 - MA_5 - MA_6$$

M represents a very large number. A $-M$ coefficient will force each artificial variable to be zero. For example, $-MA_5$ has its maximized (highest) value when $A_5 = 0$, or $-MA_5 = -M(0) = 0$. Zero is higher than the product of $-M$ multiplied by any positive nonzero number.

STEP 2 Construct the simplex tables to find the optimum solution.

The first simplex table based on the preceding information is shown in Table 10-8. This table is identical with the first simplex table of Example 2, Table 10-4, page 199, except that the signs in the C_i column, and C_j, Z_j, and $C_j - Z_j$ rows are changed (from + to − and vice versa). The second, third, and fourth simplex tables of this example are also identical with the simplex tables (Tables 10-5, 10-6, and 10-7) of Example 2 respectively except for the signs. The second and third simplex tables are omitted here since they are only repetitious. The fourth (final)

Table 10-8 **Initial Simplex Table (Example 3)**
   (Also see Table 10-4, except signs for C_i, C_j, Z_j, and $C_j - Z_j$)

C_j	−8	−5	0	0	−M	−M	Feasible Solution		$\dfrac{V}{X_1}$
C_i	X_1	X_2	S_3	S_4	A_5	A_6	Basic Variable	Value V	
−M	4	2	0	0	1	0	A_5	100	$100 \div 4 = 25$
0	①	0	1	0	0	0	$\rightarrow S_3$	12	$12 \div 1 = 12$ (Smallest)
−M	0	1	0	−1	0	1	A_6	20	$20 \div 0 =$ Undefined
Z_j	−4M	−3M	0	M	−M	−M	$G = -120M$		
$C_j - Z_j$	−8 + 4M	−5 + 3M	0	−M	0	0			

Highest positive value

Table 10-9   **Fourth Simplex Table – Optimal (Example 3)**
                (Also see Table 10-7, except signs for C_i, C_j, Z_j, and $C_j - Z_j$)

C_j	-8	-5	0	0	$-M$	$-M$	Feasible Solution	
C_i	X_1	X_2	S_3	S_4	A_5	A_6	Basic Variable	Value V
0	0	0	-2	1	$\frac{1}{2}$	-1	S_4	6
-8	1	0	1	0	0	0	X_1	12
-5	0	1	-2	0	$\frac{1}{2}$	0	X_2	26
Z_j	-8	-5	2	0	$-\frac{5}{2}$	0	$G = -226$	
$C_j - Z_j$	0	0	-2	0	$-M + \frac{5}{2}$	$-M$		

(The final table – all values in the $C_j - Z_j$ row are negative in the table.)

simplex table is shown in Table 10-9. The optimum solution for the table is:

$X_1 = 12$, $X_2 = 26$

and the maximized objective function:

$G = -226$

Since $G = -F$, $F = -G = -(-226) = 226$. The answer to the given problem thus is:

Minimized $F = \$226$ (cost)

Note that in a given problem if we convert both the objective function (from minimize F to maximize F in a minimization problem or from maximize F to minimize F in a maximization problem) and the signs in the constraints (from $\geq$ to $\leq$ and vice versa), the solution to the converted problem may or may not have the same answer to the given problem. (See Problems 3, 4, 5, and 7 in Exercise 10-1.)

*10.3   ALTERNATIVE SIMPLEX TABLE METHOD

The alternative simplex table method presented in Section 9.4 can also be used for a minimization problem. This is illustrated in Example 4.

EXAMPLE 4   Refer to the information given in Example 1. The initial simplex table, Table 10-1, is rearranged as shown in Table 10-10. Use the alternative simplex table method to find the optimum solution.

SOLUTION   Since there are no negative values in the F row in the minimization problem, it appears that we could take $A_5 = 100$ and $A_6 = 20$ as the optimum solution. However, M represents a very large number. We therefore must force the positive M's out of the F equation row. This can be done by pivoting the A_5 and A_6 columns with ① as the pivot in each case.

Table 10-10 **Initial Simplex Table (Example 4)
(Rearranged from Table 10-1)**

X_1	X_2	S_3	S_4	A_5	A_6	Basic Variable	V
4	2	−1	0	①	0	A_5	100
0	1	0	−1	0	①	A_6	20
8	5	0	0	M	M		F

1. Pivot the A_5 column: Convert $\begin{smallmatrix}1\\M\end{smallmatrix}$ to $\begin{smallmatrix}1\\0\end{smallmatrix}$.

Old F row	8	5	0	0	M	M		F
(−): $M \times (A_5$ row)	$4M$	$2M$	$-1M$	$0M$	$1M$	$0M$		$100M$
New F row	$8-4M$	$5-2M$	$1M$	0	0	M		$F-100M$

2. Pivot the A_6 column: Convert $\begin{smallmatrix}0\\M\end{smallmatrix}$ to $\begin{smallmatrix}0\\0\end{smallmatrix}$.

Old F row	$8-4M$	$5-2M$	$1M$	0	0	M		$F-100M$
(−): $M \times (A_6$ row)	$0M$	$1M$	$0M$	$-1M$	$0M$	$1M$		$20M$
New F row	$8-4M$	$5-3M$	M	M	0	0		$F-120M$

The last new F row is used to adjust the F row in Table 10-10 as shown in Table 10-11.

Table 10-11 **Adjusted Initial Simplex Table (Example 4)
(See Table 10-10 for A_5 and A_6 rows and the second step for the new F row.)**

X_1	X_2	S_3	S_4	A_5	A_6	Basic Variable	V
④	2	−1	0	1	0	A_5	100
0	1	0	−1	0	1	A_6	20
$8-4M$	$5-3M$	M	M	0	0		$F-120M$

Largest Negative Value

The values in the F equation row in Table 10-11 are now in complete agreement with the values on the $C_j - Z_j$ row in Table 10-1. If we perform all necessary row operations, beginning with pivoting the X_1 column in Table 10-11, we should have the optimal simplex table in agreement with the values shown in Table 10-3. The optimum solution should be:

$X_1 = 15$, $X_2 = 20$, and $F = \$220$ (the minimized cost)

Table 10-12 **Summary—Simplex Method Procedure**

Procedure	Maximization Problem (See Example 2 of Chapter 9, page 174, and Example 3 of Chapter 10, page 203)	Minimization Problem (See Example 2 of Chapter 10, page 197)
I. To convert constraints to equations for initial solution.		
1. Constraint has a "≤" sign.	Add a slack variable, such as converting $2X_1 + 6X_2 \le 12$ to $2X_1 + 6X_2 + S_3 = 12$ in Example 2 of Ch. 9.	Same as in maximization, such as converting $X_1 \le 12$ to $X_1 + S_3 = 12$ in Example 2.
2. Constraint has a "≥" sign.	Subtract a surplus variable and add an artificial variable, such as converting $X_2 \ge 20$ to $X_2 - S_4 + A_6 = 20$ in Example 3 of Ch. 10.	Same as in maximization, such as converting $X_2 \ge 20$ to $X_2 - S_4 + A_6 = 20$ in Example 2.
3. Constraint has a "=" sign.	Add an artificial variable, such as converting $4X_1 + 2X_2 = 100$ to $4X_1 + 2X_2 + A_5 = 100$ in Example 3 of Ch. 10.	Same as in maximization, such as converting $4X_1 + 2X_2 = 100$ to $4X_1 + 2X_2 + A_5 = 100$ in Example 2.
II. To complete the objective function by including slack, surplus, and artificial variables.	Assign a zero coefficient to each slack or surplus variable. Assign $-M$ coefficient to each artificial variable, such as in Example 3 of Ch. 10.	Assign a zero coefficient to each slack or surplus variable. Assign $+M$ coefficient to each artificial variable.
III. To select a column to replace a row in a simplex table—pivoting.	Select the pivot column: column with highest positive $C_j - Z_j$ value. Select the pivot row: row with smallest quotient ($= V$ value in solution column ÷ corresponding row value in the pivot column.) Transform the pivot to 1 and all other numbers in pivot column to 0's.	Select the pivot column: column with highest negative $C_j - Z_j$ value. Select the pivot row: same as in maximization. Transformation: same as in maximization.
IV. To recognize an optimum solution.	All values in $C_j - Z_j$ row are negative or zero.	All values in $C_j - Z_j$ row are positive or zero.

10.4 SUMMARY OF SIMPLEX METHOD PROCEDURE

The details of solving linear programming problems (maximization and minimization) by the simplex method are summarized in Table 10–12 on the preceding page.

EXERCISE 10–1   **Reference: Chapter 10**

Use the simplex table method to find the solution for each of the following problems:

1. Minimize the objective function:

$$F = 4X_1 + 8X_2$$

Subject to restraints: $2X_1 + 5X_2 \geq 20$
$$X_1 \geq 5$$
$$X_1, X_2 \geq 0$$

2. Minimize the objective function:

$$F = 2X_1 + 5X_2$$

Subject to restraints: $3X_1 + X_2 \geq 12$
$$X_1 + X_2 = 10$$
$$X_1 + 3X_2 \geq 18$$
$$X_1, X_2 \geq 0$$

3. Minimize the objective function:

$$F = 2X_1 + 7X_2$$

Subject to restraints: $X_1 \geq 5$
$$X_2 \geq 10$$
$$X_1, X_2 \geq 0$$

Compare this problem with Problem 5 of Exercise 9–2, page 189. Here the objective function is changed (from maximize to minimize) and the signs in the constraints are reversed (from $\leq$ to $\geq$). Are the answers to X_1, X_2, and F different for the two problems? (The answer to Problem 5 is: $X_1 = 5$, $x_2 = 10$, and $F = 80$.)

4. Minimize the objective function:

$$F = 32X_1 + 12X_2$$

Subject to restraints: $\frac{1}{2}X_1 + \frac{1}{4}X_2 \geq 3$
$$2X_1 + 3X_2 \geq 24$$
$$X_1, X_2 \geq 0$$

Compare this problem with Problem 2 of Exercise 9–2. Here the objective function is changed (from maximize to minimize) and the signs in the restraints are reversed (from $\leq$ to $\geq$). Are the answers to X_1, X_2, and F different for the two problems? (The answer to Problem 2 is: $X_1 = 6$, $X_2 = 0$ and $F = 192$.)

5. Use the maximization simplex method to find the values of X_1 and X_2 from the given information. (Same as Example 1, page 191.)

   Minimize: $F = 8X_1 + 5X_2$
   Subject to: $4X_1 + 2X_2 \geq 100$
   $$X_2 \geq 20$$
   $$X_1, X_2 \geq 0$$

   Compare this problem with Problem 4 of Exercise 9–2. Here the objective function is changed (from maximize to minimize) and the signs in the restraints are reversed (from $\leq$ to $\geq$). Are the answers to X_1, X_2, and F different for the two problems? (The answer to Problem 4 is: $X_1 = 15$, $X_2 = 20$, and $F = 220$.)

6. Refer to Example 3, page 203. Work out the second and third simplex tables.

7. Maximize the objective function:
   $$F = 8X_1 + 5X_2$$

   Subject to restraints: $4X_1 + 2X_2 = 100$
   $$X_1 \geq 12$$
   $$X_2 \leq 20$$
   $$X_1, X_2 \geq 0$$

   Compare this problem with Example 3 of this chapter. Here the objective function is changed (from minimize to maximize) and the signs in the constraints are reversed (from $\leq$ to $\geq$, and from $\geq$ to $\leq$). Are the answers to X_1, X_2, and F different for the two problems? (The answer to Example 3 is: $X_1 = 12$, $X_2 = 26$, and $F = 226$. See Table 10–9.)

8. A company plans to combine materials H and N to make a product. The product must weigh at least 80 pounds. At least 50 pounds of H and no more than 60 pounds of N should be used. The cost of H is \$10 per pound and that of N is \$15 per pound. Determine the amounts of material H and material N that should be used to minimize the cost.

9. A man plans to buy two types of stocks: A and B. He finds:

   (a) The anticipated dividend rate per year on stock A is 6% and that on stock B is 2%.
   (b) The anticipated increase in market value in one year is \$1 per dollar invested in stock A and is \$2 per dollar invested in stock B.

   He wishes:

   (1) To have at least \$300 dividend income each year.
   (2) To have at least \$10,000 increase on his investment in one year.

   Determine the minimum amount that he will use to buy the stocks to meet his wishes.

10. Assume that a man is planning to buy certain types of food: meat and fish. Both types can provide nutrient units for his minimum requirements during a given period as follows:

Type of nutrient	Units of nutrient provided in		Minimum units required
	1 pound of meat	1 pound of fish	
A	1	3	9
B	2	2	10
C	4	2	12
Price per pound	$2.50	$1.50	

How many pounds of meat and fish should he buy in order to meet the minimum nutrient requirements at the minimum cost?

11. Refer to Table 10–11, page 206. Use the alternative simplex table method to construct the second simplex table for Example 4. What is the value of F?

12. In this transportation problem:

C = the total transportation cost.

X_{11} = the number of units shipped by route No. 11; the unit transportation cost by the route is $8.

X_{12} = the number of units shipped by route No. 12; the unit transportation cost by the route is $4.

Minimize: $C = \$8X_{11} + \$4X_{12} + \$160$

Subject to restraints:　　　　$X_{11} \leq 30$ (units)
$$X_{12} \leq 10 \text{ (units)}$$
$$X_{11} + X_{12} \leq 25 \text{ (units)}$$
$$X_{11} + X_{12} \geq 5 \text{ (units)}$$
$$X_{11}, X_{12} \geq 0 \text{ (units shipped)}$$

Part Three   Mathematics For Managerial Decisions

Chapter
11 Probability and Applications

This chapter introduces the concept and operations of probability and the applications of probability theory to decision making under conditions of uncertainty. Before the work of probability is presented, however, the counting procedures which are basic to probability calculations are introduced.

11.1   COUNTING PROCEDURES

A *counting procedure* is designed for finding the number of possible arrangements of the objects in a set or sets. A simple procedure is to list all the possible arrangements in the form of a *tree diagram,* as shown in Example 1, then counting the arrangements.

EXAMPLE 1   A mother has three boys (Abe, Chuck, and Eddie) and two girls (Gail and Mary). How many ways can she take one boy and one girl at a time to see the movies?

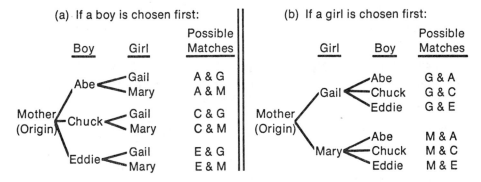

(a) If a boy is chosen first:

Boy	Girl	Possible Matches
Abe	Gail	A & G
Abe	Mary	A & M
Chuck	Gail	C & G
Chuck	Mary	C & M
Eddie	Gail	E & G
Eddie	Mary	E & M

(b) If a girl is chosen first:

Girl	Boy	Possible Matches
Gail	Abe	G & A
Gail	Chuck	G & C
Gail	Eddie	G & E
Mary	Abe	M & A
Mary	Chuck	M & C
Mary	Eddie	M & E

There are 6 ways of taking one boy and one girl at a time by actually counting the possible ways either in diagram (a) or in diagram (b).

Obviously, the tree diagram method is inconvenient in counting when the number of objects to be arranged is large. The following three methods provide systematic ways of counting without listing all possible arrangements.

A. The Multiplication Principle

Observe the diagrams in Example 1. Any one of the three boys may be teamed with any one of the two girls. Thus, the total number of possible ways of taking one boy and one girl at a time may be computed in the following manner:

3(ways of picking a boy) × 2(ways of picking a girl) = 6(total number of ways)

This expression gives the *multiplication principle* as follows:

> If one thing can be done in *a* ways, a second thing in *b* ways, a third thing in *c* ways, and so on for *n* things, then the *n* things can be done together in:
>
> $$a \times b \times c \times \ldots \text{ (n factors) ways}$$

EXAMPLE 2   A student in City *W* wishes to drive his car to City *Z* by passing cities *X* and *Y*. There are 4 highways from *W* to *X*, 3 from *X* to *Y*, and 2 from *Y* to *Z*. How many different ways can the trip be made by the student?

The total number of ways based on the multiplication principle is:

$$a \times b \times c = 4 \times 3 \times 2 = 24 \text{ ways}$$

B. Permutations

The multiplication principle provides a general method for counting the number of possible arrangements of objects within a single set or among several sets. For a *single set* of objects, however, the formulas developed for permutations and combinations are more convenient in counting the number of possible arrangements. A *permutation* is an arrangement of all or part of the objects within a single set of objects in a *definite order*. The total number of permutations of a set of objects depends on the number of objects taken at a time for each permutation. The number of objects taken at a time for each permutation may be all of the objects or part of the objects.

1. PERMUTATIONS OF DIFFERENT OBJECTS TAKEN ALL AT A TIME. The total number of permutations of a set of objects taken all at a time may be obtained by a tree diagram as shown in Example 3.

EXAMPLE 3   Find the total number of permutations of the set of letters {*a, b, c*} taken all at a time.

The possible orderly arrangements of the three letters taken three at a time are diagrammed on page 213.

By actual accounting, the diagram shows 6 permutations. Observe that the arrangement *abc* is different from *acb*, although both arrangements have the same letters. The order of each letter is important in a permutation.

The total number of permutations in Example 3 can also be obtained by applying the multiplication principle as follows:

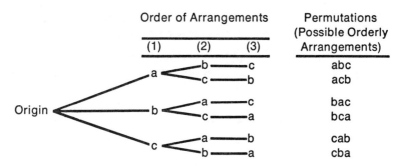

Order of Arrangements			Permutations (Possible Orderly Arrangements)
(1)	(2)	(3)	
	b	c	abc
a	c	b	acb
	a	c	bac
b	c	a	bca
	a	b	cab
c	b	a	cba

There are three ways to fill the first place in each permutation. Each of the three letters (a, b, c) can be used for the first place as shown in column (1) in the diagram.

There are two ways to fill the second place in each permutation. For example, after the first place has been filled by the letter a, the second place can be filled by either b or c as shown in column (2).

There is only one way to fill the third place in each permutation. For example, after the first and second places have been filled by letters a and b respectively, the third place can be filled by only the letter c as shown in column (3).

Thus, the total number of permutations is:

$$3 \times 2 \times 1 = 6$$

In general let:

$n =$ the number of objects in the given set; also:

$n =$ the number of objects taken at a time for each permutation

$_nP_n =$ the total number of permutations of n objects taken n at a time

Based on the illustration in Example 3, we have:

$$_nP_n = n(n-1)(n-2)(n-3) \ldots (3)(2)(1) \tag{11-1A}$$

The right side of this formula can be written as $n!$ (Read n *factorial* or *factorial n*.) Thus the formula may be simplified as:

$$_nP_n = n! \tag{11-1B}$$

Note: If $n = 0$, then we define $0! = 1$.

By using formula (11-1B), Example 3 may be computed as follows:

$$n = 3$$
$$_3P_3 = 3! = 3 \times 2 \times 1 = 6$$

EXAMPLE 4 Find the total number of permutations of the set of digits $\{6, 7, 8, 9\}$ taken all at a time.

$$n = 4 \text{(digits in the given set)}$$
$$_4P_4 = 4! = 4 \times 3 \times 2 \times 1 = 24$$

2. PERMUTATIONS OF DIFFERENT OBJECTS TAKEN PART AT A TIME.
The total number of permutations of a set of objects taken part at a time may also be obtained by a tree diagram. The diagram is similar to that

illustrated in Example 3 except that the number of columns in the present case is equal to the number of objects taken for each permutation. In general, let:

$r =$ the number of objects taken at a time for each permutation
$_nP_r =$ the total number of permutations of n objects taken r at a time

Then:

$$_nP_r = n(n-1)\ (n-2)\ (n-3) \ldots (n-r+1) \text{ for } r \text{ factors} \quad \textbf{(11-2}A\textbf{)}$$

Or:

$$_nP_r = \frac{n!}{(n-r)!} \quad \textbf{(11-2}B\textbf{)}$$

The last factor $(n-r+1)$ is equal to $[n-(r-1)]$. When $r = n$, it becomes $(n - n + 1) = 1$. In other words, when $r = n$, formula (11–2) is identical to formula (11–1).

EXAMPLE 5 Find the total number of permutations of the set of letters $\{a, b, c, d, e\}$ taken (a) three at a time and (b) two at a time.

(a) $n = 5$ (number of letters in the given set)
    $r = 3$ (number of letters taken at a time for each permutation)

$$_nP_r = {}_5P_3 = 5 \times 4 \times 3 = 60$$

Or:

$$_nP_r = \frac{n!}{(n-r)!} = \frac{5 \times 4 \times 3 \times 2 \times 1}{2 \times 1} = 60$$

(b) $n = 5$, and $r = 2$
$$_nP_r = {}_5P_2 = 5 \times 4 = 20$$

Or:

$$_nP_r = \frac{n!}{(n-r)!} = \frac{5 \times 4 \times 3 \times 2 \times 1}{3 \times 2 \times 1} = 20$$

EXAMPLE 6 Three officials—president, vice-president, and secretary—are to be elected from 12 members of a club. In how many ways can the three officials be elected?

This is a permutation problem. The order of arrangement is taken into consideration. For example, the set of president A, vice-president B, and secretary C is different from the set of president B, vice-president C, and secretary A.

$n = 12$ and $r = 3$
$$_nP_r = {}_{12}P_3 = 12 \times 11 \times 10 = 1320 \text{ ways}$$

C. Combinations

A *combination* is an arrangement of all or part of the objects of a single set without regarding the order of the objects. The total number of possible combinations of a set of objects taken all at a time is 1. For example, the possible arrangements from the set of letters $\{a, b\}$ are ab and ba. When the order of arrangements is disregarded, arrangement ab is not different from the arrangement ba. Thus, there is only one combination (a and b) possible from the set of the two letters.

The total number of possible combinations of a set of different objects taken part at a time may be obtained by first finding the total number of permutations then counting the permutations with the same objects as one combination. This procedure is illustrated in Example 7.

EXAMPLE 7   What is the total number of combinations of the set of letters $\{a, b, c\}$ taken 2 at a time?

The total number of permutations of 3 letters taken 2 letters at a time is:

$$_3P_2 = 3 \times 2 = 6$$

The total number of permutations of the same 2 letters taken all at a time is:

$$_2P_2 = 2! = 2 \times 1 = 2$$

The 2 permutations having the same two letters are considered as only one combination. Thus, the total number of combinations is:

$$\frac{_3P_2}{2!} = \frac{6}{2} = 3$$

The result of Example 7 is diagrammed as follows:

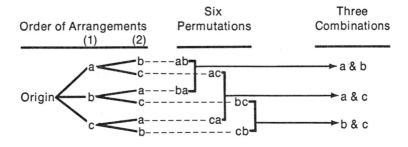

In general, let:

   $n =$ the number of objects in a given set
   $r =$ the number of objects taken at a time for each combination
   $_nC_r =$ the total number of combinations of n objects taken r at a time

Based on the illustration in Example 7, we have:

$$_nC_r = \frac{_nP_r}{r!} \qquad\qquad (11\text{--}3A)$$

Or: $$_nC_r = \frac{n!}{r!(n-r)!} \qquad\qquad (11\text{--}3B)$$

Example 7 may be computed by using formula (11–3A) in the following manner:

$$_3C_2 = \frac{3 \times 2}{2 \times 1} = 3$$

Or by using formula (11–3B):

$$_3C_2 = \frac{3!}{2!(3-2)!} = \frac{3 \times 2 \times 1}{(2 \times 1)(1)} = 3$$

EXAMPLE 8   Three out of 12 members are to be selected to form a committee. In how many ways can the committee be formed?

The committee formed by members A, B, C is the same as the committee formed by B, C, A, or other arrangements consisting of the three members. Since we disregard the order of arrangement, the given statement is a combination problem.

$$n = 12, r = 3$$
$$_nC_r = {}_{12}C_3 = \frac{{}_{12}P_3}{3!} = \frac{12 \times 11 \times 10}{3 \times 2 \times 1} = 220$$

Formula (11–4) can simplify the computation of a combination with a large value of r.

$$_nC_r = {}_nC_{n-r} \qquad\qquad (11\text{–}4)$$

EXAMPLE 9   Compute (a) $_5C_2$ and (b) $_{1,000}C_{998}$.

(a) $_5C_2 = {}_5C_{5-2} = {}_5C_3 = \dfrac{5 \times 4 \times 3}{3 \times 2 \times 1} = 10$

(b) $_{1,000}C_{998} = {}_{1,000}C_{1,000-998} = {}_{1,000}C_2 = \dfrac{1,000 \times 999}{2 \times 1} = 499,500$

EXERCISE 11–1 Reference: Section 11.1

1. Compute:
   - **a.** $_4P_4$
   - **b.** $_7P_3$
   - **c.** $_3C_3$
   - **d.** $_6C_2$
   - **e.** $_6C_4$
   - **f.** $_{100}C_{97}$

2. Compute:
   - **a.** $_3P_3$
   - **b.** $_6P_4$
   - **c.** $_4C_4$
   - **d.** $_8C_2$
   - **e.** $_8C_6$
   - **f.** $_{300}C_{298}$

3. What are the total number of permutations of a set of five letters $\{a, b, c, d, \text{and } e\}$ taken (a) all at a time, and (b) four at a time?

4. What are the total number of combinations of a set of five letters $\{a, b, c, d, \text{and } e\}$ taken (a) all at a time, and (b) four at a time?

5. How many 3-digit numbers can be formed from the ten digits (0 through 9) if the digits (a) are repeated, and (b) are not repeated?

6. There are 3 freshmen, 5 sophomores, 4 juniors, 6 seniors, and 1 graduate student in a room. How many ways can we select one student from each group in the room to form a committee of five students?

7. A housing development company provides washing machines, central air conditioning, wall to wall carpeting, musical devices, freezers, and garages as optional units to be added to new houses. In how many ways can the houses be sold?

8. In how many ways can we take a sample of the wages of 6 workers from a group of 100 workers?

9. A club was organized by 20 students. (a) In how many ways can the offices of president, secretary, and program director be filled by the 20 students? (b) In how many ways can a committee of three members be formed by the students?

10. In how many ways can a coach select a team of 5 basketball players from a group of 10 players if (a) two certain players must be included in the team, and (b) no restrictions are required in the selection?

11. A business college offers five courses: typing, shorthand, bookkeeping, marketing, and advertising. How many different ways can a student take the courses if he is required to take at least two of them?

12. How many four-letter words can possibly be formed from the letters of the word *formulas?*

11.2 INTRODUCTION TO PROBABILITY

Probability is a number expressing the likelihood of occurrence of a specific event. The probability of success or failure of an event in a trial or experiment is usually expressed in a ratio form.

Let P denote "the probability of."

Then:

$$P(\text{the event of success}) = \frac{\text{number of successful outcomes}}{\text{number of possible outcomes}}$$

$$P(\text{the event of failure or unsuccess}) = \frac{\text{number of unsuccessful outcomes}}{\text{number of possible outcomes}}$$

Where:

$$\begin{pmatrix} \text{Number of} \\ \text{successful} \\ \text{outcomes} \end{pmatrix} + \begin{pmatrix} \text{Number of} \\ \text{unsuccessful} \\ \text{outcomes} \end{pmatrix} = \begin{pmatrix} \text{Number of} \\ \text{possible} \\ \text{outcomes} \end{pmatrix}$$

Probabilities can be grouped into two types: (A) the theoretical probability and (B) the empirical probability. The two types of probabilities are derived in different manners. However, they can be supplemental to each other in solving actual business and economic problems.

A. Theoretical Probability

The *theoretical probability,* also called *mathematical probability,* of a given event in a trial is obtained under two conditions: (1) we can count exactly the different ways in which the given event may or may not happen and (2) we can assume that all the possible ways will occur on an equally likely basis.

Let A = the event of success from a trial
   n = the number of possible outcomes on an equally likely basis
   h = the number of successful outcomes

Then, the probability of the occurrence of the event A (success) is:

$$P(A) = \frac{h}{n}$$

(11–5)

The probability of the nonoccurrence of event A (failure) is:

$$P(\text{Not } A) = \frac{n-h}{n} = 1 - \frac{h}{n}$$

Or:

$$P(\text{Not } A) = 1 - P(A)$$

(11–6)

Note that the sum of the probability of success and the probability of failure is always equal to 1, or:

$$P(A) + P(\text{Not } A) = 1$$

EXAMPLE 10   Find the probability that one throw of a die will (a) have "4" on top and (b) not have "4" on top.

Assume that the die is well-balanced; that is, each of the six sides will have an equal chance to turn on the top when the die stops in a throw.

(a) $P(\text{having a "4"}) = \dfrac{1 \ (\text{number of successful outcomes})}{6 \ (\text{number of possible outcomes})}$

(b) $P(\text{not having a "4"}) = \dfrac{6-1}{6} = \dfrac{5 \ (\text{number of unsuccessful outcomes})}{6 \ (\text{number of possible outcomes})}$

B. Empirical Probability

The *empirical probability,* also called *statistical probability,* of a given event is obtained from statistical data which is recorded from our experiences or experiments. In some experiments or trials, although we may count exactly the different ways in which the given event may or may not happen, we cannot assume that all the possible ways will happen or not happen on an equally likely basis. The computation of the probability of the event thus must be based on our experiences or experiments of what has happened on similar occasions in the past. Since our experiences are subjective and the results of various experiments may be different, we must evaluate the empirical probability of an event periodically in actual applications.

EXAMPLE 11 Find the probability that a person who is now 65 years old will (a) be alive at age 66 and (b) die before reaching age 66.

Although we can count exactly the different ways in which the person will be at age 66, dead or alive, we cannot assume that the two ways are equally likely to happen. Thus, statistical data concerning similar occasions in the past must be used as a basis to predict the probability of the event that the person who is now 65 will be living at 66.

1. According to the Commissioners 1941 Standard Ordinary Mortality Table, based on the experience of life insurance companies for the years from 1930 to 1940, 554,975 out of 577,882 persons now aged 65 will be alive at age 66. Thus:

(a) P(the person will be alive at age 66) $= \dfrac{554,975}{577,882} = .96036$

(b) P(the person will die before age 66) $= \dfrac{577,882 - 554,975}{577,882}$

$$= .03964$$

Note that (a) + (b) = .96036 + .03964 = 1, the total probability.

2. According to the Commissioners 1958 Standard Ordinary Mortality Table, based on the experience of life insurance companies for the years from 1950 to 1954, 6,584,614 out of 6,800,531 persons now age 65 will be alive at age 66. Thus:

(a) P(the person will be alive at age 66) $= \dfrac{6,584,614}{6,800,531} = .96825$

(b) P(the person will die before age 66) $= \dfrac{6,800,531 - 6,584,614}{6,800,531}$

$$= .03175$$

(a) + (b) = .96825 + .03175 = 1

Example 11 indicates that in each of the two cases, (a) and (b), the answers to the same question are different because the experiences of life insurance companies are based on the different years. More people will live at age 66 based on a 1958 table than a 1941 table.

Note that in many cases empirical probability can be used in verifying the result of theoretical probability if we are in doubt that each possible outcome in a trial will occur on an equally likely basis. Example 10 has shown us that the theoretical probability of throwing a die having "4" on top is $\frac{1}{6}$. We may wish to perform an experiment to see if the die is well-balanced or loaded dishonestly.

11.3 COMPUTING PROBABILITY OF TWO OR MORE EVENTS

A trial may involve only a single event or two or more events. The computation of the probability of a single event has been illustrated in Example 10.

This section will introduce the method of computing the probability of two or more events. In the following problems, we shall assume that each of the possible outcomes of a trial will occur equally likely. In other words, we shall compute only the theoretical or mathematical probability. For convenience, the idea of sets can be used in computing probabilities. We first state the information in a given problem in the language of sets as follows:

> All possible outcomes of the trial or experiment are the elements of the universal set. For instance, the universal set in Example 10 has six elements {1, 2, 3, 4, 5, 6}.
>
> All outcomes of a given event are the elements of the event set which is a subset of the universal set. The event set in Example 10 (a) has only one element (4). The event set is a subset of the universal set since 4 is also an element of the universal set.

Let us denote a universal set by the letter U and subsets (events) by letters A, B, C, . . . and so on in the following illustrations. Then the probability of having event A can be expressed:

$$P(A) = \frac{h}{n} = \frac{\text{number of elements in set } A}{\text{number of elements in set } U}$$

Two or more events may be (A) mutually exclusive, (B) partially overlapping, (C) independent, and (D) dependent. The formulas for computing the probability of each of the four different types of events are presented in the following paragraphs.

A. Mutually Exclusive Events – Disjoint Events

Two or more events are known as *mutually exclusive events* if the events cannot occur together – that is, the occurrence of any one of them excludes the occurrences of the others. They are also called *disjoint events*. Let A and B be disjoint events as shown in the Venn diagram which follows:

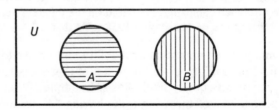

Then, the probability of event A or B, denoted by $P(A$ or $B)$ or $P(\cup B)$, is:

$$P(A \cup B) = P(A) + P(B) \tag{11-7}$$

Observe in the diagram that events A and B are subsets of universal set U. The two subsets have no elements in common; that is, they are *disjoint subsets*. Thus, the probability that both events A and B will occur is zero since the intersection of the two event subsets is the empty set or $P(A$ and $B) = 0$.

EXAMPLE 12 Find the probability that one throw of a die will produce a 4 or a 3.

A = the event having 4, and
B = the event having 3

Since one throw of a die can not have both sides on the top, events A and B are mutually exclusive.

$U = \{1, 2, 3, 4, 5, 6\}$, or $n = 6$ (elements or possible outcomes)
$A = \{4\}$, or $h = 1$ (element)

$$P(A) = \frac{h}{n} = \frac{1}{6}$$

$B = \{3\}$, or $h = 1$ (element)

$$P(B) = \frac{h}{n} = \frac{1}{6}$$

Thus:

$$P(A \cup B) = P(A) + P(B) = \frac{1}{6} + \frac{1}{6} = \frac{2}{6} = \frac{1}{3}$$

B. Partially Overlapping Events–Intersected Events

Two or more events are known as *partially overlapping events* if part of one event and part of another occur together. They are also called *intersected events*. Let A and B be intersected events; that is, part of event A is also a part of event B as shown in the following Venn diagram.

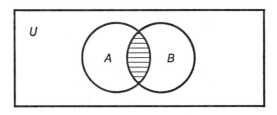

Then, the probability of event A or B is:

$$P(A \cup B) = P(A) + P(B) - P(A \cap B) \qquad (11\text{–}8)$$

where $P(A \cap B)$, or written $P(A \text{ and } B)$, denotes the probability of having both elements A and B.

Observe the diagram in which the elements in the intersection of subsets A and B appear twice–the shaded area belongs to A and B. Thus, $P(A \cap B)$, or $P(A \text{ and } B)$ represented by the shaded area, must be subtracted from the sum of $P(A)$ and $P(B)$ to avoid duplication.

EXAMPLE 13 In drawing one card from a deck of 52 cards, find the probability that a single draw will be either a face card or a spade.

There are 13 cards in each of the 4 suits: spade, heart, diamond, and club. Thus, the U set contains 52 elements (cards), or $n = 52$.

Let A = a face card. There are three face cards (jack, queen, king) in each of the 4 suits.

Then $A = \{J_s, J_h, J_d, J_c, Q_s, Q_h, Q_d, Q_c, K_s, K_h, K_d, K_c\}$, or $h = 12$ elements.

Let B = a spade. There are 13 cards with spades.

Then $B = \{Ace_s, 2_s, 3_s, 4_s, 5_s, 6_s, 7_s, 8_s, 9_s, 10_s, J_s, Q_s, K_s\}$, or $h = 13$.

Notice that the three elements (J_s, Q_s, K_s) of set B are also elements of set A, or set:

$A \cap B = \{J_s, Q_s, K_s\}$, and $h = 3$

Thus:

$$P(A \cup B) = P(A) + P(B) - P(A \cap B) = \frac{12}{52} + \frac{13}{52} - \frac{3}{52} = \frac{22}{52} = \frac{11}{26}$$

C. Independent Events

Two or more events are said to be *independent* if the events in no way influence each other. For example, A and B are independent events when the occurrence of event A has no effect on the occurrence of event B and vice versa. The probability that both events A and B will occur independently is:

$$P(A \cap B) = P(A) \cdot P(B) \tag{11-9}$$

EXAMPLE 14   A bag contains 5 balls: 2 white and 3 red. One ball is drawn from the bag and is then replaced. Another ball is drawn after the replacement. Find the probability that both balls of the two drawings will be white.

Let A = the event of the first drawing that will have a white ball, and
B = the event of the second drawing that will have a white ball.

The second ball is drawn after the first ball is replaced. Thus, the result of the first drawing certainly has nothing to do with the second drawing. The two drawings are independent events.

Since there are 2 white balls in the bag of 5 in each drawing:

$$P(A \cap B) = P(A) \cdot P(B) = \frac{2}{5} \times \frac{2}{5} = \frac{4}{25}$$

Check: The answer in Example 14 agrees with the result obtained by the multiplication principle. Let W_1 and W_2 represent the two white balls and R_1, R_2, and R_3 the three red balls. Then, there are 5 ways to get the first ball (W_1, W_2, R_1, R_2, and R_3) and 5 ways to get the second ball (again W_1, W_2, R_1, R_2, and R_3). The two drawings can be made in $5 \times 5 = 25$ possible outcomes. Four of the 25 are successful outcomes—white and white (W_1W_1, W_1W_2, W_2W_1, and W_2W_2). Thus, the probability of having the first white and the second white is $\frac{4}{25}$.

D. Dependent Events

If events A and B are so related that the occurrence of B depends on the occurrence of A, then A and B are called dependent events. The probability of event B depending on the occurrence of event A is called *conditional probability* and is written:

$P(B|A)$ or $P_A(B)$, which may be read, "the probability of B, given A"

The probability that both the dependent events A and B will occur is:

$$P(A \cap B) = P(A) \cdot P(B|A) \qquad\qquad (11\text{--}10)$$

EXAMPLE 15 Refer to Example 14. Assume that the first ball is not returned to the bag when the second ball is drawn. Find the probability that both balls of the two drawings are white.

Let $A =$ the event of the first drawing that will have a white ball, and
$B =$ the event of the second drawing that will have a white ball.

Then, $P(A) = \dfrac{2}{5}$ (since there are 2 white balls in the bag of 5)

Event B depends on the occurrence of event A. If the first ball is white, the probability of the second ball being white is:

$P(B|A) = \dfrac{1}{4}$ (since there is only 1 white ball left in the bag of 4 after the first white ball is drawn.)

The probability that both balls of the two drawings are white is:

$$P(A \cap B) = P(A) \cdot P(B|A) = \frac{2}{5} \times \frac{1}{4} = \frac{2}{20} = \frac{1}{10}$$

Check: The answer in Example 15 also agrees with the result obtained by the multiplication principle. There are 5 ways to get the first ball (W_1, W_2, R_1, R_2, and R_3), but only 4 ways to get the second ball. (For example, if the first is W_1, the second will not be W_1 since W_1 is not replaced for the second drawing). The two drawings can be made in $5 \times 4 = 20$ possible outcomes. Only two of the 20 outcomes are successful ($W_1\, W_2$, and $W_2\, W_1$). Thus, the probability is $\dfrac{2}{20}$.

EXERCISE 11-2 Reference: Sections 11.2 and 11.3

1. A room has 25 students: 9 boys and 16 girls. One person is selected at random from the room. Find the probability that the person is (a) a boy, (b) not a boy, (c) a girl, (d) a boy or a girl, and (e) a professor.

2. A box contains 7 red chips and 11 blue chips. One chip is drawn randomly from the box. Find the probability that it is (a) red, (b) not red, (c) blue, (d) red or blue, (e) red and blue.

3. According to the Commissioners 1958 Standard Ordinary Mortality Table, 9,647,694 out of 9,664,994 persons now age 20 will be alive at age 21. Find the probability that a person who is now 20 will (a) be living at age 21, and (b) die before reaching age 21. (Compute to three decimal places.)

4. Answer the two questions listed in Problem 3 based on the information given in the Commissioner's 1941 Standard Ordinary Mortality Table: 949,171 out of 951,483 persons now age 20 will be alive at age 21.

5. Among the 100 employees in a company, 45 read the morning newspaper and 30 read the evening newspaper. Among the readers, 10 employees read both morning and evening newspapers. An employee is chosen at random from the company. Find the probability that the employee reads either morning or evening newspapers.

6. A deck of 7 cards includes 3 kings (1 printed in red and 2 in yellow) and 4 queens (1 in red and 3 in yellow). One card is drawn from the deck. Find the probability that it is a (a) king, (b) queen, and (c) king or yellow card.

7. A bag contains 5 balls: 2 white and 3 red. One ball is drawn from the bag and is then replaced. Another ball is drawn after the replacement. Find the probability that (a) the first is white and the second is red, and (b) the first is red and the second is also red.

8. Refer to Problem 7. Assume that the first ball is not returned to the bag when the second ball is drawn. Find the probabilities listed in (a) and (b).

9. Find the probability that one throw of two dice (one white and one black) will have 2 or less on the white die and 3 or more on the black die.

10. Find the probability that one throw of two dice (one white and one black) will have 3 or less on the white and the sum of 7 or more.

11. A committee is to be formed by three out of a group of eight girls. (a) What is the probability that Mary is on the committee? Not on the committee? (b) What is the probability that Mary or Nancy or both are on the committee?

12. Refer to Problem 11. Given that Mary is on the committee, what is the probability that Nancy is also on the committee?

11.4 REPEATED TRIALS AND BINOMIAL DISTRIBUTION

Let $P = P(A)$, the probability that event A will happen in a single trial
$Q = P(\text{not } A)$, the probability that event A will not happen or:
$Q = 1 - P$

Then, the probability that event A will happen exactly X times in n repeated trials with the probability of success in a single trial being P, denoted by the *combined symbol $P(X;n,P)$*, is:

$$P(X;n,P) = {}_nC_X \cdot P^X Q^{n-X} \qquad (11\text{--}11)$$

The following example illustrates the derivation and use of the formula.

EXAMPLE 16 A bag contains 5 balls: 2 white and 3 red. One ball is drawn and then replaced after each drawing. In three repeated drawings find the probability that the drawings will have (a) exactly 3 white balls, (b) exactly 2 white balls, (c) exactly 1 white ball, and (d) no white ball.

Let A = the event of drawing a white ball, and
B = the event of not drawing a white ball.

Then $P(A) = P = \dfrac{2}{5} = .4.$

$$P(B) = P(\text{not } A) = Q = 1 - \frac{2}{5} = \frac{3}{5} = .6$$

Since the result of each drawing has nothing to do with each successive drawing, the three drawings are independent events.

When formula (11–9) for independent events is used, the probability that the first ball is white, the second white, and the third white is:

$$P(A \text{ and } A \text{ and } A) = P(A \cap A \cap A) = P(A) \cdot P(A) \cdot P(A) = P^3 = (.4)^3 = .064$$

The probability that the first ball is white, the second white, and the third red is:

$$P(A \cap A \cap B) = P(A) \cdot P(A) \cdot P(B) = P \cdot P \cdot Q = P^2 Q = (.4)^2(.6) = .096$$

The possible outcomes and their probabilities are shown in the following tree diagram.

	Order of Drawings		Combination of Outcomes	Probability of Combination
First	Second	Third		
		A ------ AAA	PPP = .064
	A	B ------ AAB	PPQ = .096
A (white ball)	B	A ------ ABA	PQP = .096
		B ------ ABB	PQQ = .144
Origin		A ------ BAA	QPP = .096
	A	B ------ BAB	QPQ = .144
B (red ball)	B	A ------ BBA	QQP = .144
		B ------ BBB	QQQ = .216
			Total 1.000

The probabilities of the combinations with the same number of occurrences of event A in the diagram can be added. The answers to the questions of Example 16 thus are:

(a) The probability that event A will occur exactly 3 times (or 3 white balls) in 3 repeated drawings is:

$$P(A \cap A \cap A) = P^3 = (.4)^3 = .064$$

(b) The probability that event A will occur exactly 2 times (or 2 white balls) in 3 repeated drawings is:

$$P(A \cap A \cap B) + P(A \cap B \cap A) + P(B \cap A \cap A) = P^2 Q + P^2 Q$$
$$+ P^2 Q = 3(P^2 Q) = 3(.4)^2(.6) = 3(.096) = .288$$

(c) The probability that event A will occur exactly one time (or one white ball) in the three repeated drawings is:

$$P(A \cap B \cap B) + P(B \cap A \cap B) + P(B \cap B \cap A) = PQ^2 + PQ^2$$
$$+ PQ^2 = 3(PQ^2) = 3(.4)(.6)^2 = 3(0.144) = 0.432$$

(d) The probability that event A will occur exactly zero times (or no white ball) in the three repeated drawings is:

$$P(B \cap B \cap B) = Q^3 = (.6)^3 = 0.216$$

The answers are shown in Table 11–1.

Table 11–1   Answers to Example 16

Number of White Balls (or successes) X	Probability (or theoretical frequency)
0	0.216
1	0.432
2	0.288
3	0.064
Total	1.000

The X values, called X *random variable,* represent all possible successful outcomes (0, 1, 2, and 3 white balls) in three repeated trials. The sum of the probabilities of all possible X values is always equal to 1. A set of probabilities obtained for all possible successful outcomes represented by X variable in a repeated trial is called a *probability distribution.* Here the X's are integers and thus are *discrete values.* Table 11–1 therefore shows *discrete probability distribution.* However, this discrete probability distribution can be shown graphically in either *discrete form* (Figure 11–1 (a)) or *continuous form* (Figure 11–1 (b)). Note that the continuous form of this distribution is also shown in Figure 11–2 for comparison purpose.

Figure 11-1   A Discrete Probability Distribution (Example 16)

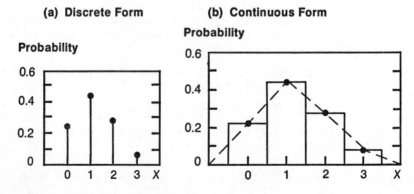

(a) **Discrete Form**   (b) **Continuous Form**

Observe the computation in Example 16 (b). The coefficient 3, which is the number of combinations with the same number of A's (AAB, ABA, BAA), corresponds to the number computed by the combination formula:

$$_nC_X = \frac{_nP_X}{X!}$$

Where:

$n = 3$ (drawings or repeated trials)
$X = 2$ (number of occurrences of event A or exactly two white balls in 3 drawings)

$$_3C_2 = \frac{3 \times 2}{2 \times 1} = 3$$

The probability in Example 16 (b) that event A will occur exactly two times in the three repeated trials with $P(A) = .4$ thus can be written as:

$$P(2; 3, .4) = _3C_2 \cdot P^2Q = 3(.4)^2(.6) = 3(0.096) = 0.288$$

In general, the probability that event A will happen exactly X times in n repeated trials with the probability of a single trial being P is:

$$P(X; n, P) = _nC_X \cdot P^X \cdot Q^{n-X}, \text{ which is formula (11–11)}$$

and is also called the *binomial probability density function* since it represents the general form of each term in a binomial expansion for computing a probability distribution.

An expression consisting of two terms is known as a binomial, such as $a + b$. By actual multiplication, the product of $(a + b)(a + b)$ may be written:

$$a + b$$
$$(\times)\ a + b$$
$$\overline{ab + b^2}$$

$$\frac{a^2 +\ ab}{a^2 + 2ab + b^2}$$

Likewise if the binomial $(a + b)$, where a and b are real numbers, is raised to successive positive integer powers, the expanded expressions are:

$$(a + b)^1 = a + b$$
$$(a + b)^2 = a^2 + 2ab + b^2$$
$$(a + b)^3 = a^3 + 3a^2b + 3ab^2 + b^3$$
$$(a + b)^4 = a^4 + 4a^3b + 6a^2b^2 + 4ab^3 + b^4$$

In general, when the binomial is raised to the nth power, n being a positive integer, the expansion can be written:

$$(a + b)^n = _nC_0a^n + _nC_1a^{n-1}b + _nC_2a^{n-2}b^2 + \cdots\cdots + _nC_{n-1}ab^{n-1} + _nC_nb^n$$

which is known as the *binomial theorem*. Observe the binomial theorem:

1. The number of terms in the expansion is $(n + 1)$.

2. The letter a in the first term of the expansion has the exponent $n;$ in the second term the exponent $(n - 1)$; in the third term the exponent $n - 2$; and so on.
3. The letter b in the second term of the expansion has the exponent 1; in the third term the exponent 2; in the fourth term the exponent 3; and so on.
4. The sum of the exponents of a and b in any term is n.
5. The coefficients of the terms which are equidistant from the ends are the same. For example, the coefficient of the first term is $1(= {}_nC_0)$ and the coefficient of the last term is also $1(= {}_nC_n)$.
6. The value of ${}_nC_r$ is the coefficient of the $(r + 1)$th term of the expansion.

Now let $b = P$ (the probability of an event in a single trial), and $a = 1 - P = Q$. The binomial theorem becomes:

$$(Q+P)^n = {}_nC_0Q^n + {}_nC_1Q^{n-1}P + {}_nC_2Q^{n-2}P^2 + \ldots + {}_nC_XQ^{n-X}P^X \ldots + {}_nC_nP^n \quad (11\text{--}12)$$

Observe the terms at the right side of the equation. Each term is identical to that of the probability formula $P(X;n,P)$; that is, $X = 0$ in the first term, $X = 1$ in the second term, and so on. Here, X again represents the exact number of times that an event, having probability P, will happen in n trials. Also, since $1 - P = Q$, or $Q + P = 1$, the left side becomes $1^n = 1$. Thus, formula (11–12) may be written in the form:

$$1 = P(0;n,P) + P(1;n,P) + P(2;n,P) + \cdots + P(X;n,P) + \cdots P(n;n,P)$$

which indicates that the sum of the probabilities of all the possible combinations $(X = 0, 1, 2, 3, \ldots, n)$ must be equal to 1. The probabilities obtained by the binomial theorem thus form a discrete probability distribution.

When Example 16 is computed by the binomial formula (11–12), the probabilities of all possible numbers of successes (X's) in the three repeated drawings may be computed systematically as follows:

$$P = \tfrac{2}{5} = .4, \ Q = \tfrac{3}{5} = .6, \ X = 0, 1, 2, 3, \text{ and } n = 3$$

$$\begin{aligned}
(Q + P)^3 &= {}_3C_0Q^3 + {}_3C_1Q^2P + {}_3C_2QP^2 + {}_3C_3P^3 \\
&= 1(\tfrac{3}{5})^3 + 3(\tfrac{3}{5})^2(\tfrac{2}{5}) + 3(\tfrac{3}{5})(\tfrac{2}{5})^2 + 1(\tfrac{2}{5})^3 \\
&= (\tfrac{27}{125}) + (\tfrac{54}{125}) + (\tfrac{36}{125}) + (\tfrac{8}{125}) \\
&= \tfrac{125}{125} = 1
\end{aligned}$$

Or:

$$\begin{aligned}
&= 1(.6)^3 + 3(.6^2)(.4) + 3(.6)(.4^2) + 1(.4)^3 \\
&= 0.216 + 3(0.144) + 3(0.096) + 0.064 \\
&= 0.216 + 0.432 + 0.288 + 0.064 = 1
\end{aligned}$$

Thus, the probabilities that event A will occur exactly 0, 1, 2, and 3 times (white balls) in the 3 repeated drawings are 0.216, 0.432, 0.288, and 0.064, respectively. The preceding answers are similar to those obtained in Example 16.

EXAMPLE 17 Refer to Example 16 concerning the five balls (two white and three red). Find the probabilities of having the various possible numbers of white balls in (a) 10 repeated drawings, and (b) 20 repeated drawings. Again, assume that one ball is drawn and is replaced after each drawing.

SOLUTION $P = \frac{2}{5}$ (two white out of 5 balls) $= 0.4$
$Q = 1 - \frac{2}{5} = \frac{3}{5} = 0.6$

(a) $X = 0, 1, 2, 3, \ldots, 8, 9, 10$ (exact number of successes or white balls)

$n = 10$ (repeated trials)

The probabilities are obtained by using formula $P(X;n,P)$ and are shown in Table 11–2 and Figure 11–2.

When $X = 3$ (3 white balls in 10 repeated drawings), for example:

$P(3,10,0.4) = {}_{10}C_3(0.4)^3 \ (0.6)^7$.
$$= \frac{10 \times 9 \times 8}{3 \times 2 \times 1} (.064)(.0279936)$$
$$= 0.2149908480, \text{ rounded to } 0.2150$$

(b) $X = 0, 1, 2, 3, \cdots, 20$ (exact number of white balls)
$n = 20$ (repeated trials)

The probabilities are shown in Table 11–3 and Figure 11–2. Example of the computation of the probabilities.

When $X = 8$, or exactly 8 white balls in twenty repeated drawings, then:

$P(8;20,0.4) = {}_{20}C_8(0.4)^8(0.6)^{12} = 0.1797057878$, rounded to 0.1797

Table 11–2 Answers to Example 17a

Number of white balls (or successes) X	Probability (or theoretical frequency) $P(X;10,0.4)$
0	.0060
1	.0403
2	.1209
3	.2150
4	.2508
5	.2007
6	.1115
7	.0425
8	.0106
9	.0016
10	.0001
Total	1.0000

Table 11–3   **Answers to Example 17b**

Number of white balls (or successes) X	Probability (or theoretical frequency) $P(X;20,0.4)$	Number of white balls (or successes) X	Probability (or theoretical frequency) $P(X;20,0.4)$
0	.0000	11	0.0710
1	.0005	12	0.0355
2	.0031	13	0.0146
3	.0123	14	0.0049
4	.0350	15	0.0013
5	.0746	16	0.0003
6	.1244	17	0.0000
7	.1659	18	0.0000
8	.1797	19	0.0000
9	.1597	20	0.0000
10	.1171		

Note that the binomial probability distribution for selected P and n values are tabulated in the appendix for easy reference.

Figure 11–2 illustrates two facts: (1) The three curves are skewed to the right because the value of $P(= .40)$ is smaller than the value of $Q(= .60)$. (If P is larger than Q, the curves would skew to the left.) (2) When the value of n

Figure 11-2   Binomial Probability Distributions

For n (Number of Trials) = 3, 10, and 20, where $P = 2/5 = .40$

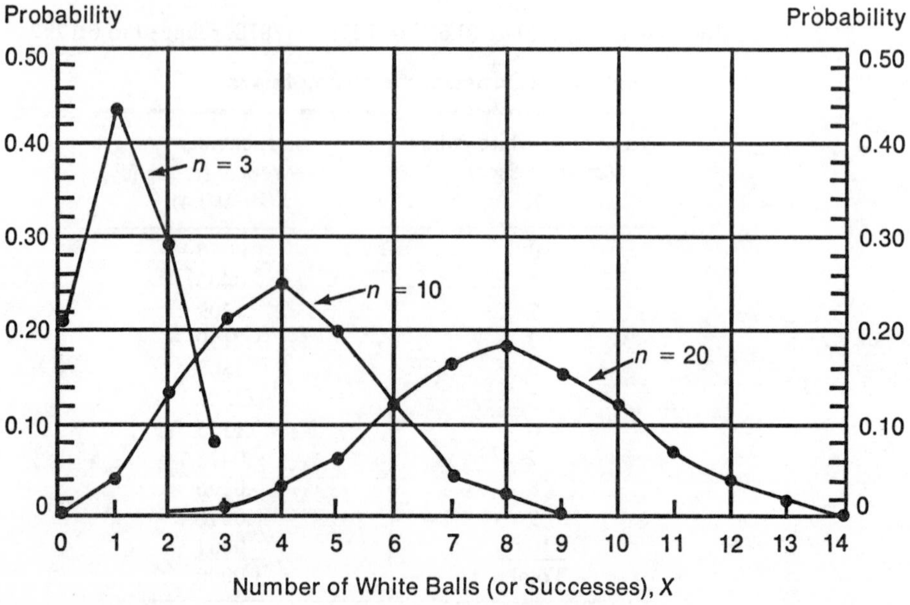

Number of White Balls (or Successes), X

gets larger and larger (increased from 3 to 10 to 20), the curve representing the probability distribution tends to be less and less skewed.

In general, when $P = Q = .5$ (n is either small or large) the binomial distribution curve is symmetrical. If the value of n becomes infinitely large, either $P = .5$ or P deviates from .5, the curve will approach a symmetrical or bell-shaped smooth curve known as the *normal curve*. The normal curve, which is shown in Figure 11-3, is extremely important to numerous types of decision-making problems.

Figure 11-3 **Areas Under the Normal Curve For** $z = 0, \pm 1, \pm 2,$ **and** ± 3
From Table 11-5

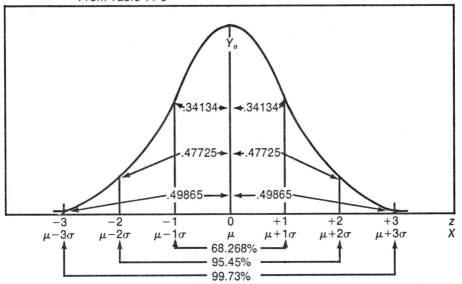

The binomial probability distribution has an arithmetic mean and a standard deviation. The *arithmetic mean,* or simply the *mean,* of a set of values is the quotient of the sum of the values divided by the number of values in the set. It is a type of average and can be used to measure the central tendency of a distribution. The *standard deviation* of a set of values is the square root of the arithmetic mean of the individual deviations squared. A deviation is the difference between an individual value and the arithmetic mean of the values in the set. A standard deviation can be used to measure the degree of variation or dispersion among values in a distribution. We usually use X to represent a set of values. The method of computing the mean (μ) and the standard deviation (σ) of a set of X values thus can be written symbolically:

The mean (of n X values),

$$\mu = \frac{\Sigma X}{n}$$

The standard deviation (of n X values),

$$\sigma = \sqrt{\frac{\Sigma(X - \mu)^2}{n}}$$

The frequency of each X value in the above formula is 1. When the frequency is different from 1, we must weigh each X value. (See Table 11-4 where frequency $= f$.)

The basic method of computing the mean and the standard deviation of the binomial probability distribution in Example 16, Table 11-1, is illustrated as follows:

Table 11-4 Computation of Mean and Standard Deviation From the Probability Distribution of Example 16

Number of successes (white balls) X	$P(X;n,P)$, or theoretical frequency, f	fX	$X - \mu$ $(\mu = 1.2)$	$f(X - \mu)$	$f(X - \mu)^2$
0	0.216	0	−1.2	−0.2592	0.31104
1	0.432	0.432	−0.2	−0.0864	0.01728
2	0.288	0.576	0.8	0.2304	0.18432
3	0.064	0.192	1.8	0.1152	0.20736
Total	1.000	1.200		0.0000	0.72000

The mean (of numbers of successes), $\mu = \dfrac{\Sigma fX}{\Sigma f} = \dfrac{1.200}{1.000} = 1.2$

The standard deviation (of numbers of successes):

$$\sigma = \sqrt{\frac{\Sigma f(X - \mu)^2}{\Sigma f}} = \sqrt{\frac{0.72000}{1.000}} = \sqrt{0.72} = 0.85$$

Here $\Sigma f =$ the sum of the frequencies $= n$

Note: The Greek symbols μ (mu) and σ (sigma) are customarily used to denote the mean and the standard deviation of a set of values in a population. From a population, samples of different sizes can be drawn.

However, a simple way to compute the mean and the standard deviation of a binomial probability distribution is to use the formulas: (Here $n =$ number of repeated trials)

$$\mu = nP \tag{11-13}$$

$$\sigma = \sqrt{nPQ} \tag{11-14}$$

When these two formulas are used, the means and the standard deviations of the distributions in Example 16, 17(a), and 17(b) can be obtained in the following manner:

In Example 16:

$$\mu = nP = 3(0.4) = 1.2$$
$$\sigma = \sqrt{nPQ} = \sqrt{3(0.4)(0.6)} = \sqrt{0.72} = 0.85$$

In Example 17(a):

$$\mu = 10(0.4) = 4.0$$
$$\sigma = \sqrt{10(0.4)(0.6)} = \sqrt{2.4} = 1.55$$

In Example 17(b):

$$\mu = 20(0.4) = 8.0$$
$$\sigma = \sqrt{20(0.4)(0.6)} = \sqrt{4.8} = 2.19$$

Note that when n is large, the work of computing a binomial probability distribution is very tedious. For practical purposes we usually use the normal distribution to estimate the binomial distribution since, as stated previously, the latter approaches the former when n becomes large. The method of computing the probability for the normal distribution is presented in the next section.

EXERCISE 11-3   Reference: Section 11.4

A. *Expand each of the following by the binomial formula and simplify:*

    1. $(a - b)^3$ **5.** $(.2 + .8)^4$

    2. $(x + y)^4$ **6.** $(1 + i)^6$

    3. $(3a + 2b)^5$ **7.** $(Q + P)^5$

    4. $(x - 2y)^3$ **8.** $(\frac{1}{2} + \frac{1}{2})^6$

B. *Statement Problems*

    9. A bag contains 5 balls: 3 blue and 2 black. One ball is drawn and replaced for the next draw. Required:

        (a) Find the probability distribution of having 0, 1, 2, and 3 blue balls in three repeated drawings.

        (b) Graph the probability distribution in (1) the discrete form, and (2) the continuous form.

        (c) Compute the mean of the distribution by the methods: (1) shown in Table 11–4, and (2) formula (11–13).

        (d) Compute the standard deviation of the distribution by the methods: (1) shown in Table 11–4 and (2) formula (11–14).

    10. A fair coin is tossed 6 times. Required:

        (a) Find the probability distribution of having 0, 1, 2, 3, 4, 5, and 6 heads in the tosses.

        (b) Graph the probability distribution in (1) the discrete form, and (2) the continuous form.

        (c) Compute the mean of the distribution by the methods: (1) shown in Table 11–4, and (2) formula (11–13).

        (d) Compute the standard deviation of the distribution by the methods: (1) shown in Table 11–4 and (2) formula (11–14).

11. Refer to Problem 9. Find the probability of having exactly 4 blue balls in 8 repeated drawings. Also, find the mean and standard deviation of the probability distribution resulting from the repeated drawings.
12. Refer to Problem 10. Find the probability of having exactly 8 heads in 10 repeated tosses. Also, find the mean and standard deviation of the probability distribution resulting from the repeated tosses.

★11.5 THE NORMAL DISTRIBUTION

Many types of quantitative information concerning business and economic problems are displayed in the form of normal distribution. The *normal distribution,* also called the *normal probability distribution* is thus regarded as the most important type among various probability distributions. When the normal distribution is shown graphically in Figure 11–3, the curve representing the distribution, called the *normal curve,* is symmetrical or bell-shaped. The maximum ordinate, denoted by Y_0 under the curve, is at the midpoint which is the mean (μ or the arithmetic mean) of the distribution. The shape of the normal curve indicates that the frequencies in a normal distribution are concentrated in the center portion of the distribution and the values above and below the mean are equally distributed.

The normal distribution is a continuous distribution. The probability of the occurrence of a certain event is therefore measured according to the size of the area representing the event under the normal curve. The method of finding the size of an area under the normal curve is based on the "Areas Under the Normal Curve Table" as shown in Table 11–5.

Each value in column $A(z)$ of the table represents the area under the curve between the maximum ordinate at the mean (Y_0) and the ordinate at a point on the X-axis, the distance being expressed in units of z ($= x/\sigma$), where:

$x = X - \mu$, the distance between an X value and the mean μ on the X-axis
σ = the standard deviation of the distribution

The table gives z values at an interval of .25, from $z = 0$ to $z = 4$. The total area under the normal curve and above the X-axis is the total probability which is equal to 1 or 100%. Figure 11–3 shows that approximately 68% ($= .34134(2) = .68268$ or 68.268%) of the area is within the range $\mu \pm 1\sigma$, 95% ($= .47725(2) = .95450$ or 95.45%) is within the range $\mu \pm 2\sigma$, and 99.73% ($= .49865(2) = .99730$) is within the range of $\mu \pm 3\sigma$. If the total area is used to represent the total absolute frequency N of a distribution, a portion of the area expressed in terms of probability (or relative frequency) obtained from Table 11–5 must be multiplied by N to find the absolute frequency of that portion.

When the z values are employed, the curve is known as the *standard normal curve,* and z is referred to as the *standard normal deviate.* Because any shape of a normal curve can be converted to the shape of a standard normal curve simply by using the expression $z = \dfrac{x}{\sigma}$, the table (Table 11–5) showing the areas

Table 11-5 **Areas Under the Normal Curve Between the Ordinate At Mean (Y_0) and the Ordinate At Z**

$z\left(\text{or } \dfrac{x}{\sigma}\right)$	$A(z)$ (Area)	$z\left(\text{or } \dfrac{x}{\sigma}\right)$	$A(z)$ (Area)
0.00	.00000	2.25	.48778
0.25	.09871	2.50	.49379
0.50	.19146	2.75	.49702
0.75	.27337	3.00	.49865
1.00	.34134	3.25	.49942
1.25	.39435	3.50	.49977
1.50	.43319	3.75	.49991
1.75	.45994	4.00	.49997
2.00	.47725		

Note: For more detailed values of $A(z)$, see Table 6 in the Appendix.

under the standard normal curve is the only one required in finding the probability of a certain area under a normal curve. The method of computing the probability based on Table 11–5 is illustrated in the following two cases.

A. The area A(z) is to be determined since the X value is given.

EXAMPLE 18 Assume that the average monthly income of 10,000 families in a city is $1,000, the standard deviation is $200, and the distribution of income amounts is normal. Find the number of families having a monthly income (a) below $1,000, (b) above $1,000 but below $1,200, and (c) above $1,200.

The related information is shown in Figure 11–4. Before using the normal curve area table (Table 11–5), the value of X must be converted to z by using the formula:

$$z = \frac{x}{\sigma} = \frac{X - \mu}{\sigma}$$

(11–15)

This example shows that $\mu = \$1,000$ and $\sigma = \$200$.

(a) When $X = \$1,000$ and $\mu = \$1,000$, or $X = \mu$:

$$z = \frac{X - \mu}{\sigma} = \frac{1,000 - 1,000}{200} = 0$$

The required area is below $1,000 or below the point $z = 0$. Since the maximum ordinate Y_0 is located at the point where $z = 0$, the area representing below $z = 0$ is the entire area to the left side of Y_0, or .5 (= 50%). (Note: When $z = -\infty$, the value of $A(z) = .50000$ in Table 6 in the Appendix.)

The number of families having monthly incomes below $1,000 is:

$10,000(.5) = 5,000$ (families)

Figure 11-4 **Areas Under the Normal Curve**
(For Values of Example 18)

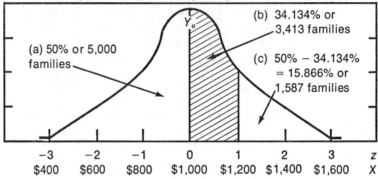

(b) When $X = \$1,000$, $z = 0$

When $X = \$1,200$, $z = \dfrac{(1,200 - 1,000)}{200} = 1$

In Table 11–5, the area (or probability) between $z = 0$ and $z = 1$ is .34134 or 34.134%.

$A(1) = .34134$ or 34.134%

The number of families having a monthly income above $1,000 but below $1,200 is:

$10,000 \times .34134 = 3,413.4$ or 3,413 (families)

(c) When $X = \$1,200$, $z = 1$. The area above $z = 1$ is the difference between the area above $z = 0$ (or .5) and the area between $z = 0$ and $z = 1$, (or .34134).

$.50000 - .34134 = .15866$ or 15.866%

The number of families having a monthly income above $1,200 is:

$10,000 \times .15866 = 1,586.6$ or 1,587 (families)

Note that the total probability is equal to 1 or 100%. Thus, the total number of families obtained in the above category is 10,000, which is 100% of the distribution.

Total families = $5,000 + 3,413 + 1,587 = 10,000$

Examples 19 and 20 further illustrate the use of Table 11–5. However, for simplicity, only the areas of probability are computed, not the absolute frequencies.

EXAMPLE 19 If $\mu = 600$ and $\sigma = 100$, what is the probability (area) of values (a) between 450 and 700 and (b) less than 450?

This example is analyzed in Figure 11–5 and computed as follows:

Figure 11-5 **Areas Under the Normal Curve**
(For Values of Example 19)

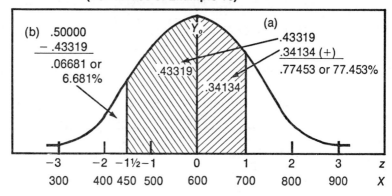

(a) The area under the curve between 450 and 700 is the sum of the area between 450 and 600 and the area between 600 and 700.

The area between 450 and 600, where $600 = \mu$, is computed as follows:

When $X = 450$, $z = \dfrac{X - \mu}{\sigma} = \dfrac{450 - 600}{100} = -1.5$

The area between Y_0 and the ordinate at $z = -1.5$ is the same as the area between Y_0 and the ordinate at $z = +1.5$ in the table since the normal curve shows a perfect symmetrical distribution.

$A(z) = A(-1.5) = .43319$

The area between 600 and 700 is computed as follows:

When $X = 700$, $z = \dfrac{700 - 600}{100} = 1$

$A(z) = A(1) = .34134$

Thus, the area between 450 and 700 is:

.43319
+.34134
.77453 or 77.453%

(b) When $X = 450$, $z = -1.5$ (See (a).)

The area below $z = -1.5$ is the difference between the area below $z = 0$ (or .50000) and the area between Y_0 and the ordinate at $z = -1.5$ (or .43319).

.50000
−.43319
.06681 or 6.681%

EXAMPLE 20 If $\mu = 800$ and $\sigma = 50$, find the probability (area) of values (a) between 700 and 725 and (b) between 850 and 900.

This example is analyzed in Figure 11–6 and is computed as follows:

(a) When $X = 700$, $z = \dfrac{700 - 800}{50} = -2$. $A(-2) = .47725$.

When $X = 725$, $z = \dfrac{725 - 800}{50} = -1.5$. $A(-1.5) = .43319$.

**Figure 11-6   Areas Under the Normal Curve
(For Values of Example 20)**

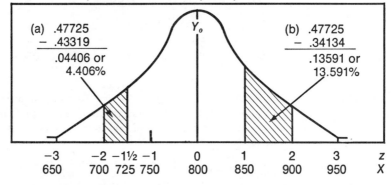

The area between 700 and 725 = the area between 700 and 800 −
the area between 725 and 800
= .47725 − .43319
= .04406 or 4.406%

(b) When $X = 850$, $z = \dfrac{850 - 800}{50} = 1$. $A(1) = .34134$.

When $X = 900$, $z = \dfrac{900 - 800}{50} = 2$. $A(2) = .47725$.

The area between 850 and 900 = the area between 800 and 900 −
the area between 800 and 850
= .47725 − .34134
= .13591 or 13.591%

B. The X value is to be determined since the area A(z) is given.

EXAMPLE 21 Assume that the average monthly income of 10,000 families in a city is $1,000 and the standard deviation is $200. If the distribution is normal, what is the amount of income above which are the earnings of 69% of the families?

This problem is analyzed in Figure 11–7 and is computed as follows:

Figure 11-7 Areas Under the Normal Curve
(For Values of Example 21)

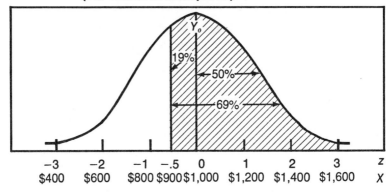

Here $N = 10,000$ (families), $\mu = \$1,000$, and $\sigma = \$200$. Since the area to the right of the maximum ordinate Y_0 is 50%, we must find the 19% (or 69% − 50%) area to the left of the maximum ordinate, or $A(-z) =$ 19%. Table 11–5 shows that,

$$A(z) = A(0.50) = .19146, \text{ or } 19.146\%$$

which is the closest value to the required 19%. Since the area required is to the left of Y_0, the value of z must be negative, or:

$$z = -0.50$$

Since $z = \dfrac{x}{\sigma}$:

$$x = z(\sigma) = (-0.50)(200) = -\$100$$
$$X = \mu + x = 1,000 - 100 = \$900$$

Thus, above $900 on the X scale will be 69% of the area under the normal curve. In other words there are:

$$10,000 \times 69\% = 6,900 \text{ families}$$

making a monthly income above $900.

EXAMPLE 22 Let $\mu = 500$ miles and $\sigma = 100$ miles. Find (a) point X_1 above which there will be 10% of the area under the curve and (b) point X_2 below which there will be 10% of the area under the curve. Assume that the curve represents the normal distribution of miles travelled by a large group of salesmen during a week.

This problem is analyzed in Figure 11–8 and is computed as follows:

Find the 40% (= 50% – 10%) area between the maximum ordinate and the ordinate at z, or $A(z) = .40$ or 40%. The area table in the Appendix (Table 11–5 is not detailed enough for this example) shows,

$$A(z) = A(1.28) = .39973 \text{ or } 39.973\%$$

which is the closest value to the required 40%.

Figure 11-8   Areas Under the Normal Curve
(For Values of Example 22)

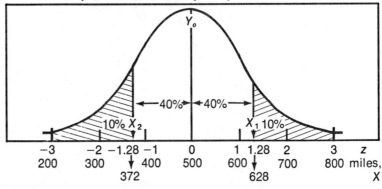

(a) Since the area required represents the top 10% of the values, the value of z is on the right side of Y_0 and is positive, or:

$z = +1.28$

Since $z = \dfrac{x}{\sigma}$:

$x = z(\sigma) = 1.28(100) = 128$ miles
$X_1 = \mu + x = 500 + 128 = 628$ miles

Thus, above 628 miles on the X scale will be the 10% area under the normal curve. That is, 10% of the salesmen in the group travelled more than 628 miles during the week.

(b) Since the required area represents the low 10% of the values, the value of z must be on the left side of Y_0 and be negative, or:

$z = -1.28$
$x = z(\sigma) = (-1.28)(100) = -128$ miles
$X_2 = \mu + x = 500 + (-128) = 372$ miles

Thus, below 372 miles on the X scale, there will be a 10% area under the normal curve. Note that there will be 80% values above 372 miles but below 628 miles (or 128 miles deviate from the mean, 500 miles, in both directions).

Table 11-6 **Summary of Formulas**

Application	Formula	Formula Number	Reference Page
Permutations	$_nP_n = n!$ $= n(n-1)(n-2)\cdots(1)$	11-1	213
	$_nP_r = \dfrac{n!}{(n-r)!}$ $= n(n-1)(n-2)\cdots$ $(n-r+1)$	11-2	214
Combinations	$_nC_r = \dfrac{_nP_r}{r!}$	11-3	215
	$= \dfrac{n!}{r!(n-r)!}$		
	$_nC_r = {}_nC_{n-r}$	11-4	216
Probability	$P(A) = \dfrac{h}{n}$	11-5	218
	$P(\text{not } A) = 1 - P(A)$ $= 1 - \dfrac{h}{n}$	11-6	218
	$P(A \cup B) = P(A) + P(B)$	11-7	220
	$P(A \cup B) = P(A) + P(B)$ $- P(A \cap B)$	11-8	221
	$P(A \cap B) = P(A) \cdot P(B)$	11-9	222
	$P(A \cap B) = P(A) \cdot P(B|A)$	11-10	223
n repeated trials	$P(X;n,P) = {}_nC_X \cdot P^X \cdot Q^{n-X}$	11-11	224
Binomial theorem	$(Q + P)^n = {}_nC_0Q^n +$ ${}_nC_1Q^{n-1}P + {}_nC_2Q^{n-2}P^2$ $+ \cdots + {}_nC_nP^n$	11-12	228
	$\mu = nP$	11-13	232
	$\sigma = \sqrt{nPQ}$	11-14	232
Normal distribution	$z = \dfrac{x}{\sigma} = \dfrac{X - \mu}{\sigma}$	11-15	235

EXERCISE 11-4   Reference: Section 11.5

1. A grocery store has 10,000 sales tickets this month. The average of the tickets is $60 and the standard deviation is $10. The store manager believes that the distribution of sales amounts on the tickets is normal. Based on the amounts, predict the number of sales for each of the amounts listed for the next month's sales. (Assume that there will be 10,000 sales tickets again next month.)

(a) above $70
(b) above $90
(c) below $60
(d) below $40

2. The average of 20,000 cash deposits in a bank is $500 and the standard deviation of the deposits is $80. Assume that the distribution of the amounts of the deposits is normal. How many deposits in the distribution are:

(a) above $500
(b) above $500 but below $580
(c) above $580
(d) below $380

3. Refer to Problem 1. Predict the number of sales that will have the amounts:

(a) between $60 and $70
(b) between $60 and $80
(c) between $60 and $90
(d) between $50 and $60
(e) between $40 and $60
(f) between $30 and $60

4. Refer to Problem 2. How many deposits in the distribution are:

(a) above $380 but below $580
(b) above $380 but below $420
(c) between $620 and $740
(d) between $260 and $340

5. Refer to Problem 1. Find the sales amount above which there will be (a) the upper 10% of the sales tickets, and (b) the upper 70% of the sales tickets. Also, find the sales amount below which there will be (c) the lower 10% of the sales tickets, and (d) the lower 80% of the sales tickets.

6. Refer to Problem 2. Find the deposited amount above which there will be (a) the upper 20% of the deposits, and (b) the upper 60% of the deposits. Also, find the deposited amount below which there will be (c) the lower 20% of the deposits, and (d) the lower 90% of the deposits.

Chapter

12 Bayes' Theorem and Applications

This chapter will introduce Bayes' theorem and its applications in the decision-making process for problems under conditions of uncertainty. In these types of problems the probabilities of all possible events are first determined by either theoretical or empirical computation. Next, the old set of probabilities is revised as additional information becomes available. The additional information can be obtained from past records of a business firm or from samples. The revision process is usually based on Bayes' theorem.

12.1 INTRODUCTION TO BAYES' THEOREM

Bayes' theorem was developed by the English clergyman and mathematician Thomas Bayes and published in 1763 after his death. The basic concept of this theorem is explained in the following example.

EXAMPLE 1 A box of 20 cards can be classified two ways based on:

(1) The colors of the cards: 6 white cards, including 2 printed with a flower and 4 with an animal; and 14 red cards, including 6 printed with a flower and 8 with an animal.

(2) The types of pictures printed on the cards: 8 flower cards, including 2 white and 6 red; and 12 animal cards, including 4 white and 8 red.

The cross classifications of the 20 cards are shown in Table 12-1.

Table 12-1 **Cross-Classified Cards**

Picture / Color	Flower	Animal	Row Total
white	2 (white and flower)	4 (white and animal)	6 (white)
red	6 (red and flower)	8 (red and animal)	14 (red)
Column Total	8 (flower)	12 (animal)	20 cards

One card is drawn randomly from the box. What are the probabilities of all possible outcomes of the drawing?

SOLUTION

The probabilities of all possible outcomes are shown in Table 12-2. Each probability is obtained by dividing the number in the cell of Table 12-1 by the total number 20. For example, the probability of drawing a white-and-flower card, P(white and flower), is $\frac{2}{20} = .10$. Likewise, the probability of drawing a white, ignoring the picture, or P(white), is $\frac{6}{20} = .30$. The probabilities, P(white and flower), P(white and animal), P(red and flower), and P(red and animal) in the center four cells of the table are known as *joint probabilities*. The sum of all joint probabilities is equal to 1. The probabilities in the right column (row totals) and the bottom row (column totals) are called *marginal probabilities*. The sum of the column marginal probabilities and that of the row marginal probabilities are also equal to 1 respectively.

Table 12-2 **Probabilities of Outcomes of One Drawing (Example 1)**

Picture / Color	Flower	Animal	Row Total
white	P(white and flower) $\frac{2}{20} = .10$	P(white and animal) $\frac{4}{20} = .20$	P(white) $\frac{6}{20} = .30$
red	P(red and flower) $\frac{6}{20} = .30$	P(red and animal) $\frac{8}{20} = .40$	P(red) $\frac{14}{20} = .70$
Column Total	P(flower) $= \frac{8}{20}$ $= .40$	P(animal) $= \frac{12}{20}$ $= .60$	$\frac{20}{20} = 1$

EXAMPLE 2

Refer to Example 1. Suppose that the card drawn has a flower. What is the probability that the card is white? red?

SOLUTION

Table 12-2 shows that the probability of drawing a white card is $\frac{6}{20}$. In the absence of any additional information, this would be the final answer. However, in the present case the additional information is available: the card drawn has a flower. Thus, the above question may be rephrased in another manner. Suppose that the box contained only 8 flower cards: 2 white and 6 red. What is the probability of drawing a white card?

The probability of drawing a white card is $\frac{2}{8}$, or .25.

Likewise, the probability of drawing a red card is $\frac{6}{8}$, or .75.

These given questions in the original problem may also be answered by the use of conditional probability. Observe the "flower" column in Table 12-2. Find the probability of drawing a white card, given the card has a flower, or:

$$P(\text{white} \mid \text{flower}) = \frac{P(\text{white and flower})}{P(\text{flower})} = \frac{\frac{2}{20}}{\frac{8}{20}} = \frac{2}{8} = .25$$

Similarly, find the probability of drawing a red card, given the card has a flower, or:

$$P(\text{red} \mid \text{flower}) = \frac{P(\text{red and flower})}{P(\text{flower})} = \frac{\frac{6}{20}}{\frac{8}{20}} = \frac{6}{8} = .75$$

Observe:

$$P(\text{flower}) = P(\text{white and flower}) + P(\text{red and flower}) = \frac{2}{20} + \frac{6}{20} = \frac{8}{20}$$

$$P(\text{white} \mid \text{flower}) + P(\text{red} \mid \text{flower}) = .25 + .75 = 1$$

EXAMPLE 3 Refer to Example 1 again. Suppose that the card drawn has an animal. What is the probability that the card is white? red?

SOLUTION This question may also be rephrased in another manner. Suppose that the box contained only 12 animal cards: 4 white and 8 red. What is the probability of drawing a white card? red card?

The probability of drawing a white card is $\frac{4}{12}$, or .33.

The probability of drawing a red card is $\frac{8}{12}$, or .67.

Again, the above questions may be answered by the use of conditional probability. Observe the "animal" column in Table 12–2.

The probability of drawing a white card, given the card has an animal, is:

$$P(\text{white} \mid \text{animal}) = \frac{P(\text{white and animal})}{P(\text{animal})} = \frac{\frac{4}{20}}{\frac{12}{20}} = \frac{4}{12} = .33$$

The probability of drawing a red card, given the card has an animal is:

$$P(\text{red} \mid \text{animal}) = \frac{P(\text{red and animal})}{P(\text{animal})} = \frac{\frac{8}{20}}{\frac{12}{20}} = \frac{8}{12} = .67.$$

Observe:

$$P(\text{animal}) = P(\text{white and animal}) + P(\text{red and animal}) = \frac{4}{20} + \frac{8}{20} = \frac{12}{20}$$
$$P(\text{white} \mid \text{animal}) + P(\text{red} \mid \text{animal}) = .33 + .67 = 1$$

In general, let:

A_1 and A_2 = the set of events which are mutually exclusive (the two events cannot occur together) and exhaustive (the combination of the two events is the universal set)

B = a single event which intersects each of the A events as shown in the following diagram:

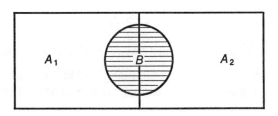

In the diagram, the part of B which is within A_1 represents the area "A_1 and B," and the part of B within A_2 represents the area "A_2 and B."

The probability of event A_1, given event B, is:

$$P(A_1|B) = \frac{P(A_1 \text{ and } B)}{P(B)}$$

The probability of event A_2, given B, is:

$$P(A_2|B) = \frac{P(A_2 \text{ and } B)}{P(B)}$$

where $P(B) = P(A_1 \text{ and } B) + P(A_2 \text{ and } B)$, and by applying the dependent events formula:

$$P(A_1 \text{ and } B) = P(A_1) \cdot P(B|A_1)$$
$$P(A_2 \text{ and } B) = P(A_2) \cdot P(B|A_2)$$

Further, let A_1, A_2, A_3, . . . , A_n be the set of n mutually exclusive and exhaustive events as shown in the following diagram ($n = 3$):

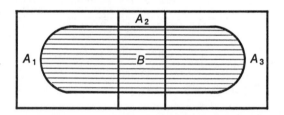

These expressions, called *Bayes' theorem,* may be summarized as follows:

The probability of event A_i ($i = 1, 2, 3, . . . n$), given event B, is:

$$P(A_i|B) = \frac{P(A_i \text{ and } B)}{P(B)}$$

where $P(B) = P(A_1 \text{ and } B) + P(A_2 \text{ and } B) + . . . + P(A_n \text{ and } B)$, and
$\quad P(A_i \text{ and } B) = P(A_i) \cdot P(B|A_i)$

When Bayes' theorem is used to revise a set of old probabilities, called *prior probabilities,* to a set of new probabilities, called *posterior probabilities,* the revision is usually done systematically in a table. Example 2 is now used to illustrate the application of such a Table.

Let A_1 = the event of drawing a white card,
$\quad A_2$ = the event of drawing a red card, and
$\quad B$ = the event of drawing a card having a flower, either white or red.

The required values for Example 2 are computed in Table 12–3 and shown in the tree diagram, Figure 12–1. The final answer is shown in column (5) of Table 12–3.

Figure 12-1 **Computation of Posterior Probabilities (Examples 2 and 3)**

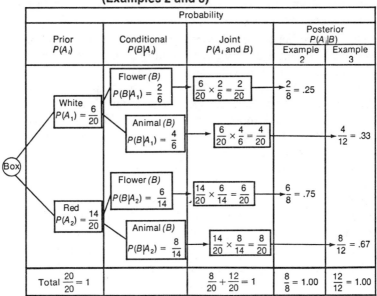

Table 12-3 **Computation of Posterior Probabilities (Example 2)**

(1) Event	(2) Prior Probability $P(A_i)$	(3) Conditional Probability of Event B (Flower) Given Event A $P(B\mid A_i)$	(4) Joint Probability $P(A_i \text{ and } B)$ (2) × (3)	(5) Posterior (Revised) Probability $P(A_i\mid B)$ (4) ÷ $P(B)$
A_1 (White)	$\frac{6}{20}$	$\frac{2}{6}$	$\frac{2}{20}$	$\frac{\frac{2}{20}}{\frac{8}{20}}=\frac{2}{8}=.25$
A_2 (Red)	$\frac{14}{20}$	$\frac{6}{14}$	$\frac{6}{20}$	$\frac{\frac{6}{20}}{\frac{8}{20}}=\frac{6}{8}=.75$
Total	$\frac{20}{20}=1$		$P(B)=\frac{8}{20}$	1.00

The individual columns in Table 12–3 are explained as follows:

(1) List the events for which the prior and posterior probabilities are computed.

(2) $P(A_i)$ represents the prior probability. Each card in the box has an equal chance to be drawn. Since 6 out of 20 cards are white, the probability of the event of drawing a white, $P(A_1)$, is $\frac{6}{20}$. Likewise, since 14 out of 20 cards are red, the probability of the event of drawing a red, $P(A_2)$, is $\frac{14}{20}$.

Prior probabilities are assigned to the events in column (1) before having the additional information that may be used to improve the quality of the probabilities. Prior probabilities may be assigned mathematically, such as the computation of $\frac{6}{20}$ and $\frac{14}{20}$ (Table 12–3), or subjectively, such as based on past records of business activities. (See Example 5.)

(3) $P(B|A_i)$ represents the conditional probability of B, given (or depending on) event A_i. Event B describes the additional information, having a flower card. For example, the additional information states that there are 2 flower-white and 4 animal-white cards. Then, the probability of the event of drawing a flower card, given the card is a white or $P(B|A_1)$, is $\frac{2}{6}$. Also, there are 6 flower-red and 8 animal-red cards. Thus, the probability of the event of drawing a flower card, given the card is red, or $P(B|A_2)$, is $\frac{6}{14}$.

(4) $P(A_i \text{ and } B)$ represents the joint probability of events A_i and B. The probability of the event of drawing a white and flower card is $P(A_1 \text{ and } B) = P(A_1) \cdot P(B|A_1) = (\frac{6}{20})(\frac{2}{6}) = \frac{2}{20}$. The probability of the event of drawing a red and flower card is $P(A_2 \text{ and } B) = P(A_2) \cdot P(B|A_2) = (\frac{14}{20})(\frac{6}{14}) = \frac{6}{20}$. The sum is $P(B) = \frac{2}{20} + \frac{6}{20} = \frac{8}{20}$.

(5) $P(A_i|B)$ represents the posterior (or revised prior) probability of A_i, given B. It is computed directly from the corresponding joint probability and the sum of the joint probabilities. For example, $P(A_1|B) = (\frac{2}{20}) \div (\frac{8}{20}) = 0.25$, which represents the probability that the white card has a flower. The sum of the posterior probabilities is equal to 1.

In a similar manner, let:

$A_1 =$ the event of drawing a white card
$A_2 =$ the event of drawing a red card
$B =$ the event of drawing a card having an animal, either white or red

Then, the required values for Example 3, by the use of Bayes' theorem, can be computed in Table 12–4 and also shown in the tree diagram in Figure 12–1. The final answer is also shown in column (5) of Table 12–4.

Table 12–4 **Computation of Posterior Probabilities (Example 3)**

(1) Event	(2) Prior Probability $P(A_i)$	(3) Conditional Probability of Event B (Animal) Given Event A $P(B|A_i)$	(4) Joint Probability $P(A_i \text{ and } B)$ (2) × (3)	(5) Posterior (Revised) Probability $P(A_i|B)$ (4) ÷ $P(B)$
A_1 (White)	$\frac{6}{20}$	$\frac{4}{6}$	$\frac{4}{20}$	$\frac{\frac{4}{20}}{\frac{12}{20}} = \frac{4}{12} = .33$
A_2 (Red)	$\frac{14}{20}$	$\frac{8}{14}$	$\frac{8}{20}$	$\frac{\frac{8}{20}}{\frac{12}{20}} = \frac{8}{12} = .67$
Total	$\frac{20}{20} = 1$		$P(B) = \frac{12}{20}$	1.00

The use of Bayes' theorem is further illustrated in Example 4.

EXAMPLE 4 A bag contains green and black balls marked 1, 2, or 3 as follows:

1	(10 balls, including 2 green and 8 black)
2	(15 balls, including 6 green and 9 black)
3	(25 balls, including 12 green and 13 black)
Total	(50 balls, including 20 green and 30 black)

A green ball is drawn randomly from the bag. Determine the probabilities that the green ball is marked 1, 2, and 3.

SOLUTION A_1 = the event of drawing a 1 ball
A_2 = the event of drawing a 2 ball
A_3 = the event of drawing a 3 ball
B = the event of drawing a green ball, marked either 1, 2, or 3

Events $A_1, A_2,$ and A_3 are mutually exclusive since one drawing of one ball cannot have two kinds of balls at one time. The events are also exhaustive since there is no ball other than 1, 2, and 3. The required values are computed in Table 12–5. The final answer is shown in column (5) of the table.

Table 12–5 **Computation of Posterior Probabilities (Example 4)**

(1) Event	(2) Prior Probability $P(A_i)$	(3) Conditional Probability of Event B (Green Ball) Given Event A $P(B\|A_i)$	(4) Joint Probability $P(A_i \text{ and } B)$ (2) × (3)	(5) Posterior (Revised) Probability $P(A_i\|B)$ (4) ÷ $P(B)$
A_1 (1 Ball)	$\frac{10}{50}$	$\frac{2}{10}$	$\frac{2}{50}$	$\frac{\frac{2}{50}}{\frac{20}{50}} = \frac{2}{20} = 0.10$
A_2 (2 Ball)	$\frac{15}{50}$	$\frac{6}{15}$	$\frac{6}{50}$	$\frac{\frac{6}{50}}{\frac{20}{50}} = \frac{6}{20} = 0.30$
A_3 (3 Ball)	$\frac{25}{50}$	$\frac{12}{25}$	$\frac{12}{50}$	$\frac{\frac{12}{50}}{\frac{20}{50}} = \frac{12}{20} = 0.60$
Total	$\frac{50}{50} = 1$		$P(B) = \frac{20}{50}$	1.00

The individual columns in Table 12–5 are explained as follows:

(1) $P(A_i)$ represents the prior probability. Each ball in the bag has an equal chance to be drawn. Since 10 out of 50 balls are marked 1, the probability of drawing a 1 ball, $P(A_1)$, is $\frac{10}{50}$. The probabilities $\frac{15}{50}$ and $\frac{25}{50}$, assigned to $P(A_2)$ and $P(A_3)$ respectively are obtained in the same manner.

(2) $P(B\|A_i)$ represents the conditional probability of event B (described by the additional information), given event A_i. For example, the additional information states that there are 2 green and 8 black balls marked 1. Then, the probability of the event of drawing a green ball, given balls marked 1, or $P(B\|A_1)$, is $\frac{2}{10}$.

(3) $P(A_i \text{ and } B)$ represents the joint probability of event A_i and event B. For example, $P(A_1 \text{ and } B) = P(A_1) \cdot P(B|A_1) = (\frac{10}{50})(\frac{2}{10}) = \frac{2}{50}$. Thus, the probability of drawing a 1 and green ball from the bag is $\frac{2}{50}$ (or 2 out of 50 balls in the bag).

(4) $P(A_i|B)$ is the posterior (or revised prior) probability of A_i, given B. For example, $P(A_1|B) = (\frac{2}{50}) \div (\frac{20}{50}) = 0.10$, which represents the probability that the green ball is marked 1. The sum of the posterior probabilities is equal to 1.

12.2 APPLICATIONS IN BUSINESS PROBLEMS

Applications of Bayes' theorem to business problems will be illustrated in this section. The original probability of a business event is revised according to additional information which can be obtained from historical records or samples. The illustrations are presented in Examples 5, 6, and 7.

EXAMPLE 5　Certain units in an automobile factory are manufactured by two machines. Machine 1 produces 30% and machine 2 produces 70% of the output. Past records showed that 12% of the units produced by machine 1 were defective and only 4% produced by machine 2 were defective. If a defective unit is drawn at random, what is the probability that the defective unit was produced by machine 1? machine 2?

SOLUTION　$A_1 =$ the event of drawing a unit produced by machine 1
　　　　　$A_2 =$ the event of drawing a unit produced by machine 2
　　　　　$B =$ the event of drawing a defective unit produced either by machine 1 or machine 2

From the first information, or prior probabilities:

$P(A_1) = 30\% = 0.30$
$P(A_2) = 70\% = 0.70$

From the second information, or conditional probabilities:

$P(B|A_1) = 12\% = 0.12$
$P(B|A_2) = 4\% = 0.04$

The given information is diagrammed below and the required values are computed in Table 12–6. The final answer is shown in column (5) of the table:

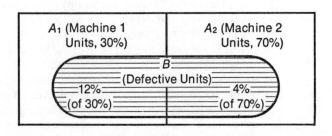

A_1 (Machine 1 Units, 30%)	A_2 (Machine 2 Units, 70%)
B (Defective Units)	
12% (of 30%)	4% (of 70%)

Table 12-6 Computation of Posterior Probabilities (Example 5)

(1) Event	(2) Prior Probability $P(A_i)$	(3) Conditional Probability of Event B (Defective Unit) Given Event A $P(B \mid A_i)$	(4) Joint Probability $P(A_i$ and $B)$ $(2) \times (3)$	(5) Posterior (Revised) Probability $P(A_i \mid B)$ $(4) \div P(B)$
A_1 (Units by Machine 1)	0.30	0.12	0.036	$\dfrac{0.036}{0.064} = 0.5625$
A_2 (Units by Machine 2)	0.70	0.04	0.028	$\dfrac{0.028}{0.064} = 0.4375$
Total	1.00		$P(B) = 0.064$	1.0000

Based on the prior probability distribution shown in Table 12-6, or without the additional information, we may believe that the defective unit is drawn from machine 2 output since $P(A_2) = 70\%$ is larger than $P(A_1) = 30\%$. With the additional information, the accuracy of the probability distribution is improved. The posterior distribution shows that the probability of having the defective unit produced by machine 1 is 0.5625 or 56.25% and that by machine 2 is only 0.4375 or 43.75%. Thus, the defective unit is most likely drawn from the output produced by machine 1.

The answer of Example 5 may be checked by using an actual number of units as follows:

Assuming that 1,000 units were produced by the two machines in a given period, the number of units produced by machine 1 is:

$1{,}000 \times 30\% = 300$

and the number of units produced by machine 2 is:

$1{,}000 \times 70\% = 700$

The number of defective units produced by machine 1 is:

$300 \times 12\% = 36$

and by machine 2 is:

$700 \times 4\% = 28$

The probability that a defective unit was produced by machine 1 is:

$$\frac{36}{36 + 28} = \frac{36}{64} = 0.5625$$

and that by machine 2 is:

$$\frac{28}{36+28} = \frac{28}{64} = 0.4375$$

EXAMPLE 6   Refer to Example 5. Assume that the products of machine 1 and machine 2 are packed separately into different boxes. Two defective units instead of only one are drawn at random from an unidentified box. What is the probability that the two defective units were produced by machine 1? By machine 2?

SOLUTION   Let A_1 and A_2 represent the same events as defined in Example 5, and:

$B =$ the event of drawing two defective units produced either by machine 1 or machine 2.

The conditional probabilities based on the additional information are computed as follows:

$P(B|A_1) = (.12)^2 = .0144$
$P(B|A_2) = (.04)^2 = .0016$

The required values are computed in Table 12–7. The final answer is shown in column (5) of the table. The final answer indicates that the probability of drawing the defective units produced by machine 1 increases to .794 from .5625 (Table 12–6). The increase is expected since machine 1 produces more defective units (12% of 30% of the total production) than machine 2 (4% of 70% of the total).

Table 12–7   Computation of Posterior Probabilities (Example 6)

(1) Event	(2) Prior Probability $P(A_i)$	(3) Conditional Probability of Event B (2 Defective Units), Given Event A $P(B|A_i)$	(4) Joint Probability $P(A_i \text{ and } B)$ (2) × (3)	(5) Posterior (Revised) Probability $P(A_i|B)$ (4) ÷ $P(B)$
A_1 (Units by Machine 1)	0.30	0.0144	0.00432	$\frac{.00432}{.00544} = .794$
A_2 (Units by Machine 2)	0.70	0.0016	0.00112	$\frac{.00112}{.00544} = .206$
Total	1.00		$P(B) = 0.00544$	1.000

The additional information in Examples 5 and 6 was obtained from past records. The additional information in Example 7 is obtained from a sample. The basic procedure of applying Bayes' theorem for various sample sizes is the same. The binomial formula $P(X;n,P)$ is used in computing the conditional probabilities in such cases.

EXAMPLE 7 A factory has five machines. The production distribution and the defective production of the five machines are:

Machine Number	Production Distribution $P(A_i)$	Defective Units Produced by Each Machine (P)
1	20%	1%
2	32%	2%
3	30%	4%
4	10%	5%
5	8%	20%
	100%	

A sample of 20 units is taken from a lot produced by an unidentified machine. Three of the 20 units are defective. Find the probability that the defective units were produced by each machine.

SOLUTION Let $A_1, A_2 \ldots A_5$ represent the event of drawing a unit produced by machine 1, 2, . . . 5 respectively, and

B = the event of drawing three defective units from 20 units produced by one of the five machines.

The conditional probabilities based on the sample are computed according to the binomial formula

$P(B|A_i) = P(X;n,P) = {}_nC_X \cdot P^X \cdot Q^{n-X}$ (or from the Binomial Probability Distribution Table, Table 5 in the Appendix)

Where $X = 3$ (defective units), $n = 20$ (sample size), P = percent of defective units produced by each machine, and $Q = 1 - P$.

Substitute $P = 1\%, 2\%, 4\%, 5\%$, and 20% in the formula. The conditional probabilities of the sample result for given events are:

$$P(3;20,.01) = {}_{20}C_3 \cdot (.01)^3 \cdot (.99)^{17} = \frac{20 \cdot 19 \cdot 18}{3 \cdot 2 \cdot 1}(.01)^3 (.99)^{17}$$

$$= .0010 \text{ (machine 1 with } P = 1\% = .01)$$
$$P(3;20,.02) = .0065 \text{ (machine 2 with } P = 2\% = .02)$$
$$P(3;20,.04) = .0364 \text{ (machine 3 with } P = 4\% = .04)$$
$$P(3;20,.05) = .0596 \text{ (machine 4 with } P = 5\% = .05)$$
$$P(3;20,.20) = .2054 \text{ (machine 5 with } P = 20\% = .20)$$

The required values are computed in Table 12–8. The final answer is shown in column (5) of the table. The final answer indicates that the probability of drawing the defective units produced by machine 5 is .4617, the highest probability among the revised probabilities in the column. According to the prior probabilities, however, the probability of drawing a unit produced by machine 2 is the highest, .32. Thus we may state that the given sample is most likely drawn from the lot produced by machine 5, but very unlikely by machine 2.

Table 12-8   **Computation of Posterior Probabilities (Example 7)**

(1) Event	(2) Prior Probability $P(A_i)$	(3) Conditional Probability of Event B (3 Defective Units Out of 20), Given Event A $P(B\|A_i) = P(X = 3; n = 20,P)$	(4) Joint Probability $P(A_i \text{ and } B)$ (2) × (3)	(5) Posterior (Revised) Probability $P(A_i\|B)$ (4) ÷ $P(B)$
$A_1(P = .01)$	.20	.0010	.000200	.0056
$A_2(P = .02)$	.32	.0065	.002080	.0584
$A_3(P = .04)$	.30	.0364	.010920	.3068
$A_4(P = .05)$	.10	.0596	.005960	.1675
$A_5(P = .20)$	.08	.2054	.016432	.4617
Total	1.00		$P(B) = .035592$	1.0000

12.3 COMPUTING THE MEAN AND THE STANDARD DEVIATION OF THE PRIOR AND POSTERIOR PROBABILITY DISTRIBUTIONS

The mean of the defective units, expressed in percent rates in Example 7, can be computed from either the prior probability distribution or the posterior probability distribution. The computation is based on the formula on page 232.

The mean $= \mu = \dfrac{\Sigma fX}{\Sigma f}$, where $f =$ the relative frequency or the probability,
$X =$ defective rate $= P$ in Example 7, and
$\Sigma f = 1$, the sum of the probabilities.

Thus: $\mu = \Sigma fX = \Sigma fP$

The mean of the P values is also called the *expected value* of P, denoted by $E(P)$, and $E(P) = \Sigma fP$. The expected value of the prior probability is shown in column (3) and that of the posterior probability is shown in column (5) of Table 12-9. The expected values of P based on the posterior probability distribution (.114211) is greater than that based on the prior probability distribution (.0414). This reflects the greater weight that the posterior probability distribution gives to the larger values of P. The weight factors are determined by the sample. Observe that the posterior expected value is closer to the sample proportion .15 (or 3 defective units out from 20 units in the sample) than the expected value based on the prior probability distribution.

The posterior expected value of .114211 is obtained from the result of the sample of 20 units with a defective proportion of .15. The posterior expected value will be closer to .15 if the sample size is increased and the defective proportion remains at .15.

Table 12-9 Computation of Expected Values of Prior and Posterior Probability Distributions (Example 7)

(1) Values P	(2) Prior Probability $P(A_i) = f$	(3) Expected Value of P based on Prior Probability (1) × (2) = fP	(4) Posterior Probability $P(A_i\|B) = f$	(5) Expected Value of P based on Posterior Probability (1) × (4) = fP
A_1: .01 A_2: .02 A_3: .04 A_4: .05 A_5: .20	.20 .32 .30 .10 .08	.0020 .0064 .0120 .0050 .0160	.0056 .0584 .3068 .1675 .4617	.000056 .001168 .012272 .008375 .092340
Total	1.00	$E(P) = .0414$	1.0000	$E(P) = .114211$

The standard deviation of the posterior probability distribution is computed in Table 12-10. (The standard deviation of the prior probability distribution is omitted here since the prior distribution is replaced by the updated posterior probability distribution.) The computation is based on the formula on page 232, or:

$$\text{the standard deviation} = \sigma = \sqrt{\frac{\Sigma f(X - \mu)^2}{\Sigma f}} = \sqrt{\Sigma f(X - \mu)^2}$$

or, written in the symbols used in Example 7:

$$\sigma = \sqrt{\Sigma f(P - E(P))^2}$$

Table 12-10 Computation of the Standard Deviation of a Posterior Probability Distribution (Example 7) (Sample Size = 20 Units)

Event Value P	Posterior Probability f	$x = P - E(P)$ $= P - .114211$	fx	fx^2
.01 .02 .04 .05 .20	.0056 .0584 .3068 .1675 .4617	−.104211 −.094211 −.074211 −.064211 .085789	−.000584 −.005502 −.022768 −.010755 .039609	.000061 .000518 .001690 .000691 .003398
Total	1.0000		.000000	.006358

$$\sigma = \sqrt{.006358} = 0.08$$

The value of the standard deviation obtained can be used to measure the dispersion of the set of P values away from the expected value of P.

EXERCISE 12–1   Reference: Sections 12.1–12.3

1. A bag contains 40 ping pong balls as follows:

> 10 white balls, including 3 marked X and 7 marked Y
> 30 yellow balls, including 12 marked X and 18 marked Y

One ball is drawn randomly from the bag. (a) What are the probabilities of all possible outcomes of the drawing? (b) Suppose that the ball drawn has an X mark. What is the probability that the ball is white? yellow?

2. Refer to problem 1. Suppose that the ball drawn has a Y mark. What is the probability that the ball is white? yellow?

3. A company has 60 employees grouped into three departments as follows:

> Accounting Department – 12 employees with college education
>   8 employees without college education
> Purchasing Department – 3 employees with college education
>   17 employees without college education
> Marketing Department – 6 employees with college education
>   14 employees without college education

An employee selected randomly from the company is college educated. From which department was he most likely employed?

4. Refer to problem 3. Assume that the employee selected from the company has no college education. From which department was he most likely employed?

5. A grocery store regularly receives large shipments of apples. A good shipment contains 2% spoiled apples and a poor shipment, 5%. The store's past experience is that 90% of the shipments have been good and only 10% have been poor. A large shipment has just been received. A sample of 30 apples drawn randomly from the shipment has 3 spoiled ones. What is the probability that it is a good shipment? a poor shipment?

6. An electric company has two plants producing light bulbs. Plant 1 produces 40% and plant 2, 60% of the total output. Past records showed that 1 in 10 bulbs produced by plant 1 was defective and 1 in 20 bulbs produced by plant 2 was defective. If a defective bulb is drawn at random, what is the probability that the defective bulb was produced by plant 1? by plant 2?

7. Refer to problem 6. Assume that a nondefective bulb is drawn at random. What is the probability that the nondefective bulb was produced by plant 1? by plant 2? Also, assume that the company has produced 10,000 bulbs during a day. Check your answers obtained in this problem and in problem 6 by using the actual number of products.

8. Refer to problem 6. Assume that the bulbs produced in each plant are packed into cartons of two bulbs each. A carton is drawn randomly from the company's warehouse and it is found that both bulbs in the carton are defective. What is the probability that the two defective bulbs were produced by plant 1? by plant 2?

9. A company has four plants. The production distribution and defective production of the four plants are as follows:

Plant Number	Production Distribution	Defective Units Produced by Each Plant
1	30%	1%
2	20%	5%
3	40%	10%
4	10%	20%
Total	100%	

A sample of 25 units is taken from a lot produced by an unidentified plant. Four of the 25 units are defective. Find the probability that the defective units were produced by each plant.

10. Refer to problem 9. Compute the expected values of the prior and posterior probability distributions and the standard deviation of the posterior distribution.

⋆12.4 COMPUTATION BASED ON THE NORMAL CURVE TABLE

The prior probabilities and conditional probabilities in the examples presented in previous sections were obtained by exact computations. The probabilities may also be obtained by using the Areas Under the Normal Curve Table, Table 6 in the Appendix. The method involving the use of the normal curve table is illustrated in Example 8.

EXAMPLE 8 Bee Company manufactures television sets. Based on past experience, the sales manager believes that there is a 50% probability that the company's share of sales this year will be between 18% and 22% (or 20% ± 2%) of the national market. In addition, he requested that a sample of 1,600 customers be drawn and interviewed. The sample indicated the following results:

X = the number of Bee television sets to be purchased by each customer. If a customer says that he will not buy, $X = 0$; if he will buy one, $X = 1$; if two, $X = 2$; and so on.

The mean $\overline{X}$ and the standard deviation s of the 1,600 X's of the sample are computed as follows: (Details of the 1,600 X's are omitted here for simplicity.) The sum of 1,600 X's, or ΣX, is 352. Thus:

$$\overline{X} = \frac{\Sigma X}{n} = \frac{352}{1,600} = 0.22 \text{ set per customer}$$

$$s = \sqrt{\frac{\Sigma(X - \overline{X})^2}{n}} = \sqrt{\frac{256}{1,600}} = \sqrt{0.16} = .40 \text{ set (Details of 256 are also omitted here.)}$$

The manager wishes to combine his past experience and sample results in the decision-making analysis. Find the probability distribution of the company's share of sales this year in the national market based on the analysis. Also, find the company's expected sales for this year.

Note that the Greek symbols μ (mu) and σ (sigma), with or without subscripts, are customarily used to denote the mean and the standard deviation of the population or a complete set of data, whereas the symbols $\overline{X}$ and s are used to denote the two values of a sample (drawn from a population).

The decision-making analysis of Example 8 is presented in parts A to D which follow. The probability distribution of the company's share of sales this year in the national market based on the analysis is shown in Table 12–13, column (5). The expected sales are 21.768% as shown in column (6) of the table.

A. Finding the Prior Probability Distribution

The prior probability is derived from the sales manager's past experience.

$P =$ sales proportion, or the company's share of sales this year in the national market expressed in percent rates

The value of P will vary. Thus, P is a random variable. The information based on the manager's past experience suggests that the expected value of P, or $E(P)$, is $20\% = .20$, the midpoint of 18% to 22%. Assume that the P values are normally distributed. Then, the probability of sales from 18% to 20% is 25% and that of sales from 20% to 22% is also 25% since the total between 18% and 22% is 50%. This information gives the prior probability distribution as shown in Figure 12–2.

Figure 12-2 **Prior Probability Distribution (Example 8)**

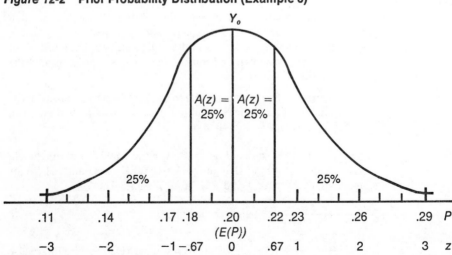

The standardized normal deviate z is:

$$z = \frac{P - E(P)}{\sigma}$$

where σ is the standard deviation of the distribution of P values.

When $P = 22\% = .22$, $E(P) = 20\% = .20$, and the area under the normal curve $A(z) = 25\% = .25$ are given as shown in Figure 12–2, the value of z can be determined as follows:

The Areas Under the Normal Curve Table shows:

$$A(z) = .24857, \qquad z = .67$$
$$A(z) = .25175, \qquad z = .68$$

Since $A(z) = .25$ is closer to .24857 than to .25175, we select the nearest value, or $z = .67$. Substitute the z value in the above z formula and solve for σ:

$$.67 = \frac{.22 - .20}{\sigma}$$

$$\sigma = \frac{.02}{.67} = .03$$

As stated in Chapter 11, Section 11.5, when the mean and the standard deviation of a normally distributed random variable are known, the normal probability distribution of the variable is completely determined. Here the random variable is P, the mean, or the expected value of $P = E(P) = .20$, and the standard deviation of P values $= \sigma = .03$.

When:

$$z = \ \ 1, P = .20 + .03(1) = .23$$
$$z = \ \ 2, P = .20 + .03(2) = .26$$
$$\cdots$$
$$z = -1, P = .20 - .03(1) = .17$$
$$z = -2, P = .20 - .03(2) = .14 \text{ and so on}$$

The z and P values are labeled under the normal curve.

Thus, the area representing the probability between any two points under the curve can be obtained by using the normal curve table.

However, most decision problems are concerned with discrete variables. In the present example, the manager is interested in knowing the sales proportion P expressed only in a single number, such as 18%, 20%, or 22% of the national market, not in a range specified by two numbers. The normal distribution is a continuous distribution. The probability of a single value of P, or at a point on the continuous P scale, is always equal to zero. Therefore, when the normal curve is used to compute the probabilities of a discrete variable, each individual value of the variable must be considered as the midpoint of a selected interval on the continuous scale. The upper limit of an interval will be the lower limit of the next interval on the scale.

Although all of the intervals need not be of equal size, equal size is generally preferred for convenience. In Table 12–11, which is designed to compute the prior probability distribution of variable P, each interval is arbitrarily assigned to be 2% = .02. Also, in selecting the intervals, care should be taken so that the mean and the standard deviation of the discrete prior probability distribution are the same as those for the original continuous normal distribution; that is, mean $(E(P))$ is .20 and standard deviation (σ) is .03. The final answer, prior probability, is shown in column (6) of Table 12–11.

Further comments refer to the construction of Table 12–11: Columns (1) and (2) P variable. First, determine in column (1) the upper and lower limits of

Table 12-11 Computation of Prior Probabilities Using Areas Under the Normal Curve Table (Example 8)

P(Event)		$P - E(P)$ = Col. (1) − .20	$z = \dfrac{P - E(P)}{\sigma}$ = $\dfrac{\text{Col. (3)}}{.03}$	Area Between Y_o and z	Prior Probability
Interval Limits	Midpoint				
(1)	(2)	(3)	(4)	(5)	(6)
· · · · ·					· · · · ·
.17		−.03	−1.00	.34134	
	.18				.21204
.19		−.01	−0.33	.12930	
	.20 (mean)			+ ⟩	.25860
.21		.01	0.33	.12930	
	.22				.21204
.23		.03	1.00	.34134	
	.24				.11120
.25		.05	1.67	.45254	
	.26				.03756
.27		.07	2.33	.49010	
· · · · ·					· · · · ·
Total					1.00000

the interval located in the center of the prior probability distribution. Since the distribution is normal, the midpoint of the central interval is the mean of the distribution and is equal to .20. The size of each interval is arbitrarily assigned to be .02. The upper and lower limits of each interval thus must be $\frac{1}{2}$ of .02, or .01, above and below each midpoint respectively. For the central interval:

$$\text{the upper limit} = .20 + .01 = .21$$
$$\text{the lower limit} = .20 - .01 = .19$$

Next, determine the limits of other intervals in column (1). The other limits can be obtained consecutively by either subtracting .02 from .19 (for the intervals with values smaller than that of the central interval) or adding .02 to .21 (for the intervals with values larger than that of the central interval).

The range in column (1) is from .17 to .27. This range agrees with the range used in Table 12–12 for computing conditional probabilities. The conditional probabilities of P values outside the range will have 0 values. Thus, it is unnecessary to include them in the tables.

Table 12-12   Computation of Conditional Probabilities Using Areas Under the Normal Curve Table (Example 8)

P (Event)		$\overline{X} - P$ $= .22 -$ Col. (1)	$z = \dfrac{\overline{X} - P}{s_{\overline{X}}}$ $= \dfrac{\text{Col. (3)}}{.01}$	Area Between Y_o and z	Conditional Probability
Interval limits	Midpoint				
(1)	(2)	(3)	(4)	(5)	(6)
	.16				.00000
.17		.05	5.00	.50000	
	.18				.00135
.19		.03	3.00	.49865	
	.20				.15731
.21		.01	1.00	.34134	
	.22 ($\overline{X}$)			+	.68268
.23		−.01	−1.00	.34134	
	.24				.15731
.25		−.03	−3.00	.49865	
	.26				.00135
.27		−.05	−5.00	.50000	
	.28				.00000

In column (3), the P values are the interval limits shown in column (1). The expected value of P, or $E(P)$, is the mean of the prior distribution, .20. Thus, in the 1st line: $P - E(P) = .17 - .20 = -.03$.

In column (4), each z value is the quotient of the difference obtained from column (3) divided by .03. The value of .03 is the standard deviation of P values of the prior distribution.

In column (5), the areas are obtained from the Areas Under the Normal Curve Table. Each number represents the probability of the sales proportion for the areas between a given limit (at z) and the mean (at Y_o).

In column (6), each prior probability is the difference between two adjoining values shown in column (5). For example, the prior probability of the interval .17 to .19 with the midpoint .18 is:

$$.34134 - .12930 = .21204.$$

One exception is the prior probability of the central interval, which is the sum of the two adjoining values in column (5), or:

$$.12930 + .12930 = .25860.$$

The total of the prior probabilities, including the probabilities not shown or the probabilities of the P values outside the range listed in column (1), is 1.

B. Finding the Conditional Probability

The conditional probability for each of the intervals listed in the prior probability distribution of Table 12–11 is derived from the sample information. Each conditional probability is the probability of obtaining the mean $\bar{X} = .22$ of a sample drawn from a population which has a given mean P.

The conditional probabilities are systematically computed in Table 12–12. For example, the probability of obtaining a sample with the mean .22 from a population with the given mean $P = .18$ is .00135. The population mean $P = .18$ is specified by the interval limits .17 and .19 in the table. The difference between the sample mean and each limit, or $X - P$ in column (3) in the table, is measured by the unit of standard error of the mean $s_{\bar{x}}$ which is an estimate of the standard deviation (also called standard error) of the means ($\bar{X}$'s) of all possible samples of the same size drawn from the population. When the sample size is large, preferably 100 or more, the formula of $s_{\bar{x}}$ is:

$$s_{\bar{x}} = \frac{s}{\sqrt{n)}} \text{ where } s = \text{the standard deviation of}$$
$$\text{the sample}$$
$$n = \text{the sample size.}$$

Substitute the s and n values from the given sample information:

$$s_{\bar{x}} = \frac{.40}{\sqrt{1,600}} = .01$$

When the difference $\bar{X} - P$ is expressed in the units of $s_{\bar{x}}$, the result is the value of z, or:

$$z = \frac{\bar{X} - P}{s_{\bar{x}}} = \frac{.22 - P}{.01} \quad \begin{array}{l}\text{(shown in column (4)}\\ \text{in the table)}\end{array}$$

The area between Y_o and z in column (5) is obtained from the Areas Under the Normal Curve Table. Each conditional probability is the difference between two adjoining values shown in column (5). One exception is the probability of the central interval, which is the sum of the two adjoining values in column (5).

C. Finding the Posterior Probability Distribution

The computation of the posterior probabilities is shown in Table 12–13. The prior probabilities obtained from Table 12–11 are used in the computation. The highest prior probability in the table is .25860 for $P = .20$. However, when the prior probabilities are revised according to the sample information, the highest posterior probability is .71215 for $P = .22$.

D. Computing the Expected Value and the Standard Deviation of the Posterior Probability Distribution

The expected value of the posterior probability distribution (also shown in Table 12–13) is .21768 which is very close to the sample mean .22. The standard deviation of the posterior distribution is .01 as computed in Table 12–14.

Table 12–13 Computation of Posterior Probability and Expected Value (Example 8)

(1) P (Event)		(2) Prior Probability (Table12–11)	(3) Conditional Probability (Table 12–12)	(4) Joint Probability (2) × (3)	(5) Posterior Probability (4) ÷ .20327	(6) Posterior Expected Value of P (1) × (5)
Interval Limits	Mid- point					
.17 − .19	.18	.21204	.00135	.00029	.00143	.00026
.19 − .21	.20	.25860	.15731	.04068	.20013	.04003
.21 − .23	.22	.21204	.68268	.14476	.71215	.15667
.23 − .25	.24	.11120	.15731	.01749	.08604	.02065
.25 − .27	.26	.03756	.00135	.00005	.00025	.00007
Total		1.00000		.20327	1.00000	.21768

Table 12–14 Computation of the Standard Deviation of Posterior Probability Distribution (Example 8)

Event P (Midpoint)	Posterior Probability	$x = P - E(P)$ $= P - .21768$	fx	fx^2
.18	.00143	−.03768	−.000054	.000002
.20	.20013	−.01768	−.003538	.000063
.22	.71215	.00232	.001652	.000004
.24	.08604	.02232	.001920	.000043
.26	.00025	.04232	.000011	.000000
Total	1.00000		−.000009 (near zero)	.000112

$\sigma = \sqrt{.000112} = .01$

EXERCISE 12–2   Reference: Section 12.4

1. Refer to Example 8, page 257. Assume that the past experience of the sales manager in Bee Company states that there is a 95% probability that the company's share of sales this year will be between 18% and 22% of the

national market. Find the prior probability distribution for the sales proportions (P values) from 19% to 23%. Let the size of each interval be $1\% = .01$. (Thus, the first interval is $.185 - .195$ and the midpoint is $.19 = 19\%$.)

2. Refer to Example 8. Assume that the sales manager took a sample of 2,500 customers. The mean and the standard deviation of the sample are: $\overline{X} = 0.21$ television set per customer, and $s = .25$ set. Find the conditional probabilities for the interval limits specified in problem 1.

3. Find the posterior probability distribution based on the answers obtained from problems 1 and 2.

4. Compute (a) the expected value and (b) the standard deviation of the posterior probability distribution obtained in problem 3.

Chapter
13 Expected Values and Decision Rules

This chapter continues to illustrate the process of decision making for problems under conditions of uncertainty. We shall first introduce the method of constructing the payoff table since the table provides essential information for making decisions on these types of problems.

13.1 THE PAYOFF TABLE AND CONDITIONAL PROFITS

In analyzing a problem under conditions of uncertainty, the decision maker should list the following related items in a table:

1. All possible *states of nature,* called *events,* obtained from the given problem. The events are mutually exclusive since the occurrence of any one of them excludes the occurrences of the others; that is, two or more of the events cannot occur together. The events are uncertain since we do not know which one of the events will occur.
2. All possible acts that he may take. The uncertainty of individual events provides many alternative actions available to the decision maker.
3. All consequences; each consequence is the outcome of an act under a given event.

A table which systematically shows the relationships among all possible events, all possible acts, and all related consequences, or sometimes the values associated with the consequences, is called a *payoff table, decision matrix,* or *payoff matrix.* A "payoff" is simply a consequence, regardless of whether it is favorable or unfavorable. For example, the payoffs may be profits, either positive numbers (representing profits) or negative numbers (representing losses) as shown in Table 13–1.

EXAMPLE 1   A drug store buys news magazines for resale. According to past experience, the numbers of copies sold in a week vary from 21 to 24. The cost of each copy is $2 and the selling price, $3. If the copy is not sold during the week, it cannot be returned; thus it has no value. Construct a payoff table to show all possible profits.

SOLUTION   See Table 13-1.

Table 13-1   **A Payoff Table (Example 1)**

Possible Weekly Sales (Events)	Possible Number of Copies Stocked (Acts)			
	21	22	23	24
	Conditional Profit by Each Act (Consequences)			
21 (Copies)	(21×3) $-(21 \times 2) = \$21$	(21×3) $-(22 \times 2) = \$19$	(21×3) $-(23 \times 2) = \$17$	(21×3) $-(24 \times 2) = \$15$
22	(21×3) $-(21 \times 2) = 21$	(22×3) $-(22 \times 2) = 22$	(22×3) $-(23 \times 2) = 20$	(22×3) $-(24 \times 2) = 18$
23	(21×3) $-(21 \times 2) = 21$	(22×3) $-(22 \times 2) = 22$	(23×3) $-(23 \times 2) = 23$	(23×3) $-(24 \times 2) = 21$
24	(21×3) $-(21 \times 2) = 21$	(22×3) $-(22 \times 2) = 22$	(23×3) $-(23 \times 2) = 23$	(24×3) $-(24 \times 2) = 24$

The profit by each act is computed as follows:

$$\left(\begin{array}{c}\text{Number of Copies Possibly Sold} \\ \text{(Limited by Number Stocked)} \\ \times \\ \text{Selling Price Per Copy}\end{array}\right) - \left(\begin{array}{c}\text{Number of Copies Stocked} \\ \times \\ \text{Cost Per Copy}\end{array}\right)$$

For example, if the demand event (in the possible weekly sales column) is 23 copies and the number of copies stocked is 22, the profit is computed as follows:

$$[22 \text{ (Copies Sold)} \times \$3] - [22 \text{ (Copies Stocked)} \times \$2] = \$66 - \$44$$
$$= \$22$$

When 22 are stocked, only 22 can be sold, even if the demand event is 23 or more copies. On the other hand, if the demand event, such as 21, is less than the number stocked, such as 22, the profit is reduced by the cost of overstocked copies, at $2 each.

The profits shown in the payoff table are also called the *conditional profits* since each profit is computed under the condition that a specific number of copies is sold (certain event has occurred) when a specific number of copies is stocked (certain act is taken).

13.2 FINDING EXPECTED PROFITS

The exact occurrence of each event is not known. A decision maker thus cannot choose the best or optimum act based only on the conditional profits. However, the probability of occurrence of each event can be assigned. A general way to assign the probability is based on historical data as shown in Example 2. The sum of the probabilities of all events is 1. The conditional profit of each event under a given act is weighted by the probability of the event occurring. The sum of the weighted conditional profits is called the *expected profit* for the act. The *optimum act* is the one with the highest expected value or profit.

Note that the term *expected value* is generally used to represent an average of a set of values. The average used in this chapter is a weighted arithmetic mean, the weights being the probabilities of the individual values. Thus, the term will be used not only for conditional profits but also for other types of values, such as sales (Example 2) and conditional losses (Section 13.4).

EXAMPLE 2   Refer to Example 1. The following additional information concerning weekly sales during the past 50 weeks is obtained from the drug store.

Weekly Sales (Number of Copies)	Number of Weeks
21	5
22	15
23	20
24	10
Total	50

Compute: (a) the probability of each event of weekly sales, (b) the expected sales for each week, and (c) the expected profit for each act.

SOLUTION   (a) The probability of each event is obtained by dividing the number of weeks for the event (such as in 5 weeks 21 copies per week were sold), by the total number of weeks (50). Thus, the probability of the event of 21 copies is $5 \div 50 = .10$. The probabilities of other events are computed in the same manner and are shown in column (3) of Table 13–2.

Table 13–2   Solution to Example 2(a) and (b).

(1) Weekly Sales (Event)	(2) Number of Weeks	(3) Probability of Each Event Col. (2) ÷ 50	(4) Expected Weekly Sales Col. (1) × Col. (3)
21 Copies	5	.10	2.10
22	15	.30	6.60
23	20	.40	9.20
24	10	.20	4.80
Total	50	1.00	22.70

(b) The expected sales for each week based on the given information are computed in column (4) in Table 13–2. The weekly sales for each event are multiplied by the probability of the event. The sum of the products, 22.70 copies, is the expected sales per week.

(c) The expected profit for each act is computed in Table 13-3. The probability of each event is obtained from Table 13-2 and the profits by each act are obtained from Table 13-1. The expected profit of the act of stocking 22 copies is the highest value, $21.70. Thus, the optimum act is to stock 22 copies each week in the drug store.

Table 13-3 **Computing Expected Profits (Example 2c)**

		Act (Number of Copies Stocked)							
		21		22		23		24	
		Calculations of Expected Profit (Conditional Profit × Probability of Event) *							
		Profit		Profit		Profit		Profit	
Event (Copies)	Probability of Event	Condi-tional	Expec-ted	Condi-tional	Expec-ted	Condi-tional	Expec-ted	Condi-tional	Expec-ted
21	.10	$21	$ 2.10	$19	$ 1.90	$17	$ 1.70	$15	$ 1.50
22	.30	21	6.30	22	6.60	20	6.00	18	5.40
23	.40	21	8.40	22	8.80	23	9.20	21	8.40
24	.20	21	4.20	22	4.40	23	4.60	24	4.80
Total	1.00		$21.00		$21.70		$21.50		$20.10

Highest
(Optimum Act)

* For example, the expected profit of the act of stocking 22 copies is computed as follows:

$$\begin{aligned}
\$19 \times .10 &= \$\ 1.90 \\
22 \times .30 &= \ 6.60 \\
22 \times .40 &= \ 8.80 \\
22 \times .20 &= \ \underline{4.40} \\
&\quad\ \ \$21.70
\end{aligned}$$

13.3 EXPECTED VALUE OF PERFECT INFORMATION

Perfect information, also called *perfect predictor*, can be used by a business manager to change occurrences of events from uncertainty to certainty. The manager is usually interested in the cost of obtaining the perfect information. If the cost is higher than the additional profit derived from the information, he should not seek the information. The decision of whether or not to seek more information is commonly based on the expected value of perfect information.

The expected profits in Table 13-3 were computed under the conditions of uncertainty; the exact occurrence of each event is unknown to the manager of the drug store. The expected profit of the optimum act (stocking 22 copies) under uncertainty is $21.70. If the perfect information is known in advance, the manager can predict with certainty the demand of copies in each week. The best act for each event is to stock the number of copies in each week exactly equal to the demand. The profit of each act thus is maximized. For example, if the manager knew that the demand was 22 copies, he would stock 22 copies for the highest possible profit of $22. If he took other actions, such as only stocking 21 or 23 copies, he would receive a smaller profit. The highest conditional profit of each act can be obtained from Tables 13-3 and 13-4.

When each conditional profit in Table 13-4 is multiplied by the probability of a corresponding event and the products are added, the *expected profit under certainty* is obtained. This is the highest average profit that the manager can receive from the varying demand (21, 22, 23, and 24 copies) if he can get the perfect information a week in advance. The difference between the expected profit under certainty and the expected profit of the optimum act under uncertainty sets the limit for the manager to pay for the perfect information. The limit, or the difference, is called *expected value of perfect information.*

EXAMPLE 3   Refer to Example 2. Compute: (a) the expected profit under certainty, and (b) the expected value of perfect information.

(a) The expected profit under certainty is $22.70. The computation is shown in the last column of Table 13-4.

(b) Expected profit under certainty $22.70
Less: Expected profit of the optimum act
(stocking 22 copies) under uncertainty...................... 21.70
Expected value of perfect information......................... $ 1.00
(See Tables 13-3 and 13-4.)

Table 13-4   **Conditional and Expected Profit Under Certainty (Example 3)**

(1) Event (Copies Demand)	(2) Probability of Event	Act (Number of Copies Stocked)				(4) Expected Profit Under Certainty (2) × (3)
		21	22	23	24	
		(3) Conditional Profit Under Certainty (See Table 13-3)				
21	.10	$21				$ 2.10
22	.30		$22			$ 6.60
23	.40			$23		$ 9.20
24	.20				$24	$ 4.80
Total	1.00					$22.70

13.4 FINDING CONDITIONAL AND EXPECTED LOSSES

The same optimum act and the same expected value of perfect information may be found by using the conditional losses. In finding the optimum act by using conditional profits, we maximize the expected profit. Conversely, by using *conditional losses*, we must minimize the expected loss. The optimum act is the act with the highest expected profit or with the smallest expected loss. The function of perfect information is to reduce the loss for any act under uncertainty to zero. Thus, the expected loss of the optimum act equals the expected value of perfect information.

The conditional losses are incurred under the conditions of uncertainty of event occurrences when a given action is taken. The losses can be classified into two groups: (See Table 13–5.)

1. Cost losses — caused by stocking more units than demanded. A cost loss = unit cost × number of units overstocked.
2. Opportunity losses — caused by stocking fewer units than demanded. An opportunity loss = unit profit × number of units understocked.

The conditional loss is zero if the action is the best for that event, or number of units stocked = number of units demanded.

EXAMPLE 4   Use the information given in Examples 1 and 2. (a) Construct a conditional loss table, (b) compute the expected loss for each act, and (c) find the optimum act based on the lowest expected loss.

(a) The conditional loss table is shown in Table 13–5. A simple way to construct the loss table is first to place zeros in the diagonal row. Each zero indicates the best action for a given event, such as stocking 21 copies if the demand is 21 and stocking 22 copies if the demand is 22. (The loss in either case is zero.)

The losses above the diagonal row of zeros are cost losses. Each cost loss = $2 (unit cost) × units overstocked. For example, if we stocked 23 copies and the demand for the week is 21, the cost loss = $2 × (23 − 21) = $2 × 2 = $4.

The losses below the diagonal row of zeros are opportunity losses. Each opportunity loss = $1 (unit profit: selling price, $3, − cost, $2) × units understocked. For example, if we stocked 21 copies and the demand for the week is 24, the opportunity loss = $1 × (24 − 21) = $1 × 3 = $3. We have lost the opportunity to sell 3 copies, or $3 profit, since we have understocked 3 copies more than the demand.

Table 13-5   **The Conditional Loss Table (Example 4a)**

Event (Copies Demand)	Act (Number of Copies Stocked)			
	21	22	23	24
	Conditional Loss			
21	0	2	4	6
22	1	0	2	4
23	2	1	0	2
24	3	2	1	0

Cost Losses: $2 (Unit Cost) × Units Overstocked

Opportunity Losses: $1 (Unit Profit: Selling Price, $3 − Cost, $2) × Units Understocked.

(b) The expected loss for each act is computed in Table 13–6. The probability of each event is again obtained from Table 13–2 and the conditional losses by each act are obtained from Table 13–5. The conditional loss by a given act for each event is multiplied by the probability of the event for obtaining the expected loss.

(c) The expected loss of the act of stocking 22 copies is the lowest value, $1.00. Thus, the optimum act is to stock 22 copies of the news magazine each week in the drug store. The answer is the same as that of Example 2. Also, notice that the expected loss of the optimum act is equal to the expected value of perfect information obtained in Example 3(b), $1.

Table 13–6 **Computing Expected Losses (Example 4b)**

Event (Copies Demand)	Probability of Event	Act (Number of Copies Stocked)							
		21		22		23		24	
		Calculations of Expected Loss (Conditional Loss × Probability of Event)							
		Loss		Loss		Loss		Loss	
		Condi-tional	Ex-pected	Condi-tional	Ex-pected	Condi-tional	Ex-pected	Condi-tional	Ex-pected
21	.10	$0	$0.00	$2	$0.20	$4	$0.40	$6	$0.60
22	.30	1	0.30	0	0.00	2	0.60	4	1.20
23	.40	2	0.80	1	0.40	0	0.00	2	0.80
24	.20	3	0.60	2	0.40	1	0.20	0	0.00
Total	1.00		$1.70		$1.00		$1.20		$2.60

Lowest
(Optimum Act)

Special notes to Tables 13–1 through 13–6:

1. For a given event, the sum of each conditional profit in Table 13–3 and a corresponding conditional loss in Table 13–5 under uncertainty is equal to the conditional profit of the best act under certainty in Table 13–4. For example, for the event of selling 21 copies a week, the following values may be checked:

	Act (Number of Copies Stocked)			
	21	22	23	24
Conditional profit under uncertainty (Table 13–3)	$21	$19	$17	$15
Add: Conditional loss under uncertainty (Table 13–5)	0	2	4	6
Conditional profit under certainty (Table 13–4)	$21	$21	$21	$21

2. The sum of expected profit and expected loss of any act under uncertainty is equal to the expected profit under certainty:

	Act (Number of Copies Stocked)			
	21	22	23	24
Expected profit under uncertainty (Table 13–3)	$21.00	$21.70	$21.50	$20.10
Add: Expected loss under uncertainty (Table 13–6)	1.70	1.00	1.20	2.60
Expected profit under certainty (Table 13–4)	$22.70	$22.70	$22.70	$22.70

3. In the previous illustrations, it was assumed that the overstocked copies at the end of each week were completely worthless. However, if the overstocked copies have some salvage value, such as $0.50 for each copy, the value must be included in computing the conditional profits and losses for each action. For example, when the demand event is 21 copies and the act is stocking 23 copies, the conditional profit is:

    21 (copies sold) × $3 (price per copy)   = $63
Less: 23 (copies stocked) × $2 (cost per copy)   = $\underline{46}$
    Conditional profit without salvage value   17 (Table 13–1)
Add: 2 (copies overstocked) × $0.50 (salvage
    value per copy) $\underline{1}$ (+)
    Conditional profit with salvage value $18

An alternative way to compute the conditional profit is:

    Profit per copy = $3 − $2 = $1
Net loss per copy overstocked = $2 − $0.50 = $1.50

    21 (copies sold) × $1   = $21
Less: 2 (copies overstocked) × $1.50 = $\underline{3}$ (−)
    Conditional profit $18

The conditional loss is:

  $1.50 (net loss per copy) × 2 (overstocked) = $3

	Event 21, Act 23
Check:	
Conditional profit under uncertainty	$18
Add: Conditional loss under uncertainty	3
Conditional profit under certainty (of the best act: stocking 21 since the demand event is 21 — Table 13–4)	$21

The procedures of computing various expected values involving salvage values are the same as those illustrated above involving no salvage values.

EXERCISE 13–1   Reference: Sections 13.1–13.4

1. AA Candy Store purchases doughnuts for resale. The following information is obtained from the record of recent daily sales in the store:

Daily Sales (Number of Dozens of Doughnuts)	Number of Days
10	20
11	30
12	50
Total	100

The cost of a dozen doughnuts is $1.00 and the selling price, $1.60. If the doughnuts are not sold during the day, they have no value.

a. Construct a payoff table to show all possible profits.
b. Find the probability of each event of daily sales.
c. Compute the expected sales (in the number of dozens) for a day.
d. Compute the expected profit for each possible act and indicate the optimum act.
e. Compute the expected profit under certainty.
f. Compute the expected value of perfect information.

2. Use the information in problem 1.
a. Prepare a conditional loss table.
b. Compute the expected loss for each possible act.
c. Find the optimum act and the expected value of perfect information based on the expected losses.

3. Refer to problem 1. Assume that the leftover doughnuts have a value of $0.80 per dozen. What are the answers to the questions listed in this problem?

4. Use the assumptions given in problem 3. Find the answers to the questions listed in problem 2.

5. Use the information in problem 1, except that the record of recent daily sales in the AA Candy Store shows the following:

Daily Sales (Number of Dozens of Doughnuts)	Number of Days
10	60
20	50
30	40
40	30
50	20
Total	200

Find the answers to the questions listed in problem 1.

6. Use the information in problem 5 to find the answers to the questions in problem 2.

7. Refer to problem 5. Assume that the leftover doughnuts have a value of $0.80 per dozen. What are the answers to the questions listed in problem 1?

8. Use the assumption in problem 7. Find the answers to the questions listed in problem 2.

⋆13.5 DECISIONS BASED ON UTILITY VALUES

The decisions made in the previous discussions were based on monetary values. The rule for finding the optimum act from all possible acts is to choose the act that has either the highest expected monetary profit or the lowest expected monetary loss. An expected value is the arithmetic mean of a set of values. The decision maker thus considers each dollar being equally important under this rule. However, there are cases in which dollars are not treated with equal importance. Under those cases the expected monetary value is not an appropriate guide for making decisions. Instead, an expected index number, called *utility value,* can be used as a basis for making decisions.

Utility is a measurement of money based on the decision maker's opinion of monetary value under conditions of uncertainty. In general a higher utility value represents a higher monetary value. However, utility values may or may not be proportional to monetary values represented. When a set of utility values is proportional to a set of monetary values, the relationship is said to be *linear.* A linear relationship suggests that every dollar is equally important, such as the representation listed below and shown in Figure 13–1.

Monetary Value (Dollar Unit)	Utility Value (Utility Unit)
$ 0	0
100	10
200	20
300	30
400	40
500	50

Figure 13-1 **Linear Relationship Between Monetary Values and Their Utility Values**

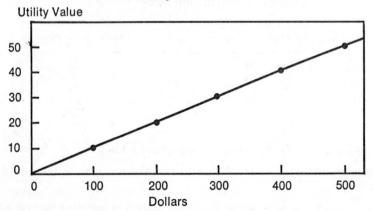

When all dollars are not equally important, monetary values and their corresponding utility values will not be related linearly. The procedure of constructing a set of utility values from a set of monetary values for nonlinear relationship is illustrated as follows:

(1) Select two monetary values and assign their utility values correspondingly. The two monetary values should have a fairly wide range, such as $0 and $6,000. Their utility values are assigned arbitrarily, such as letting zero be the utility for $0 and 100 the utility for $6,000. For convenience, let the symbol:

U(a monetary value) = the utility value of the given monetary value

Thus, the two selected monetary values and their assigned utility values may be written symbolically as follows:

$$U(\$0) = 0$$
$$U(\$6{,}000) = 100$$

(2) Compute the third utility for the decision maker. The utility must reflect the maker's opinion of indifference between an amount of money with certainty and a lottery which will offer him the two selected monetary values in (1) with uncertainty but known probability distribution. The amount with certainty should be larger than $0 and smaller than $6,000, preferably close to the lower value $0.

Assume that we are constructing a set of utility values for the decision maker, Mr. Robertson. He is indifferent between (A) $1,000 for certain and (B) a lottery which will give $0 with .5 probability and $6,000 with .5 probability. The sum of the two probabilities must be equal to 1, or .5 + .5 = 1. Since the amount in alternative (A) is indifferent from the amounts in alternative (B), their corresponding utilities must also be equal or:

$U(\$1{,}000) = U(\$0$ with .5 probability and $\$6{,}000$ with .5 probability)

The utility of a lottery is the expected utility of its component amounts. Thus:

$$U(\$1{,}000) = .5 \times U(\$0) + .5 \times U(\$6{,}000) \qquad \text{(See (1))}$$
$$= (.5 \times 0) + (.5 \times 100) = 0 + 50 = 50.$$

(3) Compute the fourth utility for the decision maker. The basic steps of computing the fourth utility are similar to those illustrated in (2). However, in designing a new lottery, we must follow the two guidelines:

(a) The amounts included in the lottery must be selected from the amounts whose utilities are already available.

(b) The utility of a higher monetary value, such as $U(\$2{,}000)$, should be numerically larger than the utility of a lower monetary value, such as $U(\$1{,}000)$.

Assume that Mr. Robertson is indifferent between (A) $2,000 for certain and (B) a lottery which will give $1,000 with .5 probability and $6,000 with .5 probability. What is the utility value of $2,000?

$$U(\$2{,}000) = .5 \times U(\$1{,}000) + .5 \times U(\$6{,}000)$$
$$= (.5 \times 50) + (.5 \times 100)$$
$$= 25 + 50 = 75$$

Observe that $U(\$1,000) = 50$ is obtained from (2) and $U(\$6,000) = 100$ is obtained from (1). Also, the new lottery in (3) is based on the old lottery in (2) with the same probability distribution (.5 and .5). Notice that the amount $0 in the old lottery is increased to $1,000 in the new lottery.

(4) Compute the fifth utility for the decision maker.

Assume that Mr. Robertson is indifferent between (A) $4,000 for certain and (B) a lottery having $1,000 with .2 probability and $6,000 with .8 probability. What is the utility of $4,000?

$$U(\$4,000) = .2 \times U(\$1,000) + .8 \times U(\$6,000)$$
$$= (.2 \times 50) + (.8 \times 100) = 10 + 80 = 90$$

The lottery in (4) is based on the lottery in (3) with the same amounts, $1,000 and $6,000. Notice that the probability of the larger amount, $6,000, has been increased from .5 to .8. The increase is consistent with the expectation that the utility of $4,000 is larger than the utility of $2,000. When the probability of the larger amount is increased, the utility is also increased, from 75 in (3) to 90 in (4).

(5) Compute the sixth utility.

The utilities from (2) to (4) were computed for the amounts within the range of the two selected monetary values, $0 and $6,000. However, it is also possible to compute the utilities of the amounts either below $0 or above $6,000. The utility of an amount above $6,000 is computed in this step.

Assume that Mr. Robertson is indifferent between (A) $6,000 for certain and (B) a lottery having $1,000 with .2 probability and $9,000 with .8 probability. What is the utility of $9,000?

$$U(\$6,000) = .2 \times U(\$1,000) + .8 \times U(\$9,000)$$
$$100 = (.2 \times 50) + (.8 \times U(\$9,000))$$
$$100 = 10 + (.8 \times U(\$9,000))$$
$$U(\$9,000) = (100 - 10)/.8 = 112.5$$

Note: The lottery in (5) is based on the lottery in (4) with the same probability distribution (.2 and .8). Notice that the amount, $6,000, in the old lottery is increased to $9,000 in the new lottery. The increase is consistent with the fact that the utility of $6,000 is larger than the utility of $4,000 ($U(\$6,000) = 100$ and $U(\$4,000) = 90$). When the utility is increased, the amount (or amounts) in the new lottery should also be increased, to a reasonable extent. We have assumed that Mr. Robertson is a reasonable man.

(6) Compute the seventh utility.

The utility of an amount below $0 (a negative amount representing a loss) is computed in a similar manner.

Assume that Mr. Robertson is indifferent between (A) $0 for certain and (B) a lottery which will give a loss of $1,500 (or −$1,500) with a .6 probability and a gain of $4,000 with .4 probability. Find the utility of $1,500 loss.

$$U(\$0) = (.6 \times U(-\$1,500)) + (.4 \times U(\$4,000))$$
$$0 = (.6 \times U(-\$1,500)) + (.4 \times 90)$$
$$U(-\$1,500) = (0 - 36)/.6 = -60$$

Note: The lottery in (6) is different from any lottery illustrated in the previous steps (2)–(5). It has a different probability distribution (.6 and .4) and different amounts (−$1,500 and $4,000). However, the answer must be consistent with the basic concept that a higher (or lower) utility value represents a higher (or lower) monetary value. The example in (6) is used to illustrate another way to design a lottery for the decision maker's approval.

(7) Construct a utility curve (Figure 13–2).

The monetary values and their corresponding utility values, either assigned or computed in (1) through (6), are listed below and plotted in Figure 13–2. The points plotted in the figure give us enough indication in drawing a smoothed curve to represent the utility function of the decision maker, Mr. Robertson. From this utility curve, we may (a) read off the utility representing a desired monetary value (See Example 5.), and (b) find the probability distribution of two monetary values in a lottery, which is equivalent to an amount for certain in the decision maker's opinion. (See Example 6.)

Monetary Value (Dollar Unit)	Utility Value (Utility Unit)
−$1,500	−60
0	0
1,000	50
2,000	75
4,000	90
6,000	100
9,000	112.5

Figure 13-2 **Mr. Robertson's Utility Curve**

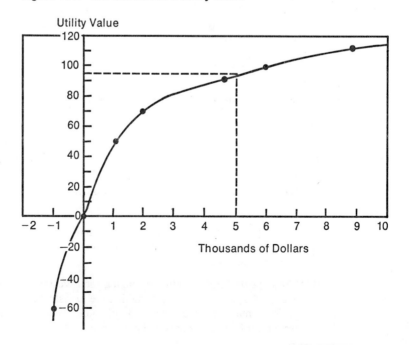

EXAMPLE 5   Find the utility of $5,000 from Figure 13–2.

The utility of $5,000 is about 96 as shown by the broken line intersecting the curve in the figure.

EXAMPLE 6   A lottery, which is equivalent to $5,000 for certain, has two amounts: $1,000 (which is smaller than $5,000) and $6,000 (which is larger than $5,000). Find the probability distribution of the two amounts.

Let P be the probability of $1,000. Then, the probability of $6,000 must be $(1 - P)$.

$$U(\$5,000) = (P \times U(\$1,000)) + ((1 - P) \times U(\$6,000)),$$
$$96 = (P \times 50) + ((1 - P) \times 100),$$
$$96 = 50P + 100 - 100P$$
$$50P = 4, \qquad \text{(See Example 5 for}$$
$$P = \tfrac{4}{50} = .08 \text{ or } 8\% \qquad U(\$5,000) = 96)$$
$$1 - P = 1 - 8\% = 92\%$$

Thus, Mr. Robertson is indifferent between (A) $5,000 for certain and (B) a lottery giving $1,000 with 8% probability and $6,000 with 92% probability.

As stated in the beginning of this section, there are cases in which the expected monetary value is not an appropriate guide for making decisions. Examples 7 and 8 are used to illustrate the use of expected utility values in those cases.

EXAMPLE 7   Suppose that Mr. Robertson received $1,500 from his mother in April of this year. He may keep this money until September for the payment of wedding expenses; or, he may invest his money in a restaurant at a beach during the summer. If the weather is good, he will be able to make a net profit of $6,000. If the weather is bad, he will lose $1,500. The chance of having good weather during the summer is 30% and that of having bad weather is 70%. What would be his decision if he uses (a) expected monetary value, and (b) expected utility values as a basis for decision making?

SOLUTION   (a) The expected monetary value of this investment is (also, see Table 13–7):

$$(\$6,000 \times .30) + (-1,500 \times .70) = \$1,800 - \$1,050$$
$$= \$750 \text{ (higher profit)}$$

The expected monetary value of not making the investment is $0 (lower profit). Thus, Mr. Robertson should invest his money in the restaurant business.

(b) The expected utility value of making the investment is −12 and the expected utility value of not making the investment is 0. (See Table 13–8.) Thus, Mr. Robertson should not invest his money in the restaurant business. (Zero is larger than −12.)

Table 13-7 **Computing Expected Monetary Values (Example 7(a))**

Event	Probability of Event	Invest		Not Invest	
		Conditional Profit	Expected Profit	Conditional Profit	Expected Profit
Good Weather	.30	$6,000	$1,800	$0	$0
Bad Weather	.70	−$1,500	−$1,050	$0	$0
Total	1.00		$ 750		$0

↑
Higher

Table 13-8 **Computing Expected Utility Values (Example 7(b))**

Event	Probability of Event	Invest			Not Invest		
		Conditional Profit		Expected Utility	Conditional Profit		Expected Utility
		Dollars	Utility Value		Dollars	Utility Value	
Good Weather	.30	$6,000	100	30	$0	0	0
Bad Weather	.70	−$1,500	−60	−42	$0	0	0
Total	1.00			−12			0

↑
Higher

The decision obtained in (a) conflicts with that in (b). However, the expected utility value is more appropriate for decision making in our case. Mr. Robertson is most likely not to make the investment since he does not wish to risk his marriage for the uncertain profit.

EXAMPLE 8   Mr. Patterson owns a house valued at $50,000. Statistical records showed that the probability of having a fire in his type of house is 1 out of 1,000 (or $\frac{1}{1,000} = .001$). However, the insurance company said that the fire insurance premium on his house should be $100 per year. Assume that the home owner's utility values assigned to represent monetary values are as follows:

$$U(-\$50,000) = -5,000$$
$$U(-\$100) = -1$$
$$U(\$0) = 0$$

Should Mr. Patterson buy the fire insurance policy? State the home owner's decision based on (a) expected monetary value and (b) expected utility value.

SOLUTION   (a) The expected monetary value of buying the insurance policy is −$100
(or $100 loss) and that of not buying is −$50 (or $50 loss). (See Table
13–9.)

Table 13–9   **Computing Expected Monetary Values (Example 8(a))**

| | | Act | | | |
| | | Buy Insurance | | Not Buy Insurance | |
Event	Probability of Event	Conditional Value	Expected Value	Conditional Value	Expected Value
No Fire	.999	−$100	−$99.90	$0	$ 0
Fire	.001	−$100	−$ .10	−$50,000	−$50
Total	1.000		−$100.00		−$50

↑
Higher Value
(Smaller Loss)

Thus, Mr. Patterson should not buy the insurance.

(b) The expected utility value of buying the fire insurance is −1 and that
of not buying the insurance is −5. Since −1 is larger than −5, the
home owner should buy the fire insurance. See the computation in
Table 13–10.

Table 13–10   **Computing Expected Utility Values (Example 8(b))**

		Act					
		Buy Insurance			Not Buy Insurance		
		Conditional Value			Conditional Value		
Event	Probability of Event	Dollars	Utility Value	Expected Utility	Dollars	Utility Value	Expected Utility
No Fire	.999	−$100	−1	−.999	$0	0	0
Fire	.001	−$100	−1	−.001	−$50,000	−5,000	−5
Total	1.000			−1.000			−5

↑
Higher

Again, the decision obtained in (a) conflicts with that in (b). However,
the expected utility value is also more appropriate for decision making in
this case. In practice, Mr. Patterson will buy the fire insurance since he
cannot afford to risk his house being destroyed by fire and recovering nothing.

The results obtained in Examples 7(b) and 8(b) are consistent with the
practical cases where the expected monetary values were not appropriate for

decision making. The rule of finding the optimum act from all possible acts thus is to choose the act that has the highest *expected utility* value. However, care must be exercised in using utility as a basis for decision making. Utilities are derived from personal opinions on monetary values under conditions of uncertainty. A person's opinion may change from time to time and from one situation to another. Thus, the utility constructed for a person at one time may not be suitable to the same type of problem at another time. Different types of problems usually require the construction of different sets of utility values for the same person in making decisions. For example, a person's opinion on money used for college education, although the outcome is uncertain – graduate or not graduate, is usually different from that used for gambling purposes. Furthermore, the scale of utility values of a person is arbitrarily selected. To compare the utility values of one person with those of another thus is meaningless.

★13.6 DECISION RULES

Various rules have been developed for making decisions. The decision rules commonly employed by decision makers are:

(1) Choose the act with the *highest expected value*. This rule is also called the *Bayesian decision rule*. It applies to subjective probability distributions as illustrated in the previous sections. The expected values can be expressed either in terms of dollars or utilities. However, utilities are generally preferred. (See Examples 7 and 8.) The Bayesian decision rule is generally regarded as the superior one among available decision rules.

(2) Choose the act with the *maximum possible profit*. Here the decision maker ignores both the probability distribution and the fact of possible losses. Refer to the information given in Table 13–7. The decision maker will choose the act of investing since the act will yield the highest possible profit of $6,000. The fact of possible loss of $1,500 is not taken into consideration.

(3) Choose the act with the *minimum possible loss*. Here the decision maker again ignores the probability distribution of event occurrences. This rule tends to lead a decision maker to do nothing since the minimum possible loss in a given problem is usually zero dollar. The fact of making possible profit, either very high or small, is also ignored. By using the information given in Table 13–7, the decision is to choose the act of not investing since the act has the minimum possible loss of $0.

(4) Choose the act with the *highest average profit*. Under this rule, the decision maker considers that the occurrences of events are equally likely. Thus, if there are 2 possible events, the probability of the occurrence of each event is $\frac{1}{2}$; for 3 possible events, the probability is $\frac{1}{3}$; and so on. The method of computing the average profit for each event is the same as that used for computing the expected values. The computed average profit may be positive or negative. If the highest average profit is positive, the rule maximizes the average profit; if it is negative, the rule minimizes the average loss.

Refer to Table 13–7. The average profit of the act of investing is:

$$\frac{1}{2}(\$6,000) + \frac{1}{2}(-\$1,500) = 3,000 - 750 = \$2,250$$

The average profit of the act of not investing is:

$$\frac{1}{2} (\$0) = \frac{1}{2} (\$0) = \$0.$$

The highest average profit among the two possible acts is $2,250. Thus, the act of investing should be chosen under this rule.

(5) Choose the act with the *highest profit of the event occurring most likely.* Under this rule, the consequences of the events occurring at lesser degrees are ignored. The profit may also be positive (gain) or negative (loss). Refer to Table 13–7. The event which is most likely to occur is "bad weather", with .70 probability. The profit of not investing is $0 which is higher than the profit of investing −$1,500 or losing $1,500. Thus, the act of not investing is chosen under this decision rule.

EXERCISE 13–2   Reference: Sections 13.5 and 13.6

1. Mr. Anderson has a utility value of 30 for a gain of $100 and 80 for a gain of $4,000. He is indifferent between (A) having $700 for certain and (B) a lottery which will gain $100 with a .5 chance and $4,000 with a .5 chance. Find his utility value for $700.

2. Miss Virginia has a utility value of 10 for a loss of $300 and 160 for a gain of $5,000. She is indifferent between (A) receiving $2,000 for certain and (B) a loss of $300 at a .2 chance and a gain of $5,000 at a .8 chance. Find her utility value for $2,000.

3. Refer to problem 1. Assume that Mr. Anderson is indifferent between (A) having $1,000 for certain and (B) a lottery which will gain $100 with a .3 chance and $4,000 with a .7 chance. Find his utility value for $1,000.

4. Refer to problem 2. Assume that Miss Virginia is indifferent between (A) receiving $3,000 for certain and (B) a loss of $300 at a .1 chance and a gain of $5,000 at a .9 chance. What is her utility value for $3,000?

5. Refer to problem 1. If Mr. Anderson is indifferent between (A) having $4,000 for certain and (B) gaining $100 at a .3 chance and $9,000 at a .7 chance, what is his utility value for $9,000?

6. Refer to problem 2. If Miss Virginia is indifferent between (A) receiving nothing for sure and (B) a loss of $300 at a .9 chance and a gain of $2,000 at a .1 chance, what is her utility value for $0?

7. Refer to problem 1. If Mr. Anderson is indifferent between (A) having $100 for certain and (B) a loss of $2,500 at a .4 chance and a gain of $4,000 at a .6 chance, what is his utility value for a loss of $2,500?

8. Refer to problem 2. If Miss Virginia is indifferent between (A) receiving a loss of $300 for certain and (B) a loss of $3,500 at a .3 chance and a gain of $5,000 at a .7 chance, what is her utility value for a loss of $3,500?

9. Refer to problem 1. Assume that Mr. Anderson is indifferent between (A) having $1,200 for certain and (B) a lottery which will gain $100 with a .7 chance and $4,000 with a .3 chance. Find his utility value for $1,200. Would you consider his opinion consistent? Why?

10. Refer to problem 2. Assume that Miss Virginia is indifferent between (A) receiving $4,000 for certain and (B) a loss of $300 at a .4 chance and a gain of $5,000 at a .6 chance. What is her utility value for $4,000? Based on your finding, would you consider Miss Virginia's opinion consistent? If not, why?

11. A lottery, which is equivalent to $4,000 for certain, will give either $1,000 or $6,000 by chance. Based on the information given in Figure 13-2, find the probability distribution of the chance.

12. A lottery, which is equivalent to $2,000 for certain, will give either nothing or $9,000 by chance. Based on the information given in Figure 13-2, find the probability distribution of the chance.

13. Using the dollar values given in Table 13-10, state the decisions concerning whether or not to buy the fire insurance if Mr. Patterson chooses:
    a. the act with the maximum possible profit
    b. the act with the minimum possible loss
    c. the act with the highest average profit
    d. the act with the highest profit of the event occurring most likely

14. Use the utility values given in Table 13-10. State your answers to the questions listed in problem 13.

Chapter
14 Mathematics of Finance

Mathematics of finance affects the life of practically every family in the United States. This subject includes the financial problems ranging from making a simple personal loan to financing the huge social security program. The basic mathematical devices for solving those problems are the methods of computing simple interest, compound interest, and annuities. Simple interest is usually charged for short-term borrowing, whereas compound interest is commonly employed in long-term obligations. Examples of annuity applications are rental payments, purchases of houses or home appliances on installment payment plans, life insurance premiums, interest payments on bonds, and social security programs.

This chapter will introduce the methods of computing simple interest, compound interest, annuities, and some selected applications of these methods. The computation of different types of annuities is based on the method of computing compound interest, which in turn is based on the method of computing simple interest. We thus begin with the computation of simple interest.

14.1 SIMPLE INTEREST

Interest is a fee for the use of the money a person borrowed. The value of interest (I) is determined by three factors: The *principal* borrowed (P), the *rate of interest* charged (i), and the number of interest periods (n) during the borrowing time. The basic formula for computing simple interest is:

$$\text{Interest} = \text{Principal} \times \left(\begin{array}{c}\text{Interest Rate} \\ \text{Per Period}\end{array}\right) \times \left(\begin{array}{c}\text{Number of Interest} \\ \text{Periods (or Time)}\end{array}\right)$$

Or, simply written:

$$I = Pin \qquad\qquad (14\text{--}1)$$

The period in computing simple interest problems is usually expressed in the unit of a year. Thus, in formula (14–1):

$i =$ interest rate per year, and
$n =$ number of years or a fraction of a year

The sum of the principal and the interest is called the *amount*. Let $S =$ the amount. Then:

$$S = P + I, \text{ or:}$$
$$S = P + Pin$$
$$S = P(1 + in) \qquad\qquad (14\text{-}2)$$

EXAMPLE 1 What are the amount and the simple interest on $1,000 at 6% for (a) two years, and (b) three months?

(a) $P = \$1,000$, $i = 6\%$ (per year)
$n = 2$ (years)

Substitute the values in formula (14-1):

$I = Pin = 1,000 \times 6\% \times 2 = \$120$ (Interest)

Substitute the above values in formula (14-2):

$S = P + I = 1,000 + 120 = \$1,120$ (Amount)

Or:

$S = P(1 + in) = 1,000(1 + 6\%(2))$
$\quad = 1,000(1.12) = \$1,120$ (Amount)

(b) $n = \dfrac{3}{12}$, since three months $= \dfrac{3}{12}$ years

$I = Pin = 1,000 \times 6\% \times \dfrac{3}{12} = \$15$ (Interest)

$S = P + I = 1,000 + 15 = \$1,015$ (Amount)

There are two methods of determining the number of days in a year—the exact method and the approximate method. Under the *exact method,* each year has 365 days except leap years, which have 366 days. When the *approximate method* is used, it is assumed that each of the 12 months in a year has 30 days, and thus there are 360 days in a year. The value of interest computed by using 360 as the divisor in the time factor:

$$\frac{\text{Exact Number of Days}}{360}$$

is called *ordinary interest.* Ordinary interest is commonly used in commercial practice and the method of computing ordinary interest is known as the *Banker's Rule.* When 365 is used as the divisor in the time factor:

$$\frac{\text{Exact Number of Days}}{365}$$

the result is called *exact interest.* The exact interest method is usually used in calculating interest payments on government obligations, in foreign trade, and in rediscounting notes for member banks by the Federal Reserve Banks. The Banker's Rule will be used in this chapter in computing simple interest.

EXAMPLE 2 What is the simple interest on $500 at 8% for 90 days? What is the amount?

$$P = \$500, \ i = 8\%, \ n = \frac{90}{360} \text{ years}$$

Thus:

$$I = Pin = 500 \times 8\% \times \frac{90}{360} = \$10 \text{ (Interest)}$$

$$S = P + I = 500 + 10 = \$510 \text{ (Amount)}$$

Or:

$$S = P(1 + in) = 500\left(1 + 8\%\left(\frac{90}{360}\right)\right)$$
$$= 500(1.02)$$
$$= \$510.00 \text{ (Amount)}$$

14.2 COMPOUND INTEREST

The compound interest method is generally used in long-term borrowing. There is usually more than one period for computing interest during the borrowing time. The interest for each period is added (compounded or converted) to the principal before the interest for the next period is computed. The final sum at the end of the period of borrowing is called the *compound amount*.

A. Computing Compound Amount and Compound Interest

Compound interest is the difference between the original principal and the compound amount. The period for computing interest, usually at regularly stated intervals such as annually, semiannually, quarterly, or monthly, is called the *conversion period*, or the *interest period*. The interest rate per conversion period is equal to the stated annual interest rate divided by the number of conversion periods in one year. The stated annual interest rate is called the *nominal annual rate*, or simply, the *nominal rate*. Thus, if the nominal rate is 8%, the interest rate for the annual conversion period is also 8%; if the conversion period is semiannual, the interest rate for a period of six months is $\frac{8\%}{2}$, or 4%.

The basic method for computing compound interest for each conversion period is the same as the method for computing simple interest. Thus, if there is only one conversion period, compound interest is the same as simple interest. Example 3 illustrates the basic method for computing compound interest for three conversion periods.

EXAMPLE 3 What are the compound amount and the compound interest at the end of three months if $10,000 is borrowed at 12% compounded monthly?

P = original principal = $10,000
i = interest rate per conversion period

$$= \frac{\text{stated annual interest rate (nominal rate)}}{\text{number of conversion periods in one year}}$$

$$= \frac{12\%}{12} = 1\% \text{ per month}$$

$$n = \text{number of conversion periods} = \frac{3 \text{ months}}{1 \text{ month}} = 3$$

S = compound amount (or the sum), or the principal at the end of the nth period

The computation by the basic method is arranged in actual values and symbols as follows:

Original principal .... $10,000.00   = P
1st month interest .... (+) 100.00 = $10,000(1%) = Pi
Principal, end of 1st month ..... $10,100.00   = P(1 + i)
2d month interest .... (+) 101.00 = $10,100(1%) = P(1 + i)i
Principal, end of 2d month $10,201.00   = P(1 + i)²
3d month interest .... (+) 102.01 = $10,201(1%) = P(1 + i)²i
Principal, end of 3d month,
   or the compound amount ..... $10,303.01 =   S = P(1 + i)³

Compound interest = Compound amount − Original principal
                  = 10,303.01 − 10,000.00
                  = $303.01

Note that the simple interest at 12% on $10,000 for three months is:

$$10,000 \times 12\% \times \frac{3}{12} = \$300$$

The compound interest is greater than the simple interest for the problem in Example 3 by:

$$303.01 - 300.00 = \$3.01$$

Observe the symbolical expression in Example 3. The result shows that the compound amount,

$S = P(1 + i)^3$ when the number of conversion periods is 3.

When the computation is extended, the compound amount at the end of the nth period can be written:

$$S = P(1 + i)^n \tag{14-3}$$

Example 3 may be computed in the following manner when formula (14-3) is used:

$$\begin{aligned} S = P(1 + i)^n &= 10,000(1 + 1\%)^3 \\ &= 10,000(1.030301) \\ &= \$10,303.01 \end{aligned}$$

For convenience, selected values of $(1 + i)^n$, which is frequently referred to as the *accumulation factor*, are tabulated in Table 14-1 for computing the problems in this chapter.

Table 14–1 **Compound Amount When Principal Is 1**
$$S = (1 + i)^n$$

n	$\frac{1}{2}\%$	1%	$1\frac{1}{2}\%$	2%	3%	n
1	1.0050 00	1.0100 00	1.0150 00	1.0200 00	1.0300 00	1
2	1.0100 25	1.0201 00	1.0302 25	1.0404 00	1.0609 00	2
3	1.0150 75	1.0303 01	1.0456 78	1.0612 08	1.0927 27	3
4	1.0201 51	1.0406 04	1.0613 64	1.0824 32	1.1255 09	4
5	1.0252 51	1.0510 10	1.0772 84	1.1040 81	1.1592 74	5
6	1.0303 78	1.0615 20	1.0934 43	1.1261 62	1.1940 52	6
7	1.0355 29	1.0721 35	1.1098 45	1.1486 86	1.2298 74	7
8	1.0407 07	1.0828 57	1.1264 93	1.1716 59	1.2667 70	8
9	1.0459 11	1.0936 85	1.1433 90	1.1950 93	1.3047 73	9
10	1.0511 40	1.1046 22	1.1605 41	1.2189 94	1.3439 16	10
11	1.0563 96	1.1156 68	1.1779 49	1.2433 74	1.3842 34	11
12	1.0616 78	1.1268 25	1.1956 18	1.2682 42	1.4257 61	12
13	1.0669 86	1.1380 93	1.2135 52	1.2936 07	1.4685 34	13
14	1.0723 21	1.1494 74	1.2317 56	1.3194 79	1.5125 90	14
15	1.0776 83	1.1609 69	1.2502 32	1.3458 68	1.5579 67	15
16	1.0830 71	1.1725 79	1.2689 86	1.3727 86	1.6047 06	16
17	1.0884 87	1.1843 04	1.2880 20	1.4002 41	1.6528 48	17
18	1.0939 29	1.1961 47	1.3073 41	1.4282 46	1.7024 33	18
19	1.0993 99	1.2081 09	1.3269 51	1.4568 11	1.7535 06	19
20	1.1048 96	1.2201 90	1.3468 55	1.4859 47	1.8061 11	20
21	1.1104 20	1.2323 92	1.3670 58	1.5156 66	1.8602 95	21
22	1.1159 72	1.2447 16	1.3875 64	1.5459 80	1.9161 03	22
23	1.1215 52	1.2571 63	1.4083 77	1.5768 99	1.9735 87	23
24	1.1271 60	1.2697 35	1.4295 03	1.6084 37	2.0327 94	24
25	1.1327 96	1.2824 32	1.4509 45	1.6406 06	2.0937 78	25
26	1.1384 60	1.2952 56	1.4727 10	1.6734 18	2.1565 91	26
27	1.1441 52	1.3082 09	1.4948 00	1.7068 86	2.2212 89	27
28	1.1498 73	1.3212 91	1.5172 22	1.7410 24	2.2879 28	28
29	1.1556 22	1.3345 04	1.5399 81	1.7758 45	2.3565 66	29
30	1.1614 00	1.3478 49	1.5630 80	1.8113 62	2.4272 62	30
31	1.1672 07	1.3613 27	1.5865 26	1.8475 89	2.5000 80	31
32	1.1730 43	1.3749 41	1.6103 24	1.8845 41	2.5750 83	32
33	1.1789 08	1.3886 90	1.6344 79	1.9222 31	2.6523 35	33
34	1.1848 03	1.4025 77	1.6589 96	1.9606 76	2.7319 05	34
35	1.1907 27	1.4166 03	1.6838 81	1.9998 90	2.8138 62	35
36	1.1966 81	1.4307 69	1.7091 40	2.0398 87	2.8982 78	36
37	1.2026 64	1.4450 76	1.7347 77	2.0806 85	2.9852 27	37
38	1.2086 77	1.4595 27	1.7607 98	2.1222 99	3.0747 83	38
39	1.2147 21	1.4741 23	1.7872 10	2.1647 45	3.1670 27	39
40	1.2207 94	1.4888 64	1.8140 18	2.2080 40	3.2620 38	40
41	1.2268 98	1.5037 52	1.8412 29	2.2522 00	3.3598 99	41
42	1.2330 33	1.5187 90	1.8688 47	2.2972 44	3.4606 96	42
43	1.2391 98	1.5339 78	1.8968 80	2.3431 89	3.5645 17	43
44	1.2453 94	1.5493 18	1.9253 33	2.3900 53	3.6714 52	44
45	1.2516 21	1.5648 11	1.9542 13	2.4378 54	3.7815 96	45
46	1.2578 79	1.5804 59	1.9835 26	2.4866 11	3.8950 44	46
47	1.2641 68	1.5962 63	2.0132 79	2.5363 44	4.0118 95	47
48	1.2704 89	1.6122 26	2.0434 78	2.5870 70	4.1322 52	48
49	1.2768 42	1.6283 48	2.0741 30	2.6388 12	4.2562 19	49
50	1.2832 26	1.6446 32	2.1052 42	2.6915 88	4.3839 06	50

Table 14-1   **Compound Amount When Principle Is 1**
$S = (1 + i)^n$ **(Continued)**

n	4%	5%	6%	7%	8%	n
1	1.0400 00	1.0500 00	1.0600 00	1.0700 00	1.0800 00	1
2	1.0816 00	1.1025 00	1.1236 00	1.1449 00	1.1664 00	2
3	1.1248 64	1.1576 25	1.1910 16	1.2250 43	1.2597 12	3
4	1.1698 59	1.2155 06	1.2624 77	1.3107 96	1.3604 89	4
5	1.2166 53	1.2762 82	1.3382 26	1.4025 52	1.4693 28	5
6	1.2653 19	1.3400 96	1.4185 19	1.5007 30	1.5868 74	6
7	1.3159 32	1.4071 00	1.5036 30	1.6057 81	1.7138 24	7
8	1.3685 69	1.4774 55	1.5938 48	1.7181 86	1.8509 30	8
9	1.4233 12	1.5513 28	1.6894 79	1.8384 59	1.9990 05	9
10	1.4802 44	1.6288 95	1.7908 48	1.9671 51	2.1589 25	10
11	1.5394 54	1.7103 39	1.8982 99	2.1048 52	2.3316 39	11
12	1.6010 32	1.7958 56	2.0121 96	2.2521 92	2.5181 70	12
13	1.6650 74	1.8856 49	2.1329 28	2.4098 45	2.7196 24	13
14	1.7316 76	1.9799 32	2.2609 04	2.5785 34	2.9371 94	14
15	1.8009 44	2.0789 28	2.3965 58	2.7590 32	3.1721 69	15
16	1.8729 81	2.1828 75	2.5403 52	2.9521 64	3.4259 43	16
17	1.9479 01	2.2920 18	2.6927 73	3.1588 15	3.7000 18	17
18	2.0258 17	2.4066 19	2.8543 39	3.3799 32	3.9960 20	18
19	2.1068 49	2.5269 50	3.0256 00	3.6165 28	4.3157 01	19
20	2.1911 23	2.6532 98	3.2071 35	3.8696 84	4.6609 57	20
21	2.2787 68	2.7859 63	3.3995 64	4.1405 62	5.0338 34	21
22	2.3699 19	2.9252 61	3.6035 37	4.4304 02	5.4365 40	22
23	2.4647 16	3.0715 24	3.8197 50	4.7405 30	5.8714 64	23
24	2.5633 04	3.2251 00	4.0489 35	5.0723 67	6.3411 81	24
25	2.6658 36	3.3863 55	4.2918 71	5.4274 33	6.8484 75	25
26	2.7724 70	3.5556 73	4.5493 83	5.8073 53	7.3963 53	26
27	2.8833 69	3.7334 56	4.8223 46	6.2138 68	7.9880 61	27
28	2.9987 03	3.9201 29	5.1116 87	6.6488 38	8.6271 06	28
29	3.1186 51	4.1161 36	5.4183 88	7.1142 57	9.3172 75	29
30	3.2433 98	4.3219 42	5.7434 91	7.6122 55	10.0626 57	30
31	3.3731 33	4.5380 39	6.0881 01	8.1451 13	10.8676 69	31
32	3.5080 59	4.7649 41	6.4533 87	8.7152 71	11.7370 83	32
33	3.6483 81	5.0031 89	6.8405 90	9.3253 40	12.6760 50	33
34	3.7943 16	5.2533 48	7.2510 25	9.9781 14	13.6901 34	34
35	3.9460 89	5.5160 15	7.6860 87	10.6765 81	14.7853 44	35
36	4.1039 33	5.7918 16	8.1472 52	11.4239 42	15.9681 72	36
37	4.2680 90	6.0814 07	8.6360 87	12.2236 18	17.2456 26	37
38	4.4388 13	6.3854 77	9.1542 52	13.0792 71	18.6252 76	38
39	4.6163 66	6.7047 51	9.7035 07	13.9948 20	20.1152 98	39
40	4.8010 21	7.0399 89	10.2857 18	14.9744 58	21.7245 22	40
41	4.9930 61	7.3919 88	10.9028 61	16.0226 70	23.4624 83	41
42	5.1927 84	7.7615 88	11.5570 33	17.1442 57	25.3394 82	42
43	5.4004 95	8.1496 67	12.2504 55	18.3443 55	27.3666 40	43
44	5.6165 15	8.5571 50	12.9854 82	19.6284 60	29.5559 72	44
45	5.8411 76	8.9850 08	13.7646 11	21.0024 52	31.9204 49	45
46	6.0748 23	9.4342 58	14.5904 87	22.4726 23	34.4740 85	46
47	6.3178 16	9.9059 71	15.4659 17	24.0457 07	37.2320 12	47
48	6.5705 28	10.4012 70	16.3938 72	25.7289 07	40.2105 73	48
49	6.8333 49	10.9213 33	17.3775 04	27.5299 30	43.4274 19	49
50	7.1066 83	11.4674 00	18.4201 54	29.4570 25	46.9016 13	50

EXAMPLE 4   Find the compound amount and compound interest of $2,000 invested at 8% compounded semiannually and due at the end of 10 years.

$P = \$2,000$, $i = 8\%/2 = 4\%$ per 6 months or per semiannual period, and $n = 10 \times 2 = 20$ semiannual periods. $S =$ Compound amount:

$$S = P(1 + i)^n = 2,000(1 + 4\%)^{20}$$
$$= 2,000(2.191123) = \$4,382.25$$

Compound interest $= 4,382.25 - 2,000.00 = \$2,382.25$

EXAMPLE 5   A note having a face value of $1,000 and bearing interest at 6% compounded quarterly will mature in 20 years. Find the maturity value.

$P = \$1,000$, $i = 6\%/4 = 1\frac{1}{2}\%$ per quarter, and $n = 20 \times 4 = 80$ quarters

The maturity value is the compound amount S:

$$S = 1,000(1 + 1\tfrac{1}{2}\%)^{80} = 1,000(1 + 1\tfrac{1}{2}\%)^{40} (1 + 1\tfrac{1}{2}\%)^{40}$$
$$= 1,000(1.814018) (1.814018)$$
$$= \$3,290.66$$

When the number of conversion periods is greater than that given in Table 14–1, the number (n) may be divided into several smaller numbers which are listed in the table. Here the factor $(1 + 1\frac{1}{2}\%)^{80}$ is converted to $(1 + 1\frac{1}{2}\%)^{40} \times (1 + 1\frac{1}{2}\%)^{40}$ based on the law of exponents.

Note that the computation of $(1 + 1\frac{1}{2}\%)^{80}$ may also be performed by using logarithms as shown in Example 12 of Chapter 16.

B. Computing Present Value

There are numerous occasions in business when it becomes necessary to discount an amount which is due on a future date. The discounted value is called the *present value* of the amount. The present value is obtained by solving for P from formula (14–3), or:

$$P = \frac{S}{(1 + i)^n}$$

and:

$$P = S(1 + i)^{-n} \qquad\qquad (14\text{–}4)$$

The difference between the value of the compound amount and its present value is called *compound discount*. From the above expression, we can see that the compound discount on the compound amount is the same as the compound interest on the present value.

For convenience, selected values of $(1 + i)^{-n}$, which is frequently called the *discount factor*, are tabulated in Table 14–2 for computing the problems in this chapter.

EXAMPLE 6   Find the present value of $4,382.25 due at the end of 10 years if money is worth 8% compounded semiannually.

Table 14-2 **Present Value When Compound Amount Is 1**

$$P = (1 + i)^{-n}$$

n	½%	1%	1½%	2%	3%	n
1	0.9950 25	0.9900 99	0.9852 22	0.9803 92	0.9708 74	1
2	0.9900 75	0.9802 96	0.9706 62	0.9611 69	0.9425 96	2
3	0.9851 49	0.9705 90	0.9563 17	0.9423 22	0.9151 42	3
4	0.9802 48	0.9609 80	0.9421 84	0.9238 45	0.8884 87	4
5	0.9753 71	0.9514 66	0.9282 60	0.9057 31	0.8626 09	5
6	0.9705 18	0.9420 45	0.9145 42	0.8879 71	0.8374 84	6
7	0.9656 90	0.9327 18	0.9010 27	0.8705 60	0.8130 92	7
8	0.9608 85	0.9234 83	0.8877 11	0.8534 90	0.7894 09	8
9	0.9561 05	0.9143 40	0.8745 92	0.8367 55	0.7664 17	9
10	0.9513 48	0.9052 87	0.8616 67	0.8203 48	0.7440 94	10
11	0.9466 15	0.8963 24	0.8489 33	0.8042 63	0.7224 21	11
12	0.9419 05	0.8874 49	0.8363 87	0.7884 93	0.7013 80	12
13	0.9372 19	0.8786 63	0.8240 27	0.7730 33	0.6809 51	13
14	0.9325 56	0.8699 63	0.8118 49	0.7578 75	0.6611 18	14
15	0.9279 17	0.8613 49	0.7998 52	0.7430 15	0.6418 62	15
16	0.9233 00	0.8528 21	0.7880 31	0.7284 46	0.6231 67	16
17	0.9187 07	0.8443 77	0.7763 85	0.7141 63	0.6050 16	17
18	0.9141 36	0.8360 17	0.7649 12	0.7001 59	0.5873 95	18
19	0.9095 88	0.8277 40	0.7536 07	0.6864 31	0.5702 86	19
20	0.9050 63	0.8195 44	0.7424 70	0.6729 71	0.5536 76	20
21	0.9005 60	0.8114 30	0.7314 98	0.6597 76	0.5375 49	21
22	0.8960 80	0.8033 96	0.7206 88	0.6468 39	0.5218 92	22
23	0.8916 22	0.7954 42	0.7100 37	0.6341 56	0.5066 92	23
24	0.8871 86	0.7875 66	0.6995 44	0.6217 21	0.4919 34	24
25	0.8827 72	0.7797 68	0.6892 06	0.6095 31	0.4776 06	25
26	0.8783 80	0.7720 48	0.6790 21	0.5975 79	0.4636 95	26
27	0.8740 10	0.7644 04	0.6689 86	0.5858 62	0.4501 89	27
28	0.8696 62	0.7568 36	0.6590 99	0.5743 75	0.4370 77	28
29	0.8653 35	0.7493 42	0.6493 59	0.5631 12	0.4243 46	29
30	0.8610 30	0.7419 23	0.6397 62	0.5520 71	0.4119 87	30
31	0.8567 46	0.7345 77	0.6303 08	0.5412 46	0.3999 87	31
32	0.8524 84	0.7273 04	0.6209 93	0.5306 33	0.3883 37	32
33	0.8482 42	0.7201 03	0.6118 16	0.5202 29	0.3770 26	33
34	0.8440 22	0.7129 73	0.6027 74	0.5100 28	0.3660 45	34
35	0.8398 23	0.7059 14	0.5938 66	0.5000 28	0.3553 83	35
36	0.8356 45	0.6989 25	0.5850 90	0.4902 23	0.3450 32	36
37	0.8314 87	0.6920 05	0.5764 43	0.4806 11	0.3349 83	37
38	0.8273 51	0.6851 53	0.5679 24	0.4711 87	0.3252 26	38
39	0.8232 35	0.6783 70	0.5595 31	0.4619 48	0.3157 54	39
40	0.8191 39	0.6716 53	0.5512 62	0.4528 90	0.3065 57	40
41	0.8150 64	0.6650 03	0.5431 16	0.4440 10	0.2976 28	41
42	0.8110 09	0.6584 19	0.5350 89	0.4353 04	0.2889 59	42
43	0.8069 74	0.6519 00	0.5271 82	0.4267 69	0.2805 43	43
44	0.8029 59	0.6454 45	0.5193 91	0.4184 01	0.2723 72	44
45	0.7989 64	0.6390 55	0.5117 15	0.4101 97	0.2644 39	45
46	0.7949 89	0.6327 28	0.5041 53	0.4021 54	0.2567 37	46
47	0.7910 34	0.6264 63	0.4967 02	0.3942 68	0.2492 59	47
48	0.7870 98	0.6202 60	0.4893 62	0.3865 38	0.2419 99	48
49	0.7831 83	0.6141 19	0.4821 30	0.3789 58	0.2349 50	49
50	0.7792 86	0.6080 39	0.4750 05	0.3715 28	0.2281 07	50

Table 14–2   **Present Value When Compound Amount Is 1**
$$P = (1 + i)^{-n}$$

n	4%	5%	6%	7%	8%	n
1	0.9615 38	0.9523 81	0.9433 96	0.9345 79	0.9259 26	1
2	0.9245 56	0.9070 29	0.8899 96	0.8734 39	0.8573 39	2
3	0.8889 96	0.8638 38	0.8396 19	0.8162 98	0.7938 32	3
4	0.8548 04	0.8227 02	0.7920 94	0.7628 95	0.7350 30	4
5	0.8219 27	0.7835 26	0.7472 58	0.7129 86	0.6805 83	5
6	0.7903 15	0.7462 15	0.7049 61	0.6663 42	0.6301 70	6
7	0.7599 18	0.7106 81	0.6650 57	0.6227 50	0.5834 90	7
8	0.7306 90	0.6768 39	0.6274 12	0.5820 09	0.5402 69	8
9	0.7025 87	0.6446 09	0.5918 98	0.5439 34	0.5002 49	9
10	0.6755 64	0.6139 13	0.5583 95	0.5083 49	0.4631 93	10
11	0.6495 81	0.5846 79	0.5267 88	0.4750 93	0.4288 83	11
12	0.6245 97	0.5568 37	0.4969 69	0.4440 12	0.3971 14	12
13	0.6005 74	0.5303 21	0.4688 39	0.4149 64	0.3676 98	13
14	0.5774 75	0.5050 68	0.4423 01	0.3878 17	0.3404 61	14
15	0.5552 65	0.4810 17	0.4172 65	0.3624 46	0.3152 42	15
16	0.5339 08	0.4581 12	0.3936 46	0.3387 35	0.2918 90	16
17	0.5133 73	0.4362 97	0.3713 64	0.3165 74	0.2702 69	17
18	0.4936 28	0.4155 21	0.3503 44	0.2958 64	0.2502 49	18
19	0.4746 42	0.3957 34	0.3305 13	0.2765 08	0.2317 12	19
20	0.4563 87	0.3768 89	0.3118 05	0.2584 19	0.2145 48	20
21	0.4388 34	0.3589 42	0.2941 55	0.2415 13	0.1986 56	21
22	0.4219 55	0.3418 50	0.2775 05	0.2257 13	0.1839 41	22
23	0.4057 26	0.3255 71	0.2617 97	0.2109 47	0.1703 15	23
24	0.3901 21	0.3100 68	0.2469 79	0.1971 47	0.1576 99	24
25	0.3751 17	0.2953 03	0.2329 99	0.1842 49	0.1460 18	25
26	0.3606 89	0.2812 41	0.2198 10	0.1721 95	0.1352 02	26
27	0.3468 17	0.2678 48	0.2073 68	0.1609 30	0.1251 87	27
28	0.3334 77	0.2550 94	0.1956 30	0.1504 02	0.1159 14	28
29	0.3206 51	0.2429 46	0.1845 57	0.1405 63	0.1073 28	29
30	0.3083 19	0.2313 77	0.1741 10	0.1313 67	0.0993 77	30
31	0.2964 60	0.2203 59	0.1642 55	0.1227 73	0.0920 16	31
32	0.2850 58	0.2098 66	0.1549 57	0.1147 41	0.0852 00	32
33	0.2740 94	0.1998 73	0.1461 86	0.1072 35	0.0788 89	33
34	0.2635 52	0.1903 55	0.1379 12	0.1002 19	0.0730 45	34
35	0.2534 15	0.1812 90	0.1301 05	0.0936 63	0.0676 35	35
36	0.2436 69	0.1726 57	0.1227 41	0.0875 35	0.0626 25	36
37	0.2342 97	0.1644 36	0.1157 93	0.0818 09	0.0579 86	37
38	0.2252 85	0.1566 05	0.1092 39	0.0764 57	0.0536 90	38
39	0.2166 21	0.1491 48	0.1030 56	0.0714 55	0.0497 13	39
40	0.2082 89	0.1420 46	0.0972 22	0.0667 80	0.0460 31	40
41	0.2002 78	0.1352 82	0.0917 19	0.0624 12	0.0426 21	41
42	0.1925 75	0.1288 40	0.0865 27	0.0583 29	0.0394 64	42
43	0.1851 68	0.1227 04	0.0816 30	0.0545 13	0.0365 41	43
44	0.1780 46	0.1168 61	0.0770 09	0.0509 46	0.0338 34	44
45	0.1711 98	0.1112 97	0.0726 50	0.0476 13	0.0313 28	45
46	0.1646 14	0.1059 97	0.0685 38	0.0444 99	0.0290 07	46
47	0.1582 83	0.1009 49	0.0646 58	0.0415 87	0.0268 59	47
48	0.1521 95	0.0961 42	0.0609 98	0.0388 67	0.0248 69	48
49	0.1463 41	0.0915 64	0.0575 46	0.0363 24	0.0230 27	49
50	0.1407 13	0.0872 04	0.0542 88	0.0339 48	0.0213 21	50

The amount $S = \$4,382.25$, $i = 8\%/2 = 4\%$, and $n = 10 \times 2 = 20$ semi-annual periods

$$P = S(1 + i)^{-n} = 4,382.25(1 + 4\%)^{-20}$$
$$= 4,382.25(0.456387) = \$2,000.00$$

Compound discount $= 4,382.25 - 2,000.00$
$$= \$2,382.25$$

which is the same as the compound interest on the present value $2,000.00. Also see the solution of Example 4.

EXAMPLE 7   A note of $1,000 dated January 1, 1973, at 8% compounded quarterly for ten years, was discounted in a bank on January 1, 1977. Find the proceeds and the compound discount if the note was discounted at 10% compounded semiannually.

This type of problem requires two steps in finding the solution:

STEP 1   Find the maturity value on January 1, 1983 (ten years from the date of the note), according to the rate and the time stipulated on the note.

$P = \$1,000$ (face value), $i = 8\%/4 = 2\%$ (per quarter), $n = 10 \times 4 = 40$ (quarters)

Maturity value $S = P(1 + i)^n = 1,000(1 + 2\%)^{40}$
$$= 1,000(2.208040)$$
$$= \$2,208.04$$

STEP 2   Find the present value (as of January 1, 1977, the date of discounting, or six years before the maturity date) of the maturity value, according to the rate and the time of discount.

$S = 2,208.04$, $i = 10\%/2 = 5\%$ per semiannual period, $n = 6 \times 2 = 12$ semiannual periods.

Proceeds (the value received by the seller of the note) = the present value P:

$$P = S(1 + i)^{-n} = 2,208.04(1 + 5\%)^{-12}$$
$$= 2,208.04(0.556837) = \$1,229.52$$

Compound discount $=$ Maturity value $-$ Proceeds
$$= 2,208.04 - 1,229.52$$
$$= \$978.52$$

C. Effective Annual Interest Rate

The effective annual interest rate, simply called the *effective rate*, is the compound interest earned for a one-year period on the principal of $1. In other words, an effective rate is an interest rate compounded annually. The effective rate is frequently used as a device to compare one interest rate with another rate compounded at different time intervals. It is especially useful to those who invest or borrow money from various sources. By comparing the effective rates of the sources, a person may select the one having the lowest effective rate for borrowing and the one having the highest effective rate for investing.

EXAMPLE 8   If $1 is invested at 8% compounded quarterly for one year, what is the effective rate?

$P = \$1, i = 8\%/4 = 2\%$ (per quarter), $n = 4$ (quarters)

The compound amount is:

$S = P(1 + i)^n = 1(1 + 2\%)^4 = \$1.082432$

The compound interest for one year is:

$1.082432 - 1 = \$0.082432$

Thus, the effective rate is 0.082432, or rounded, 8.24%.
    We may state that 8.24% compounded annually is equivalent to an interest rate of 8% compounded quarterly.

In general, let:

j = nominal rate
m = number of conversion periods for one year
$i = j/m$ (interest rate per conversion period)
f = effective rate

Substitute the values in the compound amount formula (14–3). The compound amounts for a one-year period are:

Based on the nominal rate, $S = P(1 + j/m)^m$.
Based on the effective rate, $S = P(1 + f)$.

Equate the right sides of the two equations above:

$$P(1 + f) = P(1 + j/m)^m$$

Divide both sides by P:

$$1 + f = (1 + j/m)^m$$

Thus:

$$f = \left(1 + \frac{j}{m}\right)^m - 1$$

or:

$$f = (1 + i)^m - 1 \qquad (14\text{–}5)$$

Example 8 now may be computed by using formula (14–5) directly without mentioning the values of S and P as follows:

$j = 8\%$, and $m = 4$ (quarters in one year)

$$f = \left(1 + \frac{8\%}{4}\right)^4 - 1 = (1 + 2\%)^4 - 1 = 1.082432 - 1$$
$$= 0.082432, \text{ or } 8.24\%$$

EXAMPLE 9   Bank X offers its depositors an interest rate of 6% compounded monthly, while Bank Y offers its depositors an interest rate of 8% compounded semiannually. Which of the two banks makes the better offer?

The effective rate based on the interest rate offered by Bank X is:

$$f = \left(1 + \frac{6\%}{12}\right)^{12} - 1 = (1 + \tfrac{1}{2}\%)^{12} - 1$$
$$= 1.061678 - 1 = 0.061678, \text{ or rounded, 6.17\%}$$

The effective rate based on the interest rate offered by Bank Y is:

$$f = \left(1 + \frac{8\%}{2}\right)^{2} - 1 = (1 + 4\%)^{2} - 1$$
$$= 1.081600 - 1 = 0.0816, \text{ or 8.16\%}$$

Bank Y offers a better interest rate since the bank has a higher effective rate.

D. Continuously Compounded Interest

In previous discussions, compound interest was computed at regularly stated intervals, such as annually, semiannually, quarterly, or monthly. Compound interest may be computed more frequently such as every minute, every second, or continuously. Continuous compounding is not commonly used in the actual investment market. However, its concept is theoretically important in analyzing financial problems.

In computing an interest at a nominal rate j compounded continuously, first find the equivalent effective rate based on formula (14–5):

$$f = \left(1 + \frac{j}{m}\right)^{m} - 1$$

Next, compute the compound interest based on the found effective rate.

The formula for finding the effective rate of the nominal rate j compounded continuously is derived as follows:

Let $k = \dfrac{m}{j}$,   and $\dfrac{1}{k} = \dfrac{j}{m}$. Then, the term in formula (14–5):

$$\left(1 + \frac{j}{m}\right)^{m} = \left[\left(1 + \frac{j}{m}\right)^{\frac{m}{j}}\right]^{j} = \left[\left(1 + \frac{1}{k}\right)^{k}\right]^{j}$$

When j is compounded continuously, m becomes infinitely large, and the value of $k = m/j$ also becomes infinitely large. When k approaches an infinitely large value ($k \to \infty$), the limit of $(1 + 1/k)^k$, usually denoted by the letter e, is an irrational number and is approximately 2.71828. It can be written:

$$e = \lim_{k \to \infty} \left(1 + \frac{1}{k}\right)^{k} = 2.71828, \text{ approximately (Also see Examples 23 and 27 of Chapter 15.)}$$

The effective rate of the nominal rate j compounded continuously is:

$$f = \left(1 + \frac{j}{m}\right)^m - 1 = \left[\left(1 + \frac{1}{k}\right)^{k}\right]^{j} - 1 = e^j - 1$$

or:

$$f = e^j - 1 \qquad (14\text{--}6)$$

The value of e^j can be approximated by using logarithms (Chapter 16) or obtained directly from a table such as Table 14–3. Table 14–3 shows the values of e^j for selected values of j computed to eight decimal places. The value of e^j for other values of j may also be obtained from the table after applying the laws of exponents as illustrated in Example 11.

EXAMPLE 10 Find the effective rate if money is worth 8% compounded continuously.

$j = 8\% = .08$

$e^j = 1.08328707$ (Table 14–3)

$f = e^j - 1 = 1.08328707 - 1 = 0.08328707$, or rounded to 8.33%
(formula 14–6)

Note that the effective rate of 8% compounded quarterly is 8.24% (Example 8.) In general, the effective rate (f) increases as the number of conversion periods (m) increases. However, the increases of the rate are moderate.

EXAMPLE 11 Find the effective rate if money is worth $6\frac{1}{2}\%$ compounded continuously.

$j = 6\frac{1}{2}\% = .065$

$e^j = e^{.065} = e^{.06}(e^{.005}) = 1.06183655(1.00501252)$
$= 1.06715903$ (Table 14–3)

Substitute e^j value in formula (14–5).

$f = e^j - 1 = 1.06715903 - 1 = 0.06715903$, or rounded to 6.72%

Substitute the effective rate $f(= e^j - 1)$ for i in the compound amount formula:

$$S = P(1 + i)^n = P(1 + f)^n = P(1 + e^j - 1)^n$$
$$= P(e^j)^n = P(e^{jn})$$

Thus, the formula for finding the compound amount S of the principal P at the nominal rate j compounded continuously for n years is:

$$S = P(e^{jn}) \qquad (14\text{--}7)$$

EXAMPLE 12 Find the compound amount and the compound interest if $1,000 is invested at 6% compounded continuously for (a) one year, and (b) five years.

$P = \$1,000, j = 6\% = .06$

(a) $n = 1$ year, and $jn = .06(1) = .06$

Substitute the above values in formula (14–7):

$$S = 1,000(e^{.06}) = 1,000(1.06183655)$$
$$= \$1,061.84$$

(b) $n = 5$ years, and $jn = .06(5) = .30$

$$S = 1,000(e^3) = 1,000(1.34985881)$$
$$= \$1,349.86$$

EXAMPLE 13 Find the present value of $10,000 due at the end of 10 years if money is worth 5% compounded continuously. What is the compound discount?

The compound amount $S = \$10,000$, $j = 5\% = .05$, $n = 10$ years, and $jn = .05(10) = .5$.

Solve for P from formula (14–7):

$$P = S(e^{-jn})$$

Substitute the values in the above formula:

$$P = 10,000(e^{-.5}) = 10,000(0.60653066) \qquad \text{(Table 14–3)}$$
$$= \$6,065.31 \text{ (present value)}$$

Compound discount $= 10,000 - 6,065.31 = \$3,934.69$.

Table 14–3 **Values of e^j and e^{-j} for Selected Values of j**

j	e^j	e^{-j}	j	e^j	e^{-j}
.000	1.0000 0000	1.0000 0000			
.001	1.0010 0050	.9990 0050	.06	1.0618 3655	.9417 6453
.002	1.0020 0200	.9980 0200	.07	1.0725 0818	.9323 9382
.003	1.0030 0450	.9970 0450	.08	1.0832 8707	.9231 1635
.004	1.0040 0801	.9960 0799	.09	1.0941 7428	.9139 3119
.005	1.0050 1252	.9950 1248	.1	1.1051 7092	.9048 3742
.006	1.0060 1804	.9940 1796	.2	1.2214 0276	.8187 3075
.007	1.0070 2456	.9930 2444	.3	1.3498 5881	.7408 1822
.008	1.0080 3209	.9920 3191	.4	1.4918 2470	.6703 2005
.009	1.0090 4062	.9910 4038	.5	1.6487 2127	.6065 3066
.01	1.0100 5017	.9900 4983	.6	1.8221 1880	.5488 1164
.02	1.0202 0134	.9801 9867	.7	2.0137 5271	.4965 8530
.03	1.0304 5453	.9704 4553	.8	2.2255 4093	.4493 2896
.04	1.0408 1077	.9607 8944	.9	2.4596 0311	.4065 6966
.05	1.0512 7110	.9512 2942	1.0	2.7182 8183	.3678 7944

For more detailed values, see Table 4 in the Appendix.

EXERCISE 14–1 Reference: Sections 14.1 and 14.2

1. What are the amount and simple interest on $1,000 at 5% for (a) 3 years and (b) 6 months?

2. What are the amount and simple interest on $2,000 at 7% for (a) 5 years and (b) 4 months?

3. Find (a) the simple interest on $4,000 at 12% for 60 days and (b) the amount.

4. Find (a) the simple interest on $600 at 6% for 30 days and (b) the amount.

5. What are (a) the amount and (b) the interest at the end of 3 years if $1,000 is borrowed at 5% compounded annually?

6. What are (a) the amount and (b) the interest at the end of 5 years if $2,000 is invested at 7% compounded annually?

7. Find the compound amount and the compound interest of $1,000 due at the end of 4 years invested at 8% compounded (a) semiannually and (b) quarterly.

8. Find the compound amount and the compound interest of $3,000 invested at 6% compounded monthly and due at the end of (a) 10 months and (b) 2 years.

9. A note having a face value of $5,000 and bearing interest at 10% compounded semiannually will mature in 6 years. Find the maturity value.

10. Find the maturity value of a 2-year note with a face value of $2,000 and interest at 8% compounded quarterly.

11. Find the present value of $1,500 due at the end of 8 years if money is worth 7% compounded annually.

12. Find the present value of $1,000 due at the end of 2 years if money is worth 6% compounded monthly.

13. A 6-year note having a face value of $1,000 charged an interest rate of 5% compounded annually. The note was discounted in a bank $1\frac{1}{2}$ years before the due time at 8% compounded quarterly. What are the proceeds and the compound discount?

14. A note of $2,000, dated July 1, 1975, at 6% compounded quarterly for 5 years, was discounted in a bank on July 1, 1978. Find the proceeds and the compound discount if the note was discounted at 12% compounded monthly.

15. Find the effective rate if $1 is invested for 1 year at 6% compounded (a) monthly, (b) quarterly, (c) semiannually, (d) annually, and (e) continuously.

16. Find the effective rate if money is worth 12% compounded (a) continuously, (b) monthly, (c) quarterly, (d) semiannually, and (e) annually.

17. Find the compound amount and the compound interest if $1,000 is invested at 5% compounded continuously for (a) 1 year and (b) 4 years.

18. What are the compound amount and the compound interest if $2,000 is invested at 12% compounded continuously for (a) 1 year and (b) 5 years?

19. Find the present value of $1,000 due at the end of 5 years if money is worth 6% compounded continuously. (b) Find the compound discount.

20. (a) What is the present value of $2,000 due at the end of 10 years if money is worth 8% compounded continuously? (b) What is the compound discount?

14.3 ORDINARY ANNUITY

An *annuity* is a series of periodic payments, usually made in equal amounts. The period of time between two successive payment dates is called the *payment interval*. The time between the beginning of the first payment interval and the end of the last payment interval is called the *term* of the annuity.

There are various types of annuities. However, the most important and basic type is the ordinary annuity. In an *ordinary annuity*, periodic payments are made at the end of each payment interval. The interest on each payment is computed by the compound interest method. Hereafter, unless otherwise specified, the word *annuity* means an ordinary annuity and the payment interval of an annuity will coincide with the interest conversion period. Thus, the *payment date* is the interest computing date. For example, when the payment interval is one month, the interest is compounded monthly; when each of the payments is made at the end of each month, the interest is also computed and compounded at the end of each month.

A. Amount of An Annuity

The *amount of an annuity* is the final value at the end of the term of the annuity. The amount includes all of the periodic payments and the compound interest. Basically, the amount of an annuity can be obtained by totaling the compound amounts of the individual periodic payments. Each of the compound amounts is computed by the formula $S = P(1 + i)^n$, as illustrated in Example 14.

EXAMPLE 14   Find the amount of an annuity if the size of each payment is $100, payable at the end of each quarter for one year at an interest rate of 4% compounded quarterly. Assume that the term of the annuity begins on January 1, 1976, and ends on January 1, 1977.

SOLUTION   The first payment is made at the end of the first quarter (April 1, 1976), which is 3 quarters before the end of the term of the annuity. Thus, interest is accumulated on the first payment for 3 interest periods; on the second payment, for 2 interest periods; and on the third payment, for 1 interest period. The fourth payment is not entitled to interest since it is paid at the end of the term (January 1, 1977).

This analysis is diagrammed on page 300.

The amount of the annuity, $406.0401, which is the final value at the end of the fourth quarter, is computed as follows, using Table 14–1:

$$
\begin{aligned}
100 \quad &= 100(1) \\
100(1 + 1\%) &= 100(1.01) \\
100(1 + 1\%)^2 &= 100(1.0201) \\
\underline{100(1 + 1\%)^3} &= \underline{100(1.030301)} \\
\text{Total amount} &= 100(4.060401) = \$406.0401 \\
&\qquad\qquad\quad \text{or rounded to } \$406.04
\end{aligned}
$$

Total compound interest on the four payments is:

$$\$406.04 - (\$100 \times 4) = \$6.04$$

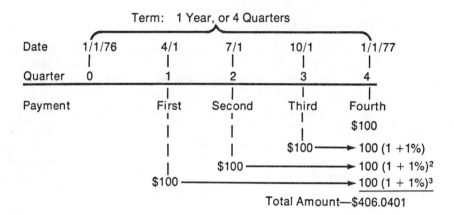

In general, let:

S_n = the amount of an ordinary annuity
R = the size of each regular payment (or periodic rent)
i = interest rate per conversion period
n = number of payments during the term of an annuity
    (It is also the number of payment intervals, or the number of interest conversion periods.)

Then, the formula for the amount of an ordinary annuity is:

$$S_n = R \cdot \frac{(1 + i)^n - 1}{i}$$

(14–8)

The derivation of formula (14–8) follows.

Based on the diagram in Example 14, the amount of an annuity of R payable at the end of each period for 4 periods at interest rate i per period, denoted by S_4, can be written:

$$S_4 = R + R(1 + i) + R(1 + i)^2 + R(1 + i)^3$$

(1)

Multiply both sides of equation (1) by $(1 + i)$:

$$S_4(1 + i) = R(1 + i) + R(1 + i)^2 + R(1 + i)^3 + R(1 + i)^4$$

(2)

Subtract equation (2) from equation (1):

$$S_4 - S_4(1 + i) = R - R(1 + i)^4$$

Factor both sides:

$$S_4[1 - (1 + i)] = R[1 - (1 + i)^4]$$

$$S_4 = \frac{R[1 - (1 + i)^4]}{-i} = \frac{R[1 - (1 + i)^4]}{-i} \cdot \frac{-1}{-1}$$

$$S_4 = R \cdot \frac{(1 + i)^4 - 1}{i}$$

Extend this idea by letting $n =$ number of periods or payments. Then, the amount of an annuity of R payable at the end of each period for n periods at interest rate i per period is:

$$S_n = R \cdot \frac{(1 + i)^n - 1}{i}$$

which is formula (14–8).

Note: The right side of equation (1) is a *geometric progression* and may also be solved by the geometric progression formula as shown in Chapter 15.

The use of the formula is illustrated in Example 15.

EXAMPLE 15   A man deposits $100 in a bank at the end of each quarter for one year. If the money earns interest at 4% compounded quarterly, how much does he have in his account at the end of the year after the last payment is made?

$R = \$100$ (per quarter), $i = 4\%/4 = 1\%$ (per quarter), and $n = 4$ (quarterly payments)

Substitute the values in formula (14–8):

$$S_n = 100 \cdot \frac{(1 + 1\%)^4 - 1}{1\%}$$

$$= 100 \cdot \frac{1.040604 - 1}{.01}$$

$$= 100(4.0604) = \$406.04. \text{ (Table 14–1)}$$

For convenience, let the simple symbol:

$$s_{\overline{n}|i} \text{ (read: } s \text{ angle } n \text{ at } i) = \frac{(1 + i)^n - 1}{i}$$

The values of $s_{\overline{n}|i}$ for selected interest rates (i) and numbers of payments (n) are tabulated in Table 14–4 on pages 302 and 303. When Table 14–4 is employed, the computation of the amount of an annuity is simplified. Formula (14–8) now can be written:

$$S_n = R s_{\overline{n}|i} \tag{14–8}$$

Example 15 can also be computed in the following manner:

$$S_n = R s_{\overline{n}|i} = 100 s_{\overline{4}|1\%} = 100(4.060401) = \$406.04 \text{ (Table 14–4)}$$

EXAMPLE 16   If $50 is deposited at the end of each month for three years in a fund which earns 6% interest compounded monthly, what will be the final value at the end of the third year? What is the total interest?

$R = \$50$ (per month), $i = 6\%/12 = \frac{1}{2}\%$ (per month), $n = 3 \times 12 = 36$ (monthly payments)

Substitute the values in formula (14–8):

Table 14–4   **Amount of Annuity When Periodic Payment Is 1**

$$s_{\overline{n}|i} = \frac{(1+i)^n - 1}{i}$$

n	$\frac{1}{2}\%$	1%	$1\frac{1}{2}\%$	2%	3%	n
1	1.0000 00	1.0000 00	1.0000 00	1.0000 00	1.0000 00	1
2	2.0050 00	2.0100 00	2.0150 00	2.0200 00	2.0300 00	2
3	3.0150 00	3.0301 00	3.0452 25	3.0604 00	3.0909 00	3
4	4.0301 00	4.0604 01	4.0909 03	4.1216 08	4.1836 27	4
5	5.0502 51	5.1010 05	5.1522 67	5.2040 40	5.3091 36	5
6	6.0755 02	6.1520 15	6.2295 51	6.3081 21	6.4684 10	6
7	7.1058 79	7.2135 35	7.3229 94	7.4342 83	7.6624 62	7
8	8.1414 09	8.2856 71	8.4328 39	8.5829 69	8.8923 36	8
9	9.1821 16	9.3685 27	9.5593 32	9.7546 28	10.1591 06	9
10	10.2280 26	10.4622 13	10.7027 22	10.9497 21	11.4638 79	10
11	11.2791 67	11.5668 35	11.8632 62	12.1687 15	12.8077 96	11
12	12.3355 62	12.6825 03	13.0412 11	13.4120 90	14.1920 30	12
13	13.3972 40	13.8093 28	14.2368 30	14.6803 32	15.6177 90	13
14	14.4642 26	14.9474 21	15.4503 82	15.9739 38	17.0863 24	14
15	15.5365 48	16.0968 96	16.6821 38	17.2934 17	18.5989 14	15
16	16.6142 30	17.2578 64	17.9323 70	18.6392 85	20.1568 81	16
17	17.6973 01	18.4304 43	19.2013 55	20.0120 71	21.7615 88	17
18	18.7857 88	19.6147 48	20.4893 76	21.4123 12	23.4144 35	18
19	19.8797 17	20.8108 95	21.7967 16	22.8405 59	25.1168 68	19
20	20.9791 15	22.0190 04	23.1236 67	24.2973 70	26.8703 74	20
21	22.0840 11	23.2391 94	24.4705 22	25.7833 17	28.6764 86	21
22	23.1944 31	24.4715 86	25.8375 80	27.2989 84	30.5367 80	22
23	24.3104 03	25.7163 02	27.2251 44	28.8449 63	32.4528 84	23
24	25.4319 55	26.9734 65	28.6335 21	30.4218 62	34.4264 70	24
25	26.5591 15	28.2432 00	30.0630 24	32.0303 00	36.4592 64	25
26	27.6919 11	29.5256 32	31.5139 69	33.6709 06	38.5530 42	26
27	28.8303 70	30.8208 88	32.9866 79	35.3443 24	40.7096 34	27
28	29.9745 22	32.1290 97	34.4814 79	37.0512 10	42.9309 23	28
29	31.1243 95	33.4503 88	35.9987 01	38.7922 35	45.2188 50	29
30	32.2800 17	34.7848 92	37.5386 81	40.5680 79	47.5754 16	30
31	33.4414 17	36.1327 40	39.1017 62	42.3794 41	50.0026 78	31
32	34.6086 24	37.4940 68	40.6882 88	44.2270 30	52.5027 59	32
33	35.7816 67	38.8690 09	42.2986 12	46.1115 70	55.0778 41	33
34	36.9605 75	40.2576 99	43.9330 92	48.0338 02	57.7301 77	34
35	38.1453 78	41.6602 76	45.5920 88	49.9944 78	60.4620 82	35
36	39.3361 05	43.0768 78	47.2759 69	51.9943 67	63.2759 44	36
37	40.5327 85	44.5076 47	48.9851 09	54.0342 55	66.1742 23	37
38	41.7354 49	45.9527 24	50.7198 85	56.1149 40	69.1594 49	38
39	42.9441 27	47.4122 51	52.4806 84	58.2372 38	72.2342 33	39
40	44.1588 47	48.8863 73	54.2678 94	60.4019 83	75.4012 60	40
41	45.3796 42	50.3752 37	56.0819 12	62.6100 23	78.6632 98	41
42	46.6065 40	51.8789 89	57.9231 41	64.8622 23	82.0231 96	42
43	47.8395 72	53.3977 79	59.7919 88	67.1594 68	85.4838 92	43
44	49.0787 70	54.9317 57	61.6888 68	69.5026 57	89.0484 09	44
45	50.3241 64	56.4810 75	63.6142 01	71.8927 10	92.7198 61	45
46	51.5757 85	58.0458 85	65.5684 14	74.3305 64	96.5014 57	46
47	52.8336 64	59.6263 44	67.5519 40	76.8171 76	100.3965 01	47
48	54.0978 32	61.2226 08	69.5652 19	79.3535 19	104.4083 96	48
49	55.3683 21	62.8348 34	71.6086 98	81.9405 90	108.5406 48	49
50	56.6451 63	64.4631 82	73.6828 28	84.5794 01	112.7968 67	50

Table 14-4 **Amount of Annuity When Periodic Payment Is 1**

$$s_{\overline{n}|i} = \frac{(1+i)^n - 1}{i}$$

n	4%	5%	6%	7%	8%	n
1	1.0000 00	1.0000 00	1.0000 00	1.0000 00	1.0000 00	1
2	2.0400 00	2.0500 00	2.0600 00	2.0700 00	2.0800 00	2
3	3.1216 00	3.1525 00	3.1836 00	3.2149 00	3.2464 00	3
4	4.2464 64	4.3101 25	4.3746 16	4.4399 43	4.5061 12	4
5	5.4163 23	5.5256 31	5.6370 93	5.7507 39	5.8666 01	5
6	6.6329 75	6.8019 13	6.9753 19	7.1532 91	7.3359 29	6
7	7.8982 94	8.1420 08	8.3938 38	8.6540 21	8.9228 03	7
8	9.2142 26	9.5491 09	9.8974 68	10.2598 03	10.6366 28	8
9	10.5827 95	11.0265 64	11.4913 16	11.9779 89	12.4875 58	9
10	12.0061 07	12.5778 93	13.1807 95	13.8164 48	14.4865 62	10
11	13.4863 51	14.2067 87	14.9716 43	15.7835 99	16.6454 87	11
12	15.0258 05	15.9171 27	16.8699 41	17.8884 51	18.9771 26	12
13	16.6268 38	17.7129 83	18.8821 38	20.1406 43	21.4952 97	13
14	18.2919 11	19.5986 32	21.0150 66	22.5504 88	24.2149 20	14
15	20.0235 88	21.5785 64	23.2759 70	25.1290 22	27.1521 14	15
16	21.8245 31	23.6574 92	25.6725 28	27.8880 54	30.3242 83	16
17	23.6975 12	25.8403 66	28.2128 80	30.8402 17	33.7502 26	17
18	25.6454 13	28.1323 85	30.9056 53	33.9990 33	37.4502 44	18
19	27.6712 29	30.5390 04	33.7599 92	37.3789 65	41.4462 63	19
20	29.7780 79	33.0659 54	36.7855 91	40.9954 92	45.7619 64	20
21	31.9692 01	35.7192 52	39.9927 27	44.8651 77	50.4229 21	21
22	34.2479 70	38.5052 14	43.3922 90	49.0057 39	55.4567 55	22
23	36.6178 89	41.4304 75	46.9958 28	53.4361 41	60.8932 96	23
24	39.0826 04	44.5019 99	50.8155 77	58.1766 71	66.7647 59	24
25	41.6459 08	47.7270 99	54.8645 12	63.2490 38	73.1059 40	25
26	44.3117 45	51.1134 54	59.1563 83	68.6764 70	79.9544 15	26
27	47.0842 14	54.6691 26	63.7057 66	74.4838 23	87.3507 68	27
28	49.9675 83	58.4025 83	68.5281 12	80.6976 91	95.3388 30	28
29	52.9662 86	62.3227 12	73.6397 98	87.3465 29	103.9659 36	29
30	56.0849 38	66.4388 48	79.0581 86	94.4607 86	113.2832 11	30
31	59.3283 35	70.7607 90	84.8016 77	102.0730 41	123.3458 68	31
32	62.7014 69	75.2988 29	90.8897 78	110.2181 54	134.2135 37	32
33	66.2095 27	80.0637 71	97.3431 65	118.9334 25	145.9506 20	33
34	69.8579 09	85.0669 59	104.1837 55	128.2587 65	158.6266 70	34
35	73.6522 25	90.3203 07	111.4347 80	138.2368 78	172.3168 04	35
36	77.5983 14	95.8363 23	119.1208 67	148.9134 60	187.1021 48	36
37	81.7022 46	101.6281 39	127.2681 19	160.3374 02	203.0703 20	37
38	85.9703 36	107.7095 46	135.9042 06	172.5610 20	220.3159 45	38
39	90.4091 50.	114.0950 23	145.0584 58	185.6402 92	238.9412 21	39
40	95.0255 16	120.7997 74	154.7619 66	199.6351 12	259.0565 19	40
41	99.8265 36	127.8397 63	165.0476 84	214.6095 70	280.7810 40	41
42	104.8195 98	135.2317 51	175.9505 45	230.6322 40	304.2435 23	42
43	110.0123 82	142.9933 39	187.5075 77	247.7764 97	329.5830 05	43
44	115.4123 77	151.1430 06	199.7580 32	266.1208 51	356.9496 46	44
45	121.0293 92	159.7001 56	212.7435 14	285.7493 11	386.5056 17	45
46	126.8705 68	168.6851 64	226.5081 25	306.7517 63	418.4260 67	46
47	132.9453 90	178.1194 22	241.0986 12	329.2243 86	452.9001 52	47
48	139.2632 06	188.0253 93	256.5645 29	353.2700 93	490.1321 64	48
49	145.8337 34	198.4266 63	272.9584 01	378.9990 00	530.3427 37	49
50	152.6670 84	209.3479 96	290.3359 05	406.5289 29	573.7701 56	50

$$S_n = 50s_{\overline{36}|\frac{1}{2}\%} = 50(39.336105) = \$1,966.81.$$
$$\text{(Table 14-4)} \qquad \text{(Final value)}$$

Total deposits $= 50 \times 36 = \$1,800.00$
Total interest $= 1,966.81 - 1,800.00 = \$166.81$

EXAMPLE 17 In Example 16, how much will the amount be $2\frac{1}{2}$ years after the last deposit is made?

The final value at the end of the 3 years now becomes the principal which is accumulated for $2\frac{1}{2}$ years at the same interest rate for the deposits.

Substitute $P = \$1,966.81$, $i = \frac{1}{2}\%$ (per month), $n = 2\frac{1}{2} \times 12 = 30$ (months) in the formula.

$$S = P(1 + i)^n = 1,966.81(1 + \tfrac{1}{2}\%)^{30}$$
$$= 1,966.81(1.161400) = \$2,284.25 \text{ (Amount) (Table 14-1)}$$

B. Present Value of An Annuity

The present value of an annuity is the value at the beginning of the term of the annuity. The computation of the present value is based on either one of the following two different interpretations of the value:

1. It is the sum of the present values of the periodic payments of an annuity.
2. It is the single principal which will accumulate to the amount of an annuity at a given compound interest rate by the end of the term of the annuity.

The two interpretations are illustrated in Example 18.

EXAMPLE 18 Find the present value of an annuity if the size of each payment is $100, payable at the end of each quarter for one year at an interest rate of 4% compounded quarterly. Assume that the term of the annuity begins on January 1, 1976 and ends on January 1, 1977.

METHOD 1 When the present value of an annuity is considered as the sum of the present values of the periodic payments, each of the present values (P) is computed by the compound discount formula $P = S(1 + i)^{-n}$.

$S = \$100$ in each case, $i = 4\%/4 = 1\%$ per quarter

The first payment of $100 is made at the end of the first quarter, which is one quarter after the beginning of the term of the annuity (April 1, 1976). The present value of the first payment is thus obtained by discounting the payment for one quarter, or one interest period. The present value of the second payment is obtained by discounting the payment for two periods, and so on.

This analysis is diagrammed on page 305.

The present value of the annuity, or the total of the present values of the 4 payments, $390.20, is computed as follows, using Table 14-2:

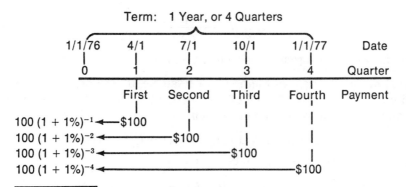

$390.20 (Total, Present Value of the Annuity)

$$100(1 + 1\%)^{-1} = 100(0.990099)$$
$$100(1 + 1\%)^{-2} = 100(0.980296)$$
$$100(1 + 1\%)^{-3} = 100(0.970590)$$
$$100(1 + 1\%)^{-4} = 100(0.960980)$$
$$\text{Total} = 100(3.901965) = \$390.1965, \text{ or } \$390.20$$

METHOD 2   When the present value of an annuity is considered as the single principal of the amount of the annuity, the principal (P) is obtained by using the compound discount formula $P = S(1 + i)^{-n}$.

$S =$ the amount of the annuity, S_n. Use formula (14–8) to find the value of S_n:

$R = \$100$ (per quarter), $i = 4\%/4 = 1\%$ (per quarter), and $n = 4$ (quarterly payments)

$S_n = Rs_{\overline{n}|i} = 100s_{\overline{4}|1\%} = 100(4.060401) = \$406.0401$. (Also see Example 15.)

Substitute these values in the compound discount formula:

$$P = S_n(1 + i)^{-n} = 406.0401(1 + 1\%)^{-4}$$
$$= 406.0401(0.960980) = \$390.20. \text{ (Table 14–2)}$$

The sum of compound discounts on the four payments is:

$$(\$100 \times 4) - \$390.20 = \$9.80$$

In general, let:

$A_n =$ the present value of an ordinary annuity
$R =$ the size of each regular payment (or periodic rent)
$i =$ interest rate per conversion period
$n =$ number of payments during the term of an annuity. (It is also the number of payment intervals, or the number of interest conversion periods.)

Then, the formula for the present value of an ordinary annuity is:

$$A_n = R \cdot \frac{1 - (1 + i)^{-n}}{i} \qquad (14\text{-}9)$$

The derivation of formula (14-9) follows.

Based on the diagram in Method 1 of Example 18, the present value of an annuity of R payable at the end of each period for 4 periods at interest rate i per period, denoted by A_4, can be written:

$$A_4 = R(1 + i)^{-4} + R(1 + i)^{-3} + R(1 + i)^{-2} + R(1 + i)^{-1}$$

Multiply both sides of the equation by $(1 + i)^4$:

$$A_4(1 + i)^4 = R + R(1 + i) + R(1 + i)^2 + R(1 + i)^3 = S_4$$
(See equation (1), page 300.)

Extend this idea by letting n = number of periods or payments. Then:

$$A_n(1 + i)^n = S_n$$

$$A_n = S_n(1 + i)^{-n} = R \cdot \frac{(1 + i)^n - 1}{i}(1 + i)^{-n}$$

$$A_n = R \cdot \frac{1 - (1 + i)^{-n}}{i}$$

which is formula (14-9).

The use of the formula is illustrated in Example 19.

EXAMPLE 19 A man wishes to borrow money from a bank which charges interest at 4% compounded quarterly. If he agrees to pay $100 at the end of each quarter for one year, how much money should he receive from the bank at the time of borrowing?

$R = \$100$ (per quarter), $i = \dfrac{4\%}{4} = 1\%$ (per quarter), and $n = 4$ (quarterly payments).

Substitute the values in formula (14-9):

$$A_n = 100 \cdot \frac{1 - (1 + 1\%)^{-4}}{1\%}$$

$$= 100 \cdot \frac{1 - 0.960980}{.01}$$

$$= 100(3.9020) = \$390.20 \text{ (Table 14-2)}$$

For convenience, let the simple symbol:

$$a_{\overline{n}|i} \text{ (read: } a \text{ angle } n \text{ at } i) = \frac{1 - (1 + i)^{-n}}{i}$$

The values of $a_{\overline{n}|i}$ for selected interest rates (i) and numbers of payments (n) are tabulated in Table 14-5 on pages 307 and 308. When Table 14-5 is employed, the computation of the present value of an annuity is simplified. Formula (14-9) now can be written:

Table 14-5 **Present Value of Annuity When Periodic Payment Is 1**

$$a_{\overline{n}|i} = \frac{1 - (1 + i)^{-n}}{i}$$

n	½%	1%	1½%	2%	3%	n
1	0.9950 25	0.9900 99	0.9852 22	0.9803 92	0.9708 74	1
2	1.9850 99	1.9703 95	1.9558 83	1.9415 61	1.9134 70	2
3	2.9702 48	2.9409 85	2.9122 00	2.8838 83	2.8286 11	3
4	3.9504 96	3.9019 66	3.8543 85	3.8077 29	3.7170 98	4
5	4.9258 66	4.8534 31	4.7826 45	4.7134 60	4.5797 07	5
6	5.8963 84	5.7954 76	5.6971 87	5.6014 31	5.4171 91	6
7	6.8620 74	6.7281 95	6.5982 14	6.4719 91	6.2302 83	7
8	7.8229 59	7.6516 78	7.4859 25	7.3254 81	7.0196 92	8
9	8.7790 64	8.5660 18	8.3605 17	8.1622 37	7.7861 09	9
10	9.7304 12	9.4713 05	9.2221 85	8.9825 85	8.5302 03	10
11	10.6770 27	10.3676 28	10.0711 18	9.7868 48	9.2526 24	11
12	11.6189 32	11.2550 77	10.9075 05	10.5753 41	9.9540 04	12
13	12.5561 51	12.1337 40	11.7315 32	11.3483 74	10.6349 55	13
14	13.4887 08	13.0037 03	12.5433 82	12.1062 49	11.2960 73	14
15	14.4166 25	13.8650 53	13.3432 33	12.8492 64	11.9379 35	15
16	15.3399 25	14.7178 74	14.1312 64	13.5777 09	12.5611 02	16
17	16.2586 32	15.5622 51	14.9076 49	14.2918 72	13.1661 18	17
18	17.1727 68	16.3982 69	15.6725 61	14.9920 31	13.7535 13	18
19	18.0823 56	17.2260 09	16.4261 68	15.6784 62	14.3237 99	19
20	18.9874 19	18.0455 53	17.1686 39	16.3514 33	14.8774 75	20
21	19.8879 79	18.8569 83	17.9001 37	17.0112 09	15.4150 24	21
22	20.7840 59	19.6603 79	18.6208 24	17.6580 48	15.9369 17	22
23	21.6756 81	20.4558 21	19.3308 61	18.2922 04	16.4436 08	23
24	22.5628 66	21.2433 87	20.0304 05	18.9139 26	16.9355 42	24
25	23.4456 38	22.0231 56	20.7196 11	19.5234 56	17.4131 48	25
26	24.3240 18	22.7952 04	21.3986 32	20.1210 44	17.8768 42	26
27	25.1980 28	23.5596 08	22.0676 17	20.7068 98	18.3270 31	27
28	26.0676 89	24.3164 43	22.7267 17	21.2812 72	18.7641 08	28
29	26.9330 24	25.0657 85	23.3760 76	21.8443 85	19.1884 55	29
30	27.7940 54	25.8077 08	24.0158 38	22.3964 56	19.6004 41	30
31	28.6508 00	26.5422 85	24.6461 46	22.9377 02	20.0004 28	31
32	29.5032 84	27.2695 89	25.2671 39	23.4683 35	20.3887 66	32
33	30.3515 26	27.9896 93	25.8789 54	23.9885 64	20.7657 92	33
34	31.1955 48	28.7026 66	26.4817 28	24.4985 92	21.1318 37	34
35	32.0353 71	29.4085 80	27.0755 95	24.9986 19	21.4872 20	35
36	32.8710 16	30.1075 05	27.6606 84	25.4888 42	21.8322 53	36
37	33.7025 04	30.7995 10	28.2371 27	25.9694 53	22.1672 35	37
38	34.5298 54	31.4846 63	28.8050 52	26.4406 41	22.4924 62	38
39	35.3530 89	32.1630 33	29.3645 83	26.9025 89	22.8082 15	39
40	36.1722 28	32.8346 86	29.9158 45	27.3554 79	23.1147 72	40
41	36.9872 91	33.4996 89	30.4589 61	27.7994 89	23.4124 00	41
42	37.7983 00	34.1581 08	30.9940 50	28.2347 94	23.7013 59	42
43	38.6052 74	34.8100 08	31.5212 32	28.6615 62	23.9819 02	43
44	39.4082 32	35.4554 54	32.0406 22	29.0799 63	24.2542 74	44
45	40.2071 96	36.0945 08	32.5523 37	29.4901 60	24.5187 13	45
46	41.0021 85	36.7272 36	33.0564 90	29.8923 14	24.7754 49	46
47	41.7932 19	37.3536 99	33.5531 92	30.2865 82	25.0247 08	47
48	42.5803 18	37.9739 59	34.0425 54	30.6731 20	25.2667 07	48
49	43.3635 00	38.5880 79	34.5246 83	31.0520 78	25.5016 57	49
50	44.1427 86	39.1961 18	34.9996 88	31.4236 06	25.7297 64	50

Table 14–5 Present Value of Annuity When Periodic Payment is 1

$$a_{\overline{n}|i} = \frac{1 - (1 + i)^{-n}}{i}$$

n	4%	5%	6%	7%	8%	n
1	0.9615 38	0.9523 81	0.9433 96	0.9345 79	0.9259 26	1
2	1.8860 95	1.8594 10	1.8333 93	1.8080 18	1.7832 65	2
3	2.7750 91	2.7232 48	2.6730 12	2.6243 16	2.5770 97	3
4	3.6298 95	3.5459 51	3.4651 06	3.3872 11	3.3121 27	4
5	4.4518 22	4.3294 77	4.2123 64	4.1001 97	3.9927 10	5
6	5.2421 37	5.0756 92	4.9173 24	4.7665 40	4.6228 80	6
7	6.0020 55	5.7863 73	5.5823 81	5.3892 89	5.2063 70	7
8	6.7327 45	6.4632 13	6.2097 94	5.9712 99	5.7466 39	8
9	7.4353 32	7.1078 22	6.8016 92	6.5152 32	6.2468 88	9
10	8.1108 96	7.7217 35	7.3600 87	7.0235 82	6.7100 81	10
11	8.7604 77	8.3064 14	7.8868 75	7.4986 74	7.1389 64	11
12	9.3850 74	8.8632 52	8.3838 44	7.9426 86	7.5360 78	12
13	9.9856 48	9.3935 73	8.8526 83	8.3576 51	7.9037 76	13
14	10.5631 23	9.8986 41	9.2949 84	8.7454 68	8.2442 37	14
15	11.1183 87	10.3796 58	9.7122 49	9.1079 14	8.5594 79	15
16	11.6522 96	10.8377 70	10.1058 95	9.4466 49	8.8513 69	16
17	12.1656 69	11.2740 66	10.4772 60	9.7632 23	9.1216 38	17
18	12.6592 97	11.6895 87	10.8276 03	10.0590 87	9.3718 87	18
19	13.1339 39	12.0853 21	11.1581 16	10.3355 95	9.6035 99	19
20	13.5903 26	12.4622 10	11.4699 21	10.5940 14	9.8181 47	20
21	14.0291 60	12.8211 53	11.7640 77	10.8355 27	10.0168 03	21
22	14.4511 15	13.1630 03	12.0415 82	11.0612 41	10.2007 44	22
23	14.8568 42	13.4885 74	12.3033 79	11.2721 87	10.3710 59	23
24	15.2469 63	13.7986 42	12.5503 58	11.4693 34	10.5287 58	24
25	15.6220 80	14.0939 45	12.7833 56	11.6535 83	10.6747 76	25
26	15.9827 69	14.3751 85	13.0031 66	11.8257 79	10.8099 78	26
27	16.3295 86	14.6430 34	13.2105 34	11.9867 09	10.9351 65	27
28	16.6630 63	14.8981 27	13.4061 64	12.1371 11	11.0510 78	28
29	16.9837 15	15.1410 74	13.5907 21	12.2776 74	11.1584 06	29
30	17.2920 33	15.3724 51	13.7648 31	12.4090 41	11.2577 83	30
31	17.5884 94	15.5928 11	13.9290 86	12.5318 14	11.3497 99	31
32	17.8735 52	15.8026 77	14.0840 43	12.6465 55	11.4349 99	32
33	18.1476 46	16.0025 49	14.2302 30	12.7537 90	11.5138 88	33
34	18.4111 98	16.1929 04	14.3681 41	12.8540 09	11.5869 34	34
35	18.6646 13	16.3741 94	14.4982 46	12.9476 72	11.6545 68	35
36	18.9082 82	16.5468 52	14.6209 87	13.0352 08	11.7171 93	36
37	19.1425 79	16.7112 87	14.7367 80	13.1170 17	11.7751 79	37
38	19.3678 64	16.8678 93	14.8460 19	13.1934 73	11.8288 69	38
39	19.5844 85	17.0170 41	14.9490 75	13.2649 28	11.8785 82	39
40	19.7927 74	17.1590 86	15.0462 97	13.3317 09	11.9246 13	40
41	19.9930 52	17.2943 68	15.1380 16	13.3941 20	11.9672 35	41
42	20.1856 27	17.4232 08	15.2245 43	13.4524 49	12.0066 99	42
43	20.3707 95	17.5459 12	15.3061 73	13.5069 62	12.0432 40	43
44	20.5488 41	17.6627 73	15.3831 82	13.5579 08	12.0770 74	44
45	20.7200 40	17.7740 70	15.4558 32	13.6055 22	12.1084 02	45
46	20.8846 54	17.8800 67	15.5243 70	13.6500 20	12.1374 09	46
47	21.0429 36	17.9810 16	15.5890 28	13.6916 08	12.1642 67	47
48	21.1951 31	18.0771 58	15.6500 27	13.7304 74	12.1891 36	48
49	21.3414 72	18.1687 22	15.7075 72	13.7667 99	12.2121 63	49
50	21.4821 85	18.2559 25	15.7618 61	13.8007 46	12.2334 85	50

$$A_n = Ra_{\overline{n}|i} \qquad\qquad (14\text{-}9)$$

Example 19 can also be computed in the following manner:

$$A_n = Ra_{\overline{n}|i} = 100a_{\overline{4}|1\%} = 100(3.901966) = \$390.20 \text{ (Table 14-5)}$$

EXAMPLE 20   What is the cash value of a car that can be bought for $500 down and $150 a month for 3 years if money is worth 12% compounded monthly?

After the down payment is made, the first of the regular payments will be made at the end of the month following the date of purchase. Thus, this is an ordinary annuity problem.

$R = \$150$ (per month), $i = \dfrac{12\%}{12} = 1\%$ (per month), $n = 3 \times 12 = 36$ (monthly payments)

Substitute the above values in formula (14-9):

$$A_n = 150a_{\overline{36}|1\%} = 150(30.107505) = \$4,516.13 \text{ (Table 14-5)}$$

Cash value of the car $= 500 + 4,516.13$
$\qquad\qquad\qquad\qquad = \$5,016.13$

EXAMPLE 21   A man purchased a house for $60,000. He made a down payment of $10,000 and agreed to make equal payments at the end of each quarter for 10 years. If the interest is charged at 8% compounded quarterly, what is the size of the quarterly payment?

The debt to be paid $= A_n = 60,000 - 10,000 = \$50,000, i = 8\%/4 = 2\%$ (per quarter), $n = 10 \times 4 = 40$ (quarterly payments).

Substitute these values in formula (14-9):

$$50,000 = Ra_{\overline{40}|2\%} = R(27.355479). \text{ (Table 14-5)}$$

Solve for R:

$$R = \frac{50,000}{27.355479} = \$1,827.79$$

which is the size of each quarterly payment.

14.4   ANNUITY DUE

This section and Section 14.5 will introduce two other types of annuities: annuity due and deferred annuity. The computations of the two types of annuities may conveniently be carried out if the ordinary annuity formulas are used.

When the periodic payments are made at the beginning of each payment interval, the annuity is known as an *annuity due*. The term of an annuity due begins on the date of the first payment and ends one payment interval after the last payment is made.

A. Finding the Amount of An Annuity Due

EXAMPLE 22 Mary made four quarterly deposits of $100 each in a bank that pays an interest rate of 4% compounded quarterly. Assume that she made the first deposit on January 1, 1977. How much will Mary have in her account on January 1, 1978?

$R = \$100$ (per quarter), $i = 4\%/4 = 1\%$ (per quarter), and $n = 4$ (quarterly payments)

First, find the amount of an ordinary annuity of $100 payable at the end of each quarter for $(n + 1)$ quarters at 4% compounded quarterly: (Formula 14–8)

$$S_{4+1} = 100s_{\overline{4+1}|1\%} = 100s_{\overline{5}|1\%}$$
$$= 100(5.101005) = \$510.10$$

Next, subtract the additional payment from the amount obtained:

$S_4(\text{due}) = 510.10 - 100.00 = \$410.10$ (Answer)

The entire computation in Example 22 is diagrammed as follows:

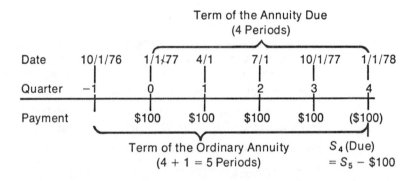

The diagram shows that if an additional payment ($100) is made at the end of the fourth quarter (1/1/78), the 5 (or $4+1$) payments form an ordinary annuity. The amount of the ordinary annuity is S_5. Since the 5th payment was actually not made, the payment must be subtracted. Thus, the amount of the annuity due of four payments is:

$$S_4(\text{due}) = S_{4+1} - 100 = 100s_{\overline{4+1}|1\%} - 100$$

In general, let $S_n(\text{due}) =$ the amount (final value) of an annuity due of R payable at the beginning of each period for n periods at interest rate i per period

Then:

$$S_n(\text{due}) = Rs_{\overline{n+1}|i} - R$$

or:

$$S_n(\text{due}) = R(s_{\overline{n+1}|i} - 1) \qquad (14\text{--}10)$$

EXAMPLE 23 What is the amount of an annuity due for one year if each payment is $100 payable at the beginning of each quarter and the interest rate is 4% compounded quarterly?

$R = \$100$ (per quarter), $i = 4\%/4 = 1\%$ (per quarter), and $n = 4$ (quarterly payments). Substitute the values in formula (14–10):

$$S_n(\text{due}) = 100s_{\overline{4+1}|1\%} - 100 = 100s_{\overline{5}|1\%} - 100$$
$$= 100(5.101005) - 100$$
$$= \$410.10$$

This answer is the same as that of Example 22.

B. Finding the Present Value of An Annuity Due

EXAMPLE 24 Mary wishes to know the selling price of a refrigerator that can be bought for $100 a quarter for 4 quarters with the first payment due on the purchase date, January 1, 1977. The interest rate is 4% compounded quarterly.

$R = \$100$ (per quarter), $i = 4\%/4 = 1\%$ (per quarter), and $n = 4$ (quarterly payments)

First, find the present value of the ordinary annuity of $(n - 1)$ quarterly payments: (Formula 14–9)

$$A_{4-1} = 100a_{\overline{4-1}|1\%} = 100a_{\overline{3}|1\%} = 100(2.940985)$$
$$= \$294.10$$

Next add the excluded payment to the present value obtained:

$$A_4(\text{due}) = 294.10 + 100 = \$394.10 \quad (\text{Answer})$$

The entire computation in Example 24 is diagrammed as follows:

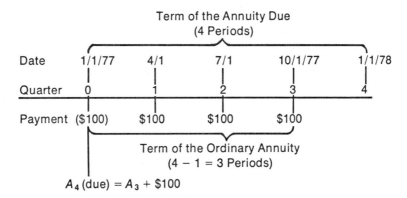

The diagram shows that if the $100 payment made at the beginning of the first quarter (1/1/77) is excluded, the remaining three (or $4 - 1$) payments form an ordinary annuity. The present value of the ordinary annuity is A_3. Since the first payment was actually made, the payment must be added to A_3. Thus, the present value of the annuity due of four payments is:

$$A_4(\text{due}) = A_{4-1} + 100 = 100a_{\overline{4-1}|1\%} + 100$$

In general, let $A_n(\text{due}) =$ the present value of an annuity due of R payable at the beginning of each period for n periods at interest rate i per period.

Then:

$$A_n(\text{due}) = Ra_{\overline{n-1}|i} + R$$

or:

$$A_n(\text{due}) = R(a_{\overline{n-1}|i} + 1) \tag{14-11}$$

EXAMPLE 25 What is the present value of an annuity due for one year if each payment is $100 payable at the beginning of each quarter and the interest rate is 4% compounded quarterly?

$R = \$100$ (per quarter), $i = 4\%/4 = 1\%$ (per quarter), and $n = 4$ (quarterly payments). Substitute the values in formula (14-11):

$$A_n(\text{due}) = 100a_{\overline{4-1}|1\%} + 100$$
$$= 100(2.940985) + 100$$
$$= \$394.10$$

This answer is the same as that of Example 24.

C. Finding the Size of Each Payment of An Annuity Due

EXAMPLE 26 A man wishes to have $5,000 six years from now. How much must he invest at the beginning of each semiannual period if the first payment starts now and the interest is 10% compounded semiannually?

The amount is given, or $S_n(\text{due}) = \$5,000$

$i = 10\%/2 = 5\%$ (per semiannual period)
$n = 6 \times 2 = 12$ (semiannual periods)
$R = ?$ (per semiannual period)

Substitute the values in formula (14-10):

$$5,000 = R(s_{\overline{12+1}|5\%} - 1) = R(s_{\overline{13}|5\%} - 1)$$
$$= R(17.712983 - 1)$$
$$R = \frac{5,000}{16.712983} = \$299.17$$

EXAMPLE 27 A boat which sells for $4,000 can be bought under the payment plan of 30 equal monthly payments starting now. If money is worth 6% compounded monthly, what is the size of each payment?

The present value is given, or:

$A_n(\text{due}) = \$4,000$, $i = 6\%/12 = \frac{1}{2}\%$ (per month)
$n = 30$ (monthly payments), $R = ?$ (per month)

Substitute the values in formula (14-11):

$$4{,}000 = R(a_{\overline{30-1}|}\tfrac{1}{2}\% + 1) = R(a_{\overline{29}|}\tfrac{1}{2}\% + 1)$$
$$= R(26.933024 + 1)$$
$$R = \frac{4{,}000}{27.933024} = \$143.20$$

★ 14.5 DEFERRED ANNUITY

As discussed in the previous sections, the first payment of an ordinary annuity is always made at the end of the first payment period. When the first payment is made on a future date, more than one payment period from now, the ordinary annuity is then called a *deferred annuity*. The amount of a deferred annuity is the final value at the end of the term of the annuity. Thus, it is the same as the amount of the ordinary annuity. However, the present value of a deferred annuity is the value for now, not at the beginning of the term of the ordinary annuity.

EXAMPLE 28   On April 1, 1976, Mary plans to make four quarterly deposits of $100 each in a bank which pays an interest rate of 4% compounded quarterly. Assume that she will make the first deposit on January 1, 1977 and the last deposit on October 1, 1977. She wishes to know (a) the amount in her account on October 1, 1977 after the deposit, and (b) the single value on April 1, 1976 which is equivalent to the four deposits.

SOLUTION   The computation of Example 28 is diagrammed as follows:

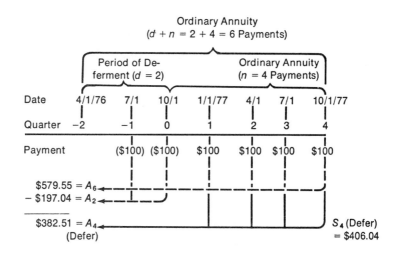

(a) $R = \$100$ (per quarter), $i = \dfrac{4\%}{4} = 1\%$ (per quarter), $n = 4$ (quarterly payments)

The amount of the deferred annuity is:

$$S_n(\text{defer}) = S_n = 100 s_{\overline{4}|1\%} = 100(4.060401)$$
$$= \$406.04. \qquad \text{(Formula 14–8)}$$

Thus, the amount in her account on October 1, 1977 is $406.04.

(b) Consider two payments were made during the period of deferment (from April 1, 1976 to October 1, 1976.) The present value of the annuity consisting of the two payments is:

$$A_2 = 100a_{\overline{2}|1\%} = 100(1.970395) = \$197.04$$

The present value of the annuity consisting of the six payments $(2 + 4$ for the entire period from April 1, 1976 to October 1, 1977) is:

$$A_{2+4} = A_6 = 100a_{\overline{6}|1\%} = 100(5.795476)$$
$$= \$579.55$$

The present value of the deferred annuity consisting of four payments (from October 1, 1976 to October 1, 1977) is:

$$A_4(\text{defer}) = A_6 - A_2 = 579.55 - 197.04$$
$$= \$382.51$$

Thus, the single value on April 1, 1976, $382.51, is equivalent to the four deposits.

Check:　The amount of the four deposits is $406.04 on October 1, 1977. (See the answer in (a).)

The compound amount of the single value for 6 quarters (from April 1, 1976 to October 1, 1977) at 1% per quarter is the same, or:

$$S = 382.51(1 + 1\%)^6$$
$$= 382.51(1.061520) = \$406.04$$

In general, let S_n (defer) = the amount of a deferred annuity. Then:

$$S_n(\text{defer}) = S_n = Rs_{n|i} \qquad (14\text{-}12)$$

Let A_n (defer) = the present value of a deferred annuity, and
　　　　　d = the number of the deferred payment intervals or payments.

Then:

$$A_n(\text{defer}) = Ra_{\overline{d+n}|i} - Ra_{\overline{d}|i}$$

Or:

$$A_n(\text{defer}) = R(a_{\overline{d+n}|i} - a_{\overline{d}|i}) \qquad (14\text{-}13)$$

When formula (14–13) is used, Example 28(b) may be computed in the following manner:

$$A_n(\text{defer}) = 100(a_{\overline{2+4}|1\%} - a_{\overline{2}|1\%}) = 100(a_{\overline{6}|1\%} - a_{\overline{2}|1\%})$$
$$= 100(5.795476 - 1.970395)$$
$$= 100(3.825081) = \$382.51$$

EXAMPLE 29　A 45-year-old man wishes to receive $1,000 per year for 25 years with the first receipt at age 65. Assume that he can invest his money at 6% effective. How much should he invest now in order to get the annual receipts?

$R = \$1,000$ (per year), $i = 6\%$ (per year), $d = 19 \ (= 64 - 45$, deferred the annual receipts from age 45 to age 64), and $n = 25$ (yearly receipts).

Substitute the above values in formula (14-13):

$$A_n(\text{defer}) = 1.000(a_{\overline{19+25}|6\%} - a_{\overline{19}|6\%})$$
$$= 1.000(a_{\overline{44}|6\%} - a_{\overline{19}|6\%})$$
$$= 1.000(15.383182 - 11.158116)$$
$$= \$4.225.07 \quad (\text{Table } 14\text{-}5)$$

Note: Example 29 can also be computed in the following way:

The present value (at age 64, one year before receiving the first $1,000) of the annuity of 25 annual receipts of $1.000 each at 6% effective is:

$$A_n = 1.000a_{\overline{25}|6\%} = 1.000(12.783356) = \$12.783.356 \quad (\text{Table } 14\text{-}5)$$

The present value (at age 45, or 19 years before reaching age 64) of the single amount at 6% effective is:

$$P = 12.783.356(1 + 6\%)^{-19} = 12.783.356(.330513)$$
$$= \$4.225.07 \quad (\text{Table } 14\text{-}3)$$

Here we have used two tables (Tables 14-3 and 14-5). The computation in Example 29 used only one table (Table 14-5).

14.6 SUMMARY OF FORMULAS

Application	Formula	Formula Number	Reference Page
Simple Interest	$I = Pin$	14-1	284
	$S = P(1 + in) = P + I$	14-2	285
Compound Interest	$S = P(1 + i)^n$	14-3	287
	$P = S(1 + i)^{-n}$	14-4	290
	$f = (1 + i)^m - 1$	14-5	294
	$f = e^j - 1$	14-6	296
	$S = P(e^{jn})$	14-7	296
Ordinary Annuity	$S_n = R \cdot \dfrac{(1 + i)^n - 1}{i}$ $= Rs_{\overline{n}|i}$	14-8	300
	$A_n = R \cdot \dfrac{1 - (1 + i)^{-n}}{i}$ $= Ra_{\overline{n}|i}$	14-9	306
Annuity Due	$S_n(\text{due}) = Rs_{\overline{n+1}|i} - R$	14-10	310
	$A_n(\text{due}) = Ra_{\overline{n-1}|i} + R$	14-11	312
Deferred Annuity	$S_n(\text{defer}) = S_n = Rs_{\overline{n}|i}$	14-12	314
	$A_n(\text{defer}) = Ra_{\overline{d+n}|i} - Ra_{\overline{d}|i}$	14-13	314

EXERCISE 14-2   Reference: Sections 14.3–14.5

A. *Ordinary Annuity*

1. Find the amount of an annuity if the size of each payment is $1,000 payable at the end of each month for 2 years at an interest rate of 12% compounded monthly.

2. What is the amount of an annuity if the payment is $500 payable at the end of every 6 months for 10 years at 8% compounded semiannually?

3. (a) If $200 is deposited at the end of each quarter for 5 years in a bank that pays 6% interest compounded quarterly, what will be the final value at the end of 5 years? (b) What is the total interest at the end of 5 years?

4. (a) If $3,000 is deposited at the end of each year for four years in a fund that earns 5% interest compounded annually, what will be the value of the fund at the end of four years? (b) What is the total interest at the end of four years?

5. Refer to problem 3. (a) How much will the amount be 3 years after the last deposit is made? (b) What is the total interest at the end of 8 years?

6. Refer to problem 4. (a) How much will the amount be 6 years after the last deposit is made? (b) What is the total interest at the end of 10 years?

7. What is the present value of an annuity if the payment is $150 payable at the end of each month for 3 years and the interest rate is 6% compounded monthly?

8. Find the present value of an annuity if the payment is $300 payable at the end of each quarter for 2 years and the interest rate is 12% compounded quarterly.

9. A man wishes to borrow money from a bank which charges interest at 8% compounded semiannually. If he agrees to pay $400 at the end of each semiannual period for 3 years, how much money will he receive from the bank at the time of borrowing?

10. What is the cash price of a truck that can be bought for $1,000 down and $200 a month for $2\frac{1}{2}$ years if money is worth 6% compounded monthly?

11. What are the amount and the present value of an annuity of $100 payable at the end of each month for 3 years if the interest rate is 12% compounded monthly?

12. Find the amount and the present value of an annuity of $250 payable at the end of each quarter for 5 years if the interest rate is 8% compounded quarterly.

13. The price of a house is $80,000. The buyer made a down payment of $20,000. The balance is to be paid in monthly installments for 4 years. If the interest rate charged is 6% compounded monthly, how much should the buyer pay each month?

14. A debt of $10,000 is to be discharged by equal payments at the end of every 6 months for 3 years. If the interest charged is 10% compounded semiannually, find the size of the periodic payment.

15. Refer to problem 13. Find the unpaid balance after the 12th monthly payment was made.

16. Refer to problem 14. Find the unpaid balance after the second payment.

B. *Annuity Due*

17. Find the amount of an annuity due if the size of each payment is $1,000 payable at the beginning of each month for 2 years at an interest rate of 12% compounded monthly.

18. What is the amount of an annuity due if the payment is $500 payable at the beginning of every 6 months for 10 years at 8% compounded semiannually?

19. (a) If $200 is deposited at the beginning of each quarter for 5 years in a bank that pays 6% interest compounded quarterly, what will be the final value at the end of 5 years? (b) What is the total interest at the end of 5 years?

20. (a) If $3,000 is deposited at the beginning of each year for four years in a fund that earns 5% interest compounded annually, what will be the value of the fund at the end of 4 years? (b) What is the total interest at the end of 4 years?

21. What is the present value of an annuity due if the payment is $150 payable at the beginning of each month for 3 years and the interest rate is 6% compounded monthly?

22. Find the present value of an annuity due if the payment is $300 payable at the beginning of each quarter for 2 years and the interest rate is 12% compounded quarterly.

23. A lady wishes to know the selling price of a mink coat that can be bought for $400 every semiannual period for 6 periods with the first payment due on the purchase date. The interest rate charged is 8% compounded semiannually.

24. What is the cash price of a truck that can be bought for $200 down and $200 a month thereafter for 30 monthly payments if money is worth 6% compounded monthly?

25. A company plans to have $10,000 3 years from now for paying a debt due then. How much must the company invest now and thereafter for 11 equal additional quarterly deposits in a fund which can earn an interest at 8% compounded quarterly?

26. A car can be purchased under the payment plan that requires 36 equal monthly payments with the first payment to be made on the date of purchase. The cash price of the car is $6,500. What is the size of each payment if the interest rate is 12% compounded monthly?

C. *Deferred Annuity*

27. Find the amount and the present value of an annuity of $1,000 payable at the end of each month for 2 years or 24 payments. The first payment is due at the end of 1 year. The interest rate is 12% compounded monthly.

28. Find the amount and the present value of 20 semiannual payments of $500 each. The first payment is due in 3 years. The interest rate is 8% compounded semiannually.

29. John purchased a store for $80,000 on July 1, 1976. He paid $20,000 cash and agreed to pay the balance plus interest at 6% compounded monthly in 36 monthly payments, with the first payment due on July 1, 1977. What is the size of each payment?

30. Steve bought a farm on May 1, 1977. He agreed to pay for it in 20 quarterly payments of $1,000 each. If the first payment is to be made on August 1, 1978, and the interest rate is 8% compounded quarterly, find the cash price of the farm.

Part Four   Calculus and Related Topics

*Chapter

15 Sequences and Series

A *sequence* is a set of numbers arranged in such a manner that there is a first number, a second number, a third number, and so on. The successive numbers are called the *terms* of the sequence. A sequence is said to be *finite* if there is an ending or final term, and *infinite* if the terms are unending. Thus, the sequence:

$$1, 2, 3, 4$$

is a *finite sequence* composed of four terms, whereas the unending sequence:

$$1, 2, 3, 4, \ldots$$

is an *infinite sequence*.

A *series* is an indicated sum of the numbers of a sequence. Thus, the indicated sum of the four terms:

$$1 + 2 + 3 + 4$$

is a *finite series* and that of the unlimited terms:

$$1 + 2 + 3 + 4 + \ldots$$

is an *infinite series*.

A sequence is said to be fully defined when one has the information needed to find any required term of the sequence. This information is usually in the form of a formula representing the nth (first, second, etc.) term, or a statement which indicates how each term can be determined, or a description of a specific pattern observed in the given terms. The nth term, also called the *general term*, represents the first term when $n = 1$; the second term when $n = 2, \ldots$; the kth term when $n = k$; and so on.

The terms of a sequence are generally increased or decreased in one of three ways: (1) the arithmetic progression, (2) the geometric progression, and (3) other specified patterns. Methods of finding the value of the nth term and the sum of the first n terms of a sequence formed by arithmetic or geometric progressions are presented in Sections 15.1 and 15.2. The methods for finding these values in sequences of other specified patterns are shown in Sections 15.3 and 15.4.

15.1 ARITHMETIC PROGRESSION (AP)

An *arithmetic progression* is a sequence of numbers in which each successive term is obtained by adding a fixed number called the *common difference* to the preceding term. Thus, the sequence 1, 2, 3, 4 is an arithmetic progression having a common difference of 1; the sequence 1, 4, 7, 10, 13 is an arithmetic progression having a common difference of 3.

A. The *n*th Term of an Arithmetic Progression

Based on the definition of an arithmetic progression, if the first number of a sequence of numbers is 1 and the common difference is 3, we may obtain the terms of the sequence in the following manner:

$$
\begin{aligned}
&\text{1st term} = 1 \\
&\text{2d term} = 1 + 3 = 4 \\
&\text{3d term} = 1 + 2 \cdot 3 = 7 \\
&\text{4th term} = 1 + 3 \cdot 3 = 10, \text{ or: } \quad 1 + (4 - 1) \cdot 3 = 10 \\
&\text{5th term} = 1 + 4 \cdot 3 = 13, \text{ or: } \quad 1 + (5 - 1) \cdot 3 = 13 \\
&\qquad \ldots = \ldots \\
&\text{the } n\text{th term} = \qquad\qquad\qquad 1 + (n - 1)3
\end{aligned}
$$

In general, let: a = the first term of an AP
d = the common difference
L = the *n*th term

Then:

$$
\begin{aligned}
&\text{1st term} = a \\
&\text{2d term} = a + d \\
&\text{3d term} = a + 2d \\
&\text{4th term} = a + 3d, \text{ or: } \quad a + (4 - 1)d \\
&\text{5th term} = a + 4d, \text{ or: } \quad a + (5 - 1)d \\
&\qquad \ldots = \ldots \\
&\text{the } n\text{th term is:}
\end{aligned}
$$

$$L = a + (n - 1)d$$

B. The Sum of an Arithmetic Progression

Let S_n = the sum of the first n terms of an arithmetic progression.

The indicated sum of the n terms expressed above may be written in both direct (1) and reverse (2) orders as follows:

$$S_n = a + (a + d) + (a + 2d) + (a + 3d) + \ldots + L \tag{1}$$
$$S_n = L + (L - d) + (L - 2d) + (L - 3d) + \ldots + a \tag{2}$$

Add equations (1) and (2):

$$
\begin{aligned}
2S_n &= (a + L) + (a + L) + (a + L) + (a + L) + \ldots + (a + L) \\
&= n(a + L)
\end{aligned}
$$

Thus:

$$S_n = \frac{n}{2}(a + L)$$

Based on the preceding L and S_n formulas, if any three of the five quantities, a, d, n, L, and S_n are given, the other two unknowns may be computed.

EXAMPLE 1 Find (a) the 20th term and (b) the sum of AP 1, 4, 7, 10, ... to 20 terms.

$n = 20$, $a = 1$, $d = 4 - 1 = 3$. Apply the formulas:

(a) $L = a + (n - 1)d = 1 + (20 - 1)3 = 58$

(b) $S_n = \frac{n}{2}(a + L) = \frac{20}{2}(1 + 58) = 590$

EXAMPLE 2 A radio manufacturing company started production in January of this year. The quantity produced each month was increased at a constant number based on the production of the preceding month. During December the company produced 640 radios and the total production of this year was 5,040 radios. How many radios were produced each month this year?

This is an arithmetic progression problem since the production in each month was increased at a constant number.

$S_n = 5{,}040$, $L = 640$, and $n = 12$ (months)

Apply the S_n formula to find a, the production of the first month or January:

$$S_n = \frac{n}{2}(a + L), \quad 5{,}040 = \frac{12}{2}(a + 640)$$
$$5{,}040 = 6a + 6(640)$$
$$6a = 5{,}040 - 3{,}840 = 1{,}200$$
$$a = 1{,}200/6 = 200$$

Apply the L formula to find d, the constant number of radios increased each month:

$$L = a + (n - 1)d, \quad 640 = 200 + (12 - 1)d$$
$$640 = 200 + 11d$$
$$11d = 640 - 200 = 440$$
$$d = 440/11 = 40$$

Thus, the number of radios produced each month is:

January	200		July	$400 + 40 = 440$
February	$200 + 40 = 240$		August	$440 + 40 = 480$
March	$240 + 40 = 280$		September	$480 + 40 = 520$
April	$280 + 40 = 320$		October	$520 + 40 = 560$
May	$320 + 40 = 360$		November	$560 + 40 = 600$
June	$360 + 40 = 400$		December	$600 + 40 = 640$

The terms between any two given terms of an AP are also called the arithmetic means. To insert a given number of arithmetic means between two terms, apply the formula $L = a + (n - 1)d$, where $n =$ the number of terms to be inserted plus two.

EXAMPLE 3   Insert six arithmetic means between 5 and 7.8.

$$a = 5, L = 7.8, n = 6 + 2 = 8$$

Apply the formula:

$$7.8 = 5 + (8 - 1)d$$
$$7d = 7.8 - 5 = 2.8$$
$$d = \frac{2.8}{7} = .4$$

Thus, the required means are:

$5 + .4 = 5.4$	$6.2 + .4 = 6.6$
$5.4 + .4 = 5.8$	$6.6 + .4 = 7.0$
$5.8 + .4 = 6.2$	$7.0 + .4 = 7.4$

EXERCISE 15–1   Reference: Section 15.1

A. *Find L and S_n for each arithmetic progression:*

  1. 2, 4, 6, . . . to 10 terms   **5.** .1, .4, .7, . . . to 7 terms
  2. 6, 4, 2, . . . to 10 terms   **6.** $-3, -7, -11,$ . . . to 9 terms
  3. $\frac{1}{5}, \frac{2}{5}, \frac{3}{5},$ . . . to 8 terms   **7.** $2a, 2a + c, 2a + 2c,$ . . . to 6 terms
  4. 1, 2, 3, . . . to 100 terms   **8.** $3f + 2g, 3f, 3f - 2g,$ . . . to n terms

B. *Three of the five elements (a, d, n, L, S_n) of an AP are given below for each problem. Find the other two elements for each.*

  9. $a = 3, d = 2, n = 6$ **13.** $a = \frac{1}{4}, n = 9, S_n = 20\frac{1}{4}$
 10. $a = -2, d = -3, L = -23$ **14.** $a = 1, L = 53, S_n = 378$
 11. $a = 4, d = .5, S_n = 62.5$ **15.** $d = \frac{2}{3}, n = 12, L = 7\frac{2}{3}$
 12. $a = -1, n = 12, L = -34$ **16.** $n = 9, L = -12, S_n = 0$

C. *Statement Problems:*

 17. Insert four arithmetic means between 2 and 17.
 18. Insert three arithmetic means between 1 and 2.2.
 19. Insert five arithmetic means between 12 and 8.4.
 20. Insert six arithmetic means between 23 and 9.
 21. A retail store has sales of $10,000 the first month, $12,000 the second month, and continues to increase its sales $2,000 per month. Find the total sales at the end of 20 months.
 22. A machine which was purchased for $5,500 has an estimated useful life of 10 years and then will be worthless. Use the sum of the years-digits method to find the depreciation charge for the first year. Hint: Depreciation rate for the first year = (10, the number of remaining years of life) ÷ (the sum of the digits that represent the years of life, or $1 + 2 + \ldots + 10$.)

15.2 GEOMETRIC PROGRESSION (GP)

A *geometric progression* is a sequence of numbers in which each successive term is obtained by multiplying the preceding term by a constant factor called the *common ratio*. Thus, the sequence 1, 2, 4, 8, is a geometric progression with a common ratio of 2; and the sequence 1, 3, 9, 27, 81, is a geometric progression with a common ratio of 3.

A. The *n*th Term of a Geometric Progression

Based on the definition of a geometric progression, if the first number of a sequence of numbers is 1 and the common ratio is 3, we may obtain the terms of the sequence in the following manner:

$$1\text{st term} = 1$$
$$2\text{d term} = 1 \cdot 3$$
$$3\text{d term} = 1 \cdot 3^2 = 9$$
$$4\text{th term} = 1 \cdot 3^3 = 27, \text{ or: } 1 \cdot 3^{4-1}$$
$$5\text{th term} = 1 \cdot 3^4 = 81, \text{ or: } 1 \cdot 3^{5-1}$$
$$\ldots = \ldots$$
$$\text{the } n\text{th term} = 1 \cdot 3^{n-1}$$

In general, let $a =$ the first term of a GP
$r =$ the common ratio
$L =$ the nth term

Then:

$$1\text{st term} = a$$
$$2\text{d term} = ar$$
$$3\text{d term} = ar^2$$
$$4\text{th term} = ar^3, \text{ or: } ar^{4-1}$$
$$5\text{th term} = ar^4, \text{ or: } ar^{5-1}$$
$$\ldots = \ldots$$
$$n\text{th term is:}$$

$$L = ar^{n-1}$$

B. The Sum of a Geometric Progression

Let $S_n =$ the sum of the first n terms of a geometric progression.

The indicated sum of the n terms expressed above is written as equation (1). Also, multiply equation (1) by the common ratio r as shown in equation (2).

$$S_n = a + ar + ar^2 + ar^3 + \ldots + ar^{n-2} + ar^{n-1} \tag{1}$$
$$rS_n = ar + ar^2 + ar^3 + \ldots + ar^{n-2} + ar^{n-1} + ar^n \tag{2}$$

Subtract (2) from (1). The result is:

$$S_n - rS_n = a - ar^n$$

Factor:

$$S_n(1 - r) = a(1 - r^n)$$

Solve for S_n:

$$S_n = \frac{a(1 - r^n)}{1 - r}$$

Based on these L and S_n formulas, if any three of the five quantities, $a, r, n, L,$ and S_n are given, the other two unknowns may be computed.

EXAMPLE 4   Find (a) the 7th term and (b) the sum of GP 4, 12, 36, . . . to 7 terms.

$n = 7, a = 4, r = \dfrac{12}{4} = 3$. Apply the formulas:

(a) $L = ar^{n-1} = 4 \cdot 3^{7-1} = 4 \cdot 3^6 = 2{,}916$

(b) $S_n = \dfrac{a(1 - r^n)}{1 - r} = \dfrac{4(1 - 3^7)}{1 - 3} = \dfrac{4(-2{,}186)}{-2}$

$\quad = 4{,}372$

EXAMPLE 5   The amount of sales of the Norfolk Hardware Store is estimated to increase 10% each year. The amount of sales of the first year was $100,000. Estimate the sales of (a) the fifth year, and (b) a period of 5 years.

$a = \$100{,}000$ (sales of the first year)
$r = 100\% + 10\% = 1.1$ (rate of sales based on the sales of each previous year)
$n = 5$ (years from the first to the fifth)

(a) $L = 100{,}000(1.1)^{5-1} = 100{,}000(1.4641)$
$\quad = \$146{,}410$ (sales of the fifth year)

(b) $S_n = \dfrac{100{,}000(1 - 1.1^5)}{1 - 1.1} = \dfrac{100{,}000(1 - 1.61051)}{-.1}$

$\quad = \dfrac{100{,}000(-.61051)}{-.1} = \$610{,}510$ (sales of 5 years)

EXAMPLE 6   Given: $r = -\dfrac{1}{2}, n = 5, L = \dfrac{1}{2}$. Find a and S_n.

(a) Substitute the given values in formula:

$L = ar^{n-1}$

$\dfrac{1}{2} = a\left(-\dfrac{1}{2}\right)^{5-1}, \quad \dfrac{1}{2} = a\left(\dfrac{1}{16}\right)$

$a = \dfrac{1}{2}\left(\dfrac{16}{1}\right) = 8$

(b) Substitute the given values r and n, and the obtained value a in the formula:

$$S_n = \frac{a(1-r^n)}{1-r} = \frac{8\left(1-\left(-\frac{1}{2}\right)^5\right)}{1-\left(-\frac{1}{2}\right)}$$

$$= \frac{8\left(1-\left(-\frac{1}{32}\right)\right)}{1+\frac{1}{2}} = \frac{8\left(1+\frac{1}{32}\right)}{1\frac{1}{2}} = 5\frac{1}{2}$$

The terms between any two given terms of a GP are also called the *geometric means*. To insert a given number of geometric means between two terms, apply the formula $L = ar^{n-1}$, where $n =$ the number of terms to be inserted plus two.

EXAMPLE 7   Insert four geometric means between 4 and 128.

$a = 4, L = 128, n = 4 + 2 = 6$
Apply the formula:

$L = ar^{n-1}$ $128 = 4 \cdot r^{6-1}$
$$r^5 = \frac{128}{4} = 32$$
$$r = 2$$

Thus, the required geometric means are:

$$4(2) = 8, \quad 8(2) = 16, \quad 16(2) = 32, \quad 32(2) = 64$$

When the number of terms of a GP is unlimited (or $n \to \infty$) and the common ratio is numerically less than 1 (or $|r| < 1$), the formula for the sum of the first n terms:

$$S_n = \frac{a(1-r^n)}{1-r}$$

may be written for the infinite sequence as follows:

$$\lim_{n \to \infty} S_n = \frac{a}{1-r}$$

which is read "the limit of the value of S_n as n approaches infinity is $\frac{a}{(1-r)}$."
This is because the value of r^n approaches zero as n approaches infinity.

$$S_n = \frac{a(1-0)}{1-r} = \frac{a}{1-r}$$

EXAMPLE 8   Find the sum of the unlimited GP $1, \frac{1}{2}, \frac{1}{4}, \ldots$

$r = \frac{1}{2} \div 1 = \frac{1}{4} \div \frac{1}{2} = \frac{1}{2}$, which is < 1;
$a = 1$, and the number of terms is unlimited.

Thus:

$$\lim_{n\to\infty} S_n = \frac{a}{1-r} = \frac{1}{1-\frac{1}{2}} = 2$$

EXAMPLE 9   Convert the repeating decimal 0.5636363 . . . into a common fraction.

The given repeating decimal may be written in the following form:

$$0.56\dot{3} = 0.5636363 \ldots = 0.5 + (0.063 + 0.00063 + 0.0000063 + \ldots)$$

In the expression $0.56\dot{3}$, the figures with dots above them represent repeating figures. The terms in the parentheses form an unlimited geometric progression: $a = 0.063$ and $r = \frac{1}{100} = .01$, which is < 1.

Thus:

$$\lim_{n\to\infty} S_n = \frac{a}{1-r} = \frac{0.063}{1-0.01} = \frac{0.063}{0.990} = \frac{7}{110}$$

and the given repeating decimal is converted into a common fraction as follows:

$$0.56\dot{3} = 0.5 + \frac{7}{110} = \frac{5}{10} + \frac{7}{110} = \frac{31}{55}$$

EXAMPLE 10   Find the sum of the unlimited GP.

$$1,500(1 + 5\%)^{-1}, \; 1,500(1 + 5\%)^{-2}$$
$$1,500(1 + 5\%)^{-3}, \ldots$$

$$a = 1,500(1 + 5\%)^{-1} = 1,500(1.05)^{-1}, \text{ and}$$
$$r = (1 + 5\%)^{-1} = \frac{1}{1+5\%} = \frac{1}{1.05}, \text{ which is } < 1.$$

Thus:

$$\lim_{n\to\infty} S_n = \frac{1,500(1.05)^{-1}}{1 - \frac{1}{1.05}} = \frac{\frac{1,500}{1.05}}{\frac{1.05-1}{1.05}} = \frac{1,500}{.05} = 30,000$$

The sum \$30,000 can be interpreted as the present value of an ordinary annuity of \$1,500 payable at the end of each year forever with the interest rate at 5% compounded annually. (See Section 14.3B.) Thus, Example 10 can be used to solve a practical problem, such as the one which follows:

A man wishes to establish a college scholarship fund which will pay \$1,500 to worthwhile students at the end of each year forever. If the fund can earn interest at 5%, how much should the man deposit into the fund now?
He should deposit \$30,000 into the fund now.

Check: If \$30,000 is invested at 5%, the interest at the end of each year is:

$$\$30,000 \times 5\% = \$1,500$$

The scholarship payment may be made from the earned interest each year as long as the original deposit to the fund, $30,000, remains intact.

EXERCISE 15–2   Reference: Section 15.2

A. *Find L and S_n for each* GP:

1. 1, 2, 4, . . . to 6 terms
2. 2, 6, 18, . . . to 5 terms
3. 1, −2, 4, . . . to 7 terms
4. −3, −6, −12, . . . to 8 terms
5. 1, .5, .25, . . . to 6 terms

6. $\frac{1}{2}, \frac{1}{6}, \frac{1}{18}, \ldots$ to 5 terms

7. 2, 1, $\frac{1}{2}$, . . .
8. $1,000(1 + 2\%)^{-1}, 1,000(1 + 2\%)^{-2},$ $1,000(1 + 2\%)^{-3}, \ldots$

B. *Three of the five elements* (a, r, n, L, S_n) *of a* GP *are given below for each problem. Find the other two elements for each.*

9. $a = 1, r = 2, L = 64$
10. $a = 2, r = 3, S_n = 728$
11. $a = 1, n = 6, L = -32$
12. $a = -2, n = 3, S_n = -114$

13. $a = 1, L = .0625, S_n = 1.9375$
14. $r = 3, n = 5, L = 81$
15. $r = -2, n = 6, S_n = -42$
16. $r = -3, L = 324, S_n = 244$

C. *Statement Problems:*

17. Insert three geometric means between 2 and 32.

18. Insert four geometric means between $\frac{1}{-3}$ and 81.

19. Convert the repeating decimal 0.3454545 . . . into a common fraction.
20. Convert the repeating decimal 0.363636 . . . into a common fraction.
21. A man saved $100 the first week. Thereafter he planned to save 10% per week more than the amount he had saved the preceding week. Find the sum of his total savings at the end of the fourth week.

22. The number of units produced by a factory is estimated to increase 20% every year geometrically. The production of the first year was 10,000 units. Estimate the production of the fifth year.

23. A married couple has 4 sons. Each son has a wife and 4 boys. Each grandson also has a wife and 4 sons who are married. How many people are in the family?

24. A man wishes to establish a college scholarship fund which will pay $3,000 to worthwhile students at the end of each year forever. If the fund can earn interest at 6%, how much should the man deposit in the fund now?

15.3 SEQUENCES

The methods for finding the *n*th term and the sum of a sequence were restricted to two special types of sequences: the arithmetic progression and the geometric progression. In the remaining portion of this chapter we shall discuss methods used on sequences which are not restricted to a specific type. However, such a sequence must have a discernible pattern.

A. The *n*th Term of a Sequence

As stated previously, a sequence is fully defined when any required term, denoted as the *n*th term, of the sequence can be found. There are generally three ways to determine the *n*th term of a sequence:

1. By a formula representing the *n*th term.

EXAMPLE 11   Find the first three terms and the $(n + 1)$th term of the sequence whose *n*th term is (a) $3n + 4$, and (b) $\dfrac{1}{2n}$.

The four terms of each sequence are listed as follows:

Term	n Equivalent	Sequence (a) $3n + 4$	Sequence (b) $\dfrac{1}{2n}$
1st	1	$3(1) + 4 = 7$	$\dfrac{1}{2 \cdot 1} = \dfrac{1}{2}$
2nd	2	$3(2) + 4 = 10$	$\dfrac{1}{2 \cdot 2} = \dfrac{1}{4}$
3rd	3	$3(3) + 4 = 13$	$\dfrac{1}{2 \cdot 3} = \dfrac{1}{6}$
$(n + 1)$th	$(n + 1)$	$3(n + 1) + 4$ $= 3n + 7$	$\dfrac{1}{2(n + 1)}$

2. By a statement describing the sequence.

EXAMPLE 12   Find the 20th term of the AP when the first term $= 1$ and the common difference $= 3$.

This problem states that the sequence is an arithmetic progression. Based on the given values, the 20th term is:

$L = 1 + (20 - 1)3 = 58$. (See Example 1(a).)

EXAMPLE 13   Find the 7th term of the GP when the first term $= 4$ and the common ratio $= 3$.

This problem states that the sequence is a geometric progression. Based on the given values, the 7th term is:

$L = 4(3^{7-1}) = 4(729) = 2{,}916$ (See Example 4(a).)

3. By a rule observed from the given terms.

The work of finding a rule by observation for describing the *n*th term of a sequence may become difficult when the pattern of the sequence is not so obvious to an observer. In such a case, repeated attempts are necessary in finding the rule. Also, there may be different answers to the nth term if the number of terms given in a problem is not sufficient for making a single rule. (See Example 16.)

EXAMPLE 14   Find the nth term of the sequence:

$$\frac{\sqrt{3}}{4}, \frac{\sqrt{5}}{6}, \frac{\sqrt{7}}{8}, \frac{\sqrt{9}}{10}, \ldots$$

Observe that the numerators of the successive terms are the square roots of the positive odd integers. To find the nth term of a positive odd integer, try the expressions $(2n \pm 1)$, $(2n \pm 3)$, $(2n \pm 5)$, and so on. We select $(2n + 1)$ as the nth term, since it is the first term when $n = 1$ (or $2 \cdot 1 + 1 = 3$), the second term when $n = 2$ (or $2 \cdot 2 + 1 = 5$), the third term when $n = 3$ (or $2 \cdot 3 + 1 = 7$), and the fourth term when $n = 4$ (or $2 \cdot 4 + 1 = 9$).

The denominators of the successive terms are the positive even integers. To find the nth term of a positive even integer, try the expressions $2n$, $(2n \pm 2)$, $(2n \pm 4)$, and so on. We select $(2n + 2)$ as the nth term, since it is the first term when $n = 1$ (or $2 \cdot 1 + 2 = 4$), the second term when $n = 2$ (or $2 \cdot 2 + 2 = 6$), and so on.

Thus, the nth term of the given sequence is:

$$\frac{\sqrt{2n + 1}}{2n + 2}$$

EXAMPLE 15   Find the nth term of the sequence:

$$\frac{1}{3}, -\frac{2}{6}, \frac{3}{11}, -\frac{4}{18}, \ldots$$

Observe that the numerators of the successive terms are the positive integers $1, 2, 3, 4, \ldots$ and the denominators are equal to $(1^2 + 2 = 3)$, $(2^2 + 2 = 6)$, $(3^2 + 2 = 11)$, $(4^2 + 2 = 18)$, $\ldots$ respectively. Since the odd-numbered terms, such as the 1st and the 3rd terms, are positive and the even-numbered terms, such as the 2nd and the 4th terms, are negative, we try the expressions $(-1)^n$ and $(-1)^{n+1}$ to determine the sign of a term.

Thus, the nth term of the given sequence is:

$$(-1)^{n+1} \cdot \frac{n}{n^2 + 2}$$

Check: The 4th term is $(n = 4)$

$$(-1)^{4+1} \cdot \frac{4}{4^2 + 2} = (-1)^5 \cdot \frac{4}{16 + 2} = -\frac{4}{18}$$

EXAMPLE 16   Find the nth term and the 4th term of the sequence:

$$1, 3, 5, \ldots$$

OBSERVATION 1   The given successive terms are equal to:

$1, 1 + 2, 1 + 2 + 2, \ldots$ respectively

Thus, it is an AP and the nth term of this sequence is:

$$1 + (n - 1)2 = 1 + 2n - 2 = 2n - 1$$

The 4th term ($n = 4$) is:

$$2n - 1 = 2(4) - 1 = 7$$

OBSERVATION 2   The expression:

$$1 + 2(n - 1) + 2(n - 1)(n - 2)(n - 3)$$

may also represent any one of the three given terms, such as the 2nd term ($n = 2$):

$$1 + 2(2 - 1) + 2(2 - 1)(2 - 2)(2 - 3)$$
$$= 1 + 2 + 0 = 3$$

Thus, the 4th term ($n = 4$) is:

$$1 + 2(4 - 1) + 2(4 - 1)(4 - 2)(4 - 3)$$
$$= 1 + 6 + 2(3)(2)(1) = 1 + 6 + 12 = 19$$

Therefore, the three terms 1, 3, 5 given in Example 16 are not sufficient for making a single rule in describing the nth term of the sequence.

B. The Sum of a Sequence—Series

Let S_n = the sum of the first n terms of the sequence:

$t_1, t_2, t_3, \ldots, t_n$, or written:

$$S_n = t_1 + t_2 + t_3 + \cdots + t_n$$

The *indicated sum* on the right side of the above equation, as defined in the beginning of this chapter, is called a *series*.

Additional illustrations:

$$S_1 = t_1,$$
$$S_2 = t_1 + t_2,$$
$$S_3 = t_1 + t_2 + t_3,$$
$$\cdots\cdots\cdots\cdots\cdots$$

A series is said to be *finite* if the number of terms is limited, such as the series in the above illustrations, and *infinite* if the number of terms is unlimited, such as:

$$t_1 + t_2 + t_3 + t_4 + \ldots + t_n + \ldots$$

The sum of the first n terms of a finite series can be computed and expressed as a *definite finite number*, such as:

the sum of AP 1, 4, 7, 10, . . . to 20 terms is $S_{20} = 590$
(see Example 1(b)) and
the sum of GP 4, 12, 36, . . . to 7 terms is $S_7 = 4,372$
(see Example 4(b)).

However, the sum of an infinite series may or may not be expressed as a definite finite number. The two cases are presented in Section 15.4B.

EXERCISE 15–3   Reference: Section 15.3

A. *Find the first two terms and the* (n + 1)*th term from each given* n*th term of a sequence:*

1. $\dfrac{2n}{3n - 1}$

2. $\dfrac{3n}{2n + 1}$

3. $\dfrac{n + 4}{n^2 - 3}$

4. $\dfrac{5n - 2}{n^2 + 1}$

5. $(-1)^n \left(\dfrac{2n^2}{2n - 1} \right)$

6. $(-1)^{n+1} \left(\dfrac{2}{n^2 + 2} \right)$

B. *Find the* n*th term and the sum of the first five terms of each given sequence:*

7. $3, 5, 7, 9, \ldots$

8. $4, 6, 8, 10, \ldots$

9. $5, 7, 9, 11, \ldots$

10. $6, 8, 10, 12, \ldots$

11. $2, -3, 4, -5, \ldots$

12. $-1, 3, -5, 7, \ldots$

13. $\frac{1}{4}, \frac{4}{6}, \frac{7}{8}, 1, \ldots$

14. $\frac{1}{2}, \frac{3}{4}, \frac{5}{6}, \frac{7}{8}, \ldots$

15. $1, -\frac{1}{8}, \frac{1}{27}, -\frac{1}{64}, \ldots$

16. $-1, \frac{1}{9}, -\frac{1}{25}, \frac{1}{49}$

15.4 LIMITS OF AN INFINITE SEQUENCE

In this section, two types of limits are discussed: the limit of the *n*th term, and the limit of the sum.

A. The Limit of the *n*th Term

When *n* approaches infinity (written "$n \to \infty$" which also represents "*n* approaches infinitely large" or "*n* increases without bound"), the *n*th term of an infinite sequence will approach one of two possible values:

INFINITELY LARGE, ∞. In this case, the value of the *n*th term increases without bound as *n* increases without bound. In other words, there is no limit to the value of the *n*th term of the infinite sequence as *n* approaches infinity. An infinite sequence which has no limit is said to *diverge* or to be *divergent*.

EXAMPLE 17 Sequence:

$$1, 2, 3, 4, \ldots, n, \ldots \text{ approaches } \infty$$

EXAMPLE 18 Sequence:

$$1, 3, 5, 7, \ldots, (2n - 1), \ldots \text{ also approaches } \infty$$

A LIMIT. In this case, the value of the *n*th term approaches a limit as *n* increases without bound. An infinite sequence which has a limit is said to *converge* or to be *convergent*.

Let M = the limit, and
t_n = the value of the nth term

Then, the expression "the limit of t_n as n approaches infinitely large is M" can be written symbolically:

$$\lim_{n \to \infty} t_n = M$$

EXAMPLE 19 The limit of the infinite sequence whose nth term is $1/n$ is 0, or:

$$\frac{1}{1}, \frac{1}{2}, \frac{1}{3}, \frac{1}{4}, \ldots, \frac{1}{n}, \ldots, \lim_{n \to \infty} t_n = \lim_{n \to \infty} \frac{1}{n} = 0$$

Observe that the terms get smaller and smaller and eventually come closer and closer to 0 as a limit, but never equal 0.

EXAMPLE 20 The limit of the infinite sequence whose nth term is $1 + \dfrac{4}{10^n}$ is 1, or:

1.4, 1.04, 1.004, 1.0004, 1.00004, . . .

$$\lim_{n \to \infty} t_n = \lim_{n \to \infty} \left(1 + \frac{4}{10^n} \right) = 1$$

Observe that the terms get smaller and smaller and eventually come closer and closer to 1 as a limit. However, t_n never equals the limit 1. In other words, there always is a difference, however small, between t_n and the limit M (=1) no matter how large n is—even when n approaches infinitely large.

Based on Examples 19 and 20, we may give the following definition of the limit of an infinite sequence:

The *infinite sequence* $t_1, t_2, t_3, \ldots, t_n, \ldots$ has the limit M, if and only if there exists a sufficiently large positive integer n such that the difference between t_n and M is less in absolute value than an arbitrarily small positive number d, or:

$|t_n - M| < d$ (however small)

for all values of n and greater than n.

This definition is further explained by using the information given in Example 20. Let us pick a number, say $d = .0002$ (or any number as small as we please.) Then, the expression:

$$|t_n - M| < d$$

is written:

$$|t_n - 1| < .0002$$

Since the difference between t_5 (not t_4 or any preceding term) and the limit 1 is less than the arbitrarily picked number, or:

$$t_5 - 1 = 1.00004 - 1 = .00004, \text{ which is smaller than } .0002$$

the required "sufficiently large" value of n as stated in the definition must be 5, or:

$$t_n = t_5$$

The difference between each subsequent term, $t_6, t_7, t_8, \ldots,$ and the limit 1 is also less than .0002. We could have found a positive integer as an answer to n if we had picked a number even smaller than .0002, such as .00000001. Thus, the sequence has a limit according to the definition of the limit.

The definition of the limit of an infinite sequence only states the precise meaning of the limit. It does not provide any method for finding the limit. Common methods for finding the limit are introduced in the following paragraphs.

In many cases we can simply inspect the formula for the nth term (t_n) to find the limit, such as in the four previous examples. When t_n is a fraction whose numerator and denominator each include powers of n, the work of inspection becomes easier if the powers are changed to the form of $1/n$, $1/n^2$, $1/n^3, \ldots$ and so on. The change is usually done by dividing the numerator and denominator by the highest power of n in the fraction. When $n \to \infty$, the value of ($1/$any powers of n) approaches 0. This fact can help us simplify the fraction for inspection.

EXAMPLE 21 Find the limit of the infinite sequence whose nth term is $\dfrac{4n}{n+1}$.

$$t_n = \frac{4n}{n+1} = \frac{\dfrac{4n}{n}}{\dfrac{n+1}{n}} = \frac{4}{1+\dfrac{1}{n}}$$

$$\lim_{n\to\infty} t_n = \lim_{n\to\infty} \frac{4}{1+\dfrac{1}{n}} = \frac{4}{1+0} = 4$$

The limit of Example 21 can also be seen in the following figure.

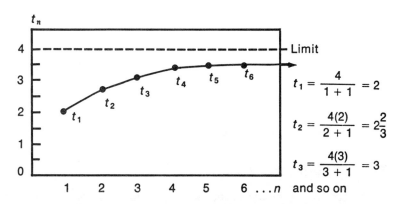

$$t_1 = \frac{4}{1+1} = 2$$

$$t_2 = \frac{4(2)}{2+1} = 2\frac{2}{3}$$

$$t_3 = \frac{4(3)}{3+1} = 3$$

and so on

EXAMPLE 22 Find the limit of the infinite sequence whose nth term is $\dfrac{3n^2 - 1}{n(n+1)}$.

Here the highest power of n in the fraction is n^2.

$$t_n = \frac{3n^2 - 1}{n(n + 1)} = \frac{\dfrac{3n^2 - 1}{n^2}}{\dfrac{n(n + 1)}{n^2}} = \frac{3 - \dfrac{1}{n^2}}{1 + \dfrac{1}{n}}$$

$$\underset{n \to \infty}{\text{limit}}\ t_n = \underset{n \to \infty}{\text{limit}}\ \frac{3 - \dfrac{1}{n^2}}{1 + \dfrac{1}{n}} = \frac{3 - 0}{1 + 0} = 3$$

EXAMPLE 23 Find the limit of the infinite sequence whose nth term is:

$$t_n = \left(1 + \frac{1}{n}\right)^n$$

Let:

$$e = \underset{n \to \infty}{\text{limit}}\ t_n = \underset{n \to \infty}{\text{limit}}\ \left(1 + \frac{1}{n}\right)^n$$

Then, $e = 2.71828$, approximately.

The following computations illustrate the fact that when the values of n get larger and larger, the values of t_n become closer and closer to e or 2.71828. The computation of the value of e is shown in Example 27.

$$t_1 = \left(1 + \frac{1}{1}\right)^1 = 2$$

$$t_2 = \left(1 + \frac{1}{2}\right)^2 = (1.5)^2 = 2.25$$

$$t_{10} = \left(1 + \frac{1}{10}\right)^{10} = (1.1)^{10} = 2.5937424601$$

$$t_{100} = \left(1 + \frac{1}{100}\right)^{100} = (1.01)^{100} = 2.705 \text{ (approximately)}$$

$$t_{1,000} = \left(1 + \frac{1}{1,000}\right)^{1,000} = (1.001)^{1,000} = 2.716 \text{ (approximately)}$$

$$t_{10,000} = \left(1 + \frac{1}{10,000}\right)^{10,000} = (1.0001)^{10,000} = 2.7164 \text{ (approximately)}$$

The number e is of fundamental importance in mathematics. For example, it is used as a base in the natural system of logarithms. This system is most convenient for theoretical purposes and will be discussed in calculus.

B. The Limit of the Sum

The sum of an infinite sequence may or may not be expressed by a definite finite number as a limit. The two cases are presented in the following paragraphs.

1. DIVERGENT SERIES. A *divergent series* is the indicated sum of the first n terms of a sequence which has no limit as n increases without bound. The value of the sum thus does not approach a definite finite number as a limit but may (a) increase without bound to infinity ∞ (see Example 24) or (b) oscillate between two numbers (see Example 25).

EXAMPLE 24 Show that the following series is divergent:

$$7 + 10 + 13 + \ldots + (3n + 4) + \ldots$$

The terms of the given series form an arithmetic progression.

$a = 7$,
$d = 10 - 7 = 13 - 10 = 3$, and
the nth term $= L = 7 + (n - 1)3$
$\qquad\qquad\qquad = 3n + 4$ (Also see Example 11(a).)

Based on the sum of an arithmetic progression formula:

$$S_n = \frac{n}{2}(a + L)$$

the required sum of the first n terms of the given series is:

$$S_n = 7 + 10 + 13 + \ldots + (3n + 4)$$
$$= \frac{n}{2}(7 + (3n + 4)) = \frac{n}{2}(3n + 11)$$

$$= \frac{3n^2}{2} + \frac{11n}{2}$$

$$\lim_{n \to \infty} S_n = \infty$$

Since the sum increases to infinitely large as n increases without bound, the given series is divergent.

EXAMPLE 25 Show that the following series is divergent:

$$1 - 1 + 1 - 1 + \ldots + (-1)^{n+1} + \ldots$$

The sum of the first n terms of the given series is:

$$S_n = (1 - 1) + (1 - 1) + \ldots + (-1)^{n+1}$$

The sum S_n is equal to 0 when n is even, and equal to 1 when n is odd. S_n does not approach a definite number as a limit but oscillates between the two numbers 0 and 1. Thus, the given series is divergent.

2. CONVERGENT SERIES. A *convergent series* is the indicated sum of the first n terms of a sequence which has a limit as n increases without bound. The value of the sum thus approaches a definite finite number as a limit.

EXAMPLE 26 Show that the following series is convergent.

$$1 + \frac{1}{2} + \frac{1}{4} + \frac{1}{8} + \ldots + \frac{1}{2^{n-1}} + \ldots$$

The terms form a geometric progression:

$a = 1$,

$r = \dfrac{1}{2} \div 1 = \dfrac{1}{4} \div \dfrac{1}{2} = \dfrac{1}{2}$, and

the nth term $= L = ar^{n-1} = 1(\frac{1}{2})^{n-1} = \dfrac{1}{2^{n-1}}$

Based on the sum of a geometric progression formula:

$$S_n = \frac{a(1 - r^n)}{1 - r}$$

the required sum of the first n terms of the given series is:

$$S_n = 1 + \frac{1}{2} + \frac{1}{4} + \frac{1}{8} + \ldots + \frac{1}{2^{n-1}}$$

$$= \frac{1(1 - (\frac{1}{2})^n)}{1 - \frac{1}{2}} = \frac{1 - \dfrac{1}{2^n}}{\frac{1}{2}} = \frac{1}{\frac{1}{2}} - \frac{\dfrac{1}{2^n}}{\frac{1}{2}}$$

$$= 2 - \frac{1}{2^{n-1}}$$

$\displaystyle \lim_{n \to \infty} S_n = 2 - 0 = 2$ (Also see Example 8.)

Since the limit exists, the given series is convergent.

EXAMPLE 27   Find the sum of the following infinite series correctly to five decimal places:

$$S_n = \frac{1}{0!} + \frac{1}{1!} + \frac{1}{2!} + \frac{1}{3!} + \ldots + \frac{1}{(n-1)!} + \ldots$$

When n in the denominator increases, the terms get smaller and smaller and eventually come closer and closer to 0 as a limit. Since the 11th term:

$$\frac{1}{(11 - 1)!} = \frac{1}{10!} = \frac{1}{3,628,800} = 0.00000028$$

has six zeros in the first six decimal places, the 11th term and the subsequent terms are omitted in computing the value of the first five decimal places for the given series.

The first ten terms are:

$\dfrac{1}{0!} = \qquad \dfrac{1}{1} = 1.0000\ 000$

$\dfrac{1}{1!} = \qquad \dfrac{1}{1} = 1.0000\ 000$

$$\frac{1}{2!} = \frac{1}{2} = 0.5000\ 000$$

$$\frac{1}{3!} = \frac{1}{6} = 0.1666\ 667$$

$$\frac{1}{4!} = \frac{1}{24} = 0.0416\ 667$$

$$\frac{1}{5!} = \frac{1}{120} = 0.0083\ 333$$

$$\frac{1}{6!} = \frac{1}{720} = 0.0013\ 889$$

$$\frac{1}{7!} = \frac{1}{5,040} = 0.0001\ 984$$

$$\frac{1}{8!} = \frac{1}{40,320} = 0.0000\ 248$$

$$\frac{1}{9!} = \frac{1}{362,880} = 0.0000\ 028$$

$$\overline{\text{Sum} = 2.7182\ 816}$$

Rounded to five decimal places:

$$S_{10} = 2.71828$$

or:

$$\underset{n \to \infty}{\text{limit}}\ S_n = 2.71828 \text{ (approximately)}$$

Note that the terms in the infinite series given in Example 27 are the expanded terms of the expression $(1 + 1/n)^n$ as n increases without bound. Thus, it is also the value of e, or $e = 2.71828$, approximately. (Also see Example 23.)

EXERCISE 15–4   Reference: Section 15.4

A. *Find the limit of each infinite sequence whose nth term is given below:*

1. $\dfrac{3n}{n+1}$

2. $\dfrac{2n}{3n+2}$

3. $\dfrac{n}{n(n+1)}$

4. $\dfrac{2n+1}{n^2-1}$

5. $\dfrac{n^2}{3n+1}$

6. $\dfrac{4n^2}{(n+1)^2}$

7. $\dfrac{(n+1)^2}{3n(n+2)}$

8. $\dfrac{n(3n-1)^2}{(n+1)^3}$

B. *Show the divergence or convergence of each of the following series:*

9. $1 + 3 + 5 + 7 + \cdots + (2n-1) + \cdots$

10. $\dfrac{1}{2} + \dfrac{1}{6} + \dfrac{1}{18} + \dfrac{1}{54} + \cdots + \dfrac{1}{2 \cdot 3^{n-1}} + \cdots$

11. $2 + 4 + 6 + 8 + \cdots + 2n + \cdots$

12. $\dfrac{1}{6}+\dfrac{2}{6}+\dfrac{3}{6}+\dfrac{4}{6}+\cdots+\dfrac{n}{6}+\cdots$

13. $1+.5+.25+.125+\cdots+.5^{n-1}+\cdots$

14. $.1+.4+.7+.10+\cdots+(.3n-.2)+\cdots$

15. $2+\dfrac{4}{3}+\dfrac{8}{9}+\dfrac{16}{27}+\cdots+2\left(\dfrac{2}{3}\right)^{n-1}+\cdots$

16. $(-3)+(-8)+(-13)+(-18)+\cdots+(2-5n)+\cdots$

17. $\dfrac{2}{5}+\dfrac{3}{5}+\dfrac{4}{5}+1+\cdots+\dfrac{n+1}{5}+\cdots$

18. $\dfrac{2}{3}+\dfrac{1}{6}+\dfrac{1}{24}+\dfrac{1}{96}+\cdots+\dfrac{2}{3\cdot4^{n-1}}+\cdots$

19. $1+2+4+8+\cdots+2^{n-1}+\cdots$

20. $2+6+18+54+\cdots+(2\cdot3^{n-1})+\cdots$

Chapter
16 Logarithms

This chapter introduces the algebraic operations involving logarithms. Logarithms can be used to simplify the basic operations of multiplication, division, raising to powers, and extracting roots. They can also be used in graphic presentations to enable a more penetrating analysis of quantitative information.

16.1 LOGARITHMS TO THE BASE 10 (COMMON LOGARITHMS)

A *logarithm* is an exponent. Let the expression b^x equal the positive number N, written in exponential form:

$$b^x = N$$

The exponent x is also called the logarithm of the number N to the base b, written in logarithmic form:

$$x = \log_b N \quad (Log \text{ represents the logarithm of.})$$

These two forms of writing state the same relationship. Thus:

if $3^2 = 9$, the exponent 2 is the logarithm of 9 to the base 3, or $\log_3 9 = 2$;

if $4^3 = 64$, the exponent 3 is the logarithm of 64 to the base 4, or $\log_4 64 = 3$; and

if $10^4 = 10,000$, the exponent 4 is the logarithm of 10,000 to the base 10, or $\log_{10} 10,000 = 4$.

Additional examples:

If $\sqrt{9} = 9^{1/2} = 3$, then $\log_9 3 = \frac{1}{2}$.
If $\sqrt[3]{64} = 64^{1/3} = 4$, then $\log_{64} 4 = \frac{1}{3}$.
If $\log_b 81 = 2$, then $b^2 = 81$, $b = \sqrt{81} = 9$.
If $\log_5 y = 3$, then $5^3 = y$, $y = 125$.
If $\log_{10} 100 = x$, then $10^x = 100$, $10^x = 10^2$, $x = 2$.

Logarithms may be computed for any base. However, logarithms to the base 10 are most convenient for computational purposes. The logarithms to the base

10 are called *common* or *Briggsian* (named after Henry Briggs, 1561–1630) system of logarithms. When the base is 10, it is usually not written and the logarithm is understood to be a common logarithm. Thus, $\log_{10} 10{,}000$ can be written simply as log 10,000. The following corresponding values expressed in exponential form and common logarithmic form give additional illustrations.

Exponential Form	Common Logarithmic Form
$10^x\ = N$ (a positive number)	$\log N \quad\ = x$
$10^4\ = 10{,}000$	$\log 10{,}000 =\ \ 4$
$10^3\ = 1{,}000$	$\log 1{,}000\ \ =\ \ 3$
$10^2\ = 100$	$\log 100 \quad =\ \ 2$
$10^1\ = 10$	$\log 10 \quad\ \ =\ \ 1$
$10^0\ = 1$	$\log 1 \quad\ \ \ =\ \ 0$
$10^{-1} = 1/10 = .1$	$\log .1 \quad\ \ \ =-1$
$10^{-2} = 1/10^2 = .01$	$\log .01 \quad\ =-2$
$10^{-3} = 1/10^3 = .001$	$\log .001 \quad =-3$
$10^{-4} = 1/10^4 = .0001$	$\log .0001\ \ =-4$

A. Characteristic and Mantissa

The above illustrations indicate that as the number (N) becomes greater, the logarithm of the number also becomes greater. Thus, if A and B are two positive numbers and A is larger than B, then $\log A$ is also larger than $\log B$. This idea is used in finding the logarithm of a number which is not an exact power of 10. For example, if the number is 350, which is larger than 100 (or 10^2) but smaller than 1,000 (or 10^3), the logarithm of 350 must be larger than the logarithm of 100 but smaller than the logarithm of 1,000.

$$\text{Numbers:} \qquad 100 < \quad 350 \quad < 1{,}000$$
$$\text{Logarithms:} \quad \log 100 < \log 350 < \log 1{,}000$$
$$\text{or:} \quad\ \ 2 < \log 350 < 3$$
$$\text{since } \log 100 = 2 \text{ and } \log 1{,}000 = 3$$

The value of log 350 is between 2 and 3, written in logarithmic form:

$$\log 350 = 2 + \text{a positive decimal}$$

which is equivalent to the exponential form:

$$10^{2\ +\ \text{a positive decimal}} = 350.$$

Likewise, we may find the logarithm of a positive number which is less than 1 and not an exact power of 10, such as .0035.

$$\text{Numbers:} \qquad\qquad .001 < \quad .0035 \quad < .01$$
$$\text{or:} \qquad\ \ 10^{-3} < \quad .0035 \quad < 10^{-2}$$
$$\text{Logarithms:} \qquad \log .001 < \log .0035 < \log .01$$
$$\text{or:} \qquad\qquad -3 < \log .0035 < -2$$
$$\text{since } \log .001 = -3 \text{ and } \log .01 = -2$$

The value of log .0035 is between −3 and −2, or written in logarithmic form:

$$\log .0035 = (-3) + \text{a positive decimal}$$

which is equivalent to the exponential form:

$$10^{-3 + \text{a positive decimal}} = .0035.$$

In general, when a number is not an exact power of 10, the logarithm of the number is the sum of a whole number and a decimal. The whole number (positive or negative) is called the *characteristic* of the logarithm, and the decimal (always positive) is called the *mantissa* of the logarithm.

Logarithm of a positive number = Characteristic + Mantissa

The characteristic of the logarithm of a number depends only on the position of the decimal point, regardless of the value of the individual digits in the number. The following rules, thus, may be used in determining characteristics.

1. If a number is greater than or equal to 1, the characteristic of its logarithm is positive and is 1 less than the number of digits in the whole-number part (or to the left of the decimal point).

Positive Numbers (1 and above)	Digits in Whole-Number Part	Characteristic
1 to 9.999999 (but less than 10)	1	0
10 to 99.99999 (but less than 100)	2	1
100 to 999.9999 (but less than 1,000)	3	2
1,000 to 9,999.999 (but less than 10,000)	4	3
10,000 to 99,999.99 (but less than 100,000)	5	4

2. If a positive number is less than 1, the characteristic of its logarithm is negative and is 1 more than the number of zeros between the decimal point and the first nonzero digit.

Positive Numbers (less than 1)	Zeros Between Decimal Point and First Nonzero Digit	Characteristic
.1 to .99999 (but less than 1)	0 (None)	−1
.01 to .09999 (but less than .1)	1	−2
.001 to .00999 (but less than .01)	2	−3
.0001 to .00099 (but less than .001)	3	−4

The mantissa of the logarithm of a number is not related to the position of the decimal point in the number. The mantissa is always positive and is determined according to the significant digits of the number. The *significant digits* in a number are the digits which do not include the zeros at the left of the first nonzero digit (such as the first two zeros in the decimal .002507), nor the zeros at the right of the last nonzero digit (such as the last three zeros in the whole number 2,507,000). The logarithms of numbers which have the same significant digits arranged in the same order will have the same mantissas. Thus, the significant digits of the numbers 2,507,000, 25,070, 2.507, and .002507 are 2, 5, 0, 7, and the logarithms of the numbers have the same mantissas, although they do not have the same characteristics.

B. Tables of Mantissas (or Tables of Logarithms)

The mantissas of the logarithms of most numbers are unending decimal fractions. For practical uses, various tables of mantissas are available. They are known as four-place tables, six-place tables, seven-place tables, and so on, according to the number of digits (decimal places) used in the mantissas. Table 2 in the Appendix contains mantissas for numbers 1–11,009. The numbers 1–9,999 have six-place mantissas and the numbers 10,000–11,009 have seven-place mantissas. The uses of Table 2 are illustrated below.

Find the logarithm of a given number. The steps are:

1. Determine the value of the characteristic based on the two rules described in Section A.
2. Find the mantissa corresponding to the significant digits of the given number from the tables of mantissas.
3. Place a decimal point before the first digit of the mantissa and add the mantissa to the characteristic. The sum is the required logarithm of the given number.

EXAMPLE 1   Find log 23.16.

STEP 1   The characteristic is +1 since there are two digits to the left of the decimal point.

STEP 2   The mantissa is 36 4739 from Table 2. It is located in the row with the first three significant digits of the given number (231 in N column) and in the column with the fourth significant digit (6).

Note: The first two digits of each mantissa are printed only in the 0 columns of the table. In order to obtain a six-place mantissa, the first two digits (such as 36 in the present example) must be prefixed to each entry on the same line and on the lines below the full entry (such as the entry 4739 in the example) until the two digits change. The first two digits of an entry marked by * are located on the line below the entry in the 0 column.

STEP 3   The logarithm of 23.16 is +1 + .364739 = 1.364739, or:

log 23.16 = 1.364739

The exponential form of this answer is:

$$10^{1.364739} = 23.16$$

EXAMPLE 2   Find log 0.002316.

STEP 1   The characteristic is −3 since there are two zeros between the decimal point and the first nonzero digit 2.

STEP 2   The mantissa is also 364739 since the significant digits of the given number are the same as those in the preceding example.

STEP 3   The logarithm of 0.002316 is −3 + .364739 = −2.635261, or:

log 0.002316 = −2.635261

The exponential form of this answer is:

$$10^{-2.635261} = 0.002316$$

The negative characteristic in Example 2 may be written in a different manner. It may be written by placing the negative sign above the characteristic, or:

log 0.002316 = $\overline{3}$.364739

It may also be written in a more convenient way by using an equivalent value, such as

log 0.002316 = 7.364739 − 10

since −3 = 7 − 10. Observe that the mantissas in the two new expressions are both positive.

EXAMPLE 3   Additional illustrations:

Use six-place mantissas in Table 2.	log 2.316 = 3.364739
	log 3.102 = 0.491642
	log 457.1 = 2.660011 (Observe that the entry 0011 has an * mark in the table.)

log 0.05942 = −2 + .773933 = −1.226067
or: log 0.05942 = $\overline{2}$.773933
= 8.773933 − 10

Use seven-place mantissas in Table 2.	log 10,492 = 4.0208583
	log 0.10504 = −1 + .0213547 = −.9786453

or: log 0.10504 = $\overline{1}$.0213547
= 9.0213547 − 10

Find the number from a given logarithm. (Find the antilogarithm.) The procedure for finding the number from a given logarithm is the inverse of the procedure used in finding the logarithm of a number. The number found is called the *antilogarithm* (abbreviated *antilog*) of the given logarithm. The steps are:

1. Find the given mantissa (decimal places) in the table of mantissas.
2. Determine the significant digits from the location of the mantissa in the table.
3. Place the decimal point with the significant digits at the place indicated by the given characteristic.

EXAMPLE 4   If log $N = 2.539202$, find N.

1. In Table 2, find the six-place mantissa 539202.
2. The significant digits are 3461.
3. The required number N is 346.1, since the given characteristic is 2.

The solution may be written:

N = antilog $2.539202 = 346.1$

EXAMPLE 5   If log $N = -1.305395$, find N.

The mantissas in the tables are all positive. Thus, first convert the given negative value of log N to its equivalent positive form having a mantissa of 694605. The conversion may be done by adding $(+10 - 10)$ to the given value as follows:

$$\log N = -1.305395 + 10 - 10$$
$$= 10 - 1.305395 - 10$$
$$= 8.694605 - 10$$

Next, find N in the usual manner:

1. In Table 2, find the six-place mantissa 694605.
2. The significant digits are 4950.
3. The required number N is 0.04950, since the characteristic is $8 - 10 = -2$. Thus:

N = antilog $(8.694605 - 10) = 0.0495$

Interpolate mantissas and numbers.

Tables of six and seven-place mantissas provide only limited numbers of digits in N columns and rows and in the bodies of the tables. However, one frequently wishes to find the mantissa for a given number whose mantissa is not listed in the tables. The unlisted mantissa and significant digits thus are approximated by using the interpolation method. This method assumes that the differences between numbers (N) and the differences between corresponding mantissas of the numbers in the tables are proportional.

EXAMPLE 6   Find log 456.38.

SOLUTION   The mantissa for the given number 456.38 cannot be found directly from Table 2. The nearest number larger than and the nearest number smaller than the given number are 456.40 and 456.30 respectively. The mantissa of 456.40 is 659346 and that of 456.30 is 659250. When the decimal points in the numbers are disregarded, the numbers and mantissas may be arranged in the following manner to facilitate calculation.

Observe that the larger number and mantissa are placed on the top line in the arrangement.

	Number	Mantissa	
	45640	659346	(1)
	45638	x	(2)
	45630	659250	(3)

Subtract line (3) from (2) $\quad \dfrac{8}{10} = \dfrac{x - 659250}{96}$ (4)

Subtract line (3) from (1) (5)

The differences between the numbers are assumed to be proportional to the differences between the corresponding mantissas: 8 is to (x − 659250) as 10 is to 96. Note that the difference 96 is also listed in the D column of Table 2.

Solve for x from the proportion formed by the differences in lines (4) and (5):

$$x - 659250 = 96\left(\frac{8}{10}\right) = 76.8, \text{ or round to } 77$$

$$x = 659250 + 77 = 659327$$

The characteristic of log 456.38 is 2. Thus:

$$\log 456.38 = 2.659327$$

Example 7 is used to illustrate the interpolation method in finding the number of a given logarithm when the mantissa is not listed in the tables.

EXAMPLE 7   If log N = 1.567196, find N to six significant digits.

SOLUTION   The mantissa 567196 cannot be found in Table 2. The significant digits thus must be approximated by the interpolation method. The nearest mantissas larger and smaller than the given mantissa are 567262 and 567144 which correspond to numbers 3,692 and 3,691, respectively. Since six significant digits are required, two zeros are added to each nearest number. Thus, the required number is between 369,200 and 369,100. The decimal point is disregarded in computing the value of N at first. The numbers and the corresponding mantissas are arranged in the following manner to facilitate calculation:

	Number	Mantissa	
	369200	567262	(1)
	x	567196	(2)
	369100	567144	(3)

Subtract: (2) − (3) $\quad \dfrac{x - 369100}{100} = \dfrac{52}{118}$ (4)

(1) − (3) (5)

Solve for x from the proportion formed by the differences in lines (4) and (5).

$$x - 369100 = 100\left(\frac{52}{118}\right) = 44.07, \text{ or round to } 44$$

$$x = 369100 + 44 = 369144$$

The given characteristic is $+1$. Thus:

$$N = 36.9144$$

EXERCISE 16–1   Reference: Section 16.1

A. *Write the following in logarithmic form:*

1. $5^2 = 25$
2. $2^3 = 8$
3. $a^4 = c$
4. $y^6 = d$

5. $10^{0.845098} = 7$
6. $10^{2.845098} = 700$
7. $10^{1.832509} = 68$
8. $10^{-1+.590173} = 0.3892$

B. *Write the following in exponential form:*

9. $\log_5 125 = 3$
10. $\log_4 16 = 2$
11. $\log_a M = x$
12. $\log_k N = y$

13. $\log 12.4 = 1.093422$
14. $\log 241.3 = 2.382557$
15. $\log 0.03219 = 8.507721 - 10$
16. $\log 0.0005432 = -4 + .734960$

C. *Find the logarithm of each number (from Table 2 in the Appendix.)*

17. 261
18. 18.24
19. 0.02146
20. 0.003678

21. 467.3
22. 0.5932
23. 1.0402
24. 10.943

D. *Find N (the number), if log N is:*

25. 2.269513
26. 1.461048
27. 0.587823
28. 3.680063

29. $8.459694 - 10$
30. $7.240549 - 10$
31. $-1 + .161967$
32. $-4 + .650405$

E. *Use the interpolation method to find each answer.*

33. $\log 0.068247 = ?$
34. $\log 4.2913 = ?$
35. $\log 0.57124 = ?$
36. $\log 2,631.84 = ?$
37. $\log N = 2.450327.$ Find N to five significant digits.
38. $\log N = 1.690482.$ Find N to five significant digits.
39. $\log N = 9.261135 - 10.$ Find N to six significant digits.
40. $\log N = 7.182654 - 10.$ Find N to six significant digits.

16.2 LOGARITHMIC COMPUTATION

Common logarithms are used in this section for the operations of multiplication, division, raising to powers, and extracting roots. Except for a few numbers, the results of logarithmic computations are approximations. In general, the results will be more accurate if the mantissas have more places included in the computation.

For simplicity, the interpolation process is omitted in the following illustrations. Instead, the nearest value in the tables of mantissas is used in the process of finding the antilog of a logarithm in the computation.

A. Multiplication

The logarithm of the product of two or more factors equals the sum of their logarithms.

The general form is:

$$\log (MN) = \log M + \log N$$

Proof   Let $x = \log_{10}M$, or $M = 10^x$;
        and $y = \log_{10}N$, or $N = 10^y$.
        Then, $MN = 10^x \cdot 10^y = 10^{x+y}$.

Write in logarithmic form:

$$\log_{10}(MN) = x + y = \log_{10}M + \log_{10}N$$

or simply by omitting the subscript 10:

$$\log (MN) = \log M + \log N$$

The applications of the above rule are illustrated in the following examples:

EXAMPLE 8   Multiply 2.481 by 0.036.

Let $MN = (2.481)(0.036)$

Take logarithms of both sides of the equation:

$\log (MN) = \log [(2.481)(0.036)]$
or: $\log (MN) = \log 2.481 + \log 0.036$

$\log 2.481 = 0.394627$
$\underline{(+) \log 0.036 = 8.556303 - 10}$
$\log MN \quad = 8.950930 - 10$

Find the antilog: $MN = 0.08932$ approximately

Notes: 1. The nearest value to the computed mantissa 950930 in the six-place table is 950949, where the number is 8932.
         2. The actual product is $(2.481)(0.036) = 0.089316$.

EXAMPLE 9   Multiply (-34.6) by (531.7).

1. Change the negative factor (-34.6) to a positive factor as follows:

$(-34.6)(531.7) = -[(34.6)(531.7)]$

2. Find the value of $[(34.6)(531.7)]$ by logarithms.

Let $MN = (34.6)(531.7)$.
$\log MN = \log 34.6 + \log 531.7$

$$\begin{array}{l} \log 34.60 = 1.539076 \\ \underline{(+) \ \log 531.7 = 2.725667} \\ \log MN \ \ \ = 4.264743 \end{array}$$

Find the antilog: $MN = 18{,}400$ approximately

3. Prefix the negative sign to the computed product to obtain the answer, or:

$$(-34.6)(531.7) = -18{,}400$$

Notes: 1. The nearest value to the computed mantissa 264743 in the six-place table is 264818, where the number is 1840.
2. The actual product is $-18{,}396.82$.

B. Division

The logarithm of a quotient equals the logarithm of the dividend (numerator) minus the logarithm of the divisor (denominator).

The general form is:

$$\log \frac{M}{N} = \log M - \log N$$

Proof   Let $x = \log_{10}M$, or $M = 10^x$;
and $y = \log_{10}N$, or $N = 10^y$.

Then, $\dfrac{M}{N} = \dfrac{10^x}{10^y} = 10^{x-y}.$

Write in logarithmic form:

$$\log_{10}\!\left(\frac{M}{N}\right) = x - y = \log_{10}M - \log_{10}N$$

or simply by omitting the subscript 10:

$$\log \frac{M}{N} = \log M - \log N$$

The application of the division rule is illustrated in the following example:

EXAMPLE 10   Divide 0.2961 by 8.46.

$$\text{Let } \frac{M}{N} = \frac{0.2961}{8.46}$$

Take logarithms of both sides of the equation:

$$\log\left(\frac{M}{N}\right) = \log\left(\frac{0.2961}{8.46}\right), \text{ or}$$

$$\log \frac{M}{N} = \log 0.2961 - \log 8.46$$

$$\log 0.2961 = 9.471438 - 10$$
$$(-)\ \log 8.4600 = 0.927370$$
$$\log \frac{M}{N} \quad = 8.544068 - 10$$

Find the antilog: $\dfrac{M}{N} = 0.035$

C. Raising to Power

The logarithm of an exponential term equals the exponent times the logarithm of its base.

The general form is:

$$\log M^p = p(\log M)$$

Proof: Let $x = \log_{10} M$, or: $M = 10^x$

Raise both sides of the equation to the pth power:

$$M^p = (10^x)^p = 10^{px}$$
and: $\qquad \log_{10}(M^p) = px = p(\log_{10} M)$
or simply: $\qquad \log M^p = p(\log M)$

EXAMPLE 11 Compute $(21)^3$

Let $N = (21)^3$
$$\log N = \log (21)^3 = 3(\log 21)$$
$$= 3(1.322219)$$
$$= 3.966657$$
$$N = \text{antilog } 3.966657 = 9{,}261$$

Thus, $(21)^3 = 9{,}261$.

EXAMPLE 12 Compute $(1.065)^{48}$.

Let $N = (1.065)^{48}$.

Then, $\log N = \log (1.065)^{48} = 48(\log 1.065)$
$$= 48(0.027350)$$
$$= 1.312800$$
$$N = \text{antilog } 1.312800 = 20.55$$

Thus, $(1.065)^{48} = 20.55$.

D. Extracting Root

If the exponent is a fraction $\frac{1}{q}$, the general form of the logarithm of an exponential term may be written:

$$\log M^{\frac{1}{q}} = \frac{1}{q} (\log M)$$

or:

$$\log \sqrt[q]{M} = \frac{1}{q} (\log M)$$

The application of the formula for extracting root is illustrated in the following two examples.

EXAMPLE 13 Compute $\sqrt[8]{256}$.

Let $N = \sqrt[8]{256} = (256)^{1/8}$.

Then, $\log N = \log (256)^{1/8} = \frac{1}{8} (\log 256)$

$$= \frac{1}{8} (2.408240) = 0.301030$$

$$N = \text{antilog } 0.301030 = 2$$

Check: $2^8 = 256$

EXAMPLE 14 Compute $\sqrt{0.1024}$.

Let $N = \sqrt{0.1024} = (0.1024)^{1/2}$.

Then, $\log N = \log (0.1024)^{1/2} = \frac{1}{2} (\log 0.1024)$
$= \frac{1}{2}(-1 + .010300) = \frac{1}{2}(-0.989700)$
$= -0.494850$

The mantissas in the tables of logarithms are all positive. The above is a negative result and therefore must be converted to its equivalent form having a positive mantissa.

Since $-0.494850 = -0.494850 + 10 - 10$
$= 9.505150 - 10,$
$\log N = -0.494850 = 9.505150 - 10$
$N = \text{antilog } (9.505150 - 10) = 0.3200.$

Check: $(0.32)^2 = 0.1024$

EXERCISE 16–2   Reference: Section 16.2

Perform each of the following indicated operations by logarithms. (Omit interpolation.)

A. *Multiplication:*
1. 3.45×0.246
2. 47.12×5.24
3. $367 \times 1,258$
4. $40,120 \times 1.453$

5. 0.5832×43.12
6. 523.4×0.02467
7. $(-35.72) \times 1.0253$
8. $1.0746 \times (-1,302)$

B. *Division:*
9. $5.62 \div 1.682$
10. $4,291 \div 25.12$
11. $32.5 \div 0.1562$
12. $642.7 \div 5.218$

13. $5.924 \div 6,349$
14. $0.3426 \div 0.0425$
15. $14.64 \div 526.8$
16. $1.0245 \div 63.14$

C. *Raising to Power:*

17. 4^2

18. 15^2

19. 34.2^3

20. 62.15^4

21. 1.045^{20}

22. 1.065^{42}

23. 524.3^{10}

24. 436.7^{50}

D. *Extracting Root:*

25. $\sqrt[4]{128}$

26. $\sqrt[5]{243}$

27. $\sqrt[10]{5,628}$

28. $\sqrt[8]{7,835}$

29. $\sqrt{20.25}$

30. $\sqrt{0.3721}$

31. $\sqrt[3]{623.4}$

32. $\sqrt[4]{1.0254}$

⋆16.3 LOGARITHMS TO BASES OTHER THAN 10

There are times when it is desirable to change a logarithm in one base to another base. The relationship between the logarithms of the same number N to different bases a and b is indicated by the following theorem:

The logarithm of number N to the base b is equal to the logarithm of N to the base a divided by the logarithm of b to the base a.

or written:

$$\log_b N = \frac{\log_a N}{\log_a b}$$

Proof: Let $b^x = N$. Then, by definition, $x = \log_b N$.

Take logarithms of both sides of the equation $b^x = N$ to the base a:

$$\log_a b^x = \log_a N, \text{ or } x(\log_a b) = \log_a N.$$

Solve for x:

$$x = \frac{\log_a N}{\log_a b}$$

or:

$$\log_b N = \frac{\log_a N}{\log_a b}$$

Another important base of logarithms is the number e, which is an irrational number equal to approximately 2.718. Logarithms to the base e are called *natural logarithms*. Natural logarithms of numbers 1 to 10 are listed in Table 3. If:

$$e^x = N$$

the exponent x is the logarithm of the number N to the base e, or written in logarithmic form:

$$x = \log_e N = ln\ N \ (ln \text{ represents } \log_e, \text{ the logarithm to base } e.)$$

Refer to the above theorem. Let $b = e$ and $a = 10$:

$$\log_e N = \frac{\log_{10} N}{\log_{10} e} = \frac{1}{\log_{10} e} \log_{10} N$$

From Table 2, the table of logarithms to the base 10, we have:

$$\log_{10} e = \log 2.718 = 0.434249, \text{ and } \frac{1}{\log_{10} e} = \frac{1}{0.434249} = 2.302826$$

Thus:

$$\log_e N = 2.302826(\log_{10} N)$$

or simply:

$$\textit{ln } N = 2.302826(\log N). \qquad \text{(Approximately)}$$

The above derived equation may be used to find logarithms of numbers to the base e from logarithms of numbers to the base 10.

EXAMPLE 15 Find $\textit{ln } 235$ (or $\log_e 235$).

(a) Use Table 2.

$$\begin{aligned}
\textit{ln } 235 &= 2.302826(\log 235) \\
&= 2.302826(2.371068) \\
&= 5.460157 \text{ (Approximately)}
\end{aligned}$$

(b) Use Table 3.

$$\begin{aligned}
\textit{ln } 235 &= \textit{ln } (2.35 \times 10^2) \\
&= \textit{ln } 2.35 + 2(\textit{ln } 10) \\
&= 0.8514 + 2(2.3026) \\
&= 5.4566
\end{aligned}$$

Note that the content in Table 2 is more comprehensive than that in Table 3. Method (a) presented in Example 15 is useful for finding $\log_e$ of a number of 4 or more digits. However, Method (b) is preferred for finding $\log_e$ of a number of 3 or fewer digits.

In general, by applying the given theorem and a table of common logarithms, the logarithms of numbers to bases other than 10 can be found as shown in the following example.

EXAMPLE 16 Find $\log_5 125$.

In the formula $\log_b N = \dfrac{\log_a N}{\log_a b}$, we let $b = 5, N = 125$, and $a = 10$.

Then:

$$\log_5 125 = \frac{\log_{10} 125}{\log_{10} 5} = \frac{2.096910}{0.698970} = 3.$$

Check: Let $x = \log_5 125$. Then $5^x = 125$. $5^3 = 125$.
Thus, the answer is correct.

*16.4 SOLVING EQUATIONS BY LOGARITHMS

Logarithms may be used to find the unknown of an exponential or logarithmic equation. An exponential equation has the unknown in an exponent, such as $2^x = 8$ and $3^{x+1} = 27$. A logarithmic equation has the logarithm of an expression involving the unknown, such as $\log (x + 1) = 3$.

To solve an exponential equation by using logarithms, first equate the logarithms of both sides of the equation; then solve the unknown from the new logarithmic equation.

EXAMPLE 17 Solve for x: $3^{x+1} = 27$

Take the logarithms of both sides:

$$\log 3^{x+1} = \log 27$$
$$(x + 1) \log 3 = \log 27$$
$$x + 1 = \frac{\log 27}{\log 3} = \frac{1.431364}{0.477121} = 3$$
$$x = 3 - 1 = 2$$

Check: $3^{2+1} = 3^3 = 27$

EXAMPLE 18 If $(1 + i)^{41} = 4.993$, find i.

Take the logarithms of both sides:

$$\log (1 + i)^{41} = \log 4.993$$
$$41 \log (1 + i) = 0.698362$$
$$\log (1 + i) = \frac{0.698362}{41} = 0.017033$$
$$1 + i = \text{antilog } 0.017033 = 1.040$$
$$i = 1.040 - 1 = 0.04$$

EXAMPLE 19 If $(1 + i)^{-14} = 0.4423$, find i.

Take the logarithms of both sides:

$$\log (1 + i)^{-14} = \log 0.4423$$
$$(-14) \log (1 + i) = -1 + .645717 = -.354283$$
$$\log (1 + i) = \frac{-.354283}{-14} = 0.025306$$
$$1 + i = \text{antilog } 0.025306 = 1.060$$
$$i = 1.060 - 1 = 0.06$$

EXAMPLE 20 Solve for x: $\log (x + 1) = 3$

Based on the definition of a logarithm, we may write the given equation in exponential form:

$$x + 1 = 10^3 = 1,000$$
$$x = 1,000 - 1 = 999$$

EXAMPLE 21   Solve for x: $\log_2 (x + 1) = 3 - \log_2 (x + 2)$

Collect all logarithmic terms to one side:

$\log_2 (x + 1) + \log_2 (x + 2) = 3$

Use the rule of multiplication to express the left side in a single logarithm:

$\log_2 [(x + 1)(x + 2)] = 3$

Write in exponential form:

$$(x + 1)(x + 2) = 2^3 = 8$$
$$x^2 + 3x + 2 = 8$$
$$x^2 + 3x - 6 = 0$$

Solve for x by the quadratic formula:

$$x = \frac{-3 \pm \sqrt{3^2 - 4(1)(-6)}}{2(1)} = \frac{-3 \pm \sqrt{33}}{2} = \frac{-3 \pm 5.74}{2}$$

$$x = \frac{-3 + 5.74}{2} = \frac{2.74}{2} = 1.37$$

$$x = \frac{-3 - 5.74}{2} = \frac{-8.74}{2} = -4.37$$

Check: Substitute $x = -4.37$ on the left side of the given equation:

$$\log_2 (x + 1) = \log_2 (-4.37 + 1)$$
$$= \log_2 (-3.37)$$

Logarithms of negative numbers are undefined. Thus, we should not use $x = -4.37$ as a solution. The substitution of $x = 1.37$ in the given equation leads to logarithms of positive numbers. Thus, we accept $x = 1.37$ as the only solution.

⋆16.5 LOGARITHMIC AND SEMILOGARITHMIC GRAPHS

The graphs presented previously are marked with arithmetic scales on X and Y axes. Equal distances on an arithmetic scale represent equal values. The axes may also be marked with logarithmic scales. A logarithmic scale is divided according to the logarithms of the values. When both axes of a graph are marked with a logarithmic scale, it is called a *logarithmic graph*. When only one axis, usually the Y or vertical axis, is marked with a logarithmic scale and the other one with an arithmetic scale, it is called a *semilogarithmic* (abbreviated *semilog*) *graph*.

EXAMPLE 22   Graph the equation $y = \log x$ (or $x = 10^y$) on arithmetic scales.

SOLUTION   The solution is shown in Figure 16–1. The paired values of x and y $(= \log x)$ are listed below for constructing the graph. On the graph, arithmetic scales are marked on the horizontal x axis and the left vertical y axis.

x	.1	.2	.3	.4	.5	.6	.7	.8	.9	1
y	−1	−.70	−.52	−.40	−.30	−.22	−.15	−.10	−.05	0

x	1	2	3	4	5	6	7	8	9	10
y	0	.30	.48	.60	.70	.78	.85	.90	.95	1

Figure 16-1

$y = \log x$ on
Arithmetic Scale

x Values on
Logarithmic Scale
(Two Cycles)

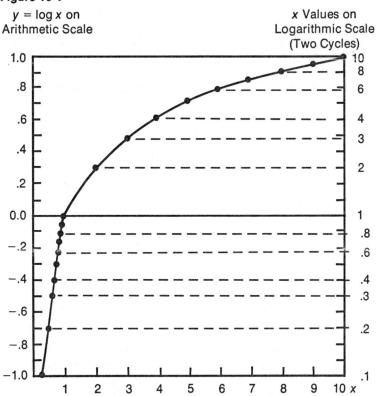

For example, when $x = .1$, $y = \log x = \log .1 = -1$;
when $x = 3$, $y = \log 3 = 0.477121$, or round to 0.48.

Also in the graph, the logarithmic scale is marked on the right vertical axis. The logarithmic scale shows the x values directly. The position of a given x value on the logarithmic scale corresponds with the logarithm of the x value plotted on the arithmetic scale (y axis).

For example, when $x = .1$ (on log scale), $\log .1 = -1$ (on y scale);
when $x = 3$ (on log scale), $\log 3 = .48$ (on y scale).

Thus, we can see that the logarithms of a group of numbers can be shown graphically on either an arithmetic scale or a logarithmic scale.

Note that the logarithmic scale marked on the graph in Example 22 has two cycles: one from .1 to 1 and the other from 1 to 10. The value at the top of a cycle on the logarithmic scale is always ten times the value at the bottom of the cycle. The first cycle, counting from the lower part of the graph, shows:

1 (at the top of the cycle) = .1 (at the bottom of the cycle) × 10;

the second cycle shows:

10 (at the top of the cycle) = 1 (at the bottom of the cycle) × 10.

Since 0 × 10 = 0, the value at the bottom of the first cycle of a logarithmic scale should not be *zero*.

The use of a semilog graph is illustrated in Examples 23 and 24. The illustration for a logarithmic graph is presented in Example 25.

EXAMPLE 23   The annual sales of Griffin Shoe Store from 1970 to 1975 are:

Year	Sales
1970	10,000
1971	20,000
1972	40,000
1973	80,000
1974	100,000
1975	200,000

Use the information to construct a semilog graph. The solution is shown in Figure 16–2.

The largest value is 20 times the smallest value (or 200,000 ÷ 10,000 = 20). Since the value at the top of a one-cycle paper is only ten times the value at the bottom, a two-cycle paper is needed to cover the given information.

The base value at the bottom of the Figure is arbitrarily assigned as 5,000. The value at the top of the first cycle thus is 50,000 (= 5,000 × 10), and that of the second cycle is 500,000 (= 50,000 × 10). Other numbers on the scale may also be computed from the base value. For example, 10,000 = 5,000 × 2, 15,000 = 5,000 × 3, and so on for the first cycle; 100,000 = 50,000 × 2, 150,000 = 50,000 × 3, and so on for the second cycle.

Observe the graph. When the ratio of two numbers is equal to the ratio of two other numbers, the distances between the points representing the pairs of numbers are the same. For example, the ratio of 20,000 to 10,000 (= 2) is the same as the ratio of 40,000 to 20,000 (= 2). The distance between 20,000 and 10,000 is the same as the distance between 40,000 and 20,000. Further, when the ratios during a period are constant, the points in the period are on a straight line. For example, the ratios of 80,000 to 40,000 (1973–1972), 40,000 to 20,000 (1972–1971), and 20,000 to 10,000 (1971–1970) are 2. The points representing the sales for the years from 1970 to 1973 are on the same straight line. A semilogarithmic figure is also called a *ratio chart* because it is useful for analyzing ratios.

Figure 16-2   A Semilog (or Ratio) Figure

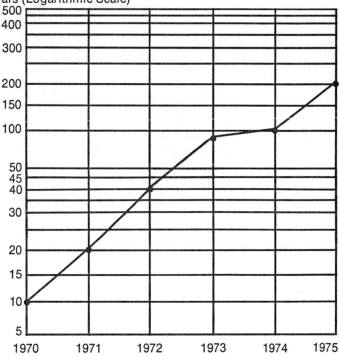
Thousands of Dollars (Logarithmic Scale)

EXAMPLE 24   Graph $y = 5(2^x)$ on a 3-cycle semilog paper.

SOLUTION

x	-2	-1	0	1	2	3	4	5	6	7
y	1.25	2.5	5	10	20	40	80	160	320	640

Observe that the resulting graph (Figure 16–3) is a straight line.

Take the logarithms of both sides of the given equation $y = 5(2^x)$:

$$\log y = \log 5 + x \log 2$$
$$= 0.698970 + 0.301030x$$

Let $\log y = s$. Since $\log y$ was plotted automatically on the logarithmic scale, the graph actually represents,

$$s = 0.698970 + 0.301030x$$

which is a linear equation.

Figure 16-3

y (Logarithmic Scale)

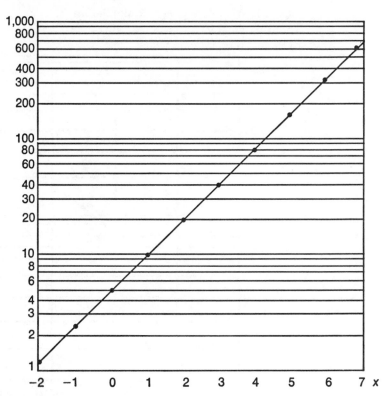

EXAMPLE 25   Graph $yx^2 = 1$ on logarithmic paper.

SOLUTION $y = \dfrac{1}{x^2}$

x	$\frac{1}{2}$	1	2	3	4	5
y	4	1	$\frac{1}{4} = .25$	$\frac{1}{9} = .11$	$\frac{1}{16} = .06$	$\frac{1}{25} = .04$

Take the logarithms of both sides of the given equation $yx^2 = 1$:

$$\log y + 2 \log x = \log 1 = 0$$

Let $\log y = s$ and $\log x = t$. Since both $\log y$ and $\log x$ were plotted automatically on the logarithmic scales, the graph actually represents:

$$s + 2t = 0, \text{ or } s = -2t$$

which is a linear equation. Observe that the given equation is represented by a straight line on the graph (Figure 16–4). Of course, not all graphs on logarithmic paper are straight lines, such as the graph in Example 23.

Figure 16-4

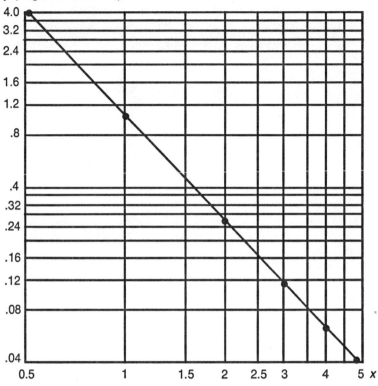

y (Logarithmic Scale)

(Logarithmic Scale)

EXERCISE 16-3   Reference: Sections 16.3–16.5

A. *Find the following logarithms by using (a) a table of common logarithms (Table 2 in the Appendix) and (b) a table of natural logarithms (Table 3 in the Appendix).*

   1. $\log_e 52$ **3.** $\log_e 186$

   2. $\log_e 5.2$ **4.** $\log_e 0.43$

B. *Find the following logarithms by using a table of common logarithms (Table 2 only).*

   5. $\log_6 36$ **8.** $\log_9 3$

   6. $\log_2 32$ **9.** $\log_5 100$

   7. $\log_8 512$ **10.** $\log_4 7$

C. *Solve for* x *by use of logarithms:*

   11. $2^{x+1} = 256$ **16.** $(1+x)^{-16} = 0.5339$

   12. $5^{2x+3} = 3{,}125$ **17.** $\log (x + 5) = 4$

   13. $(2+x)^{10} = 2{,}500$ **18.** $\log_3 (x - 51) = 6$

   14. $(1+x)^{35} = 5.5160$ **19.** $\log_2 (x^2 - 1) = 5 + \log_2 (x + 1)$

   15. $(3-x)^{-6} = 0.04621$ **20.** $\log_4 (x - 2) = 3 - \log_4 (x + 5)$

D. *Graphs*

21. Graph the equation $y = \log_2 x$ on arithmetic scales.

22. Graph the equation $y = \log_5 x$ on arithmetic scales.

23. The units produced by Begler Television Company from 1972 to 1977 are listed below. Construct a semilog graph to show the given information.

Year	Units Produced
1972	2,000
1973	5,000
1974	10,000
1975	20,000
1976	25,000
1977	40,000

24. The annual sales and operating expenses of Griffin Company from 1972 to 1978 are given below. Construct a semilog graph to show the information.

Year	Sales	Operating Expenses
1972	$120,000	$10,000
1973	240,000	20,000
1974	350,000	30,000
1975	560,000	40,000
1976	730,000	60,000
1977	870,000	50,000
1978	950,000	70,000

25. Graph the equation $y = 3(2^x)$ from $x = -1$ to $x = 8$ on a semilog graph.

26. Graph the equation $yx^2 = 5$ from $x = 1$ to $x = 6$ on a logarithmic graph.

Chapter

17 Differential Calculus: Introduction

The subject of calculus includes two major areas: differential calculus and integral calculus. This chapter introduces the concept and the basic operations of differential calculus. Other calculus topics are presented in Chapters 18, 19, and 20.

17.1 THE SLOPE OF A STRAIGHT LINE

The graph of the function:

$$y = a + bx, \text{ or conveniently written, } y = bx + a$$

is a straight line. Let $b = 2$ and $a = 3$. The graph of the equation:

$$y = 2x + 3$$

is shown in Figure 17–1.

Figure 17-1 **Basic Concept of a Straight Line and Its Slope**

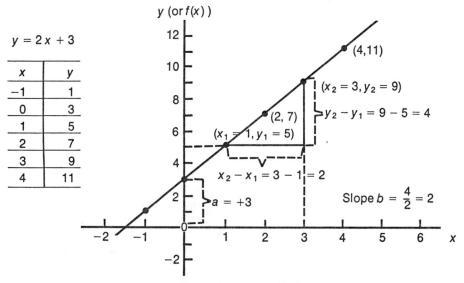

Observe the straight line in Figure 17–1. We have:

x = a point value on the straight line measured by the horizontal scale or x-axis;

$y = f(x)$, which represents the value of the function at x and is a point value on the straight line measured by the vertical scale or y-axis;

$a = y$ when $x = 0$; it is the height of the ordinate from the origin 0 to the intersection point of the straight line and the y-axis; it is also called the y-*intercept* or *constant;*

b = the *slope* of the straight line and is the coefficient of x; it represents the average amount of change in the y variable per unit change in the x variable. Let (x_1, y_1) and (x_2, y_2) be two points on the straight line. The slope is:

$$b = \frac{y_2 - y_1}{x_2 - x_1} \quad \text{or:} \quad b = \frac{y_1 - y_2}{x_1 - x_2}$$

When $(x_1 = 1, y_1 = 5)$ and $(x_2 = 3, y_2 = 9)$ represent the two points on the line as shown in the figure, the slope is:

$$b = \frac{y_2 - y_1}{x_2 - x_1} = \frac{9 - 5}{3 - 1} = \frac{4}{2} = 2$$

In general, if (x_1, y_1) and (x_2, y_2) are two points on the straight line representing the functional equation $y = f(x) = bx + a$, we can find the value of slope b, the constant a, and the equation in terms of the coordinates of the two points.

EXAMPLE 1   Refer to line I in Figure 17–2. Find the values of a and b and write the equation for the line. Given points on the line: (1, 2) and (3, 10).

SOLUTION   The slope is: $b = \dfrac{y_2 - y_1}{x_2 - x_1} = \dfrac{10 - 2}{3 - 1} = \dfrac{8}{2} = 4$

The value of a can be computed by using either one of the two given points:

1. Based on the point $(x_1 = 1, y_1 = 2)$, when $b = 4$, the equation:

$$y = bx + a$$

becomes:

$$2 = 4(1) + a$$
$$a = 2 - 4 = -2$$

2. Based on the point $(x_2 = 3, y_2 = 10)$, the straight line equation becomes:

$$10 = 4(3) + a$$
$$a = 10 - 12 = -2$$

By observing Figure 17–2, the value of a, the y-intercept, is also -2. Thus, the equation representing the straight line is:

$$y = 4x - 2$$

This equation can be used to check the points on the line, such as:

When $x = 1, y = 4(1) - 2 = 2$; and
When $x = 3, y = 4(3) - 2 = 10$

Figure 17-2   **Straight Lines I and II — (Examples 1 and 2)**

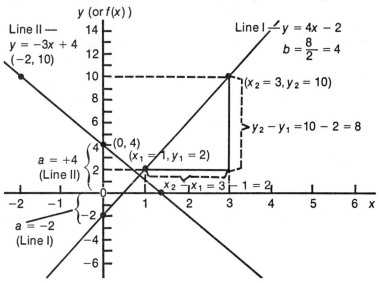

EXAMPLE 2   Find the values of a and b and write the equation for line II in Figure 17-2. Given points: $(0, 4)$ and $(-2, 10)$.

$$b = \frac{10 - 4}{-2 - 0} = \frac{6}{-2} = -3$$

Based on point $(0, 4)$, and $b = -3$, the equation $y = bx + a$ becomes:

$4 = -3(0) + a$
$a = 4 + 0 = 4$

The required equation for line II is:

$y = -3x + 4$

Check: When $x = 1\frac{1}{3}$ or $\frac{4}{3}$, $y = -3(\frac{4}{3}) + 4 = 0$.

The values of a and b may be positive or negative. Observe each line in Figure 17-2. When a line intersects the y-axis above the origin, a is positive (line II, $a = 4$); when the y-intercept is below the origin, a is negative (line I, $a = -2$). When the direction of the line is upward (or y values are increasing), the value of b is positive (line I, $b = +4$); when the direction of the line is downward (or y values are decreasing), the value of b is negative (line II, $b = -3$).

17.2 THE SLOPE OF A CURVE

The examples presented in Section 17.1 were limited only to straight lines. In this section, we wish to discuss the procedures for finding the slope of a curve at a single point on the curve. The slope at a given point on a curve is the same as the slope of the straight line drawn tangent to the curve at the given point. Thus, if we know the slope, an equation representing the line, denoted by:

$y_s = bx + a$ to avoid confusion with other function,

can be derived. The slope line at the given point, also called the *tangent line*, can then be plotted on the graph based on the derived equation.

We shall illustrate the method of finding the slope of a curve at a single point and the equation of the slope by using the curve shown in Figure 17–3. The curve plotted in the figure represents the function:

$$y = f(x) = x^2 + 1$$

Figure 17-3   The Slope of a Curve

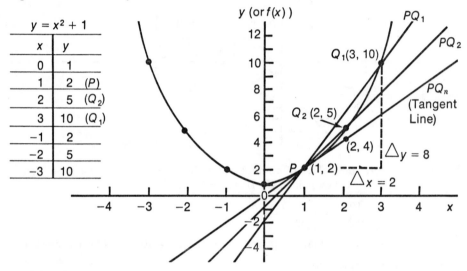

$y = x^2 + 1$	
x	y
0	1
1	2 (P)
2	5 (Q_2)
3	10 (Q_1)
−1	2
−2	5
−3	10

Suppose that we are interested in finding the slope of the curve at the point P $(x_1 = 1, y_1 = 2)$. First, take any point on the curve (either at the right or left side of P), such as point Q_1 $(x_2 = 3, y_2 = 10)$. Then the two points P and Q_1 when connected determine the straight line having a slope $b = 4$, or:

$$\text{Slope } b = \frac{y_2 - y_1}{x_2 - x_1} = \frac{10 - 2}{3 - 1} = \frac{8}{2} = 4 \text{ (Line } PQ_1)$$

Let us take other points which are nearer and nearer to the given point P on the curve, such as point Q_2 $(x_2 = 2, y_2 = 5)$. The two points P and Q_2 determine another straight line having a slope $b = 3$, or:

$$\text{Slope } b = \frac{5 - 2}{2 - 1} = \frac{3}{1} = 3 \text{ (Line } PQ_2)$$

In general, let:

$$x_2 - x_1 = \triangle x, \text{ the increment of variable } x, \text{ and}$$
$$y_2 - y_1 = \triangle y, \text{ the increment of variable } y.$$

The slope can be written in various forms:

$$b = \frac{y_2 - y_1}{x_2 - x_1} = \frac{\triangle y}{\triangle x}, \quad \text{or:} \quad b = \frac{y_2 - y_1}{\triangle x}$$

Based on the given function $y = f(x) = x^2 + 1$, we have:

$$y_1 = f(x_1) = x_1^2 + 1$$
$$y_2 = f(x_2) = f(x_1 + \Delta x) = (x_1 + \Delta x)^2 + 1$$

Since:

$$x_2 - x_1 = \Delta x \text{ and } x_2 = x_1 + \Delta x$$

The general form of the slope of line PQ on the $y = f(x)$ curve thus is:

$$b = \frac{y_2 - y_1}{\Delta x} = \frac{f(x_1 + \Delta x) - f(x_1)}{\Delta x}$$

If we substitute the values of $f(x_1 + \Delta x)$ and $f(x_1)$ in the general form, we have the slope of the line PQ on the curve for the given function $f(x) = x^2 + 1$.

$$b = \frac{[(x_1 + \Delta x)^2 + 1] - (x_1^2 + 1)}{\Delta x}$$

$$= \frac{[x_1^2 + 2x_1\Delta x + (\Delta x)^2 + 1] - (x_1^2 + 1)}{\Delta x}$$

$$= \frac{2x_1\Delta x + (\Delta x)^2}{\Delta x} = \frac{\Delta x\,(2x_1 + \Delta x)}{\Delta x} = 2x_1 + \Delta x$$

Check:

The slope of line PQ_1 is:

$$b = 2x_1 + \Delta x = 2(1) + (3 - 1) = 4$$

The slope of line PQ_2 is:

$$b = 2x_1 + \Delta x = 2(1) + (2 - 1) = 3$$

Now, let us take a point Q_n which approaches point P on the curve; that is, the distance between points P and Q_n is very small or approaches zero, but $P \neq Q_n$. The distance between P and Q_n determines the increment of the x variable, Δx. Thus, as Q_n approaches P, Δx approaches zero, written symbolically:

$$\Delta x \to 0$$

The points P and Q_n also determine a straight line. The slope of the line has already been computed above, or:

$$b = 2x_1 + \Delta x$$

When $\Delta x \to 0$, the value of slope _b_ approaches $2x_1$. In other words, the value of:

$$b = 2x_1 + \Delta x = 2x_1 + 0 = 2x_1$$

is the *limiting value* of the slope of the line through points P and Q_n as Q_n approaches P.

At point P, $x_1 = 1$, $y_1 = 2$, with the slope:

$$b = 2x_1 = 2(1) = 2$$

the equation representing the slope line:

$$y_s = bx + a$$

becomes:

$$2 = 2(1) + a$$
$$a = 2 - 2 = 0$$

Thus, the equation now is:

$$y_s = 2x + 0, \text{ or } y_s = 2x$$

According to the derived equation:

When $x = 0$, $y_s = 2(0) = 0$
When $x = 1$, $y_s = 2(1) = 2$ (See point P.)
When $x = 2$, $y_s = 2(2) = 4$

The three points obtained above are plotted in Figure 17–3. The points form a straight line PQ_n which in fact is the limit of the slope of the line through points P and Q_n as Q_n approaches P and is the tangent line to the curve at point P.

The conclusion concerning the limit of the slope of line PQ_n can be explained further. Since Q_n is very close to P, the slope of the line passing through P and Q_n should also be close to the slope of the tangent line to the curve $y = f(x)$ at point P. This fact is obvious when we observe the slopes of the three straight lines shown in Figure 17–3. Point Q_2 is nearer to P than is point Q_1 on the curve, and also the slope of the line PQ_2 is nearer to the slope of the tangent line than is the slope of the line PQ_1. Thus, as Q_n approaches P, the slope of the curve at point P is the limit of the slope of the line formed by P and Q_n.

The related values for finding the slope of the curve representing $y = x^2 + 1$ at point P are tabulated in Table 17–1 for comparison. Notice that when the increments Δx and Δy approach zero, reduced from 2 and 8 respectively, the ratio $\Delta y / \Delta x$ approaches the limit 2.

Table 17–1

Lines in Figure 17–3	Changes in x Variable			Changes in y Variable			Slope $= \dfrac{\Delta y}{\Delta x}$
	x_1	x_2	Δx	y_1	y_2	Δy	
PQ_1	1	3	2	2	10	8	4
not shown	1	2.5	1.5	2	7.25	5.25	3.5
PQ_2	1	2	1	2	5	3	3
not shown	1	1.5	.5	2	3.25	1.25	2.5
not shown	1	1.01	.01	2	2.0201	.0201	2.01
PQ_n	1	$\to 1$	$\to 0$	2	$\to 2$	$\to 0$	2

The symbol $\to$ represents "approaches."

EXERCISE 17–1   Reference: Sections 17.1 and 17.2

A. *From the given points in each of the following problems: (a) draw a straight line, (b) find the values of a and b, and (c) write the equation for the line.*

1. $(x_1 = 1, y_1 = 3)$ and $(x_2 = 2, y_2 = 5)$.
2. $(x_1 = -1, y_1 = 2)$ and $(x_2 = 4, y_2 = 7)$.
3. $(x_1 = 2, y_1 = -1)$ and $(x_2 = 1, y_2 = 1)$.
4. $(x_1 = 2, y_1 = -8)$ and $(x_2 = 0, y_2 = -2)$.
5. $(x_1 = 1, y_1 = -1)$ and $(x_2 = 3, y_2 = 7)$.
6. $(x_1 = 3, y_1 = 8)$ and $(x_2 = -1, y_2 = -4)$.

B. *Use the ratio* $\dfrac{f(x_1 + \Delta x) - f(x_1)}{\Delta x}$, $\Delta x \to 0$, *to find the slope of each of the following functions:*

7. $y = x^2 + 3$
8. $y = 2x^3$
9. $y = x^3 - 5$
10. $y = 5x^2$
11. $y = 3x^2 + 4$
12. $y = 7x^2 + x$
13. $y = 3x^2 - 5x$
14. $y = 4x^3 + 6x$
15. $y = \dfrac{3}{x + 1}$
16. $y = x^4 + 8$

C. *For the given function in each of the following problems: (a) graph the function, (b) find the slope at the indicated point, (c) find the equation of the slope, and (d) draw the line representing each slope on the graph.*

17. $y = x^2 + 3$. At the point $x = 2$. (Use the slope equation found in problem 7.)
18. $y = 2x^3$. At the point $x = 2$. (Use the slope equation found in problem 8.)
19. $y = x^3 - 5$. At the point $x = 1$. (Use the slope equation found in problem 9.)
20. $y = 5x^2$. At the point $x = -1$. (Use the slope equation found in problem 10.)

17.3 THE DERIVATIVE

The slope of the tangent to a curve at a given point (x,y) is called the *derivative* of the curve at the point. The derivative of the function $y = f(x)$ thus is the limiting value of the ratio $\Delta y/\Delta x$ as Δx approaches zero at any point on the curve representing the function. The derivative of y with respect to x can be written:

$$\frac{dy}{dx} = \lim_{\Delta x \to 0} \frac{\Delta y}{\Delta x} = \lim_{\Delta x \to 0} \frac{f(x + \Delta x) - f(x)}{\Delta x} = \begin{array}{c} \text{Slope of curve} \\ \text{at point } (x,y) \end{array}$$

It also can be written in various simpler ways, such as:

$$\frac{dy}{dx} = \frac{d}{dx}(y) = y' = f'(x) = D_x y = D_x f(x)$$

When the limit of the ratio $\Delta y/\Delta x$ exists, such as $b = 2$ in the above illustration, the function $f(x)$ is said to be *differentiable* or to *possess a derivative*. The process of finding the derivative of a function is called *differentiation*.

A function which is differentiable must be continuous. In other words, the function is differentiable at all points on the curve defined by the function, such as points P, R, and S in Figure 17–4.

Figure 17-4   Illustration of the Instantaneous Rate of Change

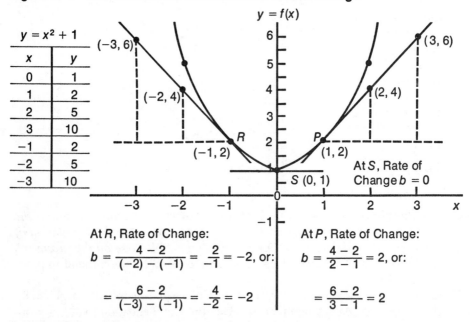

The function represented by the curve in Figure 17–4 is also $y = x^2 + 1$ (same function shown in Figure 17–3). The slope of the tangent to the curve at a given point (x_1, y_1) has already been computed on page 365, or:

$$b = 2x_1 + \Delta x$$
When $\Delta x \to 0$, $b = 2x_1$.

Since x_1 is any point on the x-axis, we may drop the subscript 1 in the above computation, or:

$$b = 2x, \quad \text{or} \quad \frac{dy}{dx} = \frac{d}{dx}(x^2 + 1) = 2x$$

At point P, $x = 1$, $y = 2$, and slope $b = 2x = 2(1) = 2$.
At point R, $x = -1$, $y = 2$, and slope $b = 2x = 2(-1) = -2$.
At point S, $x = 0$, $y = 1$, and slope $b = 2x = 2(0) = 0$.

The slope at a point on the function curve is also known as the *instantaneous rate of change* of the function at the point. The rate of change at each point on the curve can be shown graphically when the tangent line at the point is drawn.

At point P, the tangent line is represented by the equation $y_s = 2x$ (also computed on page 366.) The ratio of the amount of change in y to the amount of change in x is constant and positive: For every increase of two units in y, there is an increase of one unit in x.

At point R, the equation of the tangent line is obtained as follows:

Substitute the values $x = -1$, $y = 2$, and $b = -2$ in the equation representing the slope line:

$$y_s = bx + a$$
$$2 = (-2)(-1) + a$$
$$a = 2 - 2 = 0$$

Thus, the equation now is $y_s = (-2)x + 0 = -2x$.

The equation indicates that the ratio of the amount of change in y to the amount of change in x is constant but negative: For every increase of two units in y, there is a decrease of one unit in x.

At point S, the slope is zero since $b = 0$ and the tangent line is horizontal to the x-axis.

Observe the graphs in Figure 17–4. The point S is at the bottom of the curve and therefore is the minimum. The slopes on the points at the left of point S are negative (such as the slope at point $R = -2$) and those at the right of S are positive (such as the slope at point $P = +2$). This fact is very important in locating a minimum point on a given curve within a specified range by the use of derivatives. We remember that a positive slope is represented by a positive derivative, a negative slope by a negative derivative, and a zero slope by a zero derivative.

17.4 RULES FOR FINDING DERIVATIVES OF COMMON ALGEBRAIC FUNCTIONS

The derivative of a differentiable function can be obtained by applying the following formula as we have done in the previous section:

$$b = \frac{f(x + \Delta x) - f(x)}{\Delta x}$$

When Δx approaches 0, the value of b (the slope) is the derivative of the given function. However, the work of applying the above formula in every case is very tedious. Rules for finding the derivatives of various types of differentiable functions have been developed from the above formula. The rules can be used to find the derivatives of the given functions in order to save time and effort. Only the rules for finding derivatives of common types of algebraic functions are presented in this section. Rules for finding the derivatives of special functions are discussed in the next chapter.

For illustrative purposes, the first three rules are derived from the formula in detail. The other rules are stated and illustrated without derivations.

RULE 1   CONSTANT

The derivative of a constant is 0.

If $y = c$, where c is a constant, then:

$$\frac{dy}{dx} = 0$$

or, simply written:

$$y' = 0$$

EXAMPLE 3   Let $y = 5$. Find the derivative of y.

SOLUTION $5 = 5x^0$
Thus, the given equation can be written:

$$y = f(x) = 5x^0 = 5, \text{ and}$$
$$f(x + \Delta x) = 5(x + \Delta x)^0 = 5$$
$$b = \frac{f(x + \Delta x) - f(x)}{\Delta x} = \frac{5 - 5}{\Delta x} = \frac{0}{\Delta x}$$

When Δx approaches 0:

$$\Delta x \to 0, b = \frac{0}{\Delta x} = 0$$

Therefore:

$$\frac{dy}{dx} = 0$$

or written:

$$y' = 0, \quad \frac{d}{dx}(5) = 0,$$

$f'(x) = 0$, and so on.

Example 3 is shown graphically in Figure 17–5. The graph shows that the function is represented by a horizontal line. Thus, the slope of the function $y = 5$ is 0 for all values of x.

Additional illustrations:

(a) $y = 30$ $\qquad y' = 0$
(b) $y = 120$ $\qquad y' = 0$

RULE 2   POWER FUNCTION WITH BASE x

The derivative of a power function x^n is equal to the power n multiplied by the base x raised to the power $(n - 1)$.

Figure 17-5 (Example 3)

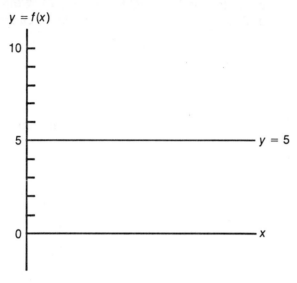

If:

$y = x^n$

then:

$$\frac{dy}{dx} = n \cdot x^{n-1}$$

or, simply written:

$y' = n \cdot x^{n-1}$

EXAMPLE 4 $y = f(x) = x^3$. Find the derivative of y and the slope at $x = 1$.

$f(x) = x^3$

$f(x + \Delta x) = (x + \Delta x)^3 = x^3 + 3x^2\Delta x + 3x(\Delta x)^2 + (\Delta x)^3$

$$b = \frac{[x^3 + 3x^2\Delta x + 3x(\Delta x)^2 + (\Delta x)^3] - x^3}{\Delta x}$$

$$= \frac{\Delta x(3x^2 + 3x(\Delta x) + (\Delta x)^2)}{\Delta x} = 3x^2 + 3x(\Delta x) + (\Delta x)^2$$

When $\Delta x \to 0$, $b = 3x^2 + 0 + 0 = 3x^2$. Thus:

$$\frac{dy}{dx} = 3x^2$$

or written:

$y' = 3x^2$, and $f'(x) = 3x^2$

The slope at $x = 1$ is:

$f'(1) = 3(1)^2 = 3$

The graph of Example 4 is shown in Figure 17–6. The graph shows that the slope at $x = 1$ is represented by the straight line $y_s = 3x - 2$ (the tangent line) with the slope $b = 3$. At $x = 1$, $y = 1$, and $b = 3$, the slope equation $y_s = bx + a$ becomes:

$$1 = 3(1) + a$$
$$a = 1 - 3 = -2$$

Thus:

$$y_s = bx + a = 3x - 2, \text{ which is plotted in the figure.}$$

Figure 17-6   (Example 4)

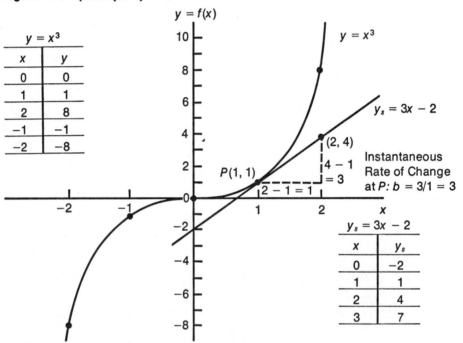

Additional illustrations:

$$y = x^7 \qquad y' = 7x^{7-1} = 7x^6$$
$$y = x^{-4} \qquad y' = (-4)x^{-4-1} = -4x^{-5}$$
$$y = x \qquad y' = x^{1-1} = x^0 = 1$$

The last illustration, if $y = x$, then $y' = 1$, indicates that the derivative of the independent variable x raised to the first power is 1.

RULE 3   PRODUCT OF A CONSTANT AND A FUNCTION

The derivative of a constant c times a function u is equal to the constant c times the derivative of the function u.

If $y = c \cdot u$, where c is a constant and u is a function of x, or $u = g(x)$, then:

$$\frac{dy}{dx} = c \cdot \frac{du}{dx}$$

or, simply written:

$$y' = c \cdot u'$$

EXAMPLE 5 $y = 4x^3$. Find the derivative of y.

$$f(x) = 4x^3$$
$$f(x + \Delta x) = 4(x + \Delta x)^3$$
$$b = \frac{4(x + \Delta x)^3 - 4x^3}{\Delta x} = 4 \cdot \frac{(x + \Delta x)^3 - x^3}{\Delta x}$$

When $\Delta x \to 0$, $b = 4(3x^2) = 12x^2$.

Let $u = g(x) = x^3$. The factor $(3x^2)$ is the derivative of $u = x^3$. (See Example 4.) Thus:

$$\frac{dy}{dx} = 4 \cdot \frac{du}{dx} = 4(3x^2) = 12x^2$$

or written:

$$y' = 12x^2$$

Additional illustrations:

$$y = 5x^7 \qquad\qquad y' = 5(7)x^{7-1} = 35x^6$$
$$y = -6x^{2/3} \qquad\qquad y' = (-6)(\tfrac{2}{3})x^{2/3-1} = -4x^{-1/3}$$
$$y = 2x^{-3} \qquad\qquad y' = 2(-3)x^{-3-1} = -6x^{-4}$$

RULE 4 SUM AND DIFFERENCE OF FUNCTIONS

The derivative of the sum (or difference) of two or more functions is equal to the sum (or difference) of the individual derivatives of the functions.

If $y = u \pm v$, where u and v are functions of x, or $u = g(x)$ and $v = h(x)$,

$$\frac{dy}{dx} = \frac{du}{dx} \pm \frac{dv}{dx}$$

or, simply written:

$$y' = u' \pm v'$$

EXAMPLE 6 $y = x^4 + x^2$. Find the derivative of y.

Here $u = x^4$ and $v = x^2$. Thus:

$$\frac{dy}{dx} = 4x^3 + 2x$$

or written:

$$y' = 4x^3 + 2x$$

EXAMPLE 7 $y = x^5 - x^3$. Find the derivative of y.

Here $u = x^5$ and $v = x^3$. Thus:

$$\frac{dy}{dx} = 5x^4 - 3x^2$$

or written:

$$y' = 5x^4 - 3x^2$$

EXAMPLE 8 $y = 3x + 2$. Find the derivative of y.

Here $u = 3x$ and $v = 2$. Thus:

$$\frac{dy}{dx} = 3 + 0 = 3$$

or written:

$$y' = 3$$

The solution of Example 8 may also be obtained in the following manner:

$$f(x) = 3x + 2$$
$$f(x + \Delta x) = 3(x + \Delta x) + 2 = 3x + 3\Delta x + 2$$
$$b = \frac{(3x + 3\Delta x + 2) - (3x + 2)}{\Delta x} = \frac{3\Delta x}{\Delta x}$$

When $\Delta x \rightarrow 0$, $b = 3$. Thus:

$$y' = 3$$

The above solution indicates that if $y = bx + a$, where a is a constant, then $y' = b$.

Thus, the derivative of a linear function is a constant.

Additional illustrations:

(a) $y = x^2 + 1$ $y' = 2x$
(b) $y = x^3 + 2x + 5$ $y' = 3x^2 + 2$
(c) $y = 3x^2 - 4x + 6$ $y' = 6x - 4$
(d) $y = 2x + 3$ $y' = 2$ (See Figure 17–1.)
(e) $y = 4x - 2$ $y' = 4$ (Example 1)
(f) $y = -3x + 4$ $y' = -3$ (Example 2)

RULE 5   PRODUCT OF FUNCTIONS

The derivative of the product of two functions $(u \cdot v)$ is equal to the first function (u) times the derivative of the second function (v') plus the second function (v) times the derivative of the first function (u').

If $y = u \cdot v$, where u and v are functions of x, or $u = g(x)$ and $v = h(x)$, then:

$$\frac{dy}{dx} = u\frac{dv}{dx} + v\frac{du}{dx}$$

or written:

$$y' = u \cdot v' + v \cdot u'$$

EXAMPLE 9   Let $y = x^5 \cdot 2x^3$. Find the derivative y'.

Here $u = x^5$ and $v = 2x^3$. Thus:

$$y' = x^5(6x^2) + 2x^3(5x^4) = 6x^7 + 10x^7 = 16x^7$$

or:

$$y = x^5 \cdot 2x^3 = 2x^8$$
$$y' = 2(8)x^7 = 16x^7 \text{ (Use Rule 3.)}$$

Additional illustrations:

(a) $y = x^4(3x^2 + 2x)$ $y' = x^4(6x + 2) + (3x^2 + 2x)(4x^3)$
$$= 6x^5 + 2x^4 + 12x^5 + 8x^4$$
$$= 18x^5 + 10x^4$$

or:

$$y = 3x^6 + 2x^5 \qquad y' = 18x^5 + 10x^4 \text{ (Use Rule 4.)}$$

(b) $y = (x^2 + 4)(x - 3)$ $y' = (x^2 + 4)(1) + (x - 3)(2x)$
$$= x^2 + 4 + 2x^2 - 6x$$
$$= 3x^2 - 6x + 4$$

or:

$$y = x^3 - 3x^2 + 4x - 12 \quad y' = 3x^2 - 6x + 4 \text{ (Use Rule 4.)}$$

RULE 6   QUOTIENT OF FUNCTIONS

The derivative of the quotient of two functions (u/v) is equal to the quotient of the product of the denominator (v) and the derivative of the numerator (u') minus the product of the numerator (u) and the derivative of the denominator (v') divided by the square of the denominator (v^2).

If $y = \dfrac{u}{v}$, where u and v are functions of x, or $u = g(x)$ and $v = h(x)$, then:

$$\frac{dy}{dx} = \frac{v\dfrac{du}{dx} - u\dfrac{dv}{dx}}{v^2}$$

or written:

$$y' = \frac{v \cdot u' - u \cdot v'}{v^2}$$

EXAMPLE 10 Let $y = \dfrac{x^5}{x^3}$. Find the derivative y'.

$u = x^5$ and $v = x^3$

Thus:

$$y' = \frac{x^3(5x^4) - x^5(3x^2)}{(x^3)^2} = \frac{5x^7 - 3x^7}{x^6} = 2x$$

or:

$$y = \frac{x^5}{x^3} = x^2 \qquad y' = 2x \text{ (Use Rule 2.)}$$

Additional illustrations:

(a) $y = \dfrac{3x^2 + 4}{2x}$ $\quad y' = \dfrac{2x(6x) - (3x^2 + 4)2}{(2x)^2} = \dfrac{12x^2 - 6x^2 - 8}{4x^4}$

$$= \frac{6x^2 - 8}{4x^4} = \frac{3x^2 - 4}{2x^4}$$

(b) $y = \dfrac{x^2 - 3x}{x + 1}$ $\quad y' = \dfrac{(x + 1)(2x - 3) - (x^2 - 3x)(1)}{(x + 1)^2}$

$$= \frac{(2x^2 - x - 3) - (x^2 - 3x)}{(x + 1)^2}$$

$$= \frac{x^2 + 2x - 3}{(x + 1)^2}$$

17.5 SUMMARY OF THE RULES FOR DIFFERENTIATION:

y is the function of x, or written:

$$y = f(x)$$

RULE 1 If $y = c$, where c is a constant:

$$\frac{dy}{dx} = 0$$

RULE 2 If $y = x^n$:

$$\frac{dy}{dx} = n \cdot x^{n-1}$$

RULE 3 If $y = c \cdot u$, where $u = g(x)$:

$$\frac{dy}{dx} = c \cdot \frac{du}{dx}$$

RULE 4 If $y = u \pm v$, where $u = g(x)$ and $v = h(x)$:

$$\frac{dy}{dx} = \frac{du}{dx} \pm \frac{dv}{dx}$$

RULE 5   If $y = u \cdot v$, where $u = g(x)$ and $v = h(x)$:

$$\frac{dy}{dx} = u\frac{dv}{dx} + v\frac{du}{dx}$$

RULE 6   If $y = \dfrac{u}{v}$, where $u = g(x)$ and $v = h(x)$:

$$\frac{dy}{dx} = \frac{v\dfrac{du}{dx} - u\dfrac{dv}{dx}}{v^2}$$

EXERCISE 17–2   Reference: Sections 17.3 and 17.4

Use the rules to find the derivative with respect to x for each of the following functions, $y = f(x)$.

1. $y = 52$
2. $y = 236$
3. $y = x^2$
4. $y = x^4$
5. $y = x^{12}$
6. $y = x^{5/2}$
7. $y = x^{8/7}$
8. $y = x^{1/3}$
9. $y = 2x^3$
10. $y = 5x^2$
11. $y = 4x^3$
12. $y = 3x^4$
13. $y = 6x^{1/2}$
14. $y = 10x^{3/5}$
15. $y = x^2 + 2$
16. $y = 2x^2 - 5$
17. $y = 4x + 3$
18. $y = x^4 - 6$
19. $y = x^3 - 5$
20. $y = x^2 + 4$

21. $y = (\tfrac{1}{5})x^{10} - 2x$
22. $y = 4x^3 + (\tfrac{1}{6})x^2$
23. $y = 3x^2 - 4x + 5$
24. $y = 7x^2 + x - 6$
25. $y = 8x^2 + 5x - 10$
26. $y = x^4 - 7x^3 + 2x$
27. $y = 3x^4 + 2x^3 - 5x^6 + 3$
28. $y = 5x^3 - 4x^5 + 6x^7 - 2$
29. $y = 2x^5 (x^3 + 4x)$
30. $y = x(x^2 + 1)$
31. $y = (x - 1)(2x^3 + 5)$
32. $y = (x^4 - 2x^3)(x^2 + 5x)$
33. $y = \dfrac{3}{x + 1}$
34. $y = \dfrac{-5}{3x^2}$
35. $y = \dfrac{2x^2 + 5}{3x}$
36. $y = \dfrac{4x^2 - 2x}{x - 3}$

37. Find the slope of the curve $y = x^{1/3}$ at the point (8,2). What is the equation of the slope?
38. What is the equation of the slope of the curve $y = x^7$ at the point (a) $x = 1$? (b) $x = 2$?

18 Differentiation for Special Functions

This chapter will illustrate the rules for differentiating several special types of functions. These are composite, exponential, logarithmic, inverse, implicit, and higher-order derivative functions. These rules, when coupled with the rules for the common algebraic functions presented in Chapter 17, will provide adequate knowledge in differential calculus for solving practical business and economic problems.

18.1 THE CHAIN RULE FOR COMPOSITE FUNCTIONS

If y is a function of u and u is a function of x, then y is a function of a function and is known as a *composite function*. The rule for finding the derivative of a composite function is called the *chain rule of differentiation* and is stated and illustrated below.

RULE 7   THE CHAIN RULE

If y is a function of u and u is a function of x, then the derivative of y with respect to x is the product of the derivative of y with respect to u and the derivative of u with respect to x.

If $y = f(u)$ and $u = g(x)$, or $y = f(g(x)) = f(x)$, then:

$$\frac{dy}{dx} = \frac{dy}{du} \cdot \frac{du}{dx}$$

EXAMPLE 1 $y = (x^2 + 1)^3$. Find the derivative $\dfrac{dy}{dx}$.

$$u = g(x) = x^2 + 1 \qquad \frac{du}{dx} = 2x$$

$$y = f(u) = u^3 \qquad \frac{dy}{du} = 3u^2$$

Thus:

$$\frac{dy}{dx} = \frac{dy}{du} \cdot \frac{du}{dx} = 3u^2 \cdot 2x = 3(x^2 + 1)^2 \, (2x)$$
$$= 6x(x^2 + 1)^2$$

Alternatively, the function can be rewritten and Rules 2 and 4 used to find its derivative:

$$y = (x^2 + 1)^3 = x^6 + 3x^4 + 3x^2 + 1$$
$$y' = 6x^5 + 12x^3 + 6x$$
$$= 6x(x^2 + 1)^2$$

The chain rule can be extended to find the derivative of a composite function involving three functions as follows:

If $y = f(u)$, $u = g(v)$, and $v = h(x)$, or $y = f\Big(g\big(h(x)\big)\Big) = f(x)$, then:

$$\frac{dy}{dx} = \frac{dy}{du} \cdot \frac{du}{dv} \cdot \frac{dv}{dx}$$

EXAMPLE 2 $\quad y = \big((x^2 + 1)^3\big)^{-2}.$ Find the derivative $\dfrac{dy}{dx}.$

$$v = x^2 + 1 \qquad\qquad \frac{dv}{dx} = 2x$$

$$u = (x^2 + 1)^3 = v^3 \qquad \frac{du}{dv} = 3v^2$$

$$y = u^{-2} \qquad\qquad \frac{dy}{du} = -2u^{-3}$$

Thus:

$$\frac{dy}{dx} = \frac{dy}{du} \cdot \frac{du}{dv} \cdot \frac{dv}{dx} = (-2u^{-3})(3v^2)(2x)$$
$$= -2\big((x^2 + 1)^3\big)^{-3}\big(3(x^2 + 1)^2\big)(2x)$$
$$= -12x(x^2 + 1)^{-7}$$

Additional illustrations:

(a) $y = \dfrac{u + 1}{u}$ and $u = \sqrt{x}.$ Find the derivative $\dfrac{dy}{dx}.$

$$u = \sqrt{x} = x^{1/2} \qquad\qquad \frac{du}{dx} = \tfrac{1}{2}(x^{-1/2}) = \frac{1}{2\sqrt{x}}$$

$$y = \frac{u + 1}{u} = 1 + \frac{1}{u} \qquad \frac{dy}{du} = 0 + \frac{0 - 1(1)}{u^2} = \frac{-1}{u^2} = \frac{-1}{x}$$

$$\frac{dy}{dx} = \frac{dy}{du} \cdot \frac{du}{dx} = \frac{-1}{x} \cdot \frac{1}{2\sqrt{x}} = \frac{-1}{2x\sqrt{x}}$$

Or alternatively, the function can be rewritten and Rule 6 used to find its derivative:

$$y = 1 + \frac{1}{u} = 1 + \frac{1}{x^{1/2}}$$

$$\frac{dy}{dx} = 0 + \frac{0 - 1(\frac{1}{2}x^{-1/2})}{(x^{1/2})^2} = \frac{-\frac{1}{2}\left(\frac{1}{\sqrt{x}}\right)}{x} = \frac{-1}{2x\sqrt{x}}$$

(b) $y = 2u + 3$, $u = v^2$, and $v = x^3$. Find the derivative $\frac{dy}{dx}$.

$$\frac{dy}{du} = 2, \quad \frac{du}{dv} = 2v, \quad \frac{dv}{dx} = 3x^2$$

$$\frac{dy}{dx} = \frac{dy}{du} \cdot \frac{du}{dv} \cdot \frac{dv}{dx} = 2(2v)(3x^2) = 2(2x^3)(3x^2) = 12x^5$$

Or alternatively, the function can be rewritten and Rule 3 used to find its derivative:

$$y = 2u + 3 = 2(v^2) + 3 = 2\left((x^3)^2\right) + 3 = 2x^6 + 3$$

$$\frac{dy}{dx} = 2(6x^5) = 12x^5$$

18.2 POWER FUNCTIONS

This section will illustrate the rule for finding the derivative of a power function whose base is a function and whose exponent is a real number.

RULE 8 POWER FUNCTION WITH A FUNCTION AS THE BASE

The derivative of a function u raised to the power n is equal to n times the function raised to the power $(n - 1)$ and times the derivative of the function.

If $y = u^n$, where u is a function of x, or $u = g(x)$, and n is a real number, then:

$$\frac{dy}{dx} = n \cdot u^{n-1} \cdot \frac{du}{dx}$$

Or, simply written:

$$y' = n \cdot u^{n-1} \cdot u'$$

Observe: I. If $y = f(u) = u^n$, then by Rule 2:

$$\frac{dy}{du} = n \cdot u^{n-1}$$

Rule 8 can be written:

$$\frac{dy}{dx} = \frac{dy}{du} \cdot \frac{du}{dx}$$

Thus, Rule 8 is a special case of the chain rule (Rule 7).

II. If $u = g(x) = x$, then:

$$y = u^n = x^n, \text{ and:}$$

$$\frac{dy}{dx} = n \cdot u^{n-1} \cdot \frac{du}{dx} = n \cdot x^{n-1} \cdot \frac{dx}{dx}$$

$$\frac{dy}{dx} = n \cdot x^{n-1}$$

Thus, Rule 2 is a special case of Rule 8.

EXAMPLE 3 $y = (x^2 + 1)^3$.   Find the derivative $\dfrac{dy}{dx}$.

$$u = x^2 + 1, \quad \frac{du}{dx} = 2x, \quad n = 3$$

Thus:

$$\frac{dy}{dx} = 3(x^2 + 1)^{3-1} \cdot 2x = 6x(x^2 + 1)^2$$

Or, rewriting the function and using Rules 2 and 4:

$$y = (x^2 + 1)^3 = x^6 + 3x^4 + 3x^2 + 1$$
$$y' = 6x^5 + 12x^3 + 6x = 6x(x^2 + 1)^2$$

Additional illustrations:

(a) $y = (x^2)^3$
$\quad u = x^2 \qquad u' = 2x$

Thus:

$$y' = 3(x^2)^{3-1} (2x) = 6x(x^4) = 6x^5$$

Or, rewriting:

$$y = (x^2)^3 = x^6$$
$$y' = 6(x^{6-1})(1) = 6x^5, \text{ which verifies Rule 2.}$$

(b) $y = (x^4 + 2x^3 - 5x^2 + 3)^5$
$\quad u = x^4 + 2x^3 - 5x^2 + 3 \qquad u' = 4x^3 + 6x^2 - 10x$

Thus:

$$y' = 5(x^4 + 2x^3 - 5x^2 + 3)^4 (4x^3 + 6x^2 - 10x)$$

(c) $y = \left(\dfrac{x + 5}{x - 1}\right)^{30}$

$$u = \frac{x + 5}{x - 1} \qquad u' = \frac{(x - 1)(1) - (x + 5)(1)}{(x - 1)^2} = \frac{-6}{(x - 1)^2}$$

Thus:

$$y' = 30 \left(\frac{x + 5}{x - 1}\right)^{29} \left(\frac{-6}{(x - 1)^2}\right)$$

18.3 EXPONENTIAL AND LOGARITHMIC FUNCTIONS

Exponential and logarithmic functions are closely related. If we know an exponential function, we can write an equivalent logarithmic function, such as:

$$\text{If } y = a^x, \text{ then } x = \log_a y.$$

Thus, the two types of functions are discussed together in this section.

RULE 9 EXPONENTIAL FUNCTIONS

If $y = a^u$, where a is a constant and u is a function of x, or $u = g(x)$, then:

$$\frac{dy}{dx} = a^u \cdot \ln a \cdot \frac{du}{dx}$$

Or, simply written:

$$y' = a^u \cdot \ln a \cdot u'$$

Special Case:

If $a = e$ and $y = e^u$, where e, a constant, $= 2.71828$ approximately, then:

$$\frac{dy}{dx} = e^u \cdot \log_e e \cdot \frac{du}{dx} = e^u \cdot (1) \cdot \frac{du}{dx}$$

$$\frac{dy}{dx} = e^u \cdot \frac{du}{dx}$$

Or, simply written:

$$y' = e^u \cdot u'$$

EXAMPLE 4 $y = 3^x$. Find the slope of the function at $x = 2$.

The slope of the function is:

$$\frac{dy}{dx} = 3^x (\ln 3) \cdot 1 = 3^x (1.0986). \quad \text{(Table 3 in the Appendix)}$$

The slope at $x = 2$ is:

$$3^2 (1.0986) = 9.8874, \text{ or round to } 9.89.$$

Example 4 is shown graphically in Figure 18–1. The figure also shows the slope (the tangent line) at $x = 2$. The tangent line is obtained as follows:

At $x = 2$, $y = 3^x = 3^2 = 9$, with the slope $b = 9.89$, the slope equation:

$$y_s = bx + a$$

becomes:

$$9 = 9.89(2) + a$$
$$a = 9 - 19.78 = -10.78$$

The tangent line is drawn from the equation $y_s = 9.89x - 10.78$.

Figure 18-1 (Example 4)

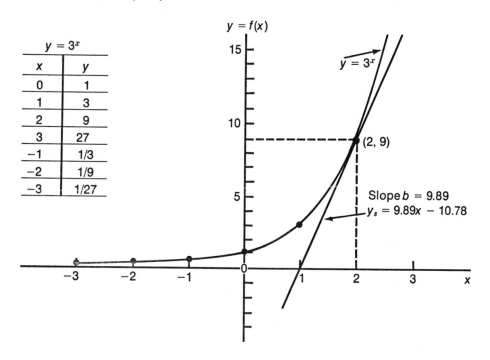

x	y
0	1
1	3
2	9
3	27
−1	1/3
−2	1/9
−3	1/27

EXAMPLE 5 $y = e^x$. Find the derivative of y.

$$\frac{dy}{dx} = e^x \cdot 1 = e^x$$

The result in Example 5 indicates that the derivative of e^x is also e^x, or:

If $y = e^x$, then $y' = e^x$.

Additional illustrations:

(a) $y = e^{2x}$ $y' = 2(e^{2x})$

(b) $y = 4e^{3x}$ $y' = 4(e^{3x})(3) = 12e^{3x}$

(c) $y = 5(2^x)$ $y' = 5(2^x)(ln\ 2)(1) = 5(2^x)(0.6931)$

                              $= 3.465\ 5\ (2^x)$

(d) $y = 10^{2x+3}$ $y' = (10^{2x+3})(ln\ 10)(2) = 2(10^{2x+3})(2.3026)$

                              $= 4.6052(10^{2x+3})$

RULE 10 LOGARITHMIC FUNCTIONS

If $y = \log_a u$, where a is a constant and u is a function of x, or $u = g(x)$, then:

$$\frac{dy}{dx} = \frac{1}{u}(\log_a e)\frac{du}{dx}$$

Or, simply written:

$$y' = \frac{u'}{u} (\log_a e)$$

Special Case:

If $a = e$ and $y = \log_e \ u = \ln u$, then:

$$\frac{dy}{dx} = \frac{1}{u} (1) \frac{du}{dx} = \frac{1}{u} \cdot \frac{du}{dx}$$

Or, simply written:

$$y' = \frac{u'}{u}$$

EXAMPLE 6 $y = \log x = \log_{10} x$. Find the slope of the function at $x = 2$.

The slope of the function is:

$$\frac{dy}{dx} = \frac{1}{x} \log_{10} e \,(1) = \frac{0.4343}{x} \qquad\qquad \text{(Table 2)}$$

The slope at $x = 2$ is:

$$\frac{0.4343}{2} = 0.21715, \text{ or round to } 0.22$$

The graph of Example 6 is shown in Figure 18–2. The tangent line at $x = 2$ is obtained as follows:

Figure 18-2   (Example 6)

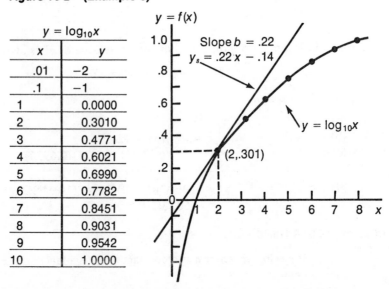

$y = \log_{10}x$	
x	y
.01	-2
.1	-1
1	0.0000
2	0.3010
3	0.4771
4	0.6021
5	0.6990
6	0.7782
7	0.8451
8	0.9031
9	0.9542
10	1.0000

At $x = 2$, $y = \log_{10} 2 = 0.3010$, with the slope $b = 0.22$, the slope equation:

$$y_s = bx + a$$

becomes:

$$0.3010 = 0.22(2) + a,$$
$$a = 0.3010 - 0.44 = -0.139, \text{ or round to } -.14$$

The tangent line equation thus is $y_s = 0.22x - .14$.

EXAMPLE 7 $y = \log(2x^3 - x^2)$. Find $\dfrac{dy}{dx}$.

$$\frac{dy}{dx} = \frac{1}{2x^3 - x^2}(\log e)(6x^2 - 2x)$$

$$= \frac{6x^2 - 2x}{2x^3 - x^2}(\log e) = \frac{6x - 2}{2x^2 - x}(\log e)$$

EXAMPLE 8 $y = \ln x$. Find $\dfrac{dy}{dx}$.

$$\frac{dy}{dx} = \frac{1}{x}(1) = \frac{1}{x}$$

Additional illustrations:

(a) $y = \log(x^2 + 3x - 5)$ $y' = \dfrac{2x + 3}{x^2 + 3x - 5}\log e$

(b) $y = (\log x^3)^2$ $y' = 2(\log x^3)\left(\dfrac{3x^2}{x^3}\right)\log e$

$$= \frac{6}{x}(\log x^3)\log e$$

(c) $y = \ln(3x^2 - 4x + 5)$ $y' = \dfrac{6x - 4}{3x^2 - 4x + 5}$

(d) $y = \ln\sqrt{x^2 - 3} = \ln(x^2 - 3)^{1/2}$ $y' = \dfrac{\frac{1}{2}(x^2 - 3)^{-1/2}(2x)}{(x^2 - 3)^{1/2}} = \dfrac{x}{x^2 - 3}$

(e) $y = \dfrac{\ln x}{x}$ $y' = \dfrac{x\left(\frac{1}{x}\right) - (\ln x)(1)}{x^2} = \dfrac{1 - \ln x}{x^2}$

RULE 11 LOGARITHMIC DIFFERENTIATION

If $y = u^v$, where u is a function of x, or $u = g(x)$, and v is also a function of x, or $v = h(x)$, then:

$$y' = y\left(v \cdot \frac{u'}{u} + v' \cdot \ln u\right), \text{ or written:}$$

$$\frac{dy}{dx} = y\left(v \cdot \frac{\dfrac{du}{dx}}{u} + \frac{dv}{dx} \cdot \ln u\right)$$

The method for finding the derivative of a function to a power function is called *logarithmic differentiation*. In developing Rule 11, we first take the natural logarithm of both sides of the given equation:

$$y = u^v$$
$$\text{Or: } ln\ y = ln\ (u^v)$$
$$ln\ y = v(ln\ u)$$

Differentiate both sides with respect to x, or:

$$\frac{d}{dx}\ (ln\ y) = \frac{d}{dx}(v \cdot ln\ u)$$

$$\frac{y'}{y} = v \cdot \frac{u'}{u} + (ln\ u)v'$$

Multiply both sides of the equation by y:

$$y' = y\left(v \cdot \frac{u'}{u} + v' \cdot ln\ u\right)$$

EXAMPLE 9 $y = (2x^3)^{5x+1}$. Find y' $\left(\text{or } \dfrac{dy}{dx}\right)$.

$$y' = (2x^3)^{5x+1}\left((5x + 1)\frac{6x^2}{2x^3} + 5(ln\ 2x^3)\right)$$

Additional illustrations:

(a) $y = x^x$ $y' = x^x\left(x \cdot \dfrac{1}{x} + 1(ln\ x)\right) = x^x(1 + ln\ x)$

(b) $y = (x + 5)^{x^3}$ $y' = (x + 5)^{x^3}\left(x^3 \cdot \dfrac{1}{x + 5} + 3x^2 \cdot ln(x + 5)\right)$

(c) $y = 3^x$ $y' = 3^x\left(x \cdot \dfrac{0}{3} + 1(ln\ 3)\right) = 3^x(ln\ 3)$

The solution to illustration (c) is the same as that for Example 4. This verifies that Rule 9 is a special case of Rule 11.

EXERCISE 18–1   Reference: Sections 18.1–18.3

Use the rules to find the derivative with respect to x for each of the following functions, y = f(x).

1. $y = (x^2 + 3)^2$
2. $y = (2x^3 + 1)^{1/3}$
3. $y = (3x^4 - 5)^{3/2}$
4. $y = (x^5 - 2x^3)^{-4}$
5. $y = (5x^2 - 4x + 7)^6$
6. $y = (2x^4 + 5x^3 - 8)^7$
7. $y = (6x^5 - 3x^4 + 2x - 3)^{-5}$

8. $y = (4x^3 + 3x^2 - 6x + 5)^{15}$

9. $y = \left(\dfrac{x + 3}{x - 2}\right)^{20}$

10. $y = \left(\dfrac{2x - 1}{x + 1}\right)^{10}$

11. $y = 3u - 4,\ u = v^3,\ v = x^2$

12. $y = 2u^3 + 5, u = v^2, v = 2x$
13. $y = 5^x$
14. $y = 3(20^x)$
15. $y = 15^{2x}$
16. $y = 100^{3x^2+2x}$
17. $y = 2e^x$
18. $y = 5e^x$
19. $y = e^{3x}$
20. $y = 3e^{4x}$
21. $y = \log_2 x$
22. $y = \log_5 x$
23. $y = \log 3x$
24. $y = \log 4x$

25. $y = \log (2x^2 - 3x)$
26. $y = \log (5x^3 + 6x^2)$
27. $y = \ln 2x$
28. $y = \ln 15x$
29. $y = \ln (4x^3 + 3x^2 - 8)$
30. $y = \ln (5x^2 - 2x + 3)$
31. $y = \ln \sqrt{x^2 + 1}$
32. $y = \dfrac{\ln 2x}{x}$
33. $y = (2x)^x$
34. $y = (3x)^{2x}$
35. $y = (x^2)^{3x+1}$
36. $y = (2x + 3)^{x^2}$

18.4 INVERSE FUNCTIONS

The functions presented in the previous sections were written in the form $y = f(x)$, where y is the dependent variable and x is the independent variable. If a value is assigned to x, then the value of y can be determined. This form expresses y explicitly in terms of x. Conversely, if we rewrite the given function $y = f(x)$ with x as the dependent variable and y as the independent variable, the new function $x = f(y)$ now expresses x explicitly in terms of y. The function $x = f(y)$ is called the *inverse* of $y = f(x)$. The relationship between the derivative of a given function and the derivative of its inverse is illustrated in the following examples:

EXAMPLE 10   Given: $y = f(x) = 2x + 3$. Find: $\dfrac{dy}{dx}, x = f(y)$, and $\dfrac{dx}{dy}$.

$$\frac{dy}{dx} = 2$$

Solve for x: $y = 2x + 3$

$$x = \frac{y - 3}{2} = \frac{1}{2}y - \frac{3}{2}$$

Thus:

$$x = f(y) = \frac{1}{2}y - \frac{3}{2}$$

$$\frac{dx}{dy} = \frac{1}{2}$$

Observe the results obtained in Example 10. They show that the derivative of y with respect to x $\left(\dfrac{dy}{dx} = 2\right)$ is the reciprocal of the derivative of x with respect to y $\left(\dfrac{dx}{dy} = \dfrac{1}{2}\right)$. This is consistent with Rule 12.

RULE 12 INVERSE FUNCTION

If $x = f(y)$ is the inverse function of $y = f(x)$, the derivative of the inverse function is the reciprocal of the derivative of $y = f(x)$, or:

$$\frac{dx}{dy} = \frac{1}{\dfrac{dy}{dx}}$$

EXAMPLE 11 Given: $y = f(x) = 3x^2 + 4x + 10$. Find $\dfrac{dx}{dy}$.

$$\frac{dy}{dx} = 6x + 4$$

$$\frac{dx}{dy} = \frac{1}{6x + 4}, \quad \begin{array}{l}\text{provided that } 6x + 4 \neq 0 \text{ if the derivative of the inverse}\\ \text{function exists}\end{array}$$

EXAMPLE 12 Given: $y = 5x^4 - 2x^3 + x^2 - 7$. Find $\dfrac{dx}{dy}$.

$$\frac{dy}{dx} = 20x^3 - 6x^2 + 2x$$

$$\frac{dx}{dy} = \frac{1}{20x^3 - 6x^2 + 2x}$$

The answers in Examples 11 and 12 are obtained directly from Rule 12, without solving for x values from the given functions.

EXAMPLE 13 Given: $y = f(x) = \ln x$.

Find $x = f(y)$, $\dfrac{dx}{dy}$, then $\dfrac{dy}{dx}$.

From the given function $y = \ln x = \log_e x$, we have the equivalent function:

$$x = e^y$$

Thus:

$$x = f(y) = e^y$$

$$\frac{dx}{dy} = e^y \text{ (See Example 5 and Rule 9.)}$$

Use Rule 12:

$$\frac{dy}{dx} = \frac{1}{e^y} = \frac{1}{x}$$

The above answer is the same as that of Example 8. This illustrates the fact that:

$$y = f(x) = \ln x \text{ and}$$
$$x = f(y) = e^y$$

are inverse functions. In other words, exponential functions and logarithmic functions are related inversely.

18.5 IMPLICIT FUNCTIONS

A function can also be expressed in an *implicit* form of equation. In this form neither y nor x is solved or is expressed explicitly on one side of the equation. For example, the equation $y - 2x = 3$ implicitly defines y in terms of x. If a value is assigned to x, the value of y is implicitly determined by the equation. Similarly, the equation implicitly defines x in terms of y. The implicit function x is determined if a value is assigned to y.

To find the derivative of an implicit function y with respect to x, it is usually easier to use the method of *implicit differentiation.* In this method y is treated as a function of x. The derivative $\dfrac{dy}{dx}$ is obtained by differentiating the implicit equation term by term and then solving the resulting equation for the derivative.

EXAMPLE 14   Given: The implicit function $y - 2x = 3$. Find the derivative of y with respect to x, or dy/dx.

Differentiate each side with respect to x, or:

$$\frac{d}{dx}(y - 2x) = \frac{d}{dx}(3)$$

Differentiate each term with respect to x (use Rule 4), or:

$$\frac{d}{dx}(y) - \frac{d}{dx}(2x) = \frac{d}{dx}(3)$$

$$\frac{dy}{dx} - 2 = 0$$

Solve the equation:

$$\frac{dy}{dx} = 2$$

Check: Solve for y from the given function $y - 2x = 3$.

$$y = 2x + 3$$

This is the explicit function given in Example 14. The answer to $\dfrac{dy}{dx}$ based on the explicit function is also equal to 2.

EXAMPLE 15   Given: The implicit function $x^4 + y^3 = 2y + 8$. Find $\dfrac{dy}{dx}$.

Differentiate each term with respect to x.

$$\frac{d}{dx}(x^4) + \frac{d}{dx}(y^3) = \frac{d}{dx}(2y) + \frac{d}{dx}(8)$$

$$4x^3 + 3y^2 \frac{dy}{dx} = 2\frac{dy}{dx} + 0$$

Solve for dy/dx.

$$3y^2 \left(\frac{dy}{dx}\right) - 2\left(\frac{dy}{dx}\right) = -4x^3$$

$$(3y^2 - 2) \cdot \frac{dy}{dx} = -4x^3$$

$$\frac{dy}{dx} = \frac{-4x^3}{3y^2 - 2}$$

Note: In differentiating the terms y^3 and $2y$, Rule 8 is used in the following manner:

Rule 8 states: If $y = u^n$, where u is a function of x, then:

$$\frac{dy}{dx} = n \cdot u^{n-1} \cdot \frac{du}{dx}$$

Now, let $z = y^n$, where y is a function of x, then:

$$\frac{dz}{dx} = n \cdot y^{n-1} \cdot \frac{dy}{dx}$$

If $z = y^3$:

$$\frac{dz}{dx} = \frac{d(y^3)}{dx} = 3y^2 \cdot \frac{dy}{dx}$$

Also, if $z = y$:

$$\frac{dz}{dx} = \frac{dy}{dx}$$

Thus:

$$\frac{d(2y)}{dx} = 2\frac{dy}{dx}$$

Check: Solve for x from the given function (we cannot solve for y or write y explicitly):

$$x = (2y + 8 - y^3)^{1/4}$$

$$\frac{dx}{dy} = \tfrac{1}{4}(2y + 8 - y^3)^{-3/4}(2 - 3y^2)$$

$$= \frac{2 - 3y^2}{4(2y + 8 - y^3)^{3/4}}$$

Use the inverse rule:

$$\frac{dy}{dx} = \frac{4(2y + 8 - y^3)^{3/4}}{2 - 3y^2}$$

Substitute the x value in the solution of Example 15:

$$\frac{dy}{dx} = \frac{-4x^3}{3y^2 - 2} = \frac{-4((2y + 8 - y^3)^{1/4})^3}{3y^2 - 2}$$

$$= \frac{4(2y + 8 - y^3)^{3/4}}{2 - 3y^2}$$

Thus, the inverse rule method and the implicit differentiation method give the same answer to the given function in the example.

Additional illustrations:

(a) $y + xy^2 - x^3 = 0$. Differentiate each term with respect to x.

$$\frac{dy}{dx} + \left(x \cdot 2y \cdot \frac{dy}{dx} + y^2(1)\right) - 3x^2 = 0$$

$$\frac{dy}{dx}(1 + 2xy) = 3x^2 - y^2$$

$$\frac{dy}{dx} = \frac{3x^2 - y^2}{1 + 2xy}$$

(b) $6xy^3 - 2y^4 = 7$. Differentiate each term with respect to x.

$$6\left(x \cdot 3y^2 \cdot \frac{dy}{dx} + y^3(1)\right) - 2\left(4y^3 \cdot \frac{dy}{dx}\right) = 0$$

$$18xy^2 \cdot \frac{dy}{dx} + 6y^3 - 8y^3 \cdot \frac{dy}{dx} = 0$$

$$\frac{dy}{dx}(18xy^2 - 8y^3) = -6y^3$$

$$\frac{dy}{dx} = \frac{-6y^3}{18xy^2 - 8y^3} = \frac{-3y}{9x - 4y}$$

18.6 HIGHER-ORDER DERIVATIVES

The derivative of a function $f(x)$ as presented in the previous sections can be called the *first derivative* of $f(x)$. If the first derivative is an algebraic function, it is obviously possible to differentiate the derivative according to the appropriate derivative rules. The derivative of the first derivative is then called the *second derivative*. If the second derivative is also an algebraic function, we can differentiate the second derivative to obtain the *third derivative*, and so on.

EXAMPLE 16   Given function: $y = f(x) = x^2 + 1$. Find the higher-order derivatives.

First derivative: $y' = 2x$

Second derivative: $y'' = 2$

Third derivative: $y''' = 0$

Additional illustrations:

(a) $y\ \ \ \ = 2x^3 - 4x^2 + 5x - 6$
$\ \ \ y'\ \ \ = 6x^2 - 8x + 5$
$\ \ \ y''\ \ = 12x - 8$
$\ \ \ y'\,''\ = 12$
$\ \ \ y''\,''\, = 0$

(b) $y\ \ \ \ = \dfrac{x^5}{x^2}$

$\ \ \ y'\ \ \ = \dfrac{x^2(5x^4) - x^5(2x)}{(x^2)^2} = \dfrac{5x^6 - 2x^6}{x^4} = 3x^2$

Or: $y = \dfrac{x^5}{x^2} = x^3.\ \ \ \ \ y' = 3x^2$

$\ \ \ y''\ \ = 6x$
$\ \ \ y'\,''\ = 6$
$\ \ \ y''\,''\, = 0$

In general, the nth derivative can be written in various ways:

$$\frac{d^n y}{dx} = \frac{d^n}{dx}(y) = y^{(n)} = f^{(n)}(x) = D_x^n y = D_x^n f(x)$$

Thus, for the second derivative, it can be written:

$$\frac{d^2 y}{dx} = \frac{d^2}{dx}(y) = y^{(2)} \text{ or } y'' = f^{(2)}(x) \text{ or } f''(x) = D_x^2 y = D_x^2 f(x)$$

Just as the first derivative represents the slope of the original function, the second derivative represents the slope of the first derivative, the third derivative represents the slope of the second derivative, and so on. Higher-order derivatives are very useful in certain theoretical problems in advanced calculus. However, we seldom use derivatives higher than the second in business and economic problems. Thus, our discussion of higher-order derivatives will be limited to the second derivatives in this text.

The graphs of the given function y, the first derivative y', and the second derivative y'' of Example 16 are shown in Figure 18–3 (a), (b), and (c) respectively. Graph (a) shows that the function y is a minimum at $x = 0$ (and $y = 1$). At $x = 0$, the slope of the function is 0 (or $y' = 2x = 2(0) = 0$ as shown in graph (b)), and the slope of the slope of the function is positive (or $y'' = +2$ as shown in graph (c)). The fact that a minimum point on the given function curve has the first derivative 0 and the second derivative positive is very important in a minimization problem to be presented in Chapter 19.

18.7 SUMMARY OF THE RULES FOR DIFFERENTIATION

The rules for finding the derivatives of various types of functions presented in Chapters 17 and 18 are summarized as follows:

Figure 18-3 **(Example 16)**

(a) Original Function:

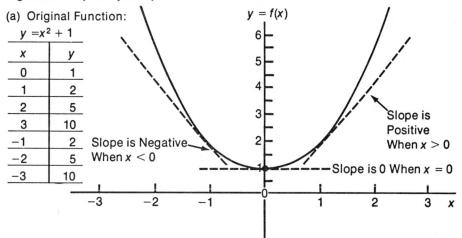

$y = x^2 + 1$

x	y
0	1
1	2
2	5
3	10
−1	2
−2	5
−3	10

$y = f(x)$

Slope is Negative When x < 0

Slope is Positive When x > 0

Slope is 0 When x = 0

(b) First Derivative:

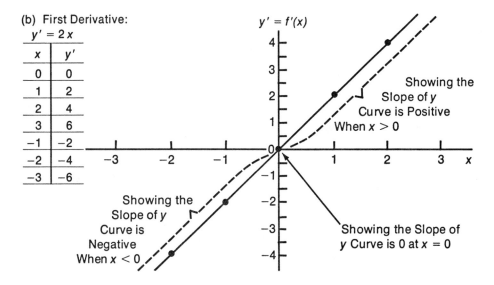

$y' = 2x$

x	y'
0	0
1	2
2	4
3	6
−1	−2
−2	−4
−3	−6

$y' = f'(x)$

Showing the Slope of y Curve is Positive When x > 0

Showing the Slope of y Curve is Negative When x < 0

Showing the Slope of y Curve is 0 at x = 0

(c) Second Derivative:

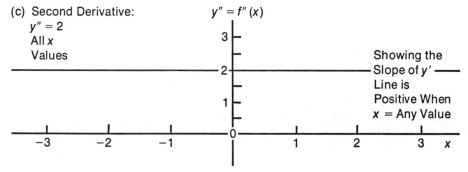

$y'' = 2$
All x
Values

$y'' = f''(x)$

Showing the Slope of y' Line is Positive When x = Any Value

A. Common Algebraic Functions

y is the function of x, or written:

$$y = f(x)$$

RULE 1 If $y = c$, where c is a constant:

$$\frac{dy}{dx} = 0$$

RULE 2 If $y = x^n$:

$$\frac{dy}{dx} = n \cdot x^{n-1}$$

RULE 3 If $y = c \cdot u$, where $u = g(x)$:

$$\frac{dy}{dx} = c \cdot \frac{du}{dx}$$

RULE 4 If $y = u \pm v$, where $u = g(x)$ and $v = h(x)$:

$$\frac{dy}{dx} = \frac{du}{dx} \pm \frac{dv}{dx}$$

RULE 5 If $y = u \cdot v$, where $u = g(x)$ and $v = h(x)$:

$$\frac{dy}{dx} = u\frac{dv}{dx} + v\frac{du}{dx}$$

RULE 6 If $y = \dfrac{u}{v}$, where $u = g(x)$ and $v = h(x)$:

$$\frac{dy}{dx} = \frac{v\dfrac{du}{dx} - u\dfrac{dv}{dx}}{v^2}$$

B. The Chain Rule for Composite Functions

RULE 7 If $y = f(u)$ and $u = g(x)$:

$$\frac{dy}{dx} = \frac{dy}{du} \cdot \frac{du}{dx}$$

If $y = f(u)$, $u = g(v)$, and $v = h(x)$:

$$\frac{dy}{dx} = \frac{dy}{du} \cdot \frac{du}{dv} \cdot \frac{dv}{dx}$$

C. Power Functions

RULE 8   If $y = u^n$, where $u = g(x)$ and n is a real number:

$$\frac{dy}{dx} = n \cdot u^{n-1} \cdot \frac{du}{dx}$$

Special case: $u = g(x) = x$

If $y = x^n$:

$$\frac{dy}{dx} = n \cdot x^{n-1} \text{ (Rule 2)}$$

D. Exponential and Logarithmic Functions

a is a constant, $u = g(x)$, and $v = h(x)$.

RULE 9   If $y = a^u$:

$$\frac{dy}{dx} = a^u \cdot \ln a \cdot \frac{du}{dx}$$

Special case: $a = e = 2.71828$

If $y = e^u$:

$$\frac{dy}{dx} = e^u \cdot \frac{du}{dx}$$

RULE 10   If $y = \log_a u$:

$$\frac{dy}{dx} = \frac{1}{u}(\log_a e)\frac{du}{dx}$$

Special case: $a = e$

If $y = \ln u$:

$$\frac{dy}{dx} = \frac{1}{u} \cdot \frac{du}{dx}$$

RULE 11   If $y = u^v$:

$$\frac{dy}{dx} = y\left(v \cdot \frac{\frac{du}{dx}}{u} + \frac{dv}{dx} \cdot \ln u\right)$$

E. Inverse Functions

RULE 12   If $x = f(y)$ is the inverse function of $y = f(x)$:

$$\frac{dx}{dy} = \frac{1}{\frac{dy}{dx}}$$

EXERCISE 18-2   Reference: Sections 18.4–18.6

A. *Inverse Functions. Use the rules to find the derivative of x with respect to y, or dx/dy, for problems 1–8.*

1. $y = 3x + 4$
2. $y = 6x - 5$
3. $y = 5x^2 - 7x + 2$
4. $y = 4x^2 + 5x - 3$
5. $y = 5x^3 + 4x^2 - 3x + 8$
6. $y = 10x^5 - 2x^3 + 5x^2 - 3$

7. $y = \dfrac{x + 2}{x - 1}$
8. $y = (x + 3)(x^2 - 2)$
9. $x = y + \frac{1}{2}y^2 + \frac{1}{4}y^4$. Find $\dfrac{dy}{dx}$.
10. $x = 2y^3 + 5y^2 - y^{1/2} + 4$. Find $\dfrac{dy}{dx}$.

B. *Implicit Functions. Find the derivative of y with respect to x, or dy/dx, for each of the following functions:*

11. $y - 3x = 4$
12. $6x - y = 5$
13. $y + 7x = 5x^2 + 2$
14. $4x^2 - y = 3 - 5x$
15. $y^2 + 4x = 3x^2$

16. $3y^2 - 5x = x^2$
17. $x^3 + y^4 = 5 - 3y$
18. $y - 5x + y^2 = 4x^3$
19. $2x^3 - 5y^2 - x^2y = 15$
20. $2xy^3 + 3y^4 = 8 - x^2$

C. *Higher-Order Derivatives. Find the higher-order derivatives with respect to x for each of the following functions in problems 21 to 28. Also, graph problems 21 and 22.*

21. $y = x^2 - 2$
22. $y = (x + 1)^2 + 2$
23. $y = 3x^2 + 5$
24. $y = (x - 1)^2 - 3$
25. $y = 4x^3 + 2x^2 - x + 7$
26. $y = 5x^3 - 4x^2 + 3x - 2$

27. $y = (x + 1)(2x - 3)$
28. $y = (3x^2 - 1)(x^2)$
29. $y = \dfrac{x - 5}{x + 1}$. Find the third derivative of y.
30. $y = \dfrac{x + 1}{x}$. Find the fourth derivative of y.

Chapter
19 Applications of Derivatives

This chapter illustrates the applications of derivatives. Derivatives are particularly useful in solving what are called optimization problems which are of either the maximization or minimization type. Maximization problems in business typically involve profit, whereas minimization problems involve cost.

19.1 FINDING MAXIMUM AND MINIMUM POINTS

The two main topics of discussion in this section are (A) Terminology, and (B) The Basic Procedure for Finding Local Maxima and Minima.

A. Terminology

The terms frequently used in finding maximum and minimum points on a curve or in a function are presented in the following paragraphs.

1. DOMAIN OF THE FUNCTION. The *domain of the function* refers to the set of possible values of an independent variable (x) of a function within a specified interval. There are three types of intervals:

Closed Interval. In a *closed interval,* the end points are included in the interval, such as $x = -1$ and $x = 5$ being included in the interval from $x = -1$ to $x = 5$, as shown in Figure 19-1, or written:

$$-1 \leq x \leq 5$$

Open Interval. In an *open interval,* the end points are not included in the interval, such as:

$$-1 < x < 5$$

Mixed Interval. In a *mixed interval,* one end point is included but the other end point is not, such as:

$$-1 \leq x < 5 \ (x = -1 \text{ is included but } x = 5 \text{ is not.})$$
$$\text{And: } \ -1 < x \leq 5 \ (x = -1 \text{ is not included but } x = 5 \text{ is.})$$

2. LOCAL MAXIMUM. A *local maximum* is also called a *relative maximum*. The function $y = f(x)$ is said to reach a local maximum at $x = a$ if $f(a)$ is greater than any value of $f(x)$ for x at points near a on both sides. Figure 19–1 shows two local maxima: $y = 3$ at $x = 1$ and $y = 4$ at $x = 3$.

Figure 19-1   **Graphical Illustration of Maximum and Minimum Points of A Function**

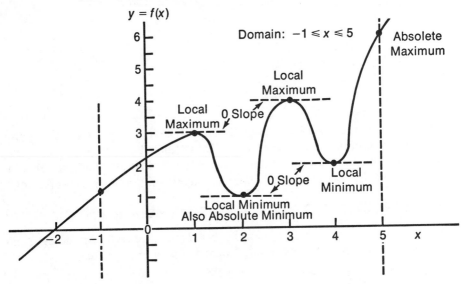

3. LOCAL MINIMUM. A *local minimum* is also called a *relative minimum*. The function $y = f(x)$ is said to reach a local minimum at $x = a$ if $f(a)$ is smaller than any value of $f(x)$ for x at points near a on both sides.

Figure 19–1 shows two local minima: $y = 1$ at $x = 2$ and $y = 2$ at $x = 4$.

End points are not considered in determining the local maxima or minima since the function is not defined at both sides of an end point.

4. ABSOLUTE MAXIMUM AND MINIMUM. An *absolute maximum* is the greatest value of $f(x)$ for x in the domain. On the other hand, an *absolute minimum* is the smallest value of $f(x)$. An absolute maximum or minimum can occur at an end point of a closed interval or at any local maximum or minimum. However, absolute maximum and minimum are not specified for a function if the domain of the function is an open interval.

Figure 19–1 shows that the absolute maximum is at the end point: $y = 6$ at $x = 5$. The absolute minimum is at the local minimum: $y = 1$ at $x = 2$.

5. CRITICAL POINTS. A *critical point* of a function is a point at which the slope of the function is 0. It is also specified by the value of the independent variable (x); that is the point at x where the slope is 0.

Figure 19–1 shows that the slopes at all local maximum and minimum points are 0; that is, slopes = 0 at $x = 1, 2, 3,$ and 4. Thus, local maxima and minima are located at critical points.

6. POINTS OF INFLECTION. A *point of inflection* of a function is a point at which the concavity of the curve changes. The change of the curve generally takes one of the following forms:

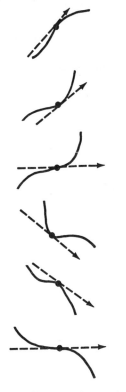

The increase at the point of inflection (•) is changed to a smaller increase. The slope at the point is positive.

The increase is changed to a greater increase. The slope at the point is positive.

The increase is changed to 0 then to another increase. The slope at the point of inflection is 0.

The decrease is changed to a smaller decrease. The slope at the point is negative.

The decrease is changed to a greater decrease. The slope at the point is negative.

The decrease is changed to 0 then to another decrease. The slope at the point of inflection is 0.

B. Basic Procedure for Finding Local Maxima and Minima

The steps for finding a maximum or minimum value of a continuous function are as follows:

1. Find the first derivative of the given function, y'.

   The first derivative represents the slope of the function. Once this step is done, the slope at any point on the curve representing the function can be found.

2. Find the critical points: Let the first derivative equal 0 and solve for x from the equation. The x values specify the locations of the critical points.

   The slope at a local maximum or minimum point on the curve is 0. This step will locate the maximum or minimum point. However, the first derivative $y' = 0$ is only a necessary condition for a maximum or minimum point; it is not sufficient evidence for the existence of such a point. A critical point may also be a point of inflection of the function. Thus, the next step is to evaluate each critical point.

3. Evaluate each critical point to determine whether the function $y = f(x)$ has a maximum or minimum value at the point. The evaluation can be done in the following manner:

First, take the second derivative of the function, y''. We recall that the first derivative represents the slope of the given function, and the second derivative represents the slope of the slope of the given function.

Substitute the x value at the critical point in the second derivative, y'' or $f''(x)$.

a. If the second derivative has a negative value, the $f(x)$ at the critical point is a maximum value. The negative derivative indicates that the slope of the function changes from positive to negative at the critical point.

   This fact is true since the slope to the left of a local maximum point must be positive and that to the right must be negative.

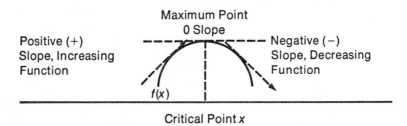

(Also see Figure 19–2 in Example 1, and Figure 19–4 (a, b, and c) for interval $-3 \leq x \leq -\frac{5}{8}$ in Example 3.)

b. If the second derivative has a positive value, the $f(x)$ at the critical point is a minimum value. The positive derivative indicates that the slope of the function changes from negative to positive at the critical point.

   This fact is true since the slope to the left of a local minimum point must be negative and that to the right must be positive.

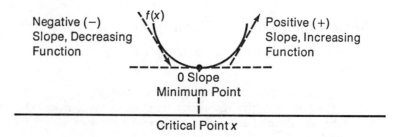

(Also see Figure 18–3 in Example 16 of Chapter 18, Figure 19–3 in Example 2, and Figure 19–4 for interval $-\frac{5}{8} \leq x \leq 2$ in Example 3.)

c. If the second derivative is 0, the $f(x)$ at the critical point is neither a maximum nor minimum value. It is a point of inflection. (See Example 5.)

The preceding steps are illustrated in Examples 1–5.

EXAMPLE 1   Find the local maxima and minima, if any, of the function:

$$y = -x^2 + 4x + 1$$

STEP 1   Find the first derivative:

$$y' = -2x + 4$$

STEP 2   Find the critical points:

Let $y' = -2x + 4 = 0$
$$x = \tfrac{4}{2} = 2$$

We have only one critical point at $x = 2$.

STEP 3   Find the second derivative:

$$y'' = -2$$

The second derivative is a negative value. Thus, $y = f(x)$ at the critical point, $x = 2$, has the maximum value:

$$y = f(x) = f(2) = -(2)^2 + 4(2) + 1 = 5$$

Check: The graph of the given function is shown in Figure 19–2. Other than by graphical observation, there are two ways to check the obtained maximum or minimum value. Both ways are based on the function $f(x)$ values at two points close to the maximum or minimum value: one at the left (denoted by x_1) and the other at the right (denoted by x_2).

Figure 19-2   (Example 1)

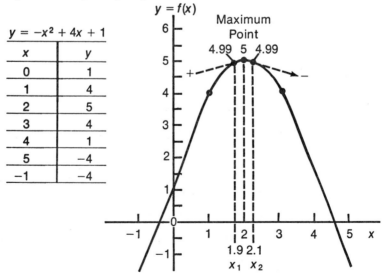

x	y
0	1
1	4
2	5
3	4
4	1
5	−4
−1	−4

1. Use $f(x)$ values directly:

The maximum value must be larger than both $f(x_1)$ and $f(x_2)$ values.
The minimum value must be smaller than both $f(x_1)$ and $f(x_2)$ values.

In Example 1: $y = -x^2 + 4x + 1$

Let $x_1 = 2 - .1 = 1.9$
$$y = f(x_1) = f(1.9) = -(1.9)^2 + 4(1.9) + 1 = 4.99$$

Let $x_2 = 2 + .1 = 2.1$
$$y = f(x_2) = f(2.1) = -(2.1)^2 + 4(2.1) + 1 = 4.99$$

Since $y = f(2) = 5$ is larger than both $f(1.9) = 4.99$ and $f(2.1) = 4.99$, the function value 5 at $x = 2$ is the maximum of the function.

2. Use the first derivative $f'(x)$:

If it is a maximum value, $f'(x_1)$ is positive and $f'(x_2)$ is negative.
If it is a minimum value, $f'(x_1)$ is negative and $f'(x_2)$ is positive.

In Example 1: $y' = f'(x) = -2x + 4$

At $x_1 = 1.9$, $y' = f'(1.9) = -2(1.9) + 4 = +0.2$
At $x_2 = 2.1$, $y' = f'(2.1) = -2(2.1) + 4 = -0.2$

Since the slope is from positive (at x_1) to negative (at x_2), the function value 5 at $x = 2$ is the maximum of the function.

EXAMPLE 2   Find the local maxima and minima, if any, of the function:

$$y = x^2 + 2x + 3$$

STEP 1   Find the first derivative:

$$y' = 2x + 2$$

STEP 2   Find the critical points:

Let $y' = 2x + 2 = 0$
$$x = -\tfrac{2}{2} = -1$$

We have only one critical point at $x = -1$.

STEP 3   Find the second derivative:

$$y'' = 2$$

The second derivative is a positive value. Thus, $y = f(x)$ at the critical point, $x = -1$, has the minimum value:

$$y = f(x) = f(-1) = (-1)^2 + 2(-1) + 3 = 2$$

Check: The graph of the given function is shown in Figure 19–3.

1. Use $f(x)$ values directly:

Let $x_1 = -1.1$. $y = f(-1.1) = (-1.1)^2 + 2(-1.1) + 3 = 2.01$
Let $x_2 = -.9$. $y = f(-.9) = (-.9)^2 + 2(-.9) + 3 = 2.01$

Since $y = f(-1) = 2$ is smaller than both $f(-1.1) = 2.01$ and $f(-.9) = 2.01$, the function value 2 at $x = -1$ is the minimum of the function.

Figure 19-3 (Example 2)

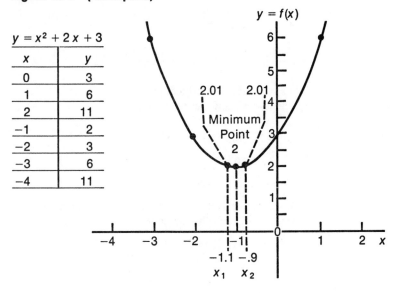

$y = x^2 + 2x + 3$

x	y
0	3
1	6
2	11
−1	2
−2	3
−3	6
−4	11

2. Use the first derivative y':

   At $x_1 = -1.1$, $y' = 2x + 2 = 2(-1.1) + 2 = -0.2$.
   At $x_2 = -.9$, $y' = 2x + 2 = 2(-.9) + 2 = 0.2$.

Since the slope is from negative (at x_1) to positive (at x_2), the function value 2 at $x = -1$ is the minimum of the function.

EXAMPLE 3   Find the local maxima and minima, if any, of the function:

$y = 2x^3 + 5x^2 - 4x - 3$

STEP 1   Find the first derivative:

$y' = 6x^2 + 10x - 4$

STEP 2   Find the critical points:

Let $y' = 6x^2 + 10x - 4 = 0$

Divide each side by 2:

$3x^2 + 5x - 2 = 0$
$(x + 2)(3x - 1) = 0$

When $x + 2 = 0$, $x = -2$.
When $3x - 1 = 0$, $x = \frac{1}{3}$.

The same answers for x can be obtained by using the quadratic formula:

$$x = \frac{-b \pm \sqrt{b^2 - 4ac}}{2a}$$

In the equation $6x^2 + 10x - 4 = 0$, $a = 6$, $b = 10$, and $c = -4$. Thus:

$$x = \frac{-10 \pm \sqrt{10^2 - 4(6)(-4)}}{2(6)} = \frac{-10 \pm \sqrt{196}}{12} = \frac{-10 \pm 14}{12}$$

$$x = \frac{-10 + 14}{12} = \frac{1}{3}, \quad x = \frac{-10 - 14}{12} = -2$$

Thus, we have two critical points:
one at $x = -2$ and the other at $x = \frac{1}{3}$.

STEP 3   Find the second derivative:

$$y'' = 12x + 10$$

At $x = -2$, $y'' = 12(-2) + 10 = -14$.

The second derivative is a negative value. Thus, $y = f(x)$ at the critical point, $x = -2$, has the maximum value:

$$y = f(-2) = 2(-2)^3 + 5(-2)^2 - 4(-2) - 3$$
$$= -16 + 20 + 8 - 3 = 9$$

At $x = \frac{1}{3}$, $y'' = 12(\frac{1}{3}) + 10 = 14$

The second derivative is a positive value. Thus, $y = f(x)$ at the critical point, $x = \frac{1}{3}$, has the minimum value:

$$y = f(\tfrac{1}{3}) = 2(\tfrac{1}{3})^3 + 5(\tfrac{1}{3})^2 - 4(\tfrac{1}{3}) - 3$$
$$= \tfrac{2}{27} + \tfrac{5}{9} - \tfrac{4}{3} - 3 = -3\tfrac{19}{27}$$

The graphs of the given function y, the first derivative y', and the second derivative y'' are shown in Figure 19–4 respectively.

EXAMPLE 4   Find the local maxima and minima, if any, of the function,

$$y = 2x^3 + 5x^2 - 4x - 3$$

in the interval $-3 \le x \le 0$.

Figure 19-4   (Example 3)

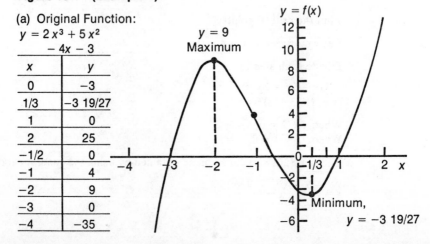

(a) Original Function:
$y = 2x^3 + 5x^2$
$\quad - 4x - 3$

x	y
0	-3
1/3	-3 19/27
1	0
2	25
-1/2	0
-1	4
-2	9
-3	0
-4	-35

(b) First Derivative:

$y' = 6x^2 + 10x - 4$

x	y
0	−4
1/3	0
1	12
2	40
−1/2	−7 1/2
−5/6	−8 1/6
−1	−8
−2	0
−3	20

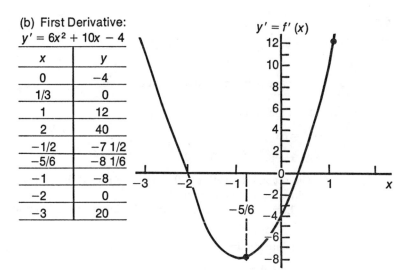

(c) Second Derivative:

$y'' = 12x + 10$

x	y
0	10
1	22
2	34
−5/6	0
−1	−2
−2	−14

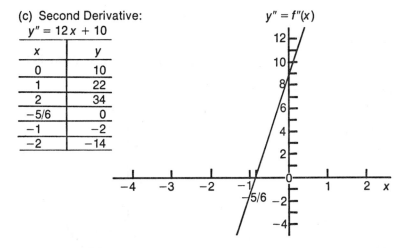

From Example 3, we have two critical points: At $x = -2$ and $x = \frac{1}{3}$. Since $\frac{1}{3}$ is larger than 0 in the interval, the critical point at $x = \frac{1}{3}$ must be disregarded.

Based on the solution in Example 3, we may state that we have only one local maximum value at $x = -2$, and the maximum value is $y = f(-2) = 9$ as shown in Figure 19–4 for the interval from −3 to 0 on the x scale.

EXAMPLE 5   Find the local maxima and minima, if any, of the function,

$y = x^3 + 4$

in the interval $-2 \le x \le 2$.

STEP 1   Find the first derivative:

$y' = 3x^2$

STEP 2   Find the critical points:

Let $y' = 3x^2 = 0$

$$x = 0$$

We have only one critical point at $x = 0$.

STEP 3   Find the second derivative:

$$y'' = 6x$$

At $x = 0, \quad y'' = 6(0) = 0.$

The second derivative is 0. Thus, the $f(x)$ at the critical point, $x = 0$, is neither maximum nor minimum.

There are no local maxima and minima in the interval $-2 \leq x \leq 2$. However, there are:

Absolute maximum at $x = 2$:

$$y = x^3 + 4 = 2^3 + 4 = 12$$

Absolute minimum at $x = -2$:

$$y = (-2)^3 + 4 = -4$$

Check: The graph of the given function is shown in Figure 19-5.

Figure 19-5   **(Example 5)**

$y = x^3 + 4$	
x	y
0	4
1	5
2	12
3	31
−1	3
−2	−4
−3	−23

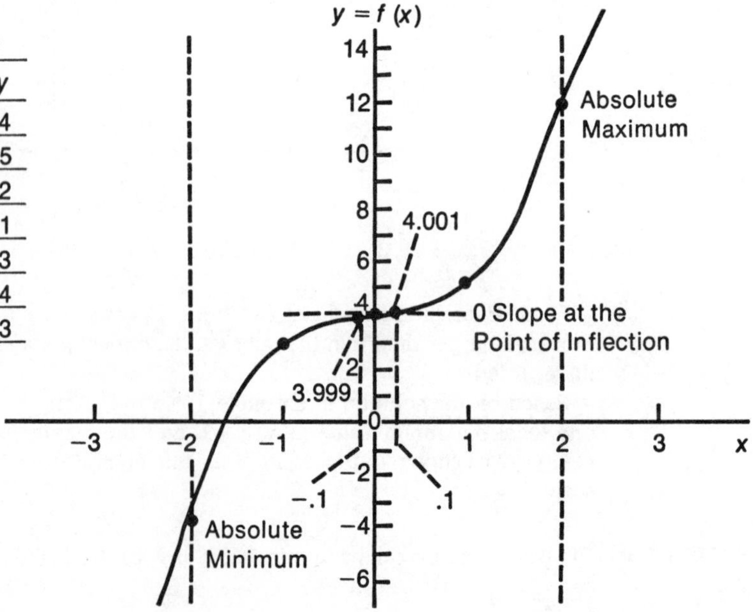

The critical point at $x = 0$ is checked as follows:

1. Use $y = x^3 + 4$ values directly:

Let $x_1 = -.1$. $y = f(-.1) = (-.1)^3 + 4 = 3.999$
Let $x_2 = +.1$. $y = f(.1) = (.1)^3 + 4 = 4.001$

At the critical point, $x = 0$, $y = f(0) = 0^3 + 4 = 4$.

Since $f(0) = 4$ is larger than $f(-.1) = 3.999$ but smaller than $f(.1) = 4.001$, the function value 4 at $x = 0$ is neither maximum nor minimum of the function.

2. Use the first derivative y':

At $x_1 = -.1$, $y' = 3x^2 = 3(-.1)^2 = +0.03$.
At $x_2 = +.1$, $y' = 3(.1)^2 = +0.03$.

Since the slopes at x_1 and x_2 are both positive, the function value 4 at $x = 0$ is neither maximum nor minimum.

Observe the critical point at $x = 0$ in Figure 19–5. We can see that the critical point is a point of inflection.

The preceding examples illustrated the procedure for finding all local maxima and minima of the functions which are continuous. Example 6 illustrates a case where the function is discontinuous.

EXAMPLE 6   Find the local maxima and minima, if any, of the function:

$$y = \frac{1}{x - 1} + 3$$

The given function y is discontinuous at $x = 1$, since the term $\frac{1}{(x - 1)}$ is undefined at $x = 1$:

$$\frac{1}{x - 1} = \frac{1}{1 - 1} = \frac{1}{0}, \text{ which is undefined.}$$

When we follow the above illustrated procedure of finding local maxima and minima for the given function, we have:

STEP 1   Find the first derivative:

$$y' = \frac{-1}{(x - 1)^2}$$

STEP 2   Find the critical points:

There is no value for x that can make $y' = 0$, since:

$$\frac{-1}{(x - 1)^2} \neq 0$$

If $x = 1$, $y' = \frac{-1}{(1 - 1)^2} = \frac{-1}{0}$, which is undefined.

Thus, y' is discontinuous at $x = 1$. We have no critical point of the given function.

STEP 3   Find the second derivative:

$$y'' = (-1)(-2)(x-1)^{-3} = \frac{2}{(x-1)^3}$$

Again, there is no value for x that can make $y'' = 0$, since:

$$\frac{2}{(x-1)^3} \neq 0$$

If $x = 1$, $y'' = \frac{2}{0}$, which is also undefined.

Thus, y'' is discontinuous at $x = 1$.

The infinite discontinuities of the functions y, y', and y'' at $x = 1$ are shown in Figure 19–6.

Figure 19-6 **(Example 6)**

(a) Original Function:

$$y = \frac{1}{x-1} + 3$$

x	y
0	2
0.9	−7
1	Undefined
1.1	13
2	4
3	3.5
4	3.3
−1	2.5
−2	2.67
−3	2.75

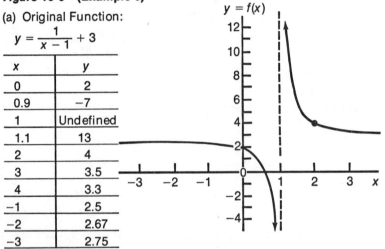

(b) First Derivative:

$$y' = \frac{-1}{(x-1)^2}$$

x	y
0	−1
0.9	−100
1	Undefined
1.1	−100
2	−1
3	−0.25
−1	−0.25
−2	−0.11
−3	−0.06

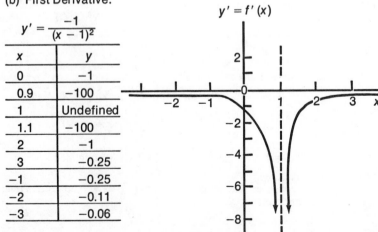

(c) Second Derivative:

$$y'' = \frac{2}{(x-1)^3}$$

x	y
0	−2
0.9	−2,000
1	Undefined
1.1	2,000
2	2
3	0.25
−1	−0.25
−2	−0.07
−3	−0.03

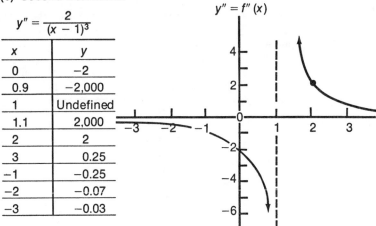

EXERCISE 19-1   Reference: Section 19.1

A. *For each of the following functions, (a) graph the function* y, *(b) find the local maxima and minima, if any, of the function, and (c) check the solution by the two methods presented in Example 1 by letting* x_1 *and* $x_2 = x$ *at a critical point* − .1 *and* + .1 *respectively.*

1. $y = -x^2 + 2x + 3$ 3. $y = 3x^2 + 12x - 10$
2. $y = 4 - 6x - 3x^2$ 4. $y = 4x^2 - 16x + 3$

B. *Find the local maxima and minima, if any, of each of the following functions:*

5. $y = 4x - x^2 + 5$ 9. $y = (\frac{1}{3})x^3 + x^2 - 3x - 4$
6. $y = 12x - 2x^2 - 2$ 10. $y = x^3 + (\frac{1}{2})x^2 - 2x + 3$
7. $y = 3x^2 - 4x + 5$ 11. $y = 2x^3 - 7x^2 + 4x - 5$
8. $y = 7x^2 + x - 6$ 12. $y = 2x^3 + 3x^2 - 12x + 6$
13. $y = x^3 - x^2 - 8x + 5$ in the interval: (a) $-3 \le x \le 3$, (b) $0 \le x \le 3$
14. $y = (\frac{2}{3})x^3 + (\frac{1}{2})x^2 - 15x + 3$ in the interval: (a) $-4 \le x \le 3$, (b) $-4 \le x \le 0$
15. $y = 2x^3 + 1$ in the interval $-3 \le x \le 1$. Also, find the absolute maximum.
16. $y = x^3 - 2$ in the interval $-4 \le x \le 3$. Also, find the absolute minimum.
17. $y = \dfrac{1}{x} + 2$
18. $y = \dfrac{1}{x+3} - 5$

19.2 MAXIMA AND MINIMA APPLICATIONS TO STATEMENT PROBLEMS

The application of the methods for finding local maxima and minima in practical business problems are illustrated in the examples which follow.

EXAMPLE 7   **MAXIMIZING TOTAL SALES (REVENUE)**

The record of Pat's Grocery Store shows that the quantity of certain types of meat sold daily varies inversely with the price of the meat as follows:

Price Per Pound of Meat, p	Number of Pounds Sold Daily, x
$2.00	4
1.90	5
1.80	6
1.70	7
1.60	8
1.50	9
1.40	10
1.30	11
1.20	12
1.10	13
1.00	14

(a) Maximize the total sales and (b) find the price per pound at which the total sales of meat will be maximum.

SOLUTION   Let $p =$ price per pound of meat
$x =$ number of pounds of meat sold.

The relationship between price and quantity is shown in Figure 19–7(a) as a straight line. The linear equation can be obtained by selecting any two points on the line, such as points $(x = 4, p = 2.00)$ and $(x = 8, p = 1.6)$.

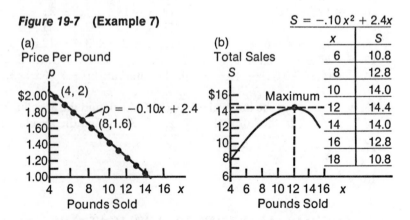

Figure 19-7   (Example 7)

(a) Price Per Pound

(b) Total Sales

$S = -.10x^2 + 2.4x$

x	S
6	10.8
8	12.8
10	14.0
12	14.4
14	14.0
16	12.8
18	10.8

The slope of the line is,

$$b = \frac{1.6 - 2.00}{8 - 4} = -0.10$$

which indicates that when the price is reduced by $0.10 (or 10¢), the quantity sold is increased by 1 pound.

Substitute the b value in the linear equation:

$y = bx + a$ where $y = p$

At the point ($x = 4$, $y = p = 2$):

$2 = -0.10(4) + a$

$a = 2 + 0.40 = 2.40$

Thus, the equation is $p = -0.10x + 2.40$.

This equation may also be obtained by writing two linear equations based on two points:

At point ($x = 4$, $p = 2$), equation $p = bx + a$ can be written: $2 = b(4) + a$.

At point ($x = 8$, $p = 1.6$), equation $p = bx + a$ can be written: $1.6 = b(8) + a$.

Solve for a and b simultaneously from the two derived equations; we have $b = -0.10$ and $a = 2.40$.

(a) The total sales (also called total revenue): S = price per pound × number of pounds sold, or:

$$S = p \cdot x = (-0.10x + 2.40)x$$
$$S = -0.10x^2 + 2.40x$$

The $S = f(x)$ is shown in Figure 19-7(b).

Find the first derivative of S with respect to x:

$$S' = \frac{dS}{dx} = -0.10(2)x + 2.40 = -0.20x + 2.40$$

Let the first derivative of $S = 0$:

$$-0.20x + 2.40 = 0$$
$$x = 2.40/0.20 = 12 \text{ (critical point)}$$

Take the second derivative of S with respect to x:

$$S'' = -0.20 \text{ (negative value)}$$

Thus, $S = f(x)$ at the critical point $x = 12$ has the maximum value:

$$S = -.10x^2 + 2.4x = -.10(12)^2 + 2.4(12) = \$14.40$$

(b) At $x = 12$, $p = -0.10x + 2.40 = -0.10(12) + 2.40 = \$1.20$ per pound

When the store sells 12 pounds of meat at $1.20 per pound, the store can maximize the total sales to $14.40.

EXAMPLE 8   **FINDING THE MARGINAL REVENUE**

Refer to Example 7. Find the marginal revenue at each level of sales.

SOLUTION   The *marginal revenue* is defined as an increase in revenue obtained from the sale of one additional unit of product. For example, if the total sale of 4 pounds of meat is $8.00 and the total sale of 5 pounds of meat is $9.50, the sale of the 5th pound contributed a marginal revenue of $1.50 (= $9.50 − $8.00.)

The total sale and the marginal revenue at each level of sales are listed in Table 19–1. Each total sale is the product of price per pound and the number of pounds sold. The total sale reaches the highest amount: $14.40 = $1.20 × 12 pounds.

Table 19–1

Price Per Pound of Meat, p	Number of Pounds Sold Daily, x	Total Sale (or Revenue), $p \cdot x$	Marginal Revenue, from $p \cdot x$ Column
$2.00	4	$ 8.00	–
1.90	5	9.50	$1.50
1.80	6	10.80	1.30
1.70	7	11.90	1.10
1.60	8	12.80	.90
1.50	9	13.50	.70
1.40	10	14.00	.50
1.30	11	14.30	.30
1.20	12	14.40 (Maximum)	.10
1.10	13	14.30	−.10
1.00	14	14.00	−.30

When we assume that the sales function is continuous, such as the smoothed curve in Figure 19–7(b), the marginal revenue is computed from the first derivative of the total sales, or $S' = -0.20x + 2.40$ as presented in Example 7. For example:

At $x = 5$ pounds, the marginal revenue $= -0.20(5) + 2.40 = \$1.40$.

Table 19–1, which gives only individual (not continuous) sales levels, shows that the marginal revenue of the 5th pound is $1.50.

At $x = 12$ pounds, where the total sale reaches the highest amount ($14.40), the marginal revenue $= -0.20(12) + 2.40 = \$0$.

Thus, we may state that when the marginal revenue is 0 (or $S' = 0$), the total revenue reaches the maximum point.

EXAMPLE 9   MAXIMIZING PROFIT

Refer to Example 7. Assume that the cost, C, is a function of the quantity sold, or can be expressed,

$C = f(x) = .60x$

which represents that the cost of each pound of meat is $.60. Our objective now is to maximize the total profit.

SOLUTION   Profit (P) = Sales (S) − Cost (C)

Thus: $P = (-0.10x^2 + 2.40x) - .60x$
$P = -0.10x^2 + 1.8x$

Find the first derivative of P with respect to x:

$$P' = \frac{dP}{dx} = -0.10(2)x + 1.8 = -0.20x + 1.8$$

Let the first derivative of $P = 0$:

$-0.20x + 1.8 = 0,$
$\qquad\qquad x = 9 \text{ (critical point)}$

Find the second derivative of P with respect to x:

$P'' = -0.20$, (negative value)

Thus, $P = f(x)$ at the critical point $x = 9$ has the maximum value:

$P = -0.10x^2 + 1.8x = -0.10(9^2) + 1.8(9) = \$8.10$

At $x = 9$, $p = -0.10x + 2.40 = -0.10(9) + 2.40$
$\qquad\qquad = \$1.50 \text{ per pound}$

When the store sells 9 pounds of meat at $1.50 per pound, the store can maximize the profit to $8.10.

Check: The total profit and the marginal profit at each level of sales are listed in Table 19–2. The total profit reaches the highest amount, $8.10. The marginal profit is an increase in profit obtained from the sale of

Table 19–2

Price Per Pound of Meat, p	Number of Pounds Sold Daily, x	Total Sale $S = p \cdot x$	Total Cost $C = \$.60x$	Total Profit $P = S - C$	Marginal Profit, From P Column
$2.00	4	$ 8.00	$2.40	$5.60	—
1.90	5	9.50	3.00	6.50	$.90
1.80	6	10.80	3.60	7.20	.70
1.70	7	11.90	4.20	7.70	.50
1.60	8	12.80	4.80	8.00	.30
1.50	9	13.50	5.40	8.10 (Maximum)	.10
1.40	10	14.00	6.00	8.00	−.10
1.30	11	14.30	6.60	7.70	−.30
1.20	12	14.40	7.20	7.20	−.50
1.10	13	14.30	7.80	6.50	−.70
1.00	14	14.00	8.40	5.60	−.90

one additional unit of product. When we assume that the profit function $P = f(x)$ is continuous, the marginal profit is 0 when the total profit is at the maximum point.

Observe Table 19–2. If the store sells 12 pounds, it will maximize the total sale, $14.40, but the profit is only $7.20. This profit is smaller than $8.10, the maximized profit when the store sells only 9 pounds.

EXAMPLE 10 MINIMIZING INVENTORY COST

The record concerning certain types of parts used in the Lynn Factory shows the following:

$n = 5,000$ units – used in the factory per year.
$a = \$50$ – cost of placing an order, such as issuing purchase order, shipping cost, and so on.
$b = \$2$ – cost of carrying one unit of inventory per year, such as warehouse men's wages, storage spaces, and so on.
$Q = $ quantity of ordering each time.
$Q/2 = $ average inventory size during a year.

Find the size of the order that should be placed each time which will result in minimum annual inventory cost.

SOLUTION Let $I = $ the total inventory cost during a year. The annual inventory cost (I) will include two items:

1. Annual cost of placing orders = number of orders in a year (n/Q) times the cost of placing a single order (a) = na/Q.
2. Annual cost of carrying inventory = cost of carrying one unit of inventory (b) times the average inventory during the year ($Q/2$) = $bQ/2$. Thus:

$$I = \frac{na}{Q} + \frac{bQ}{2}$$

Find the first derivative of I with respect to Q:

$$I' = \frac{dI}{dQ} = \frac{-na}{Q^2} + \frac{b}{2}$$

Let the first derivative of $I = 0$.

$$\frac{-na}{Q^2} + \frac{b}{2} = 0$$

$$Q^2 = \frac{2na}{b}$$

$$Q = \sqrt{\frac{2na}{b}}$$

Substitute the given values in the preceding formula:

$$Q = \sqrt{\frac{2(5,000)(\$50)}{\$2}} = \sqrt{250,000} = 500 \text{ units}$$

The second derivative of I with respect to Q is:

$$I'' = \frac{-(-na)(2Q)}{(Q^2)^2} = \frac{2na}{Q^3}$$

At $Q = 500$:

$$I'' = \frac{2(5,000)(\$50)}{500^3} = \frac{\$2}{500}, \text{ (positive value)}$$

Thus, $I = f(Q)$ at the critical point $Q = 500$ units has the minimum value:

$$I = \frac{na}{Q} + \frac{bQ}{2} = \frac{(5,000)(\$50)}{500} + \frac{(\$2)(500)}{2} = \$1,000$$

The factory should order in lots of 500 units each time. This lot size will result in minimization of total annual inventory cost.

EXAMPLE 11 **MINIMIZING A STORE SITE**

A company plans to construct a department store. The site of the store requires the following: (1) The front of the store should have 50 feet for parking; (2) each of the two sides of the store should have 20 feet for driving; (3) the back of the store should have 30 feet for delivering; and (4) the store space should be at least 10,000 square feet. What is the smallest site that the company can buy?

SOLUTION Let $w =$ the width of the store space,
$x =$ the length of the store space, and
$y =$ the area of the site of the store.

Then, $wx = 10,000$ square feet, and $w = 10,000/x$.

$$
\begin{aligned}
y &= (w + 20 + 20)(x + 50 + 30) \\
&= (w + 40)(x + 80) \\
&= wx + 40x + 80w + 3,200 \\
&= 10,000 + 40x + 80(10,000/x) + 3,200 \\
&= 13,200 + 40x + 800,000x^{-1} \\
y' &= 40 - 800,000x^{-2}
\end{aligned}
$$

Let $y' = 0$ for finding the critical point x:

$$40 - 800,000x^{-2} = 0,$$
$$x = \sqrt{20,000} = 141.42$$
$$w = \frac{10,000}{\sqrt{20,000}} = 70.71$$

Thus, the width of the site must be:

$70.71 + 20 + 20 = 110.71$ feet

The length of the site must be:

$141.42 + 50 + 30 = 221.42$ feet

The area of the site of the store is:

$y = 110.71 \times 221.42 = 24,513$ square feet

Further, $y'' = (-800,000)(-2)x^{-3} = 1,600,000x^{-3}$.
At $x = 141.42$, y'' is a positive number.
   Thus, the site "110.71 feet by 221.42 feet" is the minimum size.

19.3 PARTIAL DERIVATIVES

   In the previous discussion, we had considered only functions of a single
dependent variable y related to one independent variable x, or written, $y=f(x)$.
There are occasions, however, when two or more independent variables, such
as x and z, are involved in optimization problems.

$$y = f(x, z)$$

The functional relationship between the single dependent variable y and the
independent variables x and z is called a *multivariate function*. The derivative
of function y with respect to one of the independent variables is called a *partial
derivative* of the function. The interpretation of a partial derivative is basically
the same as that for a simple derivative. For example, the partial derivative of
function y with respect to the variable x describes the ratio of the incremental
change in y (Δy) to the incremental change in x (Δx) as Δx approaches zero, or:

$$\lim_{\Delta x \to 0} \frac{\Delta y}{\Delta x} = \lim_{\Delta x \to 0} \frac{f(x + \Delta x, z) - f(x, z)}{\Delta x}$$

A. Finding Partial Derivatives

Consider the function y of two independent variables x and z:

$$y = f(x, z)$$

The partial derivative of y with respect to x is denoted by,

$$\frac{\partial y}{\partial x} \quad \text{or} \quad f_x(x,z)$$

and is obtained by finding the derivative of y with respect to x but treating z as a
constant, or:

$$\frac{\partial y}{\partial x} = \frac{dy}{dx}, \quad \begin{array}{l}\text{where } x \text{ is treated as a variable} \\ \text{but } z \text{ is treated as a constant}\end{array}$$

The partial derivative of y with respect to z is denoted by,

$$\frac{\partial y}{\partial z} \quad \text{or} \quad f_z(x,z)$$

and is obtained by finding the derivative of y with respect to z but treating x as a
constant, or:

$$\frac{\partial y}{\partial z} = \frac{dy}{dz}, \quad \begin{array}{l}\text{where } z \text{ is treated as a variable} \\ \text{but } x \text{ is treated as a constant}\end{array}$$

EXAMPLE 12 Find the partial derivatives of the function:

$$y = 2x^3 + (xz)^2 - z^4$$

To find the partial derivative of y with respect to x, treat z as a constant:

$$\frac{\partial y}{\partial x} = 6x^2 + 2(xz)(z) - 0 = 6x^2 + 2z^2x$$

To find the partial derivative of y with respect to z, treat x as a constant:

$$\frac{\partial y}{\partial z} = 0 + 2(xz)(x) - 4z^3 = 2x^2z - 4z^3$$

EXAMPLE 13 Find the partial derivatives of the function:

$$y = (3x^2z + z^2)^5$$

$$\frac{\partial y}{\partial x} = 5(3x^2z + z^2)^4(3z(2x) + 0) = 30zx(3x^2z + z^2)^4$$

$$\frac{\partial y}{\partial z} = 5(3x^2z + z^2)^4(3x^2(1) + 2z)$$
$$= 5(3x^2z + z^2)^4(3x^2 + 2z)$$

EXAMPLE 14 Find the partial derivatives of the function:

$$z = f(x,y) = 20x^3y^2 - 5x^2y^3 + 6xy - 3$$

To find the partial derivative of z with respect to x, treat y as a constant:

$$\frac{\partial z}{\partial x} = 20y^2(3x^2) - 5y^3(2x) + 6y(1) - 0$$
$$= 60x^2y^2 - 10xy^3 + 6y$$

To find the partial derivative of z with respect to y, treat x as a constant:

$$\frac{\partial z}{\partial y} = 20x^3(2y) - 5x^2(3y^2) + 6x(1) - 0$$
$$= 40x^3y - 15x^2y^2 + 6x$$

★B. Application—The Method of Least Squares

One of many important applications of partial derivatives is to develop the normal equations required by the method of least squares. The *method of least squares* can be used to find a straight line or a curve which is considered the best fit for the scattered points representing x and y variables on a graph.

When two related variables are plotted on a graph in the form of points or dots, the graph is called a *scatter diagram*, such as the points plotted in Figure 19-7(a). Making a scatter diagram usually is the first step in investigating the functional relationship between two variables since the diagram shows visually the shape of the relationship.

When all points are on a straight line as shown in a scatter diagram, the equation representing the line can easily be obtained by selecting only two points on the line. This method was presented in Example 7. However, when the points are scattered away from a straight line or a smoothed curve, we must either arbitrarily draw a freehand graph or mathematically compute an equation for describing the functional relationship. The method of least squares used in computing an equation to represent a straight line is discussed as follows:

Let $n =$ number of paired x and y values in a given problem, or number of points in the scatter diagram,

$y =$ actual value of the dependent variable,

$y_c =$ computed value of the dependent variable based on the equation, and

$y_c = bx + a$, where x is the independent variable and a and b are constants.

Also, let $z =$ the sum of the squared deviations between the actual y values and the corresponding computed y_c values, or:

$$z = (y_1 - y_{c1})^2 + (y_2 - y_{c2})^2 + (y_3 - y_{c3})^2 + \cdots \text{ for } n \text{ squared deviations}$$

Or, simply written:

$$z = \Sigma \, (y - y_c)^2$$

The symbol Σ is the Greek letter sigma and represents "the sum of" or "the summation of."

Substitute $y_c = bx + a$ in the preceding equation:

$$z = \Sigma \, (y - bx - a)^2$$

The objective of the method of least squares is to minimize the function z. Since a and b are unknowns, we must differentiate function z with respect to a and b.

The partial derivative of z with respect to a is:

$$\frac{\partial z}{\partial a} = \Sigma \, 2(y - bx - a)(-1) = 2 \, \Sigma \, (-y + bx + a)$$

Let $\dfrac{\partial z}{\partial a} = 0$. The critical point can be derived:

$$2 \, \Sigma \, (-y + bx + a) = 0$$

Divide each side by 2 and take the sums of like terms:

$$\Sigma(-y) + \Sigma bx + \Sigma a = 0$$
$$\Sigma y = b\Sigma x + na \qquad \text{Normal Equation (1)}$$

The partial derivative of z with respect to b is:

$$\frac{\partial z}{\partial b} = \Sigma \, 2(y - bx - a)(-x) = 2 \, \Sigma \, (-xy + bx^2 + ax)$$

Let $\frac{\partial z}{\partial b} = 0$. The critical point can be derived:

$$2 \, \Sigma \, (-xy + bx^2 + ax) = 0$$
$$\Sigma \, (-xy) + \Sigma \, bx^2 + \Sigma \, ax = 0$$
$$\Sigma \, (xy) = b \, \Sigma \, x^2 + a \, \Sigma \, x. \quad \text{Normal Equation (2)}$$

Solve for a and b from the two normal equations simultaneously:

$$\begin{cases} (1) \;\; \Sigma y = b\Sigma x + na \\ (2) \;\; \Sigma(xy) = b\Sigma x^2 + a\Sigma x \end{cases}$$

We can get the required straight line equation:

$$y_c = bx + a$$

The straight line for the y dependent variable based on the method of least squares will have two mathematical properties:

1. The sum of the squared deviations is the least, or $\Sigma \, (y - y_c)^2 = $ a minimum.
2. The algebraic sum of the deviations of individual y values from (above or below) their corresponding y_c values on the line is zero, or $\Sigma \, (y - y_c) = 0$.

Property 1 can be proved by using the second partial derivatives of z. However, the proof is omitted here because of its complexity.

Property 2 can be proved in the following manner:

Linear equation: $y_c = bx + a$

Take the sum of n linear equations:

$$\Sigma y_c = b\Sigma x + na$$
$$= \Sigma y \text{ (see Normal Equation (1).)}$$

Thus:

$$\Sigma y - \Sigma y_c = 0$$

Property 2 states:

$$\Sigma(y - y_c) = 0,$$
or: $$\Sigma y - \Sigma y_c = 0,$$

which agrees with above.

The application of the method of least squares in finding a straight line to be the best fit to given data is illustrated in the following example.

EXAMPLE 15 The first three columns of Table 19-3 show the units produced (y) by a group of 3 workers in a factory during a day and the years of working experience (x) of each worker. (a) Plot a scatter diagram. (b) Compute the linear equation by the least square method. (c) Draw a straight line based on the equation. (d) Estimate the production if a worker has 5 years of experience.

Table 19-3

(1)	(2)	(3)	(4)	(5)	(6)	(7)	(8)
Worker	Units Produced y	Years of Experience x	xy	x^2	$y_c = \frac{9}{7}x + \frac{4}{7}$	$y - y_c$	$(y - y_c)^2$
A	2	1	2	1	$1\frac{6}{7}$	$\frac{1}{7}$	$\frac{1}{49}$
B	4	3	12	9	$4\frac{3}{7}$	$-\frac{3}{7}$	$\frac{9}{49}$
C	6	4	24	16	$5\frac{5}{7}$	$\frac{2}{7}$	$\frac{4}{49}$
Total (Σ)	12	8	38	26	12	0	$\frac{14}{49}$

Figure 19-8   (Example 15)

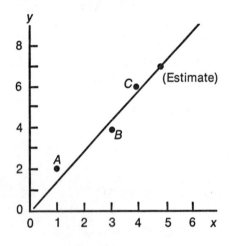

(a) A scatter diagram is plotted in Figure 19–8. Each point ($n = 3$ points) represents the units produced and a worker's years of experience.

(b) The required sums are shown in columns (2) through (5) in Table 19–3. Substitute the sums in the two normal equations:

$$12 = 8b + 3a \qquad (1)$$
$$38 = 26b + 8a \qquad (2)$$

Solve for a and b from the two equations. We have:

$$b = 1\frac{2}{7} \quad \text{and} \quad a = \frac{4}{7}$$

The linear equation is:

$$y_c = 1\frac{2}{7}x + \frac{4}{7}$$

(c) The straight line drawn in Figure 19–8 is based on the two points:

When $x = 1$,
$$y_c = 1\frac{2}{7}(1) + \frac{4}{7} = 1\frac{6}{7}$$

When $x = 4$,
$$y_c = 1\frac{2}{7}(4) + \frac{4}{7} = 5\frac{5}{7}$$

(d) When $x = 5$,

$$y_c = 1\frac{2}{7}(5) + \frac{4}{7} = 7$$

Thus, the estimated production for a worker having 5 years of experience is 7 units per day.

Check the two mathematical properties for the solution to Example 15:

1. $\Sigma(y - y_c)^2 = \frac{14}{49}$ is a minimum (See column (8) in the table.) If another straight line is drawn to fit points A, B, and C in Figure 19–8, the sum of the squared deviations must be larger than 14/49.
2. $\Sigma(y - y_c) = 0$ (See column (7) in the table.)

EXERCISE 19–2   Reference: Sections 19.2 and 19.3

A. *Maxima and Minima Applications*

1. The record of the Alan Hardware Store shows that the quantity of certain grades of paint sold per month varies linearly with the price of paint. For example, when the price was $10 per gallon in January, the total sales were 40 gallons; and when the price was $7 per gallon in May, the total sales were 100 gallons. (a) Maximize the total sales. (b) Find the price per gallon at which the store can maximize the total sales.

2. Refer to problem 1. Assume that the sales function is continuous. Find the marginal revenue if the sales per month were (a) 40 gallons, (b) 100 gallons, (c) 200 gallons, and (d) at the maximum point.

3. Refer to problem 1. Assume that the cost of each gallon of paint is $4. Find the following to maximize the profit: (a) total sales, (b) price per gallon for sale, and (c) profit.

4. Refer to problem 3. If the cost is $6 per gallon, what are the answers?

5. The Larry Manufacturing Company requires 10,000 units of machine part A each year. The cost of placing an order is $40 and the cost of carrying one unit in the inventory is $5. The average inventory size is $\frac{1}{2}$ of the quantity ordered each time. Find the minimum cost in a year and the most economic order size each time.

6. Refer to problem 5. If the company requires only 5,000 units each year, what are the answers?

7. A theater plans to construct a parking lot. The front of the lot requires 40 feet for building a gate and a small office building. The other three sides of the lot require 2 feet each for walking spaces. The parking zone should measure at least 20,000 square feet. Find the smallest size of the lot.

8. Refer to problem 7. If the theater requires only 10,000 square feet for a parking zone, what is the smallest size of the lot?

B. *Partial Derivatives*

Find the partial derivatives of each of the following functions:

9. $y = 4x^3 + (xz)^2 - z^3$
10. $y = 3x^2 - 2xz^2 + 4z^5$
11. $y = (2x^3z + z^2)^4$
12. $y = (3x^2z^3 - 2z^4)^3$

13. $z = 3xy + x + 2y$
14. $z = 5x - 2x^3y - 4y$
15. $z = 2x^3y^4 + 3x^2y^3 - 4xy^2 + 5$
16. $z = (x^2y + 5)(2xy^2 - 6)$

★C. *The Method of Least Squares*

17. The following table shows the hourly wages (x) and the weekly spendings on recreation (y) for a group of five workers in a company. (a) Plot a scatter diagram. (b) Consider y as the dependent variable and x as the independent variable. Compute the linear equation by the least square method. (c) Draw a straight line based on the equation. (d) Estimate the weekly spending on recreation for a worker if his earnings are $10 per hour.

Worker	Hourly Wage x	Weekly Spending on Recreation, y
A	$3	$2
B	4	5
C	6	3
D	8	9
E	9	6

18. Refer to problem 17. Consider x as the dependent variable and y as the independent variable. (a) Compute the linear equation by the least square method. (b) Draw a straight line based on the equation in the same graph in problem 17. (c) Estimate the hourly wage for a worker if his weekly spending on recreation is $8.

20 Integral Calculus

This chapter introduces the following topics: concepts and general operations of *indefinite* and *definite integrals;* special methods of integration; and applications of integral calculus to selected business and economic problems.

20.1 INDEFINITE INTEGRATION

Integration is the process of finding a function when its derivative is known. In other words, integration is the inverse of differentiation. The found function is called an *integral* or an *antiderivative.*

For example, in the process of finding the derivative (or rate of change) of function $F(x)$, the function is given, such as:

$$F(x) = x^2$$
$$\text{derivative: } F'(x) = 2x \text{ (also a function)}$$
$$\text{or: } f(x) = 2x$$

Inversely, in integration, the process of finding the function $F(x)$, the derivative of the function is given, such as:

$$\text{given derivative: } F'(x) = f(x) = 2x$$
$$F(x) = x^2$$

The value of x^2 is now called the integral of the value of $2x$ and can be expressed as:

$$\int (2x) \, dx = x^2 + C, \text{ where } C \text{ is any constant}$$

The integral expression at the left includes three parts. The first part is the integral sign $\int$ which is an elongated S; the second part is the *integrand* $f(x) = 2x$ which is the derivative of function $F(x) = x^2$; and the third part is dx which is the indication of the *differential* with respect to x.

The constant term C at the right of the equation is called the *constant of integration.* This constant is a necessary addition to the answer since the derivative of any constant is 0. If $y = x^2 + C$, where $C = -1, 0, 1, 2$, and 3 as shown in Figure 20–1 for illustration purposes, the derivatives of $x^2 - 1, x^2 + 0,$

Figure 20-1   **Illustration of a Family of y Functions**
              $y = x^2 + C$, Where $C = -1, 0, 1, 2, 3$

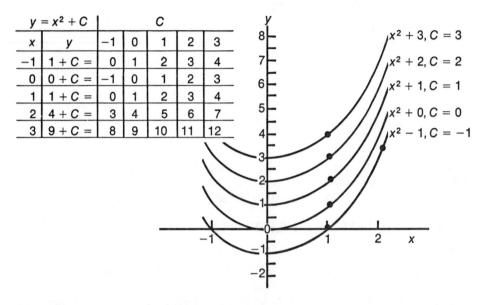

$x^2 + 1$, $x^2 + 2$, and $x^2 + 3$ are all $2x$. In other words, the slopes of all these y functions, where $C =$ any constant, are the same. The infinite number of y functions thus forms an entire family of functions, differing only by the value of C but having the same slope, $2x$. Since C does not have a definite value, we call the solution $x^2 + C$ an *indefinite integral*.

In general, we define the indefinite integral of a function $f(x)$ as:

$$\int f(x)\, dx = F(x) + C$$

where $f(x)$ is the derivative of $F(x)$. The left side is read "integral of f of x with respect to x."

Note that the constant of integration C can be determined if additional information is given. For example, if it is known that $F(x) + C = x^2 + C = 2$ for $x = 1$, then:

$$x^2 + C = 2$$
$$1^2 + C = 2$$
$$C = 1$$

The indefinite integral now is a definite function, or:

$$\int (2x)\, dx = x^2 + 1$$

This additional information is referred to as an *initial condition*.

The following examples give additional illustrations for determining indefinite integrals from given functions. Note that the examples are computed by the use of the common *rules for integration*. Those rules are obtained directly by reversing the corresponding rules for differentiation presented in Chapter 17.

RULE 1 $\int k\, dx = kx + C$, where k is any constant. If $k = 1$, the rule is written:

$$\int dx = x + C$$

EXAMPLE 1   Determine: (a) $\int 5\, dx$, and (b) $\int -10\, dx$

(a) $f(x) = 5 = 5x^0$, or $k = 5$

$$\int 5\, dx = 5x + C$$

Check: Let $y = 5x + C$. $y' = 5 = f(x)$

(b) $f(x) = -10 = -10x^0$, or $k = -10$

$$\int -10\, dx = -10x + C$$

Check: Let $y = -10x + C$. $y' = -10$

RULE 2 $\int x^n\, dx = \dfrac{x^{n+1}}{n+1} + C$, where $n \neq -1$

EXAMPLE 2   Determine: (a) $\int x\, dx$, (b) $\int x^2\, dx$, and (c) $\int x^{-3}\, dx$

(a) $f(x) = x = x^1$

$$\int x\, dx = \frac{x^{1+1}}{1+1} + C = \frac{x^2}{2} + C$$

Check: Let $y = \frac{1}{2}x^2 + C$. $y' = \frac{1}{2}(2)x^{2-1} = x$

(b) $\displaystyle \int x^2\, dx = \frac{x^{2+1}}{2+1} + C = \frac{x^3}{3} + C$

(c) $\displaystyle \int x^{-3}dx = \frac{x^{-3+1}}{-3+1} + C = \frac{x^{-2}}{-2} + C$

RULE 3 $\int k{\cdot}f(x)\, dx = k \int f(x)\, dx$

EXAMPLE 3   Determine: (a) $\int 6x^2\, dx$ (b) $\int 15x^{-4}\, dx$ and   (c) $\int \frac{1}{4}x^{-1/3}\, dx$

(a) $k = 6$ and $f(x) = x^2$

$$\int 6x^2\, dx = 6\int x^2\, dx = 6\left(\frac{x^3}{3}\right) + C = 2x^3 + C$$

(b) $\displaystyle \int 15x^{-4}\, dx = 15 \int x^{-4}\, dx = 15\left(\frac{x^{-3}}{-3}\right) + C = -5x^{-3} + C$

(c) $\displaystyle \int \frac{1}{4}x^{-1/3}\, dx = \frac{1}{4}\int x^{-1/3}\, dx = \frac{1}{4}\left(\frac{x^{2/3}}{2/3}\right) + C$

$$= \frac{3}{8}x^{2/3} + C$$

RULE 4 $\int (u \pm v)\,dx = \int u\,dx \pm \int v\,dx$, where $u = f(x)$ and $v = g(x)$

EXAMPLE 4   Determine $\int (x^2 \pm 6x)\,dx$.

$u = f(x) = x^2$ and $v = g(x) = 6x$

(a) $\int (x^2 + 6x)\,dx = \int x^2\,dx + \int 6x\,dx$

$$= \frac{x^3}{3} + 6\left(\frac{x^2}{2}\right) + C$$

$$= \frac{1}{3}x^3 + 3x^2 + C$$

Check: Let $y = \frac{1}{3}x^3 + 3x^2 + C$.

$$y' = x^2 + 6x$$

(b) $\int (x^2 - 6x)\,dx = \frac{1}{3}x^3 - 3x^2 + C$

EXAMPLE 5   Determine $y = \int (12x^3 + 15x^2 + 2)\,dx$, when $x = 1$, $y = 14$.

$y = \int (12x^3 + 15x^2 + 2)\,dx = 3x^4 + 5x^3 + 2x + C$

If $y = 14$ when $x = 1$, then:

$3(1^4) + 5(1^3) + 2(1) + C = 14$, $10 + C = 14$, $C = 4$

Thus, $y = 3x^4 + 5x^3 + 2x + 4$.

Check: $\dfrac{dy}{dx} = 12x^3 + 15x^2 + 2$

20.2 DEFINITE INTEGRATION

Integration can be interpreted in two ways. First, it is the inverse of differentiation as presented in Section 20.1. Second, it is a method of finding the area under a curve. This section will explain the second interpretation of integration.

In geometry the areas of regularly shaped figures can be computed with the use of appropriate formulas. For example, the area (A) of a rectangle is equal to the product of its width (W) and its height (H), or $A = WH$; and the area (A) of a triangle is equal to $\frac{1}{2}$ of the product of its base (B) and its height (H), or $A = \frac{1}{2}BH$. However, an area bounded by curved lines must be computed by the use of integral calculus. There is no geometrical formula for computing an area of irregular shape.

Assume that we wish to find the area under the curve $y = x^2$, above the x-axis, and between $x = 1$ and $x = 3$.

First, we may use the geometrical formula to approximate the area A by inscribing two rectangles of equal width in the area, R_1 and R_2, as shown in Figure 20-2(a).

The base of each rectangle is $\Delta x = \dfrac{3 - 1}{2} = 1$.

Figure 20-2   Finding the Area Under the Curve
y = x², Above x-axis, and Between x = 1 and x = 3

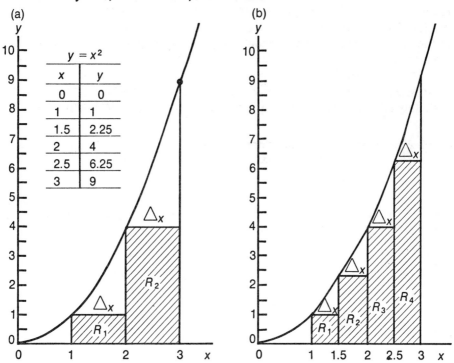

The height of R_1 is $y = f(1) = 1^2 = 1$.
The height of R_2 is $y = f(2) = 2^2 = 4$.

Let A_1 = the sum of the areas of the two rectangles R_1 and R_2. Then:

$$A_1 = f(1) \cdot \Delta x + f(2) \cdot \Delta x = 1(1) + 4(1) = 5$$

Next, we approximate the area A by inscribing four rectangles of equal width in the area, R_1, R_2, R_3, R_4, as shown in Figure 20–2(b).

The base of each rectangle is now $\Delta x = \dfrac{3-1}{4} = .5$.

The height of R_1 is $y = f(1) = 1^2 = 1$.
The height of R_2 is $y = f(1.5) = 1.5^2 = 2.25$.
The height of R_3 is $y = f(2) = 2^2 = 4$.
The height of R_4 is $y = f(2.5) = 2.5^2 = 6.25$.

Let A_2 = the sum of the areas of the four rectangles, R_1, R_2, R_3, and R_4.

Then:

$$\begin{aligned}
A_2 &= f(1) \cdot \Delta x + f(1.5) \cdot \Delta x + f(2) \cdot \Delta x + f(2.5) \cdot \Delta x \\
&= 1(.5) + 2.25(.5) + 4(.5) + 6.25(.5) \\
&= .5 + 1.125 + 2 + 3.125 = 6.750.
\end{aligned}$$

Observe graphs (a) and (b) in Figure 20–2. The gaps under the curve and above the top of the inscribed rectangles are larger in (a) than in (b). Thus, the required area A is better approximated by $A_2 = 6.75$ with four rectangles than by $A_1 = 5$ with only two rectangles. If we add more rectangles under the curve in the area, the sum of the rectangles will be increasingly closer to the actual amount of the area since the effect of the gaps becomes less important. (Note: The actual size of the area is $A = 8\frac{2}{3}$. This will be shown in Example 6.)

Now, assume that we are interested in finding the area under the curve $y = f(x)$, above the x-axis, between $x = a$ and $x = b$ as shown in Figure 20–3.

Figure 20-3 Finding the Area Under the Curve
$y = f(x)$, Above x-axis, and Between $x = a$ and $x = b$

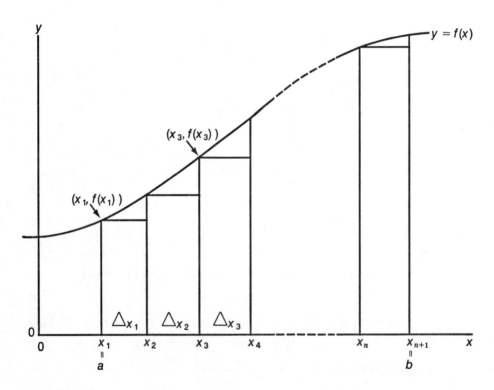

Let n = number of inscribed (or circumscribed) rectangles (of equal or unequal width) under the curve between $x = a$ and $x = b$

$i = 1, 2, 3, 4, \cdots, n$

Δx_i = base (width) of an individual rectangle

x_i = point on the x-axis which denotes the division of the rectangles, $x_1 = a$, $x_2 = a + \Delta x_1$, $x_3 = x_2 + \Delta x_2$, $x_4 = x_3 + \Delta x_3$, and so on

The sum of the areas of these inscribed rectangles is:

$$\text{Sum} = f(x_1)\cdot\Delta x_1 + f(x_2)\cdot\Delta x_2 + f(x_3)\cdot\Delta x_3 + \cdots + f(x_n)\cdot\Delta x_n$$
$$= \sum_{i=1}^{n} f(x_i)\cdot\Delta x_i$$

When the number of rectangles approaches infinity ($n \to \infty$) and the base approaches zero ($\Delta x \to 0$), the area A under the curve between $x = a$ and $x = b$ is the limit of the sum of the inscribed rectangles, or written:

$$\text{Area } A = \lim_{\substack{n\to\infty \\ (\Delta x \to 0)}} \sum_{i=1}^{n} f(x_i)\cdot\Delta x_i$$

This limit can be computed from the following expression:

$$A = \int_a^b f(x)\, dx$$

The reasoning and the operation for the preceding expression is explained as follows:

Let $A = $ the desired area *abcd* under the continuous curve $y = f(x)$ as shown in Figure 20–4.

Figure 20-4   Area A =

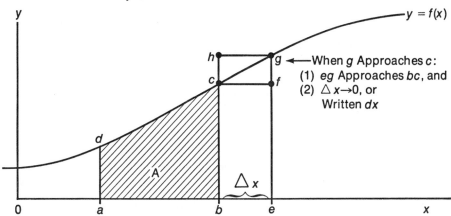

When $x = b$ is increased by Δx to e, area A is increased by $\Delta A = $ area *begc*. Observe the graph in Figure 20–4:

$$\text{Area } befc < \text{Area } begc < \text{Area } begh,$$
$$(bc \cdot \Delta x) < (\Delta A) < (eg \cdot \Delta x)$$

Divide each term by Δx:

$$bc < \frac{\Delta A}{\Delta x} < eg$$

When point g approaches point c, eg approaches bc, and:

$$\frac{\Delta A}{\Delta x} = bc = f(x) \text{ at } x = b$$

Also, when g approaches c, $\Delta x \to 0$, and:

$$\lim_{\Delta x \to 0} \frac{\Delta A}{\Delta x} = \frac{dA}{dx}$$

Thus:

$$\frac{dA}{dx} = f(x)$$

Find the antiderivative of $f(x)$ as defined in the indefinite integration:

$$\int f(x) \, dx = A$$

Let: $$A = F(x) + C$$

At $x = a$, area $A = 0$ (the area of any single point is zero), and:

$$0 = F(a) + C$$
$$C = -F(a)$$

Thus: $$A = F(x) - F(a)$$

For the area $abcd$ under the curve $f(x)$, $x = b$, and:

$$A = F(b) - F(a)$$

Now, the area formula can be written:

$$\textbf{Area } A = \int_a^b f(x) \, dx = F(b) - F(a)$$

which is known as the *fundamental theorem of integral calculus*.
The constant of integration C is not contained in the solution for A. The integral thus has a definite value. For this reason, $\int_a^b f(x) \, dx$ is called the *definite integral* of $f(x)$ from a to b. Since the definite integral can be interpreted as a summation process, the choice of the $\int$ symbol for integration does have its significance. $\int$ is an elongated S, which represents the sum.

EXAMPLE 6   Evaluate the definite integral $\int_1^3 x^2 \, dx$.

The lower limit $a = 1$, the upper limit $b = 3$, and the given function $f(x) = x^2$.

Find the function $F(x)$ whose derivative is x^2.

$$F(x) = \frac{x^3}{3} \qquad \text{(Rule 2)}$$

Find: $\qquad F(b) = \frac{3^3}{3} = \frac{27}{3}$

$$F(a) = \frac{1^3}{3} = \frac{1}{3}$$

$$F(b) - F(a) = \frac{27}{3} - \frac{1}{3} = \frac{26}{3} = 8\frac{2}{3}$$

Thus, definite integral is a number, $8\frac{2}{3}$.

The preceding steps may be combined as follows:

$$\int_1^3 x^2 \, dx = \frac{x^3}{3}\Big|_1^3 = \frac{3^3}{3} - \frac{1^3}{3} = \frac{27}{3} - \frac{1}{3} = 8\frac{2}{3}$$

Observe that the vertical line in the form $\Big|_a^b$, here $\Big|_1^3$, is used to indicate the lower and upper limits in the computation and is written at the right of the function $F(x) = x^3/3$.

The answer in Example 6 also represents the area A under the curve $y = x^2$, above the x-axis, and between $x = 1$ and $x = 3$ as shown in Figure 20–2, or $A = 8\frac{2}{3}$.

EXAMPLE 7 Evaluate the definite integral $\int_{-1}^2 (x^2 + 6x) \, dx$.

$$\int_{-1}^2 (x^2 + 6x) \, dx = \int_{-1}^2 x^2 \, dx + \int_{-1}^2 6x \, dx \qquad \text{(Rule 4)}$$

$$= \frac{x^3}{3}\Big|_{-1}^2 + 3x^2 \Big|_{-1}^2 = \left(\frac{2^3}{3} - \frac{(-1)^3}{3}\right) + (3(2^2) - 3(-1)^2)$$

$$= \left(\frac{8}{3} + \frac{1}{3}\right) + (12 - 3) = 12. \text{ (Also see Example 4.)}$$

From the above illustrations and graphs, it can be seen that definite integrals have the following three properties:

1. $\displaystyle\int_a^a f(x) \, dx = 0$

2. $\displaystyle\int_a^b f(x) \, dx = -\int_b^a f(x) \, dx$

3. $\displaystyle\int_a^c f(x) \, dx = \int_a^b f(x) \, dx + \int_b^c f(x) \, dx$

EXAMPLE 8 Illustration of property 1. Evaluate $\int_7^7 x \, dx$.

$$\int_7^7 x\,dx = \frac{x^2}{2}\Big|_7^7 = \frac{7^2}{2} - \frac{7^2}{2} = 0$$

EXAMPLE 9　Illustration of property 2.

Evaluate: (a) $\int_1^7 x\,dx$, and (b) $-\int_7^1 x\,dx$

(a) $\int_1^7 x\,dx = \dfrac{x^2}{2}\Big|_1^7 = \dfrac{7^2}{2} - \dfrac{1^2}{2} = \dfrac{49}{2} - \dfrac{1}{2} = \dfrac{48}{2} = 24$

(b) $-\int_7^1 x\,dx = -\left(\dfrac{x^2}{2}\Big|_7^1\right) = -\left(\dfrac{1^2}{2} - \dfrac{7^2}{2}\right) = -(-24) = 24$

Observe the answers in (a) and (b). They show that:

$$\int_1^7 x\,dx = -\int_7^1 x\,dx = 24$$

EXAMPLE 10　Illustration of property 3.

(a) Show that $\int_1^7 x\,dx = \int_1^4 x\,dx + \int_4^7 x\,dx$.

$\displaystyle\int_1^7 x\,dx = 24.$ (See Example 9(a).)

$\displaystyle\int_1^4 x\,dx = \frac{x^2}{2}\Big|_1^4 = \frac{4^2}{2} - \frac{1^2}{2} = \frac{15}{2}$

$\displaystyle\int_4^7 x\,dx = \frac{x^2}{2}\Big|_4^7 = \frac{7^2}{2} - \frac{4^2}{2} = \frac{33}{2}$

$\dfrac{15}{2} + \dfrac{33}{2} = \dfrac{48}{2} = 24$

Thus, the statement in (a) is true.

(b) Show that $\int_1^4 x\,dx = \int_1^7 x\,dx + \int_7^4 x\,dx$.

$\displaystyle\int_1^4 x\,dx = \frac{15}{2}$ (See (a).)

$\displaystyle\int_1^7 x\,dx = 24$ (See Example 9(a).)

$\displaystyle\int_7^4 x\,dx = \frac{x^2}{2}\Big|_7^4 = \frac{4^2}{2} - \frac{7^2}{2} = -\frac{33}{2}$

$24 + \left(-\dfrac{33}{2}\right) = \dfrac{15}{2}$

Thus, the statement in (b) is true. This solution indicates that the value of b in property 3 can be larger (or smaller) than both values of a and c.

EXERCISE 20–1   Reference: Sections 20.1 and 20.2

A. *Evaluate the following indefinite integrals:*

1. $\displaystyle \int 8\ dx$

2. $\displaystyle \int 20\ dx$

3. $\displaystyle \int -6\ dx$

4. $\displaystyle \int -15\ dx$

5. $\displaystyle \int x^3\ dx$

6. $\displaystyle \int x^{3/4}\ dx$

7. $\displaystyle \int x^{-2}\ dx$

8. $\displaystyle \int x^{-1/2}\ dx$

9. $\displaystyle \int 3x^2\ dx$

10. $\displaystyle \int 12x^{-3}\ dx$

11. $\displaystyle \int -21x^{-4}\ dx$

12. $\displaystyle \int \frac{1}{3}x^{-2/3}\ dx$

13. $\displaystyle \int (x^2 + 5)\ dx$

14. $\displaystyle \int (3x + 2)\ dx$

15. $\displaystyle \int (x^3 - 4x)\ dx$

16. $\displaystyle \int (6x^2 - 2x)\ dx$

17. $\displaystyle \int (20x^3 + 12x^2 - 3)\ dx$

18. $\displaystyle \int (18x^2 - 6x + 7)\ dx$

B. *Evaluate the following definite integrals:*

19. $\displaystyle \int_1^3 8\ dx$

20. $\displaystyle \int_2^5 -6\ dx$

21. $\displaystyle \int_0^4 x\ dx$

22. $\displaystyle \int_0^2 2x\ dx$

23. $\displaystyle \int_{-1}^2 x^3\ dx$

24. $\displaystyle \int_1^5 x^{-2}\ dx$

25. $\displaystyle \int_2^3 3x^2\ dx$

26. $\displaystyle \int_1^3 12x^{-3}\ dx$

27. $\displaystyle \int_1^4 (x^2 + 5)\ dx$

28. $\displaystyle \int_2^5 (3x^2 - 4)\ dx$

29. $\displaystyle \int_1^2 (3x^2 - 2x + 5)\ dx$

30. $\displaystyle \int_0^5 (6x^2 + 4x - 3)\ dx$

*20.3 SPECIAL METHODS OF INTEGRATION

In general, the process of integrating a given function is the reverse of the differentiation process. For example, the rules of integration presented in

Section 20.1 were derived by simply reversing the corresponding rules for differentiation. However, in many cases, computing an integral is considerably more difficult than computing a derivative. These rules are very limited in their coverage of functions. Various additional methods of integration have been developed to augment these rules. To cover all available methods is beyond the scope of this text. However, the three most popular methods for integrating complicated functions are introduced in this section. They are: (A) integration by substitution, (B) integration by parts, and (C) integration with the aid of a table of integrals. Each method has advantages and disadvantages in application.

A. INTEGRATION BY SUBSTITUTION

The method of substitution for integrating a given integral in the form of $\int f(x)\, dx$ usually involves three steps. First, substitute the variable u for a function of x in the given function $f(x)$. Second, substitute the remaining portion, if there is any after the first substitution of the given function, and dx by an equivalent value including du:

$$\int f(x)\, dx \text{ now becomes } \int f(u)\, du$$

The integral of the new form is found according to the appropriate integration rules. Third, find the answer by resubstituting the value of u in terms of x with the derived integral.

The method of integration by substitution is illustrated in Examples 11 through 14.

EXAMPLE 11   Determine the integral $\int (x^2 + 1)^3 x\, dx$.

SOLUTION $f(x) = (x^2 + 1)^3 x$

STEP 1   Substitute u for $(x^2 + 1)$ and u^3 for $(x^2 + 1)^3$.

STEP 2   Substitute the remaining portion of the given function and dx by an equivalent value including du.

Since $u = x^2 + 1$

$$\frac{du}{dx} = 2x$$

In differentiation, du/dx is considered a single symbol. In integration, du/dx may be interpreted as a quotient: du (the differential of u when $\Delta u \to 0$) is divided by dx (the differential of x when $\Delta x \to 0$). Thus, from the above derivative of u, we have:

$$du = 2x\, dx$$

$$dx = \frac{du}{2x}$$

The remaining portion of the given function is x:

$$x \, dx = x \, \frac{du}{2x} = \frac{1}{2} \, du$$

By substitution:

$$\int (x^2 + 1)^3 x \, dx = \int u^3 \, \frac{1}{2} \, du = \frac{1}{2} \int u^3 \, du$$

$$= \frac{1}{2} \left(\frac{u^{3+1}}{3+1} + C \right)$$

$$= \frac{u^4}{8} + C$$

Note: The value of C in "$\frac{u^4}{8} + C$" has an actual value of $\frac{1}{2}C$. However, it is customary to use C to represent any constant. We therefore do not distinguish between the value of a constant in the form of C and the value of a constant in the form of $\frac{1}{2}C$.

STEP 3 By resubstituting $u = x^2 + 1$ in the integral found in the second step, we have:

$$\int (x^2 + 1)^3 \, x \, dx = \frac{(x^2 + 1)^4}{8} + C$$

Check: Let $y = \dfrac{(x^2 + 1)^4}{8} + C$

$$\frac{dy}{dx} = \frac{1}{8} \, (4(x^2 + 1)^3 \, (2x)) = (x^2 + 1)^3 \, x \text{ (Use Rule 8 in Chapter 18.)}$$

EXAMPLE 12 Determine the integral $\displaystyle\int \frac{6x^2}{x^3 + 5} \, dx$.

SOLUTION $f(x) = (x^3 + 5)^{-1} \, 6x^2$

Let $u = x^3 + 5$, and $u^{-1} = (x^3 + 5)^{-1}$

$$\frac{du}{dx} = 3x^2, \text{ and } dx = \frac{du}{3x^2}$$

The remaining portion of the given function is $6x^2$.

$$6x^2 \, dx = 6x^2 \left(\frac{du}{3x^2} \right) = 2 \, du$$

By substitution:

$$\int \frac{6x^2}{x^3 + 5} \, dx = \int (x^3 + 5)^{-1} \, 6x^2 \, dx$$

$$= \int u^{-1} \, 2 \, du$$

$$= 2 \int u^{-1} \, du$$

$$= 2 \; ln \; u + C$$

$$= 2 \; ln \; (x^3 + 5) + C$$

Check: Let $y = 2 \; ln \; (x^3 + 5) + C$

$$\frac{dy}{dx} = 2 \left(\frac{1}{x^3 + 5} \right) (3x^2) = 6x^2 (x^3 + 5)^{-1} \text{ (Use Rule 10 in}$$

Chapter 18.)

EXAMPLE 13   Determine the integral $\int e^{x^3} x^2 \; dx$.

SOLUTION   Let $u = x^3$, and $e^u = e^{x^3}$.

$$\frac{du}{dx} = 3x^2, \text{ and } dx = \frac{du}{3x^2}.$$

The remaining portion of the given function is x^2.

$$x^2 \; dx = x^2 \left(\frac{du}{3x^2} \right) = \frac{1}{3} \; du$$

By substitution:

$$\int e^{x^3} x^2 \; dx = \int e^u \frac{1}{3} \; du = \frac{1}{3} \int e^u \; du$$

$$= \frac{1}{3} e^u + C = \frac{1}{3} e^{x^3} + C$$

Check: Let $y = \frac{1}{3} e^{x^3} + C$

$$\frac{dy}{dx} = \frac{1}{3} e^{x^3} (3x^2) = e^{x^3} x^2 \text{ (Use Rule 9 in Chapter 18.)}$$

EXAMPLE 14   Evaluate the integral $\int_1^2 3x^2 \; (2x^3 + 7) \; dx$.

SOLUTION   Let $u = (2x^3 + 7)$.

$$\frac{du}{dx} = 6x^2, \text{ and } dx = \frac{du}{6x^2}.$$

The remaining portion of the given function is $3x^2$.

$$3x^2 \; dx = 3x^2 \left(\frac{du}{6x^2} \right) = \frac{1}{2} \; du$$

By substitution, the indefinite integral is:

$$\int 3x^2 (2x^3 + 7) \; dx = \int u \frac{1}{2} \; du = \frac{1}{2} \int u \; du$$

$$= \frac{1}{2} \left(\frac{u^2}{2} \right) + C = \frac{u^2}{4} + C$$

$$= \frac{(2x^3 + 7)^2}{4} + C$$

The definite integral is:

$$\int_1^2 3x^2(2x^3 + 7)\, dx = \frac{(2x^3 + 7)^2}{4}\,\bigg|_1^2$$

$$= \frac{(2(2^3) + 7)^2}{4} - \frac{(2(1^3) + 7)^2}{4} = \frac{529 - 81}{4} = 112$$

Check: Let $y = \dfrac{(2x^3 + 7)^2}{4} + C.$

$$\frac{dy}{dx} = \frac{1}{4}(2(2x^3 + 7)(6x^2)) = 3x^2(2x^3 + 7)$$

B. Integration by Parts

The method of integration by parts is based on the formula for the derivative of the product of two functions. If u and v are functions of x, or $u = g(x)$ and $v = h(x)$, the derivative of the product of uv can be written:

$$(uv)' = uv' + u'v \text{ (Rule 5 of Chapter 17)}$$

The integral of this derivative is:

$$uv = \int uv'\, dx + \int u'v\, dx$$

which is rearranged as a formula:

$$\int uv'\, dx = uv - \int u'v\, dx$$

The application of the formula is illustrated in Examples 15 through 17.

EXAMPLE 15   Determine the integral $\displaystyle\int xe^x\, dx.$

SOLUTION   Let $u = x$, and $v'\, dx = e^x\, dx$, or $v' = e^x$.

Then, $u' = 1$, and $v = \displaystyle\int e^x\, dx = e^x.$

Substitute the values of the four terms, u, v', u', and v in the formula:

$$\int xe^x\, dx = xe^x - \int e^x\, dx = xe^x - e^x + C$$

Note: If we let $u = e^x$ and $v' = x$

then: $u' = e^x$ and $v = \displaystyle\int x\, dx = \frac{x^2}{2}$

The formula thus gives:

$$\int xe^x\, dx = e^x\left(\frac{x^2}{2}\right) - \int e^x\left(\frac{x^2}{2}\right) dx \text{ (See Example 19.)}$$

The integral expression at the right of the equation $\int e^x(x^2/2)\, dx$ is more complicated than the integral expression in the solution $\int e^x\, dx$. Thus, a student should be careful in selecting the terms to substitute for u and v' in the given integral.

EXAMPLE 16   Determine the integral $\int x \ln x \, dx$.

Let $u = \ln x$, and $v' = x$.

Then: $u' = \dfrac{1}{x}$, and $v = \int x \, dx = \dfrac{x^2}{2}$.

Substitute the values of u, v', u', and v in the formula:

$$\int x \ln x \, dx = (\ln x)\frac{x^2}{2} - \int \frac{1}{x}\left(\frac{x^2}{2}\right) dx$$

$$= \frac{x^2 \ln x}{2} - \frac{1}{2}\int x \, dx$$

$$= \frac{x^2 \ln x}{2} - \frac{x^2}{4} + C$$

EXAMPLE 17   Evaluate the integral $\displaystyle\int_1^5 x \ln x \, dx$.

The indefinite integral of the given integral has been determined in Example 16. Thus, the definite integral is:

$$\int_1^5 x \ln x \, dx = \left(\frac{x^2 \ln x}{2} - \frac{x^2}{4}\right)\Big|_1^5$$

$$= \left(\frac{5^2 \ln 5}{2} - \frac{5^2}{4}\right) - \left(\frac{1^2 \ln 1}{2} - \frac{1^2}{4}\right)$$
$$= 12.5 \ln 5 - .5 \ln 1 - 6$$
$$= 12.5(1.6094) - .5(0) - 6$$
$$= 14.1175$$

C. Integration by Tables of Integrals

Tables of integrals provide convenient ways to integrate given functions. Integrals can readily be determined if the forms of the integrals and their results listed in the tables can correctly be recognized. The integrals listed in Table 20–1 include the integrals used in the rules and examples of this chapter and some complicated ones selected for reference and further illustrations.

EXAMPLE 18   Determine: $\displaystyle\int \frac{3x^2 + 8x}{x^3 + 4x^2 - 3} \, dx$

Observe the list of formulas in Table 20–1. The form of formula 6:

$$\int \frac{u'}{u} \, dx = \ln u + C$$

fits this example.

Let $u = x^3 + 4x^2 - 3$, the denominator in the given fraction. Then $u' = 3x^2 + 8x$, the numerator in the given fraction.

Thus:

$$\int \frac{3x^2 + 8x}{x^3 + 4x^2 - 3} \, dx = \ln \, (x^3 + 4x^2 - 3) + C$$

EXAMPLE 19 Determine: $e^x \left(\dfrac{x^2}{2}\right) - \dfrac{1}{2} \displaystyle\int e^x(x^2) \, dx$

The form of formula 13 in Table 20–1:

$$\int u^2 e^{au} \, du = \frac{e^{au}}{a^3} \, (a^2 u^2 - 2au + 2) + C$$

fits the given integral.

Let $u = x$ and $a = 1$. Then:

$$\int x^2 e^x \, dx = e^x(x^2 - 2x + 2)$$

Thus:

$$e^x \left(\frac{x^2}{2}\right) - \frac{1}{2} \int e^x x^2 \, dx = e^x \left(\frac{x^2}{2}\right) - \frac{1}{2} e^x(x^2 - 2x + 2)$$

$$= xe^x - e^x + C$$

(Also see the solution to Example 15.)

EXAMPLE 20 Determine: $\displaystyle\int \frac{x \, dx}{\sqrt{3 + 2x}}$

The form of formula 19 in Table 20–1 fits the given integral, or:

$$\int \frac{u \, du}{\sqrt{a + bu}} = - \frac{2(2a - bu)\sqrt{a + bu}}{3b^2} + C$$

Let $u = x$, $a = 3$, and $b = 2$. Then:

$$\int \frac{x \, dx}{\sqrt{3 + 2x}} = - \frac{2(6 - 2x)\sqrt{3 + 2x}}{12} + C$$

Table 20–1   Table of Selected Integrals

A. Fundamental, Exponential, and Logarithmic Forms

1. $\displaystyle\int k \, dx = kx + C \quad \left(\text{If } k = 1, \displaystyle\int dx = x + C\right)$

2. $\displaystyle\int x^n \, dx = \frac{x^{n+1}}{n + 1} + C$, where $n \neq -1$

3. $\displaystyle\int k \, f(x) \, dx = k \displaystyle\int f(x) \, dx$

4. $\displaystyle\int (u \pm v) \, dx = \displaystyle\int u \, dx \pm \displaystyle\int v \, dx$, where $u = f(x)$ and $v = g(x)$

5. $\displaystyle\int uv' \, dx = uv - \displaystyle\int u'v \, dx$, where $v' = \dfrac{dv}{dx}$ and $u' = \dfrac{du}{dx}$

6. $\int \dfrac{u'}{u}\,dx = \int \dfrac{1}{u}\,du = \ln u + C$

7. $\int a^{kx}\,dx = \dfrac{a^{kx}}{k\,(\ln a)} + C$

8. $\int e^{kx}\,dx = \dfrac{e^{kx}}{k} + C$

9. $\int u^n\,du = \dfrac{u^{n+1}}{n+1} + C$, where $n \neq -1$

10. $\int e^u\,du = e^u + C$

11. $\int a^u\,du = \dfrac{a^u}{\ln a} + C$

12. $\int ue^{au}\,du = \dfrac{e^{au}\,(au-1)}{a^2} + C$

13. $\int u^2 e^{au}\,du = \dfrac{e^{au}}{a^3}\,(a^2u^2 - 2au + 2) + C$

14. $\int \ln u\,du = u\ln u - u + C$

15. $\int u^n \ln u\,du = u^{n+1}\left(\dfrac{\ln u}{n+1} - \dfrac{1}{(n+1)^2}\right) + C$, where $n \neq -1$

16. $\int \dfrac{du}{u\ln u} = \ln\,(\ln u) + C$

B. Forms Containing $(a + bu)$

17. $\int \dfrac{u\,du}{a+bu} = \dfrac{1}{b^2}\left((a+bu) - a\ln\,(a+bu)\right) + C$

18. $\int \dfrac{du}{u(a+bu)} = -\dfrac{1}{a}\ln\dfrac{a+bu}{u} + C$

19. $\int \dfrac{u\,du}{\sqrt{a+bu}} = -\dfrac{2(2a-bu)\sqrt{a+bu}}{3b^2} + C$

C. Forms Containing $(u^2 \pm a^2)$ and $(a^2 - u^2)$

20. $\int \dfrac{du}{a^2-u^2} = \dfrac{1}{2a}\ln\dfrac{a+u}{a-u} + C$, where $u^2 < a^2$

21. $\int \dfrac{du}{u^2-a^2} = \dfrac{1}{2a}\ln\dfrac{u-a}{u+a} + C$, where $u^2 > a^2$

22. $\int \dfrac{\sqrt{u^2+a^2}}{u}\,du = \sqrt{u^2+a^2} - a\ln\left(\dfrac{a+\sqrt{u^2+a^2}}{u}\right) + C$

★EXERCISE 20–2   Reference: **Section 20.3**

Evaluate the following integrals by the methods indicated.

A. *Integration by Substitution*

1. $\int (x^2 + 1)^4 x\ dx$

2. $\int 6x^2\ (x^3 - 5)^2\ dx$

3. $\int \dfrac{15x^4}{x^5 + 7}\ dx$

4. $\int 2x(x^2 + 5)^{-1}\ dx$

5. $\int 6xe^{x^2}\ dx$

6. $\int 8x^3 e^{x^4}\ dx$

7. $\int 2x(3x^2 + 5)\ dx$

8. $\int 30x^2(5x^3 - 2)\ dx$

9. $\int_1^5 2x(3x^2 + 5)\ dx$

10. $\int_0^3 30x^2(5x^3 - 2)\ dx$

B. *Integration by Parts* $\left(\text{Hint: } \int e^{ax}\ dx = \dfrac{e^{ax}}{a} + C\right)$

11. $\int xe^{ax}\ dx$

12. $\int x^2 e^x\ dx$ (Hint: See Example 15.)

13. $\int xe^{-x}\ dx$

14. $\int \ln x\ dx$

15. $\int (\ln x)x^2\ dx$

16. $\int_1^4 xe^{2x}\ dx$

17. $\int_0^2 xe^{-x}\ dx$

18. $\int_1^3 (\ln x)x^2\ dx$

C. *Integration by Table 20–1*

19. $\int \dfrac{2x - 3}{x^2 - 3x + 5}\ dx$

20. $\int \dfrac{15x^2 + 4x - 4}{5x^3 + 2x^2 - 4x + 6}\ dx$

21. $\int e^x\ dx$

22. $\int a^x\ dx$

23. $\int e^{5x}\ dx$

24. $\int a^{3x}\ dx$

25. $\int \ln x\ dx$

26. $\int x^3 \ln x\ dx$

27. $\int \dfrac{1}{x \ln x}\ dx$

28. $\int \dfrac{x}{5 + 2x}\ dx$

29. $\int \dfrac{1}{16 - x^2}\ dx$

30. $\int \dfrac{\sqrt{x^2 + 9}}{x}\ dx$

20.4 APPLICATIONS OF INTEGRAL CALCULUS

The applications of integral calculus to practical problems are illustrated in the examples which follow. The examples are grouped into two parts based on the two interpretations of integration: (A) the antiderivative, and (B) the area bounded by curves and straight lines.

A. Finding the Antiderivative

In this type of problem, the rate of change (or the derivative) of a function is given and the function is to be determined.

EXAMPLE 21 Let x = number of pounds of certain meat sold daily by a grocery store
$f(x)$ = marginal revenue (which is an increase in revenue obtained from the sale of one additional unit of a product), and
$f(x) = -\$.20x + \$2.40$

Find the total revenue from selling 12 pounds of meat.

SOLUTION The given function $f(x)$ is the derivative or the rate of change of the revenue with respect to the number of pounds sold. The total revenue from sales is the antiderivative of the given function, or:

$$\text{Total revenue} = \int f(x)\ dx = \int (-.20x + 2.40)\ dx$$
$$= -\$.10x^2 + \$2.40x + C$$

If $x = 0$ (or 0 pounds sold), total revenue = 0. Thus, the constant of integration $C = 0$.

At $x = 12$ pounds sold,

$$\text{the total revenue} = -.10x^2 + 2.40x = -.10(12^2) + 2.40(12)$$
$$= -14.4 + 28.8 = \$14.4$$

Observe that the process presented in this example is an inverse of the process illustrated in Examples 7 and 8 in Chapter 19.

EXAMPLE 22 The cost of producing chairs includes a fixed cost of $100 and a variable cost of $5 per chair. Find the total cost function and the cost of producing 1,000 chairs. What is the average cost per chair?

SOLUTION Let x = number of chairs produced
$f(x) = \$5$, representing the marginal cost, which is an increase in cost for producing one additional chair, or the rate of change of the cost with respect to the number of chairs produced

$$\text{Total cost} = \int 5\ dx = \$5x + C$$

Let C = the fixed cost $100.

The total cost of producing 1,000 chairs is:

$$\$5(1,000) + \$100 = \$5,100$$

The average cost per chair is:

$$\$5,100 \div 1,000 = \$5.10$$

B. Find the Area (Definite Integral)

The method for finding the area under the curve $f(x)$, above the x-axis, between $x = a$ and $x = b$, has been developed in Section 20.2 concerning definite integration. Here the examples will give additional illustrations for finding the areas bounded by curves and straight lines.

EXAMPLE 23   Find the area in Figure 20–5 bounded by the curve $y = x^3 - x^2 - 2x$, the x-axis, between $x = -1$ and $x = 2$.

Figure 20-5   Area Bounded by the Curve $y = x^3 - x^2 - 2x$, the x-Axis, and Between $x = -1$ and $x = 2$ (Example 23)

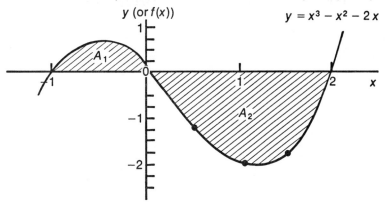

$$\int_{-1}^{2} (x^3 - x^2 - 2x)\, dx = \left(\frac{x^4}{4} - \frac{x^3}{3} - x^2\right)\Bigg|_{-1}^{2}$$

$$= \left(\frac{2^4}{4} - \frac{2^3}{3} - 2^2\right) - \left(\frac{(-1)^4}{4} - \frac{(-1)^3}{3} - (-1)^2\right)$$

$$= \left(4 - \frac{8}{3} - 4\right) - \left(\frac{1}{4} + \frac{1}{3} - 1\right)$$

$$= \left(-\frac{8}{3}\right) - \left(-\frac{5}{12}\right) = -2\frac{1}{4}$$

The answer obtained from the process of integration is the *net area*, $A = -2\frac{1}{4}$, in Example 23. If we are interested in knowing the *absolute area*, we must determine the area above the x-axis and the area under the x-axis separately. The area above the x-axis, between $x = -1$ and $x = 0$, is:

$$A_1 = \int_{-1}^{0} (x^3 - x^2 - 2x)\, dx = \left(\frac{x^4}{4} - \frac{x^3}{3} - x^2\right)\Bigg|_{-1}^{0}$$

$$= 0 - \left(\frac{(-1)^4}{4} - \frac{(-1)^3}{3} - (-1)^2\right) = 0 - \left(-\frac{5}{12}\right) = \frac{5}{12} \text{ (positive area)}$$

The area below the x-axis, between $x = 0$ and $x = 2$, is:

$$A_2 = \int_0^2 (x^3 - x^2 - 2x)\, dx = \left(\frac{x^4}{4} - \frac{x^3}{3} - x^2\right)\Big|_0^2$$

$$= \left(\frac{2^4}{4} - \frac{2^3}{3} - 2^2\right) - 0 = \left(-\frac{8}{3}\right) - 0 = -\frac{8}{3} \text{ (negative area)}$$

The absolute area, disregarding the $\pm$ signs, is:

$$|A| = |A_1| + |A_2| = \left|\frac{5}{12}\right| + \left|-\frac{8}{3}\right| = \left|\frac{37}{12}\right| = \left|3\frac{1}{12}\right|$$

Note:

1. $y = x^3 - x^2 - 2x = x(x + 1)(x - 2)$. Thus, $x = -1, 0,$ and 2 are the roots of the equation $y = 0$.

2. The net area can also be verified:

$$A = A_1 + A_2 = \frac{5}{12} + \left(-\frac{8}{3}\right) = -\frac{27}{12} = -2\frac{1}{4}$$

EXAMPLE 24 Find the area in Figure 20-6 bounded by $y = x^2 + 1$ and $y = x + 3$.

SOLUTION   It is bounded on the top by the straight line $(y = x + 3)$ and at the bottom by the parabola $(y = x^2 + 1)$. The line and the parabola are intersected at the two points: $x = -1$ and $x = 2$. The two points can be obtained in the following manner:

Figure 20-6   Area A Bounded by $y = x^2 + 1$ and $y = x + 3$
(Example 24)

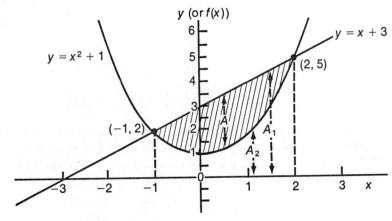

Equate the two given functions:

$x^2 + 1 = x + 3$

Solve for x from the equation:

$$x^2 + 1 - x - 3 = 0$$
$$x^2 - x - 2 = 0$$
$$(x + 1)(x - 2) = 0$$

When $x + 1 = 0$, $x = -1$.
When $x - 2 = 0$, $x = 2$.

The area under the line $y = x + 3$, above the x-axis, between $x = -1$ and $x = 2$ is denoted by A_1, or:

$$A_1 = \int_{-1}^{2} (x + 3) \, dx$$

The area under the parabola $y = x^2 + 1$, above x-axis, between $x = -1$ and $x = 2$ is denoted by A_2, or:

$$A_2 = \int_{-1}^{2} (x^2 + 1) \, dx$$

The required area, denoted by A and shaded in Figure 20–6, is:

$$A = A_1 - A_2 = \int_{-1}^{2} (x + 3) \, dx - \int_{-1}^{2} (x^2 + 1) \, dx$$

$$= \int_{-1}^{2} [(x + 3) - (x^2 + 1)] \, dx \qquad \text{(Rule 4)}$$

$$= \int_{-1}^{2} (x - x^2 + 2) \, dx$$

$$= \left(\frac{x^2}{2} - \frac{x^3}{3} + 2x \right) \Big|_{-1}^{2}$$

$$= \left(\frac{2^2}{2} - \frac{2^3}{3} + 2(2) \right) - \left(\frac{(-1)^2}{2} - \frac{(-1)^3}{3} + (2(-1)) \right)$$

$$= \left(2 - \frac{8}{3} + 4 \right) - \left(\frac{1}{2} + \frac{1}{3} - 2 \right) = 4\frac{1}{2}$$

When the A_1 area is larger than the A_2 area, the required area A is positive as shown in Figure 20–6. On the other hand, if A_1 were smaller than A_2, the sign prefixing A would have been negative.

In general, given two functions $f(x)$ and $g(x)$, the area bounded by the two function curves between $x = a$ and $x = b$ is:

$$A = \int_{a}^{b} [f(x) - g(x)] \, dx$$

EXAMPLE 25 Find the area in Figure 20–7 bounded by the curve $y = 11 - x^2$, above $p = y = 7$, and at the right of $x = 0$.

SOLUTION Substitute $y = 7$ in:

Figure 20-7   Area A Bounded By y = 11 – x², Above p = y = 7, and Right Side of x = 0 (Example 25)

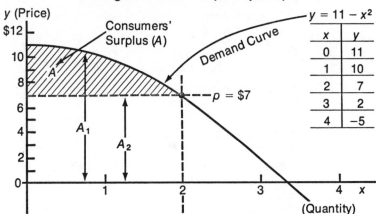

$y = 11 - x^2$
$7 = 11 - x^2$
$x^2 = 4$
$x = \pm 2$

We disregard $x = -2$ since the area required is at the right of $x = 0$.

The area under the curve $y = 11 - x^2$, above the x-axis, between $x = 0$ and $x = 2$, denoted by A_1, is:

$$A_1 = \int_0^2 (11 - x^2)\, dx = \left(11x - \frac{x^3}{3}\right) \Big|_0^2 = 11(2) - \frac{2^3}{3} = 19\frac{1}{3}$$

The area under p, above the x-axis, between $x = 0$ and $x = 2$, denoted by A_2, is:

$$A_2 = 2(7) = 14$$

The required area, denoted by A and shaded in the graph, is:

$$A = A_1 - A_2 = 19\frac{1}{3} - 14 = 5\frac{1}{3}$$

or: $A = \int_0^2 \left[(11 - x^2) - 7\right] dx = \left(4x - \frac{x^3}{3}\right)\Big|_0^2 = \left(8 - \frac{8}{3}\right) - 0 = 5\frac{1}{3}$

Example 25 can be used in analyzing the economic problem—finding the consumer's surplus—as follows:

Let $y =$ unit price of a commodity and
   $x =$ units or quantity of the commodity demanded on the market

Then, the demand quantity x is a function of the price y. The graph representing $y = 11 - x^2$ is called a *demand curve*. Generally, a demand curve indicates that the demand x will increase as the price y decreases. If the consumers in the market are willing to pay the prices described by the demand curve but in fact are paying the fixed price at $7 per unit, they have gained the amount (area A) by paying less than they are willing to pay. The amount gained is called the *consumers' surplus*.

EXAMPLE 26 Find the area in Figure 20–8 bounded by the curve $y = x^2 + 3$, below $p = y = 7$, and at the right of $x = 0$.

Figure 20-8   **Area B Bounded By y =x² + 3, Below p = y = 7, and At Right Side of x = 0 (Example 26)**

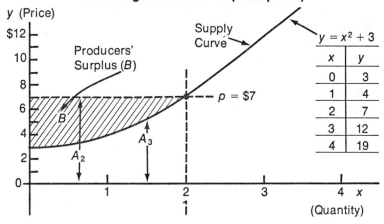

SOLUTION Substitute $y = 7$ in:

$$y = x^2 + 3$$
$$7 = x^2 + 3$$
$$x^2 = 4$$
$$x = \pm 2$$

We disregard $x = -2$ since the area required is at the right of $x = 0$.
 The area under p, above the x-axis, between $x = 0$ and $x = 2$, denoted by A_2, is:

$$A_2 = 2(7) = 14$$

The area under the curve $y = x^2 + 3$, above the x-axis, between $x = 0$ and $x = 2$, denoted by A_3, is:

$$A_3 = \int_0^2 (x^2 + 3)\, dx = \left(\frac{x^3}{3} + 3x\right)\Big|_0^2$$

$$= \frac{2^3}{3} + 3(2) = 8\frac{2}{3}$$

The required area, denoted by B and shaded in the graph, is:

$$B = A_2 - A_3 = 14 - 8\frac{2}{3} = 5\frac{1}{3}$$

Example 26 can be used in analyzing the economic problem—finding the producers' surplus—as follows:

Let $y =$ unit price of a commodity
　　　$x =$ units or quantity of the commodity supplied by producers on the market

Then, the supply quantity x is a function of the price y. The graph representing $y = x^2 + 3$ is called a *supply curve*. Generally, a supply curve indicates that the supply x will increase as the price y increases. If the producers in the market are willing to sell at the prices described by the supply curve but in fact are selling at a fixed price of \$7 per unit, they have gained the amount (area B) by selling at a price higher than those prices at which they are willing to sell. The amount gained is called the *producers' surplus*.

EXAMPLE 27　Find the area in Figure 20–9 bounded by $y = 11 - x^2$, $y = x^2 + 3$, and at the right of $x = 0$.

Figure 20-9　**Area C Bounded by y = 11 − x², y = x² + 3, and at the Right of x = 0 (Example 27)**

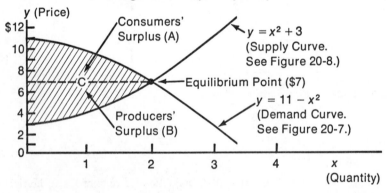

SOLUTION　Equate the two given y equations:

$$x^2 + 3 = 11 - x^2$$
$$2x^2 = 8$$
$$x = \pm 2$$

We disregard $x = -2$ since the area required is at the right of $x = 0$.
　　Thus, the required area, denoted by C, is bounded by the curve $y = 11 - x^2$, the curve $y = x^2 + 3$, and between $x = 0$ and $x = 2$.

$$C = \int_0^2 [(11 - x^2) - (x^2 + 3)] \, dx$$

$$= \int_0^2 (8 - 2x^2)\, dx$$

$$= \left(8x - \frac{2x^3}{3}\right)\Big|_0^2 = \left(8(2) - \frac{2(2)^3}{3}\right) - 0 = 10\frac{2}{3}$$

Observe that the required area C is the sum of area A (consumers' surplus) in Example 25 and area B (producers' surplus) in Example 26. Check:

$$C = A + B = 5\frac{1}{3} + 5\frac{1}{3} = 10\frac{2}{3}$$

When the supply is equal to the demand, the market reaches a point of equilibrium. The price at the equilibrium point is the market price for both producers and consumers, such as $7 in Examples 25 and 26. The equilibrium point can be obtained by solving the pair of equations representing the supply and demand curves as shown in Example 27, $x = 2$ units.

EXAMPLE 28 A company plans to establish an equipment fund by investing $1,000 at the end of each year for 4 years. Assume that the interest is 6% compounded continuously. Find the amount in the fund at the end of 4 years.

SOLUTION The formula for finding the compound amount S of the principal P at the nominal rate j compounded continuously for n years is:

$$S = P(e^{jn})$$

which is formula (14–7) as presented in Chapter 14.

(a) The exact method. Use formula (14–7) to compute the compound amount at the end of the 4th year for each annual payment.

$$P = \$1,000,\ j = 6\% = .06$$

The compound amount for the first payment:

$n = 3$ years (from the end of the first year to the end of the 4th year)
$S = 1,000(e^{.06(3)}) = 1,000(e^{.18})$
  $= 1,000(1.197217) = \$1,197.22$ (Table 4 in Appendix)

The compound amount for the second payment:

$n = 2$ years (from the end of the second year to the end of the 4th year)
$S = 1,000(e^{.06(2)}) = 1,000(e^{.12})$
  $= 1,000(1.127497) = \$1,127.50$

The compound amount for the third payment:

$n = 1$ year (from the end of the third year to the end of the 4th year)
$S = 1,000(e^{.06(1)}) = 1,000(e^{.06})$
  $= 1,000(1.061837) = \$1,061.84$

The compound amount for the fourth payment:

$S = \$1,000$, since the payment is made at the end of the 4th year.

The total amount in the fund at the end of 4 years is $4,386.56, as shown in the diagram which follows:

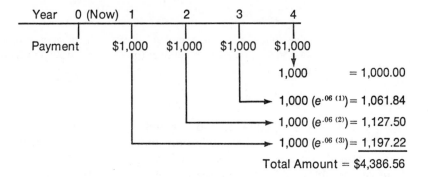

The total amount may also be computed by using the formula for the amount of an ordinary annuity, formula (14–8) in Chapter 14:

$$S_n = R \cdot \frac{(1+i)^n - 1}{i}$$

$$i = f = e^j - 1 \text{ (formula 14–6)}$$

Thus:

$$S_n = R \cdot \frac{(1 + (e^j - 1))^n - 1}{e^j - 1} = R \cdot \frac{e^{jn} - 1}{e^j - 1}$$

Substitute the values of $R = \$1,000$ (per year), $j = 6\% = .06$ (stated yearly rate), and $n = 4$ (yearly payments) in the preceding formula:

$$S_n = 1,000 \cdot \frac{e^{.06(4)} - 1}{e^{.06} - 1} = 1,000 \cdot \frac{1.271249 - 1}{1.061837 - 1}$$

$$= 1,000(4.386516) = \$4,386.52$$

The difference from $4,386.56 is due to rounding.

(b) The approximate method. Find the area under the curve $S = P(e^{jn})$ (where $y = S$ and $x = n$), above the x-axis, and between $x = 0 - .5 = -.5$ and $x = n - .5$ (where .5 represents $\frac{1}{2}$ year).

$n = 4$ (yearly payments), and $4 - .5 = 3.5$.

$$\int_{-.5}^{3.5} P(e^{jx}) \, dx = P \left(\frac{e^{jx}}{j} \right) \Big|_{-.5}^{3.5} \text{ (Use No. 8 in Table 20–1.)}$$

Substitute the values of $P = \$1,000$, and $j = .06$ in the preceding expression.

The required area, or the total amount of the 4 payments and their compound interest, is:

$$A = 1,000 \left(\frac{e^{.06(3.5)}}{.06} - \frac{e^{.06(-.5)}}{.06} \right)$$

$$= 1,000 \left(\frac{e^{.21}}{.06} - \frac{e^{-.03}}{.06}\right) = 1,000 \left(\frac{1.233678}{.06} - \frac{.970446}{.06}\right)$$

$$= 1,000 \left(\frac{.263232}{.06}\right) = \$4,387.20. \text{ (Approximated Value)}$$

The results of the exact method and the approximate method are shown in Figure 20–10.

Figure 20-10
(a) The Exact Method Result — Bars Centered at $n = 0, 1, 2,$ and $3, (n = x)$.
(b) The Approximate Method Result — Area A Under $y = P(e^{jx})$, Above x-Axis, and Between $x = -.5$ and $x = 3.5$ (Shaded)

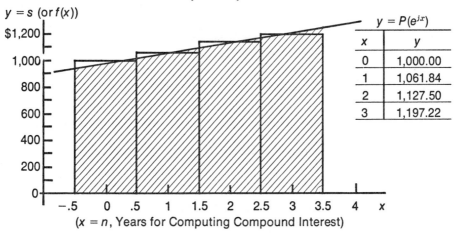

x	y
0	1,000.00
1	1,061.84
2	1,127.50
3	1,197.22

Observe the four bars in Figure 20–10. They represent the individual compound amounts of the annual payments. The width of each bar is 1 year and the height represents the respective compound amount — the first bar = $\$1,000(1) =$ $\$1,000$, the second bar = $\$1,061.84(1) = \$1,061.84$, and so on. The sum of the bars is equal to $\$4,386.56$ as computed in the exact method. The bars are plotted continuously and are centered on the discrete n values ($n = 0, 1, 2,$ and 3 years for computing compound interest). Thus, $n = 0$ covers the area from $-.5$ to $.5$, $n = 1$ covers the area from $.5$ to 1.5, and so on. The four bars continuously cover the area from $-.5$ to 3.5. The area is the basis for the computation by the approximate method.

The problem in Example 28 may also be stated as follows:

> Find the amount of an ordinary annuity if the size of each payment is $\$1,000$, payable at the end of each year for four years at an interest rate of 6% compounded continuously.

In a similar manner, the present value of the four annual payments as stated in Example 28 may be computed. This is illustrated in Example 29.

EXAMPLE 29 Find the present value of an annuity if the size of each payment is $1,000, payable at the end of each year for four years at an interest rate of 6% compounded continuously.

(a) The exact method. Use formula (14–9) to compute the present value at the beginning of the term of the annuity, or the beginning of the first year. (Also see the diagram of Example 18 in Chapter 14, page 305, for a clear understanding of the present value of an ordinary annuity.)

$$A_n = R \cdot \frac{1 - (1+i)^{-n}}{i}$$

$$i = e^j - 1$$

Thus:

$$A_n = R \cdot \frac{1 - (1 + (e^j - 1))^{-n}}{e^j - 1}$$

or, $A_n = R \cdot \dfrac{1 - e^{-jn}}{e^j - 1}$

Substitute the values of $R = \$1,000$ (per year), $j = 6\% = .06$ (stated yearly rate), and $n = 4$ (yearly payments) in the above formula:

$$A_n = 1,000 \cdot \frac{1 - e^{-.06(4)}}{e^{.06} - 1} = 1,000 \cdot \frac{1 - .786628}{1.061837 - 1}$$

$$= 1,000(3.450555) = \$3,450.56$$

(b) The approximate method. Find the area under the curve $P = S(e^{-jn})$ (where $y = P$ and $x = n$) above the x-axis, and between $x = 1 - .5 = .5$ and $x = n + .5$ (where .5 represents $\frac{1}{2}$ year).

$n = 4$ (yearly payments), and $4 + .5 = 4.5$.

$$\int_{.5}^{4.5} S(e^{-jx})\, dx = S\left(\frac{e^{-jx}}{-j}\right)\Big|_{.5}^{4.5} \quad \text{(Use No. 8 in Table 20–1.)}$$

Substitute the values of $S = \$1,000$, and $j = .06$ in the above expression.

The required area, or the total present value of the four payments, is:

$$A = 1,000\left(\frac{e^{-.06(4.5)}}{-.06} - \frac{e^{-.06(.5)}}{-.06}\right)$$

$$= 1,000\left(\frac{e^{-.27}}{-.06} - \frac{e^{-.03}}{-.06}\right)$$

$$= 1,000\left(\frac{.763379 - .970446}{-.06}\right)$$

$$= 1,000\left(\frac{-.207067}{-.06}\right) = \$3,451.12. \quad \text{(Approximated Value)}$$

The solution of Example 29 is not shown graphically. However, if the graph were constructed, the bars would be plotted continuously and centered on the discrete n values. ($n = 1, 2, 3$, and 4 years for computing compound discount.) Thus, $n = 1$ covered the area from .5 to 1.5 and $n = 4$ covered the area from 3.5 to 4.5. The four bars continuously covered the area from .5 to 4.5. The area is the basis for the computation by the approximate method.

EXERCISE 20–3 Reference: Section 20.4

A. *Graph and find the net area bounded by the y function, the x-axis, and between the x range as indicated in each problem.*
   1. $y = 5$, between (a) $x = -3$ and $x = 0$, and (b) $x = 0$ and $x = 3$.
   2. $y = -4$, between (a) $x = -2$ and $x = 1$, and (b) $x = 1$ and $x = 6$.
   3. $y = x - 5$, between (a) $x = 2$ and $x = 8$, and (b) $x = 4$ and $x = 7$.
   4. $y = 8 - 2x$, between (a) $x = 1$ and $x = 7$, and (b) $x = 0$ and $x = 6$.
   5. $y = x^3 + x^2 - 2x$, between (a) $x = -2$ and $x = 1$, and (b) $x = -1$ and $x = 1$.
   6. $y = x + 2x^2 - x^3 - 2$, between (a) $x = -1$ and $x = 2$, and (b) $x = -1$ and $x = 1$.
   7. $y = x^2 + 5$, between $x = -2$ and $x = 1$.
   8. $y = -x^2 - 4$, between $x = -2$ and $x = 2$.

B. *Graph and find the net area bounded by the two given y functions.*
   9. $y = x^2 + 3$, and $y = 5 - x$.
   10. $y = 9 - x^2$, and $y = x + 3$.
   11. $y = 2x^2 + 3$, and $y = 4 - 2x^2$.
   12. $y = 2x^2$, and $y = 20 - 3x^2$.

C. *Find the absolute area for each problem.*
   13. Use the given information in problem 3.
   14. Use the given information in problem 4.
   15. Use the given information in problem 5(a) only.
   16. Bounded by $y = 4 - x^2$ and $y = x - 2$.

D. *Statement Problems*
   17. The marginal revenue function from the sale of certain grades of paint in the Alan Hardware Store is:

   $$f(x) = -\$.1x + \$12, \text{ where } x = \text{number of gallons sold}$$

   Find the total revenue from selling (a) 100 gallons, and (b) 200 gallons. Also, find (c) the number of gallons that should be sold in order to maximize the total revenue, and (d) the maximized total revenue.

   18. The marginal revenue function from the sale of certain TV sets in Taylor Appliance Company is:

   $$f(x) = \$60{,}000 - \$20{,}000x, \text{ where}$$
   $$x = \text{units in 100 TV sets sold}$$

Find the total revenue and the average selling price per set in selling (a) 200 sets, and (b) 500 sets. Also, find (c) the number of sets that should be sold in maximizing the total revenue and (d) the maximized total revenue and the average selling price per set for the sale.

19. The cost of producing sewing machines includes a fixed cost of $2,000 and a variable cost of $50 per machine. Find the total cost function and the cost of producing 100 machines. What is the average cost per machine?

20. The marginal cost function for producing small radios in Pat Electric Company is:

$$f(x) = \$12 - \$0.008x, \text{ where } x = \text{number of radios produced}$$

The fixed cost is $300. Find the total cost and the average cost per radio in producing (a) 100, (b) 150, and (c) 200 radios.

21. If the demand function is:

$$y = \$32 - x^2, \text{ where}$$
$$y = \text{unit price of a commodity, and}$$
$$x = \text{units of the commodity demanded}$$

find the consumers' surplus at the unit price of $16.

22. If the supply function is:

$$y = x^2 + \$4x + \$4, \text{ where}$$
$$y = \text{unit price of a commodity, and}$$
$$x = \text{units of the commodity supplied}$$

find the producers' surplus at the unit price of $25.

23. If the demand function is $y = \$16 - \$2x^2$ and the supply function is $y = \$2x^2$ (see the definitions of x and y in problems 21 and 22), find (a) the consumers' surplus and (b) the producers' surplus based on the point of equilibrium.

24. Use the information given in problems 21 and 22. Compute the x and y values at the point of equilibrium. (Round to two decimal places.)

25. A man deposits $100 in a bank at the end of each year for 10 years. (The first deposit is made at the end of the first year.) If the money earns interest at 8% compounded continuously, how much does he have in his account at the end of the period after the last payment is made? Use (a) the exact method—formulas (14–6 and 14–8), and (b) the approximate method.

26. A man wishes to borrow money from a bank which charges interest at 8% compounded continuously. If he agrees to pay $100 at the end of each year for 10 years, how much money should he receive from the bank at the time of borrowing? Use (a) the exact method—formulas (14–6 and 14–9), and (b) the approximate method.

Appendix
Tables

Table 1 Squares, Square Roots, and Reciprocals

n	n^2	$\sqrt{n}$	$\sqrt{10n}$	$1/n$	n	n^2	$\sqrt{n}$	$\sqrt{10n}$	$1/n$
1.00	1.0000	1.00000	3.16228	1.000000	**1.50**	2.2500	1.22474	3.87298	.666667
1.01	1.0201	1.00499	3.17805	.990099	1.51	2.2801	1.22882	3.88587	.662252
1.02	1.0404	1.00995	3.19374	.980392	1.52	2.3104	1.23288	3.89872	.657895
1.03	1.0609	1.01489	3.20936	.970874	1.53	2.3409	1.23693	3.91152	.653595
1.04	1.0816	1.01980	3.22490	.961538	1.54	2.3716	1.24097	3.92428	.649351
1.05	1.1025	1.02470	3.24037	.952381	1.55	2.4025	1.24499	3.93700	.645161
1.06	1.1236	1.02956	3.25576	.943396	1.56	2.4336	1.24900	3.94968	.641026
1.07	1.1449	1.03441	3.27109	.934579	1.57	2.4649	1.25300	3.96232	.636943
1.08	1.1664	1.03923	3.28634	.925926	1.58	2.4964	1.25698	3.97492	.632911
1.09	1.1881	1.04403	3.30151	.917431	1.59	2.5281	1.26095	3.98748	.628931
1.10	1.2100	1.04881	3.31662	.909091	**1.60**	2.5600	1.26491	4.00000	.625000
1.11	1.2321	1.05357	3.33167	.900901	1.61	2.5921	1.26886	4.01248	.621118
1.12	1.2544	1.05830	3.34664	.892857	1.62	2.6244	1.27279	4.02492	.617284
1.13	1.2769	1.06301	3.36155	.884956	1.63	2.6569	1.27671	4.03733	.613497
1.14	1.2996	1.06771	3.37639	.877193	1.64	2.6896	1.28062	4.04969	.609756
1.15	1.3225	1.07238	3.39116	.869565	1.65	2.7225	1.28452	4.06202	.606061
1.16	1.3456	1.07703	3.40588	.862069	1.66	2.7556	1.28841	4.07431	.602410
1.17	1.3689	1.08167	3.42053	.854701	1.67	2.7889	1.29228	4.08656	.598802
1.18	1.3924	1.08628	3.43511	.847458	1.68	2.8224	1.29615	4.09878	.595238
1.19	1.4161	1.09087	3.44964	.840336	1.69	2.8561	1.30000	4.11096	.591716
1.20	1.4400	1.09545	3.46410	.833333	**1.70**	2.8900	1.30384	4.12311	.588235
1.21	1.4641	1.10000	3.47851	.826446	1.71	2.9241	1.30767	4.13521	.584795
1.22	1.4884	1.10454	3.49285	.819672	1.72	2.9584	1.31149	4.14729	.581395
1.23	1.5129	1.10905	3.50714	.813008	1.73	2.9929	1.31529	4.15933	.578035
1.24	1.5376	1.11355	3.52136	.806452	1.74	3.0276	1.31909	4.17133	.574713
1.25	1.5625	1.11803	3.53553	.800000	1.75	3.0625	1.32288	4.18330	.571429
1.26	1.5876	1.12250	3.54965	.793651	1.76	3.0976	1.32665	4.19524	.568182
1.27	1.6129	1.12694	3.56371	.787402	1.77	3.1329	1.33041	4.20714	.564972
1.28	1.6384	1.13137	3.57771	.781250	1.78	3.1684	1.33417	4.21900	.561798
1.29	1.6641	1.13578	3.59166	.775194	1.79	3.2041	1.33791	4.23084	.558659
1.30	1.6900	1.14018	3.60555	.769231	**1.80**	3.2400	1.34164	4.24264	.555556
1.31	1.7161	1.14455	3.61939	.763359	1.81	3.2761	1.34536	4.25441	.552486
1.32	1.7424	1.14891	3.63318	.757576	1.82	3.3124	1.34907	4.26615	.549451
1.33	1.7689	1.15326	3.64692	.751880	1.83	3.3489	1.35277	4.27785	.546448
1.34	1.7956	1.15758	3.66060	.746269	1.84	3.3856	1.35647	4.28952	.543478
1.35	1.8225	1.16190	3.67423	.740741	1.85	3.4225	1.36015	4.30116	.540541
1.36	1.8496	1.16619	3.68782	.735294	1.86	3.4596	1.36382	4.31277	.537634
1.37	1.8769	1.17047	3.70135	.729927	1.87	3.4969	1.36748	4.32435	.534759
1.38	1.9044	1.17473	3.71484	.724638	1.88	3.5344	1.37113	4.33590	.531915
1.39	1.9321	1.17898	3.72827	.719424	1.89	3.5721	1.37477	4.34741	.529101
1.40	1.9600	1.18322	3.74166	.714286	**1.90**	3.6100	1.37840	4.35890	.526316
1.41	1.9881	1.18743	3.75500	.709220	1.91	3.6481	1.38203	4.37035	.523560
1.42	2.0164	1.19164	3.76829	.704225	1.92	3.6864	1.38564	4.38178	.520833
1.43	2.0449	1.19583	3.78153	.699301	1.93	3.7249	1.38924	4.39318	.518135
1.44	2.0736	1.20000	3.79473	.694444	1.94	3.7636	1.39284	4.40454	.515464
1.45	2.1025	1.20416	3.80789	.689655	1.95	3.8025	1.39642	4.41588	.512821
1.46	2.1316	1.20830	3.82099	.684932	1.96	3.8416	1.40000	4.42719	.510204
1.47	2.1609	1.21244	3.83406	.680272	1.97	3.8809	1.40357	4.43847	.507614
1.48	2.1904	1.21655	3.84708	.675676	1.98	3.9204	1.40712	4.44972	.505051
1.49	2.2201	1.22066	3.86005	.671141	1.99	3.9601	1.41067	4.46094	.502513
1.50	2.2500	1.22474	3.87298	.666667	**2.00**	4.0000	1.41421	4.47214	.500000
n	n^2	$\sqrt{n}$	$\sqrt{10n}$	$1/n$	n	n^2	$\sqrt{n}$	$\sqrt{10n}$	$1/n$

Table 1 **Squares, Square Roots, and Reciprocals**

n	n²	√n	√10n	1/n	n	n²	√n	√10n	1/n
2.00	4.0000	1.41421	4.47214	.500000	2.50	6.2500	1.58114	5.00000	.400000
2.01	4.0401	1.41774	4.48330	.497512	2.51	6.3001	1.58430	5.00999	.398406
2.02	4.0804	1.42127	4.49444	.495050	2.52	6.3504	1.58745	5.01996	.396825
2.03	4.1209	1.42478	4.50555	.492611	2.53	6.4009	1.59060	5.02991	.395257
2.04	4.1616	1.42829	4.51664	.490196	2.54	6.4516	1.59374	5.03984	.393701
2.05	4.2025	1.43178	4.52769	.487805	2.55	6.5025	1.59687	5.04975	.392157
2.06	4.2436	1.43527	4.53872	.485437	2.56	6.5536	1.60000	5.05964	.390625
2.07	4.2849	1.43875	4.54973	.483092	2.57	6.6049	1.60312	5.06952	.389105
2.08	4.3264	1.44222	4.56070	.480769	2.58	6.6564	1.60624	5.07937	.387597
2.09	4.3681	1.44568	4.57165	.478469	2.59	6.7081	1.60935	5.08920	.386100
2.10	4.4100	1.44914	4.58258	.476190	2.60	6.7600	1.61245	5.09902	.384615
2.11	4.4521	1.45258	4.59347	.473934	2.61	6.8121	1.61555	5.10882	.383142
2.12	4.4944	1.45602	4.60435	.471698	2.62	6.8644	1.61864	5.11859	.381679
2.13	4.5369	1.45945	4.61519	.469484	2.63	6.9169	1.62173	5.12835	.380228
2.14	4.5796	1.46287	4.62601	.467290	2.64	6.9696	1.62481	5.13809	.378788
2.15	4.6225	1.46629	4.63681	.465116	2.65	7.0225	1.62788	5.14782	.377358
2.16	4.6656	1.46969	4.64758	.462963	2.66	7.0756	1.63095	5.15752	.375940
2.17	4.7089	1.47309	4.65833	.460829	2.67	7.1289	1.63401	5.16720	.374532
2.18	4.7524	1.47648	4.66905	.458716	2.68	7.1824	1.63707	5.17687	.373134
2.19	4.7961	1,47986	4.67974	.456621	2.69	7.2361	1.64012	5.18652	.371747
2.20	4.8400	1.48324	4.69042	.454545	2.70	7.2900	1.64317	5.19615	.370370
2.21	4.8841	1.48661	4.70106	.452489	2.71	7.3441	1.64621	5.20577	.369004
2.22	4.9284	1.48997	4.71169	.450450	2.72	7.3984	1.64924	5.21536	.367647
2.23	4.9729	1.49332	4.72229	.448430	2.73	7.4529	1.65227	5.22494	.366300
2.24	5.0176	1.49666	4.73286	.446429	2.74	7.5076	1.65529	5.23450	.364964
2.25	5.0625	1.50000	4.74342	.444444	2.75	7.5625	1.65831	5.24404	.363636
2.26	5.1076	1.50333	4.75395	.442478	2.76	7.6176	1.66132	5.25357	.362319
2.27	5.1529	1.50665	4.76445	.440529	2.77	7.6729	1.66433	5.26308	.361011
2.28	5.1984	1.50997	4.77493	.438596	2.78	7.7284	1.66733	5.27257	.359712
2.29	5.2441	1.51327	4.78539	.436681	2.79	7.7841	1.67033	5.28205	.358423
2.30	5.2900	1.51658	4.79583	.434783	2.80	7.8400	1.67332	5.29150	.357143
2.31	5.3361	1.51987	4.80625	.432900	2.81	7.8961	1.67631	5.30094	.355872
2.32	5.3824	1.52315	4.81664	.431034	2.82	7.9524	1.67929	5.31037	.354610
2.33	5.4289	1.52643	4.82701	.429185	2.83	8.0089	1.68226	5.31977	.353357
2.34	5.4756	1.52971	4.83735	.427350	2.84	8.0656	1.68523	5.32917	.352113
2.35	5.5225	1.53297	4.84768	.425532	2.85	8.1225	1.68819	5.33854	.350877
2.36	5.5696	1.53623	4.85798	.423729	2.86	8.1796	1.69115	5.34790	.349650
2.37	5.6169	1.53948	4.86826	.421941	2.87	8.2369	1.69411	5.35724	.348432
2.38	5.6644	1.54272	4.87852	.420168	2.88	8.2944	1.69706	5.36656	.347222
2.39	5.7121	1.54596	4.88876	.418410	2.89	8.3521	1.70000	5.37587	.346021
2.40	5.7600	1.54919	4.89898	.416667	2.90	8.4100	1.70294	5.38516	.344828
2.41	5.8081	1.55242	4.90918	.414938	2.91	8.4681	1.70587	5.39444	.343643
2.42	5.8564	1.55563	4.91935	.413223	2.92	8.5264	1.70880	5.40370	.342466
2.43	5.9049	1.55885	4.92950	.411523	2.93	8.5849	1.71172	5.41295	.341297
2.44	5.9536	1.56205	4.93964	.409836	2.94	8.6436	1.71464	5.42218	.340136
2.45	6.0025	1.56525	4.94975	.408163	2.95	8.7025	1.71756	5.43139	.338983
2.46	6.0516	1.56844	4.95984	.406504	2.96	8.7616	1.72047	5.44059	.337838
2.47	6.1009	1.57162	4.96991	.404858	2.97	8.8209	1.72337	5.44977	.336700
2.48	6.1504	1.57480	4.97996	.403226	2.98	8.8804	1.72627	5.45894	.335570
2.49	6.2001	1.57797	4.98999	.401606	2.99	8.9401	1.72916	5.46809	.334448
2.50	6.2500	1.58114	5.00000	.400000	3.00	9.0000	1.73205	5.47723	.333333
n	n²	√n	√10n	1/n	n	n²	√n	√10n	1/n

Table 1 **Squares, Square Roots, and Reciprocals**

n	n^2	$\sqrt{n}$	$\sqrt{10n}$	$1/n$	n	n^2	$\sqrt{n}$	$\sqrt{10n}$	$1/n$
3.00	9.0000	1.73205	5.47723	.333333	**3.50**	12.2500	1.87083	5.91608	.285714
3.01	9.0601	1.73494	5.48635	.332226	3.51	12.3201	1.87350	5.92453	.284900
3.02	9.1204	1.73781	5.49545	.331126	3.52	12.3904	1.87617	5.93296	.284091
3.03	9.1809	1.74069	5.50454	.330033	3.53	12.4609	1.87883	5.94138	.283286
3.04	9.2416	1.74356	5.51362	.328947	3.54	12.5316	1.88149	5.94979	.282486
3.05	9.3025	1.74642	5.52268	.327869	3.55	12.6025	1.88414	5.95819	.281690
3.06	9.3636	1.74929	5.53173	.326797	3.56	12.6736	1.88680	5.96657	.280899
3.07	9.4249	1.75214	5.54076	.325733	3.57	12.7449	1.88944	5.97495	.280112
3.08	9.4864	1.75499	5.54977	.324675	3.58	12.8164	1.89209	5.98331	.279330
3.09	9.5481	1.75784	5.55878	.323625	3.59	12.8881	1.89473	5.99166	.278552
3.10	9.6100	1.76068	5.56776	.322581	**3.60**	12.9600	1.89737	6.00000	.277778
3.11	9.6721	1.76352	5.57674	.321543	3.61	13.0321	1.90000	6.00833	.277008
3.12	9.7344	1.76635	5.58570	.320513	3.62	13.1044	1.90263	6.01664	.276243
3.13	9.7969	1.76918	5.59464	.319489	3.63	13.1769	1.90526	6.02495	.275482
3.14	9.8596	1.77200	5.60357	.318471	3.64	13.2496	1.90788	6.03324	.274725
3.15	9.9225	1.77482	5.61249	.317460	3.65	13.3225	1.91050	6.04152	.273973
3.16	9.9856	1.77764	5.62139	.316456	3.66	13.3956	1.91311	6.04979	.273224
3.17	10.0489	1.78045	5.63028	.315457	3.67	13.4689	1.91572	6.05805	.272480
3.18	10.1124	1.78326	5.63915	.314465	3.68	13.5424	1.91833	6.06630	.271739
3.19	10.1761	1.78606	5.64801	.313480	3.69	13.6161	1.92094	6.07454	.271003
3.20	10.2400	1.78885	5.65685	.312500	**3.70**	13.6900	1.92354	6.08276	.270270
3.21	10.3041	1.79165	5.66569	.311526	3.71	13.7641	1.92614	6.09098	.269542
3.22	10.3684	1.79444	5.67450	.310559	3.72	13.8384	1.92873	6.09918	.268817
3.23	10.4329	1.79722	5.68331	.309598	3.73	13.9129	1.93132	6.10737	.268097
3.24	10.4976	1.80000	5.69210	.308642	3.74	13.9876	1.93391	6.11555	.267380
3.25	10.5625	1.80278	5.70088	.307692	3.75	14.0625	1.93649	6.12372	.266667
3.26	10.6276	1.80555	5.70964	.306748	3.76	14.1376	1.93907	6.13188	.265957
3.27	10.6929	1.80831	5.71839	.305810	3.77	14.2129	1.94165	6.14003	.265252
3.28	10.7584	1.81108	5.72713	.304878	3.78	14.2884	1.94422	6.14817	.264550
3.29	10.8241	1.81384	5.73585	.303951	3.79	14.3641	1.94679	6.15630	.263852
3.30	10.8900	1.81659	5.74456	.303030	**3.80**	14.4400	1.94936	6.16441	.263158
3.31	10.9561	1.81934	5.75326	.302115	3.81	14.5161	1.95192	6.17252	.262467
3.32	11.0224	1.82209	5.76194	.301205	3.82	14.5924	1.95448	6.18061	.261780
3.33	11.0889	1.82483	5.77062	.300300	3.83	14.6689	1.95704	6.18870	.261097
3.34	11.1556	1.82757	5.77927	.299401	3.84	14.7456	1.95959	6.19677	.260417
3.35	11.2225	1.83030	5.78792	.298507	3.85	14.8225	1.96214	6.20484	.259740
3.36	11.2896	1.83303	5.79655	.297619	3.86	14.8996	1.96469	6.21289	.259067
3.37	11.3569	1.83576	5.80517	.296736	3.87	14.9769	1.96723	6.22093	.258398
3.38	11.4244	1.83848	5.81378	.295858	3.88	15.0544	1.96977	6.22896	.257732
3.39	11.4921	1.84120	5.82237	.294985	3.89	15.1321	1.97231	6.23699	.257069
3.40	11.5600	1.84391	5.83095	.294118	**3.90**	15.2100	1.97484	6.24500	.256410
3.41	11.6281	1.84662	5.83952	.293255	3.91	15.2881	1.97737	6.25300	.255754
3.42	11.6964	1.84932	5.84808	.292398	3.92	15.3664	1.97990	6.26099	.255102
3.43	11.7649	1.85203	5.85662	.291545	3.93	15.4449	1.98242	6.26897	.254453
3.44	11.8336	1.85472	5.86515	.290698	3.94	15.5236	1.98494	6.27694	.253807
3.45	11.9025	1.85742	5.87367	.289855	3.95	15.6025	1.98746	6.28490	.253165
3.46	11.9716	1.86011	5.88218	.289017	3.96	15.6816	1.98997	6.29285	.252525
3.47	12.0409	1.86279	5.89067	.288184	3.97	15.7609	1.99249	6.30079	.251889
3.48	12.1104	1.86548	5.89915	.287356	3.98	15.8408	1.99499	6.30872	.251256
3.49	12.1801	1.86815	5.90762	.286533	3.99	15.9201	1.99750	6.31664	.250627
3.50	12.2500	1.87083	5.91608	.285714	**4.00**	16.0000	2.00000	6.32456	.250000
n	n^2	$\sqrt{n}$	$\sqrt{10n}$	$1/n$	n	n^2	$\sqrt{n}$	$\sqrt{10n}$	$1/n$

Table 1 **Squares, Square Roots, and Reciprocals**

n	n^2	$\sqrt{n}$	$\sqrt{10n}$	$1/n$	n	n^2	$\sqrt{n}$	$\sqrt{10n}$	$1/n$
4.00	16.0000	2.00000	6.32456	.250000	**4.50**	20.2500	2.12132	6.70820	.222222
4.01	16.0801	2.00250	6.33246	.249377	4.51	20.3401	2.12368	6.71565	.221729
4.02	16.1604	2.00499	6.34035	.248756	4.52	20.4304	2.12603	6.72309	.221239
4.03	16.2409	2.00749	6.34823	.248139	4.53	20.5209	2.12838	6.73053	.220751
4.04	16.3216	2.00998	6.35610	.247525	4.54	20.6116	2.13073	6.73795	.220264
4.05	16.4025	2.01246	6.36396	.246914	4.55	20.7025	2.13307	6.74537	.219780
4.06	16.4836	2.01494	6.37181	.246305	4.56	20.7936	2.13542	6.75278	.219298
4.07	16.5649	2.01742	6.37966	.245700	4.57	20.8849	2.13776	6.76018	.218818
4.08	16.6464	2.01990	6.38749	.245098	4.58	20.9764	2.14009	6.76757	.218341
4.09	16.7281	2.02237	6.39531	.244499	4.59	21.0681	2.14243	6.77495	.217865
4.10	16.8100	2.02485	6.40312	.243902	**4.60**	21.1600	2.14476	6.78233	.217391
4.11	16.8921	2.02731	6.41093	.243309	4.61	21.2521	2.14709	6.78970	.216920
4.12	16.9744	2.02978	6.41872	.242718	4.62	21.3444	2.14942	6.79706	.216450
4.13	17.0569	2.03224	6.42651	.242131	4.63	21.4369	2.15174	6.80441	.215983
4.14	17.1396	2.03470	6.43428	.241546	4.64	21.5296	2.15407	6.81175	.215517
4.15	17.2225	2.03715	6.44205	.240964	4.65	21.6225	2.15639	6.81909	.215054
4.16	17.3056	2.03961	6.44981	.240385	4.66	21.7156	2.15870	6.82642	.214592
4.17	17.3889	2.04206	6.45755	.239808	4.67	21.8089	2.16102	6.83374	.214133
4.18	17.4724	2.04450	6.46529	.239234	4.68	21.9024	2.16333	6.84105	.213675
4.19	17.5561	2.04695	6.47302	.238663	4.69	21.9961	2.16564	6.84836	.213220
4.20	17.6400	2.04939	6.48074	.238095	**4.70**	22.0900	2.16795	6.85565	.212766
4.21	17.7241	2.05183	6.48845	.237530	4.71	22.1841	2.17025	6.86294	.212314
4.22	17.8084	2.05426	6.49615	.236967	4.72	22.2784	2.17256	6.87023	.211864
4.23	17.8929	2.05670	6.50384	.236407	4.73	22.3729	2.17486	6.87750	.211416
4.24	17.9776	2.05913	6.51153	.235849	4.74	22.4676	2.17715	6.88477	.210970
4.25	18.0625	2.06155	6.51920	.235294	4.75	22.5625	2.17945	6.89202	.210526
4.26	18.1476	2.06398	6.52687	.234742	4.76	22.6576	2.18174	6.89928	.210084
4.27	18.2329	2.06640	6.53452	.234192	4.77	22.7529	2.18403	6.90652	.209644
4.28	18.3184	2.06882	6.54217	.233645	4.78	22.8484	2.18632	6.91375	.209205
4.29	18.4041	2.07123	6.54981	.233100	4.79	22.9441	2.18861	6.92098	.208768
4.30	18.4900	2.07364	6.55744	.232558	**4.80**	23.0400	2.19089	6.92820	.208333
4.31	18.5761	2.07605	6.56506	.232019	4.81	23.1361	2.19317	6.93542	.207900
4.32	18.6624	2.07846	6.57267	.231481	4.82	23.2324	2.19545	6.94262	.207469
4.33	18.7489	2.08087	6.58027	.230947	4.83	23.3289	2.19773	6.94982	.207039
4.34	18.8356	2.08327	6.58787	.230415	4.84	23.4256	2.20000	6.95701	.206612
4.35	18.9225	2.08567	6.59545	.229885	4.85	23.5225	2.20227	6.96419	.206186
4.36	19.0096	2.08806	6.60303	.229358	4.86	23.6196	2.20454	6.97137	.205761
4.37	19.0969	2.09045	6.61060	.228833	4.87	23.7169	2.20681	6.97854	.205339
4.38	19.1844	2.09284	6.61816	.228311	4.88	23.8144	2.20907	6.98570	.204918
4.39	19.2721	2.09523	6.62571	.227790	4.89	23.9121	2.21133	6.99285	.204499
4.40	19.3600	2.09762	6.63325	.227273	**4.90**	24.0100	2.21359	7.00000	.204082
4.41	19.4481	2.10000	6.64078	.226757	4.91	24.1081	2.21585	7.00714	.203666
4.42	19.5364	2.10238	6.64831	.226244	4.92	24.2064	2.21811	7.01427	.203252
4.43	19.6249	2.10476	6.65582	.225734	4.93	24.3049	2.22036	7.02140	.202840
4.44	19.7136	2.10713	6.66333	.225225	4.94	24.4036	2.22261	7.02851	.202429
4.45	19.8025	2.10950	6.67083	.224719	4.95	24.5025	2.22486	7.03562	.202020
4.46	19.8916	2.11187	6.67832	.224215	4.96	24.6016	2.22711	7.04273	.201613
4.47	19.9809	2.11424	6.68581	.223714	4.97	24.7009	2.22935	7.04982	.201207
4.48	20.0704	2.11660	6.69328	.223214	4.98	24.8004	2.23159	7.05691	.200803
4.49	20.1601	2.11896	6.70075	.222717	4.99	24.9001	2.23383	7.06399	.200401
4.50	20.2500	2.12132	6.70820	.222222	**5.00**	25.0000	2.23607	7.07107	.200000
n	n^2	$\sqrt{n}$	$\sqrt{10n}$	$1/n$	n	n^2	$\sqrt{n}$	$\sqrt{10n}$	$1/n$

Table 1 **Squares, Square Roots, and Reciprocals**

n	n^2	$\sqrt{n}$	$\sqrt{10n}$	$1/n$	n	n^2	$\sqrt{n}$	$\sqrt{10n}$	$1/n$
5.00	25.0000	2.23607	7.07107	.200000	**5.50**	30.2500	2.34521	7.41620	.181818
5.01	25.1001	2.23830	7.07814	.199601	5.51	30.3601	2.34734	7.42294	.181488
5.02	25.2004	2.24054	7.08520	.199203	5.52	30.4704	2.34947	7.42967	.181159
5.03	25.3009	2.24277	7.09225	.198807	5.53	30.5809	2.35160	7.43640	.180832
5.04	25.4016	2.24499	7.09930	.198413	5.54	30.6916	2.35372	7.44312	.180505
5.05	25.5025	2.24722	7.10634	.198020	5.55	30.8025	2.35584	7.44983	.180180
5.06	25.6036	2.24944	7.11337	.197628	5.56	30.9136	2.35797	7.45654	.179856
5.07	25.7049	2.25167	7.12039	.197239	5.57	31.0249	2.36008	7.46324	.179533
5.08	25.8064	2.25389	7.12741	.196850	5.58	31.1364	2.36220	7.46994	.179211
5.09	25.9081	2.25610	7.13442	.196464	5.59	31.2481	2.36432	7.47663	.178891
5.10	26.0100	2.25832	7.14143	.196078	**5.60**	31.3600	2.36643	7.48331	.178571
5.11	26.1121	2.26053	7.14843	.195695	5.61	31.4721	2.36854	7.48999	.178253
5.12	26.2144	2.26274	7.15542	.195312	5.62	31.5844	2.37065	7.49667	.177936
5.13	26.3169	2.26495	7.16240	.194932	5.63	31.6969	2.37276	7.50333	.177620
5.14	26.4196	2.26716	7.16938	.194553	5.64	31.8096	2.37487	7.50999	.177305
5.15	26.5225	2.26936	7.17635	.194175	5.65	31.9225	2.37697	7.51665	.176991
5.16	26.6256	2.27156	7.18331	.193798	5.66	32.0356	2.37908	7.52330	.176678
5.17	26.7289	2.27376	7.19027	.193424	5.67	32.1489	2.38118	7.52994	.176367
5.18	26.8324	2.27596	7.19722	.193050	5.68	32.2624	2.38328	7.53658	.176056
5.19	26.9361	2.27816	7.20417	.192678	5.69	32.3761	2.38537	7.54321	.175747
5.20	27.0400	2.28035	7.21110	.192308	**5.70**	32.4900	2.38747	7.54983	.175439
5.21	27.1441	2.28254	7.21803	.191939	5.71	32.6041	2.38956	7.55645	.175131
5.22	27.2484	2.28473	7.22496	.191571	5.72	32.7184	2.39165	7.56307	.174825
5.23	27.3529	2.28692	7.23187	.191205	5.73	32.8329	2.39374	7.56968	.174520
5.24	27.4576	2.28910	7.23878	.190840	5.74	32.9476	2.39583	7.57628	.174216
5.25	27.5625	2.29129	7.24569	.190476	5.75	33.0625	2.39792	7.58288	.173913
5.26	27.6676	2.29347	7.25259	.190114	5.76	33.1776	2.40000	7.58947	.173611
5.27	27.7729	2.29565	7.25948	.189753	5.77	33.2929	2.40208	7.59605	.173310
5.28	27.8784	2.29783	7.26636	.189394	5.78	33.4084	2.40416	7.60263	.173010
5.29	27.9841	2.30000	7.27324	.189036	5.79	33.5241	2.40624	7.60920	.172712
5.30	28.0900	2.30217	7.28011	.188679	**5.80**	33.6400	2.40832	7.61577	.172414
5.31	28.1961	2.30434	7.28697	.188324	5.81	33.7561	2.41039	7.62234	.172117
5.32	28.3024	2.30651	7.29383	.187970	5.82	33.8724	2.41247	7.62889	.171821
5.33	28.4089	2.30868	7.30068	.187617	5.83	33.9889	2.41454	7.63544	.171527
5.34	28.5156	2.31084	7.30753	.187266	5.84	34.1056	2.41661	7.64199	.171233
5.35	28.6225	2.31301	7.31437	.186916	5.85	34.2225	2.41868	7.64853	.170940
5.36	28.7296	2.31517	7.32120	.186567	5.86	34.3396	2.42074	7.65506	.170649
5.37	28.8369	2.31733	7.32803	.186220	5.87	34.4569	2.42281	7.66159	.170358
5.38	28.9444	2.31948	7.33485	.185874	5.88	34.5744	2.42487	7.66812	.170068
5.39	29.0521	2.32164	7.34166	.185529	5.89	34.6921	2.42693	7.67463	.169779
5.40	29.1600	2.32379	7.34847	.185185	**5.90**	34.8100	2.42899	7.68115	.169492
5.41	29.2681	2.32594	7.35527	.184843	5.91	34.9281	2.43105	7.68765	.169205
5.42	29.3764	2.32809	7.36206	.184502	5.92	35.0464	2.43311	7.69415	.168919
5.43	29.4849	2.33024	7.36885	.184162	5.93	35.1649	2.43516	7.70065	.168634
5.44	29.5936	2.33238	7.37564	.183824	5.94	35.2836	2.43721	7.70714	.168350
5.45	29.7025	2.33452	7.38241	.183486	5.95	35.4025	2.43926	7.71362	.168067
5.46	29.8116	2.33666	7.38918	.183150	5.96	35.5216	2.44131	7.72010	.167785
5.47	29.9209	2.33880	7.39594	.182815	5.97	35.6409	2.44336	7.72658	.167504
5.48	30.0304	2.34094	7.40270	.182482	5.98	35.7604	2.44540	7.73305	.167224
5.49	30.1401	2.34307	7.40945	.182149	5.99	35.8801	2.44745	7.73951	.166945
5.50	30.2500	2.34521	7.41620	.181818	**6.00**	36.0000	2.44949	7.74597	.166667
n	n^2	$\sqrt{n}$	$\sqrt{10n}$	$1/n$	n	n^2	$\sqrt{n}$	$\sqrt{10n}$	$1/n$

Table 1 **Squares, Square Roots, and Reciprocals**

n	n^2	$\sqrt{n}$	$\sqrt{10n}$	$1/n$	n	n^2	$\sqrt{n}$	$\sqrt{10n}$	$1/n$
6.00	36.0000	2.44949	7.74597	.166667	**6.50**	42.2500	2.54951	8.06226	.153846
6.01	36.1201	2.45153	7.75242	.166389	6.51	42.3801	2.55147	8.06846	.153610
6.02	36.2404	2.45357	7.75887	.166113	6.52	42.5104	2.55343	8.07465	.153374
6.03	36.3609	2.45561	7.76531	.165837	6.53	42.6409	2.55539	8.08084	.153139
6.04	36.4816	2.45764	7.77174	.165563	6.54	42.7716	2.55734	8.08703	.152905
6.05	36.6025	2.45967	7.77817	.165289	6.55	42.9025	2.55930	8.09321	.152672
6.06	36.7236	2.46171	7.78460	.165017	6.56	43.0336	2.56125	8.09938	.152439
6.07	36.8449	2.46374	7.79102	.164745	6.57	43.1649	2.56320	8.10555	.152207
6.08	36.9664	2.46577	7.79744	.164474	6.58	43.2964	2.56515	8.11172	.151976
6.09	37.0881	2.46779	7.80385	.164204	6.59	43.4281	2.56710	8.11788	.151745
6.10	37.2100	2.46982	7.81025	.163934	**6.60**	43.5600	2.56905	8.12404	.151515
6.11	37.3321	2.47184	7.81665	.163666	6.61	43.6921	2.57099	8.13019	.151286
6.12	37.4544	2.47386	7.82304	.163399	6.62	43.8244	2.57294	8.13634	.151057
6.13	37.5769	2.47588	7.82943	.163132	6.63	43.9569	2.57488	8.14248	.150830
6.14	37.6996	2.47790	7.83582	.162866	6.64	44.0896	2.57682	8.14862	.150602
6.15	37.8225	2.47992	7.84219	.162602	6.65	44.2225	2.57876	8.15475	.150376
6.16	37.9456	2.48193	7.84857	.162338	6.66	44.3556	2.58070	8.16088	.150150
6.17	38.0689	2.48395	7.85493	.162075	6.67	44.4889	2.58263	8.16701	.149925
6.18	38.1924	2.48596	7.86130	.161812	6.68	44.6224	2.58457	8.17313	.149701
6.19	38.3161	2.48797	7.86766	.161551	6.69	44.7561	2.58650	8.17924	.149477
6.20	38.4400	2.48998	7.87401	.161290	**6.70**	44.8900	2.58844	8.18535	.149254
6.21	38.5641	2.49199	7.88036	.161031	6.71	45.0241	2.59037	8.19146	.149031
6.22	38.6884	2.49399	7.88670	.160772	6.72	45.1584	2.59230	8.19756	.148810
6.23	38.8129	2.49600	7.89303	.160514	6.73	45.2929	2.59422	8.20366	.148588
6.24	38.9376	2.49800	7.89937	.160256	6.74	45.4276	2.59615	8.20975	.148368
6.25	39.0625	2.50000	7.90569	.160000	6.75	45.5625	2.59808	8.21584	.148148
6.26	39.1876	2.50200	7.91202	.159744	6.76	45.6976	2.60000	8.22192	.147929
6.27	39.3129	2.50400	7.91833	.159490	6.77	45.8329	2.60192	8.22800	.147710
6.28	39.4384	2.50599	7.92465	.159236	6.78	45.9684	2.60384	8.23408	.147493
6.29	39.5641	2.50799	7.93095	.158983	6.79	46.1041	2.60576	8.24015	.147275
6.30	39.6900	2.50998	7.93725	.158730	**6.80**	46.2400	2.60768	8.24621	.147059
6.31	39.8161	2.51197	7.94355	.158479	6.81	46.3761	2.60960	8.25227	.146843
6.32	39.9424	2.51396	7.94984	.158228	6.82	46.5124	2.61151	8.25833	.146628
6.33	40.0689	2.51595	7.95613	.157978	6.83	46.6489	2.61343	8.26438	.146413
6.34	40.1956	2.51794	7.96241	.157729	6.84	46.7856	2.61534	8.27043	.146199
6.35	40.3225	2.51992	7.96869	.157480	6.85	46.9225	2.61725	8.27647	.145985
6.36	40.4496	2.52190	7.97496	.157233	6.86	47.0596	2.61916	8.28251	.145773
6.37	40.5769	2.52389	7.98123	.156986	6.87	47.1969	2.62107	8.28855	.145560
6.38	40.7044	2.52587	7.98749	.156740	6.88	47.3344	2.62298	8.29458	.145349
6.39	40.8321	2.52784	7.99375	.156495	6.89	47.4721	2.62488	8.30060	.145138
6.40	40.9600	2.52982	8.00000	.156250	**6.90**	47.6100	2.62679	8.30662	.144928
6.41	41.0881	2.53180	8.00625	.156006	6.91	47.7481	2.62869	8.31264	.144718
6.42	41.2164	2.53377	8.01249	.155763	6.92	47.8864	2.63059	8.31865	.144509
6.43	41.3449	2.53574	8.01873	.155521	6.93	48.0249	2.63249	8.32466	.144300
6.44	41.4736	2.53772	8.02496	.155280	6.94	48.1636	2.63439	8.33067	.144092
6.45	41.6025	2.53969	8.03119	.155039	6.95	48.3025	2.63629	8.33667	.143885
6.46	41.7316	2.54165	8.03741	.154799	6.96	48.4416	2.63818	8.34266	.143678
6.47	41.8609	2.54362	8.04363	.154560	6.97	48.5809	2.64008	8.34865	.143472
6.48	41.9904	2.54558	8.04984	.154321	6.98	48.7204	2.64197	8.35464	.143266
6.49	42.1201	2.54755	8.05605	.154083	6.99	48.8601	2.64386	8.36062	.143062
6.50	42.2500	2.54951	8.06226	.153846	**7.00**	49.0000	2.64575	8.36660	.142857
n	n^2	$\sqrt{n}$	$\sqrt{10n}$	$1/n$	n	n^2	$\sqrt{n}$	$\sqrt{10n}$	$1/n$

Table 1 **Squares, Square Roots, and Reciprocals**

n	n^2	$\sqrt{n}$	$\sqrt{10n}$	$1/n$	n	n^2	$\sqrt{n}$	$\sqrt{10n}$	$1/n$
7.00	49.0000	2.64575	8.36660	.142857	**7.50**	56.2500	2.73861	8.66025	.133333
7.01	49.1401	2.64764	8.37257	.142653	7.51	56.4001	2.74044	8.66603	.133156
7.02	49.2804	2.64953	8.37854	.142450	7.52	56.5504	2.74226	8.67179	.132979
7.03	49.4209	2.65141	8.38451	.142248	7.53	56.7009	2.74408	8.67756	.132802
7.04	49.5616	2.65330	8.39047	.142045	7.54	56.8516	2.74591	8.68332	.132626
7.05	49.7025	2.65518	8.39643	.141844	7.55	57.0025	2.74773	8.68907	.132450
7.06	49.8436	2.65707	8.40238	.141643	7.56	57.1536	2.74955	8.69483	.132275
7.07	49.9849	2.65895	8.40833	.141443	7.57	57.3049	2.75136	8.70057	.132100
7.08	50.1264	2.66083	8.41427	.141243	7.58	57.4564	2.75318	8.70632	.131926
7.09	50.2681	2.66271	8.42021	.141044	7.59	57.6081	2.75500	8.71206	.131752
7.10	50.4100	2.66458	8.42615	.140845	**7.60**	57.7600	2.75681	8.71780	.131579
7.11	50.5521	2.66646	8.43208	.140647	7.61	57.9121	2.75862	8.72353	.131406
7.12	50.6944	2.66833	8.43801	.140449	7.62	58.0644	2.76043	8.72926	.131234
7.13	50.8369	2.67021	8.44393	.140252	7.63	58.2169	2.76225	8.73499	.131062
7.14	50.9796	2.67208	8.44985	.140056	7.64	58.3696	2.76405	8.74071	.130890
7.15	51.1225	2.67395	8.45577	.139860	7.65	58.5225	2.76586	8.74643	.130719
7.16	51.2656	2.67582	8.46168	.139665	7.66	58.6756	2.76767	8.75214	.130548
7.17	51.4089	2.67769	8.46759	.139470	7.67	58.8289	2.76948	8.75785	.130378
7.18	51.5524	2.67955	8.47349	.139276	7.68	58.9824	2.77128	8.76356	.130208
7.19	51.6961	2.68142	8.47939	.139082	7.69	59.1361	2.77308	8.76926	.130039
7.20	51.8400	2.68328	8.48528	.138889	**7.70**	59.2900	2.77489	8.77496	.129870
7.21	51.9841	2.68514	8.49117	.138696	7.71	59.4441	2.77669	8.78066	.129702
7.22	52.1284	2.68701	8.49706	.138504	7.72	59.5984	2.77849	8.78635	.129534
7.23	52.2729	2.68887	8.50294	.138313	7.73	59.7529	2.78029	8.79204	.129366
7.24	52.4176	2.69072	8.50882	.138122	7.74	59.9076	2.78209	8.79773	.129199
7.25	52.5625	2.69258	8.51469	.137931	7.75	60.0625	2.78388	8.80341	.129032
7.26	52.7076	2.69444	8.52056	.137741	7.76	60.2176	2.78568	8.80909	.128866
7.27	52.8529	2.69629	8.52643	.137552	7.77	60.3729	2.78747	8.81476	.128700
7.28	52.9984	2.69815	8.53229	.137363	7.78	60.5284	2.78927	8.82043	.128535
7.29	53.1441	2.70000	8.53815	.137174	7.79	60.6841	2.79106	8.82610	.128370
7.30	53.2900	2.70185	8.54400	.136986	**7.80**	60.8400	2.79285	8.83176	.128205
7.31	53.4361	2.70370	8.54985	.136799	7.81	60.9961	2.79464	8.83742	.128041
7.32	53.5824	2.70555	8.55570	.136612	7.82	61.1524	2.79643	8.84308	.127877
7.33	53.7289	2.70740	8.56154	.136426	7.83	61.3089	2.79821	8.84873	.127714
7.34	53.8756	2.70924	8.56738	.136240	7.84	61.4656	2.80000	8.85438	.127551
7.35	54.0225	2.71109	8.57321	.136054	7.85	61.6225	2.80179	8.86002	.127389
7.36	54.1696	2.71293	8.57904	.135870	7.86	61.7796	2.80357	8.86566	.127226
7.37	54.3169	2.71477	8.58487	.135685	7.87	61 9369	2.80535	8.87130	.127065
7.38	54.4644	2.71662	8.59069	.135501	7.88	62.0944	2.80713	8.87694	.126904
7.39	54.6121	2.71846	8.59651	.135318	7.89	62.2521	2.80891	8.88257	.126743
7.40	54.7600	2.72029	8.60233	.135135	**7.90**	62.4100	2.81069	8.88819	.126582
7.41	54.9081	2.72213	8.60814	.134953	7.91	62.5681	2.81247	8.89382	.126422
7.42	55.0564	2.72397	8.61394	.134771	7.92	62.7264	2.81425	8.89944	.126263
7.43	55.2049	2.72580	8.61974	.134590	7.93	62.8849	2.81603	8.90505	.126103
7.44	55.3536	2.72764	8.62554	.134409	7.94	63.0436	2.81780	8.91067	.125945
7.45	55.5025	2.72947	8.63134	.134228	7.95	63.2025	2.81957	8.91628	.125786
7.46	55.6516	2.73130	8.63713	.134048	7.96	63.3616	2.82135	8.92188	.125628
7.47	55.8009	2.73313	8.64292	.133869	7.97	63.5209	2.82312	8.92749	.125471
7.48	55.9504	2.73496	8.64870	.133690	7.98	63.6804	2.82489	8.93308	.125313
7.49	56.1001	2.73679	8.65448	.133511	7.99	63.8401	2.82666	8.93868	.125156
7.50	56.2500	2.73861	8.66025	.133333	**8.00**	64.0000	2.82843	8.94427	.125000
n	n^2	$\sqrt{n}$	$\sqrt{10n}$	$1/n$	n	n^2	$\sqrt{n}$	$\sqrt{10n}$	$1/n$

Table 1 **Squares, Square Roots, and Reciprocals**

n	n²	√n	√10n	1/n	n	n²	√n	√10n	1/n
8.00	64.0000	2.82843	8.94427	.125000	**8.50**	72.2500	2.91548	9.21954	.117647
8.01	64.1601	2.83019	8.94986	.124844	8.51	72.4201	2.91719	9.22497	.117509
8.02	64.3204	2.83196	8.95545	.124688	8.52	72.5904	2.91890	9.23038	.117371
8.03	64.4809	2.83373	8.96103	.124533	8.53	72.7609	2.92062	9.23580	.117233
8.04	64.6416	2.83549	8.96660	.124378	8.54	72.9316	2.92233	9.24121	.117096
8.05	64.8025	2.83725	8.97218	.124224	8.55	73.1025	2.92404	9.24662	.116959
8.06	64.9636	2.83901	8.97775	.124069	8.56	73.2736	2.92575	9.25203	.116822
8.07	65.1249	2.84077	8.98332	.123916	8.57	73.4449	2.92746	9.25743	.116686
8.08	65.2864	2.84253	8.98888	.123762	8.58	73.6164	2.92916	9.26283	.116550
8.09	65.4481	2.84429	8.99444	.123609	8.59	73.7881	2.93087	9.26823	.116414
8.10	65.6100	2.84605	9.00000	.123457	**8.60**	73.9600	2.93258	9.27362	.116279
8.11	65.7721	2.84781	9.00555	.123305	8.61	74.1321	2.93428	9.27901	.116144
8.12	65.9344	2.84956	9.01110	.123153	8.62	74.3044	2.93598	9.28440	.116009
8.13	66.0969	2.85132	9.01665	.123001	8.63	74.4769	2.93769	9.28978	.115875
8.14	66.2596	2.85307	9.02219	.122850	8.64	74.6496	2.93939	9.29516	.115741
8.15	66.4225	2.85482	9.02774	.122699	8.65	74.8225	2.94109	9.30054	.115607
8.16	66.5856	2.85657	9.03327	.122549	8.66	74.9956	2.94279	9.30591	.115473
8.17	66.7489	2.85832	9.03881	.122399	8.67	75.1689	2.94449	9.31128	.115340
8.18	66.9124	2.86007	9.04434	.122249	8.68	75.3424	2.94618	9.31665	.115207
8.19	67.0761	2.86182	9.04986	.122100	8.69	75.5161	2.94788	9.32202	.115075
8.20	67.2400	2.86356	9.05539	.121951	**8.70**	75.6900	2.94958	9.32738	.114943
8.21	67.4041	2.86531	9.06091	.121803	8.71	75.8641	2.95127	9.33274	.114811
8.22	67.5684	2.86705	9.06642	.121655	8.72	76.0384	2.95296	9.33809	.114679
8.23	67.7329	2.86880	9.07193	.121507	8.73	76.2129	2.95466	9.34345	.114548
8.24	67.8976	2.87054	9.07744	.121359	8.74	76.3876	2.95635	9.34880	.114416
8.25	68.0625	2.87228	9.08295	.121212	8.75	76.5625	2.95804	9.35414	.114286
8.26	68.2276	2.87402	9.08845	.121065	8.76	76.7376	2.95973	9.35949	.114155
8.27	68.3929	2.87576	9.09395	.120919	8.77	76.9129	2.96142	9.36483	.114025
8.28	68.5584	2.87750	9.09945	.120773	8.78	77.0884	2.96311	9.37017	.113895
8.29	68.7241	2.87924	9.10494	.120627	8.79	77.2641	2.96479	9.37550	.113766
8.30	68.8900	2.88097	9.11043	.120482	**8.80**	77.4400	2.96648	9.38083	.113636
8.31	69.0561	2.88271	9.11592	.120337	8.81	77.6161	2.96816	9.38616	.113507
8.32	69.2224	2.88444	9.12140	.120192	8.82	77.7924	2.96985	9.39149	.113379
8.33	69.3889	2.88617	9.12688	.120048	8.83	77.9689	2.97153	9.39681	.113250
8.34	69.5556	2.88791	9.13236	.119904	8.84	78.1456	2.97321	9.40213	.113122
8.35	69.7225	2.88964	9.13783	.119760	8.85	78.3225	2.97489	9.40744	.112994
8.36	69.8896	2.89137	9.14330	.119617	8.86	78.4996	2.97658	9.41276	.112867
8.37	70.0569	2.89310	9.14877	.119474	8.87	78.6769	2.97825	9.41807	.112740
8.38	70.2244	2.89482	9.15423	.119332	8.88	78.8544	2.97993	9.42338	.112613
8.39	70.3921	2.89655	9.15969	.119190	8.89	79.0321	2.98161	9.42868	.112486
8.40	70.5600	2.89828	9.16515	.119048	**8.90**	79.2100	2.98329	9.43398	.112360
8.41	70.7281	2.90000	9.17061	.118906	8.91	79.3881	2.98496	9.43928	.112233
8.42	70.8964	2.90172	9.17606	.118765	8.92	79.5664	2.98664	9.44458	.112108
8.43	71.0649	2.90345	9.18150	.118624	8.93	79.7449	2.98831	9.44987	.111982
8.44	71.2336	2.90517	9.18695	.118483	8.94	79.9236	2.98998	9.45516	.111857
8.45	71.4025	2.90689	9.19239	.118343	8.95	80.1025	2.99166	9.46044	.111732
8.46	71.5716	2.90861	9.19783	.118203	8.96	80.2816	2.99333	9.46573	.111607
8.47	71.7409	2.91033	9.20326	.118064	8.97	80.4609	2.99500	9.47101	.111483
8.48	71.9104	2.91204	9.20869	.117925	8.98	80.6404	2.99666	9.47629	.111359
8.49	72.0801	2.91376	9.21412	.117786	8.99	80.8201	2.99833	9.48156	.111235
8.50	72.2500	2.91548	9.21954	.117647	**9.00**	81.0000	3.00000	9.48683	.111111
n	n²	√n	√10n	1/n	n	n²	√n	√10n	1/n

Table 1 **Squares, Square Roots, and Reciprocals**

n	n^2	$\sqrt{n}$	$\sqrt{10n}$	$1/n$	n	n^2	$\sqrt{n}$	$\sqrt{10n}$	$1/n$
9.00	81.0000	3.00000	9.48683	.111111	**9.50**	90.2500	3.08221	9.74679	.105263
9.01	81.1801	3.00167	9.49210	.110988	9.51	90.4401	3.08383	9.75192	.105152
9.02	81.3604	3.00333	9.49737	.110865	9.52	90.6304	3.08545	9.75705	.105042
9.03	81.5409	3.00500	9.50263	.110742	9.53	90.8209	3.08707	9.76217	.104932
9.04	81.7216	3.00666	9.50789	.110619	9.54	91.0116	3.08869	9.76729	.104822
9.05	81.9025	3.00832	9.51315	.110497	9.55	91.2025	3.09031	9.77241	.104712
9.06	82.0836	3.00998	9.51840	.110375	9.56	91.3936	3.09192	9.77753	.104603
9.07	82.2649	3.01164	9.52365	.110254	9.57	91.5849	3.09354	9.78264	.104493
9.08	82.4464	3.01330	9.52890	.110132	9.58	91.7764	3.09516	9.78775	.104384
9.09	82.6281	3.01496	9.53415	.110011	9.59	91.9681	3.09677	9.79285	.104275
9.10	82.8100	3.01662	9.53939	.109890	**9.60**	92.1600	3.09839	9.79796	.104167
9.11	82.9921	3.01828	9.54463	.109769	9.61	92.3521	3.10000	9.80306	.104058
9.12	83.1744	3.01993	9.54987	.109649	9.62	92.5444	3.10161	9.80816	.103950
9.13	83.3569	3.02159	9.55510	.109529	9.63	92.7369	3.10322	9.81326	.103842
9.14	83.5396	3.02324	9.56033	.109409	9.64	92.9296	3.10483	9.81835	.103734
9.15	83.7225	3.02490	9.56556	.109290	9.65	93.1225	3.10644	9.82344	.103627
9.16	83.9056	3.02655	9.57079	.109170	9.66	93.3156	3.10805	9.82853	.103520
9.17	84.0889	3.02820	9.57601	.109051	9.67	93.5089	3.10966	9.83362	.103413
9.18	84.2724	3.02985	9.58123	.108932	9.68	93.7024	3.11127	9.83870	.103306
9.19	84.4561	3.03150	9.58645	.108814	9.69	93.8961	3.11288	9.84378	.103199
9.20	84.6400	3.03315	9.59166	.108696	**9.70**	94.0900	3.11448	9.84886	.103093
9.21	84.8241	3.03480	9.59687	.108578	9.71	94.2841	3.11609	9.85393	.102987
9.22	85.0084	3.03645	9.60208	.108460	9.72	94.4784	3.11769	9.85901	.102881
9.23	85.1929	3.03809	9.60729	.108342	9.73	94.6729	3.11929	9.86408	.102775
9.24	85.3776	3.03974	9.61249	.108225	9.74	94.8676	3.12090	9.86914	.102669
9.25	85.5625	3.04138	9.61769	.108108	9.75	95.0625	3.12250	9.87421	.102564
9.26	85.7476	3.04302	9.62289	.107991	9.76	95.2576	3.12410	9.87927	.102459
9.27	85.9329	3.04467	9.62808	.107875	9.77	95.4529	3.12570	9.88433	.102354
9.28	86.1184	3.04631	9.63328	.107759	9.78	95.6484	3.12730	9.88939	.102249
9.29	86.3041	3.04795	9.63846	.107643	9.79	95.8441	3.12890	9.89444	.102145
9.30	86.4900	3.04959	9.64365	.107527	**9.80**	96.0400	3.13050	9.89949	.102041
9.31	86.6761	3.05123	9.64883	.107411	9.81	96.2361	3.13209	9.90454	.101937
9.32	86.8624	3.05287	9.65401	.107296	9.82	96.4324	3.13369	9.90959	.101833
9.33	87.0489	3.05450	9.65919	.107181	9.83	96.6289	3.13528	9.91464	.101729
9.34	87.2356	3.05614	9.66437	.107066	9.84	96.8256	3.13688	9.91968	.101626
9.35	87.4225	3.05778	9.66954	.106952	9.85	97.0225	3.13847	9.92472	.101523
9.36	87.6096	3.05941	9.67471	.106838	9.86	97.2196	3.14006	9.92975	.101420
9.37	87.7969	3.06105	9.67988	.106724	9.87	97.4169	3.14166	9.93479	.101317
9.38	87.9844	3.06268	9.68504	.106610	9.88	97.6144	3.14325	9.93982	.101215
9.39	88.1721	3.06431	9.69020	.106496	9.89	97.8121	3.14484	9.94485	.101112
9.40	88.3600	3.06594	9.69536	.106383	**9.90**	98.0100	3.14643	9.94987	.101010
9.41	88.5481	3.06757	9.70052	.106270	9.91	98.2081	3.14802	9.95490	.100908
9.42	88.7364	3.06920	9.70567	.106157	9.92	98.4064	3.14960	9.95992	.100806
9.43	88.9249	3.07083	9.71082	.106045	9.93	98.6049	3.15119	9.96494	.100705
9.44	89.1136	3.07246	9.71597	.105932	9.94	98.8036	3.15278	9.96995	.100604
9.45	89.3025	3.07409	9.72111	.105820	9.95	99.0025	3.15436	9.97497	.100503
9.46	89.4916	3.07571	9.72625	.105708	9.96	99.2016	3.15595	9.97998	.100402
9.47	89.6809	3.07734	9.73139	.105597	9.97	99.4009	3.15753	9.98499	.100301
9.48	89.8704	3.07896	9.73653	.105485	9.98	99.6004	3.15911	9.98999	.100200
9.49	90.0601	3.08058	9.74166	.105374	9.99	99.8001	3.16070	9.99500	.100100
9.50	90.2500	3.08221	9.74679	.105263	**10.00**	100.000	3.16228	10.0000	.100000
n	n^2	$\sqrt{n}$	$\sqrt{10n}$	$1/n$	n	n^2	$\sqrt{n}$	$\sqrt{10n}$	$1/n$

Table 2 Logarithms of Numbers 1,000–1,499
Six-Place Mantissas

N	0	1	2	3	4	5	6	7	8	9	D#
100	00 0000	0434	0868	1301	1734	2166	2598	3029	3461	3891	434
01	4321	4751	5181	5609	6038	6466	6894	7321	7748	8174	430
02	00 8600	9026	9451	9876	*0300	*0724	*1147	*1570	*1993	*2415	426
03	01 2837	3259	3680	4100	4521	4940	5360	5779	6197	6616	422
04	01 7033	7451	7868	8284	8700	9116	9532	9947	*0361	*0775	418
05	02 1189	1603	2016	2428	2841	3252	3664	4075	4486	4896	414
06	5306	5715	6125	6533	6942	7350	7757	8164	8571	8978	410
07	02 9384	9789	*0195	*0600	*1004	*1408	*1812	*2216	*2619	*3021	406
08	03 3424	3826	4227	4628	5029	5430	5830	6230	6629	7028	402
09	03 7426	7825	8223	8620	9017	9414	9811	*0207	*0602	*0998	399
110	04 1393	1787	2182	2576	2969	3362	3755	4148	4540	4932	395
11	5323	5714	6105	6495	6885	7275	7664	8053	8442	8830	391
12	04 9218	9606	9993	*0380	*0766	*1153	*1538	*1924	*2309	*2694	388
13	05 3078	3463	3846	4230	4613	4996	5378	5760	6142	6524	385
14	05 6905	7286	7666	8046	8426	8805	9185	9563	9942	*0320	381
15	06 0698	1075	1452	1829	2206	2582	2958	3333	3709	4083	377
16	4458	4832	5206	5580	5953	6326	6699	7071	7443	7815	374
17	06 8186	8557	8928	9298	9668	*0038	*0407	*0776	*1145	*1514	371
18	07 1882	2250	2617	2985	3352	3718	4085	4451	4816	5182	368
19	5547	5912	6276	6640	7004	7368	7731	8094	8457	8819	365
120	07 9181	9543	9904	*0266	*0626	*0987	1347	*1707	*2067	*2426	362
21	08 2785	3144	3503	3861	4219	4576	4934	5291	5647	6004	359
22	6360	6716	7071	7426	7781	8136	8490	8845	9198	9552	356
23	08 9905	*0258	*0611	*0963	*1315	*1667	*2018	*2370	*2721	*3071	353
24	09 3422	3772	4122	4471	4820	5169	5518	5866	6215	6562	350
25	09 6910	7257	7604	7951	8298	8644	8990	9335	9681	*0026	347
26	10 0371	0715	1059	1403	1747	2091	2434	2777	3119	3462	344
27	3804	4146	4487	4828	5169	5510	5851	6191	6531	6871	342
28	10 7210	7549	7888	8227	8565	8903	9241	9579	9916	*0253	339
29	11 0590	0926	1263	1599	1934	2270	2605	2940	3275	3609	337
130	3943	4277	4611	4944	5278	5611	5943	6276	6608	6940	334
31	11 7271	7603	7934	8265	8595	8926	9256	9586	9915	*0245	332
32	12 0574	0903	1231	1560	1888	2216	2544	2871	3198	3525	329
33	3852	4178	4504	4830	5156	5481	5806	6131	6456	6781	326
34	12 7105	7429	7753	8076	8399	8722	9045	9368	9690	*0012	324
35	13 0334	0655	0977	1298	1619	1939	2260	2580	2900	3219	322
36	3539	3858	4177	4496	4814	5133	5451	5769	6086	6403	319
37	6721	7037	7354	7671	7987	8303	8618	8934	9249	9564	317
38	13 9879	*0194	*0508	*0822	*1136	*1450	*1763	*2076	*2389	*2702	315
39	14 3015	3327	3639	3851	4263	4574	4885	5196	5507	5818	312
140	6128	6438	6748	7058	7367	7676	7985	8294	8603	8911	310
41	14 9219	9527	9835	*0142	*0449	*0756	*1063	*1370	*1676	*1982	308
42	15 2288	2594	2900	3205	3510	3815	4120	4424	4728	5032	306
43	5336	5640	5943	6246	6549	6852	7154	7457	7759	8061	304
44	15 8362	8664	8965	9266	9567	9868	*0168	*0469	*0769	*1068	302
45	16 1368	1667	1967	2266	2564	2863	3161	3460	3758	4055	300
46	4353	4650	4947	5244	5541	5838	6134	6430	6726	7022	297
47	16 7317	7613	7908	8203	8497	8792	9086	9380	9674	9968	296
48	17 0262	0555	0848	1141	1434	1726	2019	2311	2603	2895	293
49	3186	3478	3769	4060	4351	4641	4932	5222	5512	5802	292
N	0	1	2	3	4	5	6	7	8	9	D

*Prefix first two places on next line.
Example: The mantissa for number (N) 1072 is 03 0195.

#The *bigbest difference* between adjacent mantissas on the *individual line*. It is also the *lowest difference* between adjacent mantissas on the *preceding line* in many cases.

Table 2 Logarithms of Numbers 1,500–1,999
Six-Place Mantissas

N	0	1	2	3	4	5	6	7	8	9	D
150	17 6091	6381	6670	6959	7248	7536	7825	8113	8401	8689	290
51	17 8977	9264	9552	9839	*0126	*0413	*0699	*0986	*1272	*1558	288
52	18 1844	2129	2415	2700	2985	3270	3555	3839	4123	4407	286
53	4691	4975	5259	5542	5825	6108	6391	6674	6956	7239	284
54	18 7521	7803	8084	8366	8647	8928	9209	9490	9771	*0051	282
55	19 0332	0612	0892	1171	1451	1730	2010	2289	2567	2846	280
56	3125	3403	3681	3959	4237	4514	4792	5069	5346	5623	278
57	5900	6176	6453	6729	7005	7281	7556	7832	8107	8382	277
58	19 8657	8932	9206	9481	9755	*0029	*0303	*0577	*0850	*1124	275
59	20 1397	1670	1943	2216	2488	2761	3033	3305	3577	3848	273
160	4120	4391	4663	4934	5204	5475	5746	6016	6286	6556	272
61	6826	7096	7365	7634	7904	8173	8441	8710	8979	9247	270
62	20 9515	9783	*0051	*0319	*0586	*0853	*1121	*1388	*1654	*1921	268
63	21 2188	2454	2720	2986	3252	3518	3783	4049	4314	4579	266
64	4844	5109	5373	5638	5902	6166	6430	6694	6957	7221	265
65	21 7484	7747	8010	8273	8536	8798	9060	9323	9585	9846	263
66	22 0108	0370	0631	0892	1153	1414	1675	1936	2196	2456	262
67	2716	2976	3236	3496	3755	4015	4274	4533	4792	5051	260
68	5309	5568	5826	6084	6342	6600	6858	7115	7372	7630	259
69	22 7887	8144	8400	8657	8913	9170	9426	9682	9938	*0193	257
170	23 0449	0704	0960	1215	1470	1724	1979	2234	2488	2742	256
71	2996	3250	3504	3757	4011	4264	4517	4770	5023	5276	254
72	5528	5781	6033	6285	6537	6789	7041	7292	7544	7795	253
73	23 8046	8297	8548	8799	9049	9299	9550	9800	*0050	*0300	251
74	24 0549	0799	1048	1297	1546	1795	2044	2293	2541	2790	250
75	3038	3286	3534	3782	4030	4277	4525	4772	5019	5266	248
76	5513	5759	6006	6252	6499	6745	6991	7237	7482	7728	247
77	24 7973	8219	8464	8709	8954	9198	9443	9687	9932	*0176	246
78	25 0420	0664	0908	1151	1395	1638	1881	2125	2368	2610	244
79	2853	3096	3338	3580	3822	4064	4306	4548	4790	5031	243
180	5273	5514	5755	5996	6237	6477	6718	6958	7198	7439	241
81	25 7679	7918	8158	8398	8637	8877	9116	9355	9594	9833	240
82	26 0071	0310	0548	0787	1025	1263	1501	1739	1976	2214	239
83	2451	2688	2925	3162	3399	3636	3873	4109	4346	4582	237
84	4818	5054	5290	5525	5761	5996	6232	6467	6702	6937	236
85	7172	7406	7641	7875	8110	8344	8578	8812	9046	9279	235
86	26 9513	9746	9980	*0213	*0446	*0679	*0912	*1144	*1377	*1609	234
87	27 1842	2074	2306	2538	2770	3001	3233	3464	3696	3927	232
88	4158	4389	4620	4850	5081	5311	5542	5772	6002	6232	231
89	6462	6692	6921	7151	7380	7609	7838	8067	8296	8525	230
190	27 8754	8982	9211	9439	9667	9895	*0123	*0351	*0578	*0806	229
91	28 1033	1261	1488	1715	1942	2169	2396	2622	2849	3075	228
92	3301	3527	3753	3979	4205	4431	4656	4882	5107	5332	226
93	5557	5782	6007	6232	6456	6681	6905	7130	7354	7578	225
94	28 7802	8026	8249	8473	8696	8920	9143	9366	9589	9812	224
95	29 0035	0257	0480	0702	0925	1147	1369	1591	1813	2034	223
96	2256	2478	2699	2920	3141	3363	3584	3804	4025	4246	222
97	4466	4687	4907	5127	5347	5567	5787	6007	6226	6446	221
98	6665	6884	7104	7323	7542	7761	7979	8198	8416	8635	220
99	29 8853	9071	9289	9507	9725	9943	*0161	*0378	*0595	*0813	218
N	0	1	2	3	4	5	6	7	8	9	D

Table 2 Logarithms of Numbers 2,000–2,499
Six-Place Mantissas

N	0	1	2	3	4	5	6	7	8	9	D
200	30 1030	1247	1464	1681	1898	2114	2331	2547	2764	2980	217
01	3196	3412	3628	3844	4059	4275	4491	4706	4921	5136	216
02	5351	5566	5781	5996	6211	6425	6639	6854	7068	7282	215
03	7496	7710	7924	8137	8351	8564	8778	8991	9204	9417	214
04	30 9630	9843	*0056	*0268	*0481	*0693	*0906	*1118	*1330	*1542	213
05	31 1754	1966	2177	2389	2600	2812	3023	3234	3445	3656	212
06	3867	4078	4289	4499	4710	4920	5130	5340	5551	5760	211
07	5970	6180	6390	6599	6809	7018	7227	7436	7646	7854	210
08	31 8063	8272	8481	8689	8898	9106	9314	9522	9730	9938	209
09	32 0146	0354	0562	0769	0977	1184	1391	1598	1805	2012	208
210	2219	2426	2633	2839	3046	3252	3458	3665	3871	4077	207
11	4282	4488	4694	4899	5105	5310	5516	5721	5926	6131	206
12	6336	6541	6745	6950	7155	7359	7563	7767	7972	8176	205
13	32 8380	8583	8787	8991	9194	9398	9601	9805	*0008	*0211	204
14	33 0414	0617	0819	1022	1225	1427	1630	1832	2034	2236	203
15	2438	2640	2842	3044	3246	3447	3649	3850	4051	4253	202
16	4454	4655	4856	5057	5257	5458	5658	5859	6059	6260	201
17	6460	6660	6860	7060	7260	7459	7659	7858	8058	8257	200
18	33 8456	8656	8855	9054	9253	9451	9650	9849	*0047	*0246	200
19	34 0444	0642	0841	1039	1237	1435	1632	1830	2028	2225	199
220	2423	2620	2817	3014	3212	3409	3606	3802	3999	4196	198
21	4392	4589	4785	4981	5178	5374	5570	5766	5962	6157	197
22	6353	6549	6744	6939	7135	7330	7525	7720	7915	8110	196
23	34 8305	8500	8694	8889	9083	9278	9472	9666	9860	*0054	195
24	35 0248	0442	0636	0829	1023	1216	1410	1603	1796	1989	194
25	2183	2375	2568	2761	2954	3147	3339	3532	3724	3916	193
26	4108	4301	4493	4685	4876	5068	5260	5452	5643	5834	192
27	6026	6217	6408	6599	6790	6981	7172	7363	7554	7744	191
28	7935	8125	8316	8506	8696	8886	9076	9266	9456	9646	191
29	35 9835	*0025	*0215	*0404	*0593	*0783	*0972	*1161	*1350	*1539	190
230	36 1728	1917	2105	2294	2482	2671	2859	3048	3236	3424	189
31	3612	3800	3988	4176	4363	4551	4739	4926	5113	5301	188
32	5488	5675	5862	6049	6236	6423	6610	6796	6983	7169	187
33	7356	7542	7729	7915	8101	8287	8473	8659	8845	9030	187
34	36 9216	9401	9587	9772	9958	*0143	*0328	*0513	*0698	*0883	186
35	37 1068	1253	1437	1622	1806	1991	2175	2360	2544	2728	185
36	2912	3096	3280	3464	3647	3831	4015	4198	4382	4565	184
37	4748	4932	5115	5298	5481	5664	5846	6029	6212	6394	184
38	6577	6759	6942	7124	7306	7488	7670	7852	8034	8216	183
39	37 8398	8580	8761	8943	9124	9306	9487	9668	9849	*0030	182
240	38 0211	0392	0573	0754	0934	1115	1296	1476	1656	1837	181
41	2017	2197	2377	2557	2737	2917	3097	3277	3456	3636	180
42	3815	3995	4174	4353	4533	4712	4891	5070	5249	5428	180
43	5606	5785	5964	6142	6321	6499	6677	6856	7034	7212	179
44	7390	7568	7746	7923	8101	8279	8456	8634	8811	8989	178
45	38 9166	9343	9520	9698	9875	*0051	*0228	*0405	*0582	*0759	178
46	39 0935	1112	1288	1464	1641	1817	1993	2169	2345	2521	177
47	2697	2873	3048	3224	3400	3575	3751	3926	4101	4277	176
48	4452	4627	4802	4977	5152	5326	5501	5676	5850	6025	175
49	6199	6374	6548	6722	6896	7071	7245	7419	7592	7766	175
N	0	1	2	3	4	5	6	7	8	9	D

Table 2 **Logarithms of Numbers 2,500–2,999**
Six-Place Mantissas

N	0	1	2	3	4	5	6	7	8	9	D
250	39 7940	8114	8287	8461	8634	8808	8981	9154	9328	9501	174
51	39 9674	9847	*0020	*0192	*0365	*0538	*0711	*0883	*1056	*1228	173
52	40 1401	1573	1745	1917	2089	2261	2433	2605	2777	2949	172
53	3121	3292	3464	3635	3807	3978	4149	4320	4492	4663	172
54	4834	5005	5176	5346	5517	5688	5858	6029	6199	6370	171
55	6540	6710	6881	7051	7221	7391	7561	7731	7901	8070	171
56	8240	8410	8579	8749	8918	9087	9257	9426	9595	9764	170
57	40 9933	*0102	*0271	*0440	*0609	*0777	*0946	*1114	*1283	*1451	169
58	41 1620	1788	1956	2124	2293	2461	2629	2796	2964	3132	169
59	3300	3467	3635	3803	3970	4137	4305	4472	4639	4806	168
260	4973	5140	5307	5474	5641	5808	5974	6141	6308	6474	167
61	6641	6807	6973	7139	7306	7472	7638	7804	7970	8135	167
62	8301	8467	8633	8798	8964	9129	9295	9460	9625	9791	166
63	41 9956	*0121	*0286	*0451	*0616	*0781	*0945	*1110	*1275	*1439	165
64	42 1604	1768	1933	2097	2261	2426	2590	2754	2918	3082	165
65	3246	3410	3574	3737	3901	4065	4228	4392	4555	4718	164
66	4882	5045	5208	5371	5534	5697	5860	6023	6186	6349	163
67	6511	6674	6836	6999	7161	7324	7486	7648	7811	7973	163
68	8135	8297	8459	8621	8783	8944	9106	9268	9429	9591	162
69	42 9752	9914	*0075	*0236	*0398	*0559	*0720	*0881	*1042	*1203	162
270	43 1364	1525	1685	1846	2007	2167	2328	2488	2649	2809	161
71	2969	3130	3290	3450	3610	3770	3930	4090	4249	4409	161
72	4569	4729	4888	5048	5207	5367	5526	5685	5844	6004	160
73	6163	6322	6481	6640	6799	6957	7116	7275	7433	7592	159
74	7751	7909	8067	8226	8384	8542	8701	8859	9017	9175	159
75	43 9333	9491	9648	9806	9964	*0122	*0279	*0437	*0594	*0752	158
76	44 0909	1066	1224	1381	1538	1695	1852	2009	2166	2323	158
77	2480	2637	2793	2950	3106	3263	3419	3576	3732	3889	157
78	4045	4201	4357	4513	4669	4825	4981	5137	5293	5449	156
79	5604	5760	5915	6071	6226	6382	6537	6692	6848	7003	156
280	7158	7313	7468	7623	7778	7933	8088	8242	8397	8552	155
81	44 8706	8861	9015	9170	9324	9478	9633	9787	9941	*0095	155
82	45 0249	0403	0557	0711	0865	1018	1172	1326	1479	1633	154
83	1786	1940	2093	2247	2400	2553	2706	2859	3012	3165	154
84	3318	3471	3624	3777	3930	4082	4235	4387	4540	4692	153
85	4845	4997	5150	5302	5454	5606	5758	5910	6062	6214	153
86	6366	6518	6670	6821	6973	7125	7276	7428	7579	7731	152
87	7882	8033	8184	8336	8487	8638	8789	8940	9091	9242	152
88	45 9392	9543	9694	9845	9995	*0146	*0296	*0447	*0597	*0748	151
89	46 0898	1048	1198	1348	1499	1649	1799	1948	2098	2248	151
290	2398	2548	2697	2847	2997	3146	3296	3445	3594	3744	150
91	3893	4042	4191	4340	4490	4639	4788	4936	5085	5234	149
92	5383	5532	5680	5829	5977	6126	6274	6423	6571	6719	149
93	6868	7016	7164	7312	7460	7608	7756	7904	8052	8200	148
94	8347	8495	8643	8790	8938	9085	9233	9380	9527	9675	148
95	46 9822	9969	*0116	*0263	*0410	*0557	*0704	*0851	*0998	*1145	147
96	47 1292	1438	1585	1732	1878	2025	2171	2318	2464	2610	147
97	2756	2903	3049	3195	3341	3487	3633	3779	3925	4071	147
98	4216	4362	4508	4653	4799	4944	5090	5235	5381	5526	146
99	5671	5816	5962	6107	6252	6397	6542	6687	6832	6976	146
N	0	1	2	3	4	5	6	7	8	9	D

Table 2 **Logarithms of Numbers 3,000–3,499**
Six-Place Mantissas

N	0	1	2	3	4	5	6	7	8	9	D
300	47 7121	7266	7411	7555	7700	7844	7989	8133	8278	8422	145
01	47 8566	8711	8855	8999	9143	9287	9431	9575	9719	9863	145
02	48 0007	0151	0294	0438	0582	0725	0869	1012	1156	1299	144
03	1443	1586	1729	1872	2016	2159	2302	2445	2588	2731	144
04	2874	3016	3159	3302	3445	3587	3730	3872	4015	4157	143
05	4300	4442	4585	4727	4869	5011	5153	5295	5437	5579	143
06	5721	5863	6005	6147	6289	6430	6572	6714	6855	6997	142
07	7138	7280	7421	7563	7704	7845	7986	8127	8269	8410	142
08	8551	8692	8833	8974	9114	9255	9396	9537	9677	9818	141
09	48 9958	*0099	*0239	*0380	*0520	*0661	*0801	*0941	*1081	*1222	141
310	49 1362	1502	1642	1782	1922	2062	2201	2341	2481	2621	140
11	2760	2900	3040	3179	3319	3458	3597	3737	3876	4015	140
12	4155	4294	4433	4572	4711	4850	4989	5128	5267	5406	139
13	5544	5683	5822	5960	6099	6238	6376	6515	6653	6791	139
14	6930	7068	7206	7344	7483	7621	7759	7897	8035	8173	139
15	8311	8448	8586	8724	8862	8999	9137	9275	9412	9550	138
16	49 9687	9824	9962	*0099	*0236	*0374	*0511	*0648	*0785	*0922	138
17	50 1059	1196	1333	1470	1607	1744	1880	2017	2154	2291	137
18	2427	2564	2700	2837	2973	3109	3246	3382	3518	3655	137
19	3791	3927	4063	4199	4335	4471	4607	4743	4878	5014	136
320	5150	5286	5421	5557	5693	5828	5964	6099	6234	6370	136
21	6505	6640	6776	6911	7046	7181	7316	7451	7586	7721	136
22	7856	7991	8126	8260	8395	8530	8664	8799	8934	9068	135
23	50 9203	9337	9471	9606	9740	9874	*0009	*0143	*0277	*0411	135
24	51 0545	0679	0813	0947	1081	1215	1349	1482	1616	1750	134
25	1883	2017	2151	2284	2418	2551	2684	2818	2951	3084	134
26	3218	3351	3484	3617	3750	3883	4016	4149	4282	4415	133
27	4548	4681	4813	4946	5079	5211	5344	5476	5609	5741	133
28	5874	6006	6139	6271	6403	6535	6668	6800	6932	7064	133
29	7196	7328	7460	7592	7724	7855	7987	8119	8251	8382	132
330	8514	8646	8777	8909	9040	9171	9303	9434	9566	9697	132
31	51 9828	9959	*0090	*0221	*0353	*0484	*0615	*0745	*0876	*1007	132
32	52 1138	1269	1400	1530	1661	1792	1922	2053	2183	2314	131
33	2444	2575	2705	2835	2966	3096	3226	3356	3486	3616	131
34	3746	3876	4006	4136	4266	4396	4526	4656	4785	4915	130
35	5045	5174	5304	5434	5563	5693	5822	5951	6081	6210	130
36	6339	6469	6598	6727	6856	6985	7114	7243	7372	7501	130
37	7630	7759	7888	8016	8145	8274	8402	8531	8660	8788	129
38	52 8917	9045	9174	9302	9430	9559	9687	9815	9943	*0072	129
39	53 0200	0328	0456	0584	0712	0840	0968	1096	1223	1351	128
340	1479	1607	1734	1862	1990	2117	2245	2372	2500	2627	128
41	2754	2882	3009	3136	3264	3391	3518	3645	3772	3899	128
42	4026	4153	4280	4407	4534	4661	4787	4914	5041	5167	127
43	5294	5421	5547	5674	5800	5927	6053	6180	6306	6432	127
44	6558	6685	6811	6937	7063	7189	7315	7441	7567	7693	127
45	7819	7945	8071	8197	8322	8448	8574	8699	8825	8951	126
46	53 9076	9202	9327	9452	9578	9703	9829	9954	*0079	*0204	126
47	54 0329	0455	0580	0705	0830	0955	1080	1205	1330	1454	126
48	1579	1704	1829	1953	2078	2203	2327	2452	2576	2701	125
49	2825	2950	3074	3199	3323	3447	3571	3696	3820	3944	125
N	0	1	2	3	4	5	6	7	8	9	D

Table 2 **Logarithms of Numbers 3,500–3,999**
Six-Place Mantissas

N	0	1	2	3	4	5	6	7	8	9	D
350	54 4068	4192	4316	4440	4564	4688	4812	4936	5060	5183	124
51	5307	5431	5555	5678	5802	5925	6049	6172	6296	6419	124
52	6543	6666	6789	6913	7036	7159	7282	7405	7529	7652	124
53	7775	7898	8021	8144	8267	8389	8512	8635	8758	8881	123
54	54 9003	9126	9249	9371	9494	9616	9739	9861	9984	*0106	123
55	55 0228	0351	0473	0595	0717	0840	0962	1084	1206	1328	123
56	1450	1572	1694	1816	1938	2060	2181	2303	2425	2547	122
57	2668	2790	2911	3033	3155	3276	3398	3519	3640	3762	122
58	3883	4004	4126	4247	4368	4489	4610	4731	4852	4973	122
59	5094	5215	5336	5457	5578	5699	5820	5940	6061	6182	121
360	6303	6423	6544	6664	6785	6905	7026	7146	7267	7387	121
61	7507	7627	7748	7868	7988	8108	8228	8349	8469	8589	121
62	8709	8829	8948	9068	9188	9308	9428	9548	9667	9787	120
63	55 9907	*0026	*0146	*0265	*0385	*0504	*0624	*0743	*0863	*0982	120
64	56 1101	1221	1340	1459	1578	1698	1817	1936	2055	2174	120
65	2293	2412	2531	2650	2769	2887	3006	3125	3244	3362	119
66	3481	3600	3718	3837	3955	4074	4192	4311	4429	4548	119
67	4666	4784	4903	5021	5139	5257	5376	5494	5612	5730	119
68	5848	5966	6084	6202	6320	6437	6555	6673	6791	6909	118
69	7026	7144	7262	7379	7497	7614	7732	7849	7967	8084	118
370	8202	8319	8436	8554	8671	8788	8905	9023	9140	9257	118
71	56 9374	9491	9608	9725	9842	9959	*0076	*0193	*0309	*0426	117
72	57 0543	0660	0776	0893	1010	1126	1243	1359	1476	1592	117
73	1709	1825	1942	2058	2174	2291	2407	2523	2639	2755	117
74	2872	2988	3104	3220	3336	3452	3568	3684	3800	3915	116
75	4031	4147	4263	4379	4494	4610	4726	4841	4957	5072	116
76	5188	5303	5419	5534	5550	5765	5880	5996	6111	6226	116
77	6341	6457	6572	6687	6802	6917	7032	7147	7262	7377	116
78	7492	7607	7722	7836	7951	8066	8181	8295	8410	8525	115
79	8639	8754	8868	8983	9097	9212	9326	9441	9555	9669	115
380	57 9784	9898	*0012	*0126	*0241	*0355	*0469	*0583	*0697	*0811	115
81	58 0925	1039	1153	1267	1381	1495	1608	1722	1836	1950	114
82	2063	2177	2291	2404	2518	2631	2745	2858	2972	3085	114
83	3199	3312	3426	3539	3652	3765	3879	3992	4105	4218	114
84	4331	4444	4557	4670	4783	4896	5009	5122	5235	5348	113
85	5461	5574	5686	5799	5912	6024	6137	6250	6362	6475	113
86	6587	6700	6812	6925	7037	7149	7262	7374	7486	7599	113
87	7711	7823	7935	8047	8160	8272	8384	8496	8608	8720	113
88	8832	8944	9056	9167	9279	9391	9503	9615	9726	9838	112
89	58 9950	*0061	*0173	*0284	*0396	*0507	*0619	*0730	*0842	*0953	112
390	59 1065	1176	1287	1399	1510	1621	1732	1843	1955	2066	112
91	2177	2288	2399	2510	2621	2732	2843	2954	3064	3175	111
92	3286	3397	3508	3618	3729	3840	3950	4061	4171	4282	111
93	4393	4503	4614	4724	4834	4945	5055	5165	5276	5386	111
94	5496	5606	5717	5827	5937	6047	6157	6267	6377	6487	111
95	6597	6707	6817	6927	7037	7146	7256	7366	7476	7586	110
96	7695	7805	7914	8024	8134	8243	8353	8462	8572	8681	110
97	8791	8900	9009	9119	9228	9337	9446	9556	9665	9774	110
98	59 9883	9992	*0101	*0210	*0319	*0428	*0537	*0646	*0755	*0864	109
99	60 0973	1082	1191	1299	1408	1517	1625	1734	1843	1951	109
N	0	1	2	3	4	5	6	7	8	9	D

Table 2 **Logarithms of Numbers 4,000–4,499**
Six-Place Mantissas

N	0	1	2	3	4	5	6	7	8	9	D
400	60 2060	2169	2277	2386	2494	2603	2711	2819	2928	3036	109
01	3144	3253	3361	3469	3577	3686	3794	3902	4010	4118	109
02	4226	4334	4442	4550	4658	4766	4874	4982	5089	5197	108
03	5305	5413	5521	5628	5736	5844	5951	6059	6166	6274	108
04	6381	6489	6596	6704	6811	6919	7026	7133	7241	7348	108
05	7455	7562	7669	7777	7884	7991	8098	8205	8312	8419	108
06	8526	8633	8740	8847	8954	9061	9167	9274	9381	9488	107
07	60 9594	9701	9808	9914	*0021	*0128	*0234	*0341	*0447	*0554	107
08	61 0660	0767	0873	0979	1086	1192	1298	1405	1511	1617	107
09	1723	1829	1936	2042	2148	2254	2360	2466	2572	2678	107
410	2784	2890	2996	3102	3207	3313	3419	3525	3630	3736	106
11	3842	3947	4053	4159	4264	4370	4475	4581	4686	4792	106
12	4897	5003	5108	5213	5319	5424	5529	5634	5740	5845	106
13	5950	6055	6160	6265	6370	6476	6581	6686	6790	6895	106
14	7000	7105	7210	7315	7420	7525	7629	7734	7839	7943	105
15	8048	8153	8257	8362	8466	8571	8676	8780	8884	8989	105
16	61 9093	9198	9302	9406	9511	9615	9719	9824	9928	*0032	105
17	62 0136	0240	0344	0448	0552	0656	0760	0864	0968	1072	104
18	1176	1280	1384	1488	1592	1695	1799	1903	2007	2110	104
19	2214	2318	2421	2525	2628	2732	2835	2939	3042	3146	104
420	3249	3353	3456	3559	3663	3766	3869	3973	4076	4179	104
21	4282	4385	4488	4591	4695	4798	4901	5004	5107	5210	104
22	5312	5415	5518	5621	5724	5827	5929	6032	6135	6238	103
23	6340	6443	6546	6648	6751	6853	6956	7058	7161	7263	103
24	7366	7468	7571	7673	7775	7878	7980	8082	8185	8287	103
25	8389	8491	8593	8695	8797	8900	9002	9104	9206	9308	103
26	62 9410	9512	9613	9715	9817	9919	*0021	*0123	*0224	*0326	102
27	63 0428	0530	0631	0733	0835	0936	1038	1139	1241	1342	102
28	1444	1545	1647	1748	1849	1951	2052	2153	2255	2356	102
29	2457	2559	2660	2761	2862	2963	3064	3165	3266	3367	102
430	3468	3569	3670	3771	3872	3973	4074	4175	4276	4376	101
31	4477	4578	4679	4779	4880	4981	5081	5182	5283	5383	101
32	5484	5584	5685	5785	5886	5986	6087	6187	6287	6388	101
33	6488	6588	6688	6789	6889	6989	7089	7189	7290	7390	101
34	7490	7590	7690	7790	7890	7990	8090	8190	8290	8389	100
35	8489	8589	8689	8789	8888	8988	9088	9188	9287	9387	100
36	63 9486	9586	9686	9785	9885	9984	*0084	*0183	*0283	*0382	100
37	64 0481	0581	0680	0779	0879	0978	1077	1177	1276	1375	100
38	1474	1573	1672	1771	1871	1970	2069	2168	2267	2366	100
39	2465	2563	2662	2761	2860	2959	3058	3156	3255	3354	99
440	3453	3551	3650	3749	3847	3946	4044	4143	4242	4340	99
41	4439	4537	4636	4734	4832	4931	5029	5127	5226	5324	99
42	5422	5521	5619	5717	5815	5913	6011	6110	6208	6306	99
43	6404	6502	6600	6698	6796	6894	6992	7089	7187	7285	98
44	7383	7481	7579	7676	7774	7872	7969	8067	8165	8262	98
45	8360	8458	8555	8653	8750	8848	8945	9043	9140	9237	98
46	64 9335	9432	9530	9627	9724	9821	9919	*0016	*0113	*0210	98
47	65 0308	0405	0502	0599	0696	0793	0890	0987	1084	1181	97
48	1278	1375	1472	1569	1666	1762	1859	1956	2053	2150	97
49	2246	2343	2440	2536	2633	2730	2826	2923	3019	3116	97
N	0	1	2	3	4	5	6	7	8	9	D

Table 2 Logarithms of Numbers 4,500–4,999
Six-Place Mantissas

N	0	1	2	3	4	5	6	7	8	9	D
450	65 3213	3309	3405	3502	3598	3695	3791	3888	3984	4080	97
51	4177	4273	4369	4465	4562	4658	4754	4850	4946	5042	97
52	5138	5235	5331	5427	5523	5619	5715	5810	5906	6002	97
53	6098	6194	6290	6386	6482	6577	6673	6769	6864	6960	96
54	7056	7152	7247	7343	7438	7534	7629	7725	7820	7916	96
55	8011	8107	8202	8298	8393	8488	8584	8679	8774	8870	96
56	8965	9060	9155	9250	9346	9441	9536	9631	9726	9821	96
57	65 9916	*0011	*0106	*0201	*0296	*0391	*0486	*0581	*0676	*0771	95
58	66 0865	0960	1055	1150	1245	1339	1434	1529	1623	1718	95
59	1813	1907	2002	2096	2191	2286	2380	2475	2569	2663	95
460	2758	2852	2947	3041	3135	3230	3324	3418	3512	3607	95
61	3701	3795	3889	3983	4078	4172	4266	4360	4454	4548	95
62	4642	4736	4830	4924	5018	5112	5206	5299	5393	5487	94
63	5581	5675	5769	5862	5956	6050	6143	6237	6331	6424	94
64	6518	6612	6705	6799	6892	6986	7079	7173	7266	7360	94
65	7453	7546	7640	7733	7826	7920	8013	8106	8199	8293	94
66	8386	8479	8572	8665	8759	8852	8945	9038	9131	9224	94
67	66 9317	9410	9503	9596	9689	9782	9875	9967	*0060	*0153	93
68	67 0246	0339	0431	0524	0617	0710	0802	0895	0988	1080	93
69	1173	1265	1358	1451	1543	1636	1728	1821	1913	2005	93
470	2098	2190	2283	2375	2467	2560	2652	2744	2836	2929	93
71	3021	3113	3205	3297	3390	3482	3574	3666	3758	3850	93
72	3942	4034	4126	4218	4310	4402	4494	4586	4677	4769	92
73	4861	4953	5045	5137	5228	5320	5412	5503	5595	5687	92
74	5778	5870	5962	6053	6145	6236	6328	6419	6511	6602	92
75	6694	6785	6876	6968	7059	7151	7242	7333	7424	7516	92
76	7607	7698	7789	7881	7972	8063	8154	8245	8336	8427	92
77	8518	8609	8700	8791	8882	8973	9064	9155	9246	9337	91
78	67 9428	9519	9610	9700	9791	9882	9973	*0063	*0154	*0245	91
79	68 0336	0426	0517	0607	0698	0789	0879	0970	1060	1151	91
480	1241	1332	1422	1513	1603	1693	1784	1874	1964	2055	91
81	2145	2235	2326	2416	2506	2596	2686	2777	2867	2957	91
82	3047	3137	3227	3317	3407	3497	3587	3677	3767	3857	90
83	3947	4037	4127	4217	4307	4396	4486	4576	4666	4756	90
84	4845	4935	5025	5114	5204	5294	5383	5473	5563	5652	90
85	5742	5831	5921	6010	6100	6189	6279	6368	6458	6547	90
86	6636	6726	6815	6904	6994	7083	7172	7261	7351	7440	90
87	7529	7618	7707	7796	7886	7975	8064	8153	8242	8331	90
88	8420	8509	8598	8687	8776	8865	8953	9042	9131	9220	89
89	68 9309	9398	9486	9575	9664	9753	9841	9930	*0019	*0107	89
490	69 0196	0285	0373	0462	0550	0639	0728	0816	0905	0993	89
91	1081	1170	1258	1347	1435	1524	1612	1700	1789	1877	89
92	1965	2053	2142	2230	2318	2406	2494	2583	2671	2759	89
93	2847	2935	3023	3111	3199	3287	3375	3463	3551	3639	88
94	3727	3815	3903	3991	4078	4166	4254	4342	4430	4517	88
95	4605	4693	4781	4868	4956	5044	5131	5219	5307	5394	88
96	5482	5569	5657	5744	5832	5919	6007	6094	6182	6269	88
97	6356	6444	6531	6618	6706	6793	6880	6968	7055	7142	88
98	7229	7317	7404	7491	7578	7665	7752	7839	7926	8014	88
99	8101	8188	8275	8362	8449	8535	8622	8709	8796	8883	87
N	0	1	2	3	4	5	6	7	8	9	D

Table 2 **Logarithms of Numbers 5,000–5,499**
Six-Place Mantissas

N	0	1	2	3	4	5	6	7	8	9	D
500	69 8970	9057	9144	9231	9317	9404	9491	9578	9664	9751	87
01	69 9838	9924	*0011	*0098	*0184	*0271	*0358	*0444	*0531	*0617	87
02	70 0704	0790	0877	0963	1050	1136	1222	1309	1395	1482	87
03	1568	1654	1741	1827	1913	1999	2086	2172	2258	2344	87
04	2431	2517	2603	2689	2775	2861	2947	3033	3119	3205	86
05	3291	3377	3463	3549	3635	3721	3807	3893	3979	4065	86
06	4151	4236	4322	4408	4494	4579	4665	4751	4837	4922	86
07	5008	5094	5179	5265	5350	5436	5522	5607	5693	5778	86
08	5864	5949	6035	6120	6206	6291	6376	6462	6547	6632	86
09	6718	6803	6888	6974	7059	7144	7229	7315	7400	7485	86
510	7570	7655	7740	7826	7911	7996	8081	8166	8251	8336	86
11	8421	8506	8591	8676	8761	8846	8931	9015	9100	9185	85
12	70 9270	9355	9440	9524	9609	9694	9779	9863	9948	*0033	85
13	71 0117	0202	0287	0371	0456	0540	0625	0710	0794	0879	85
14	0963	1048	1132	1217	1301	1385	1470	1554	1639	1723	85
15	1807	1892	1976	2060	2144	2229	2313	2397	2481	2566	85
16	2650	2734	2818	2902	2986	3070	3154	3238	3323	3407	85
17	3491	3575	3659	3742	3826	3910	3994	4078	4162	4246	84
18	4330	4414	4497	4581	4665	4749	4833	4916	5000	5084	84
19	5167	5251	5335	5418	5502	5586	5669	5753	5836	5920	84
520	6003	6087	6170	6254	6337	6421	6504	6588	6671	6754	84
21	6838	6921	7004	7088	7171	7254	7338	7421	7504	7587	84
22	7671	7754	7837	7920	8003	8086	8169	8253	8336	8419	84
23	8502	8585	8668	8751	8834	8917	9000	9083	9165	9248	83
24	71 9331	9414	9497	9580	9663	9745	9828	9911	9994	*0077	83
25	72 0159	0242	0325	0407	0490	0573	0655	0738	0821	0903	83
26	0986	1068	1151	1233	1316	1398	1481	1563	1646	1728	83
27	1811	1893	1975	2058	2140	2222	2305	2387	2469	2552	83
28	2634	2716	2798	2881	2963	3045	3127	3209	3291	3374	83
29	3456	3538	3620	3702	3784	3866	3948	4030	4112	4194	82
530	4276	4358	4440	4522	4604	4685	4767	4849	4931	5013	82
31	5095	5176	5258	5340	5422	5503	5585	5667	5748	5830	82
32	5912	5993	6075	6156	6238	6320	6401	6483	6564	6646	82
33	6727	6809	6890	6972	7053	7134	7216	7297	7379	7460	82
34	7541	7623	7704	7785	7866	7948	8029	8110	8191	8273	82
35	8354	8435	8516	8597	8678	8759	8841	8922	9003	9084	82
36	9165	9246	9327	9408	9489	9570	9651	9732	9813	9893	81
37	72 9974	*0055	*0136	*0217	*0298	*0378	*0459	*0540	*0621	*0702	81
38	73 0782	0863	0944	1024	1105	1186	1266	1347	1428	1508	81
39	1589	1669	1750	1830	1911	1991	2072	2152	2233	2313	81
540	2394	2474	2555	2635	2715	2796	2876	2956	3037	3117	81
41	3197	3278	3358	3438	3518	3598	3679	3759	3839	3919	81
42	3999	4079	4160	4240	4320	4400	4480	4560	4640	4720	81
43	4800	4880	4960	5040	5120	5200	5279	5359	5439	5519	80
44	5599	5679	5759	5838	5918	5998	6078	6157	6237	6317	80
45	6397	6476	6556	6635	6715	6795	6874	6954	7034	7113	80
46	7193	7272	7352	7431	7511	7590	7670	7749	7829	7908	80
47	7987	8067	8146	8225	8305	8384	8463	8543	8622	8701	80
48	8781	8860	8939	9018	9097	9177	9256	9335	9414	9493	80
49	73 9572	9651	9731	9810	9889	9968	*0047	*0126	*0205	*0284	80
N	0	1	2	3	4	5	6	7	8	9	D

Table 2 Logarithms of Numbers 5,500–5,999
Six-Place Mantissas

N	0	1	2	3	4	5	6	7	8	9	D
550	74 0363	0442	0521	0600	0678	0757	0836	0915	0994	1073	79
51	1152	1230	1309	1388	1467	1546	1624	1703	1782	1860	79
52	1939	2018	2096	2175	2254	2332	2411	2489	2568	2647	79
53	2725	2804	2882	2961	3039	3118	3196	3275	3353	3431	79
54	3510	3588	3667	3745	3823	3902	3980	4058	4136	4215	79
55	4293	4371	4449	4528	4606	4684	4762	4840	4919	4997	79
56	5075	5153	5231	5309	5387	5465	5543	5621	5699	5777	78
57	5855	5933	6011	6089	6167	6245	6323	6401	6479	6556	78
58	6634	6712	6790	6868	6945	7023	7101	7179	7256	7334	78
59	7412	7489	7567	7645	7722	7800	7878	7955	8033	8110	78
560	8188	8266	8343	8421	8498	8576	8653	8731	8808	8885	78
61	8963	9040	9118	9195	9272	9350	9427	9504	9582	9659	78
62	74 9736	9814	9891	9968	*0045	*0123	*0200	*0277	*0354	*0431	78
63	75 0508	0586	0663	0740	0817	0894	0971	1048	1125	1202	78
64	1279	1356	1433	1510	1587	1664	1741	1818	1895	1972	77
65	2048	2125	2202	2279	2356	2433	2509	2586	2663	2740	77
66	2816	2893	2970	3047	3123	3200	3277	3353	3430	3506	77
67	3583	3660	3736	3813	3889	3966	4042	4119	4195	4272	77
68	4348	4425	4501	4578	4654	4730	4807	4883	4960	5036	77
69	5112	5189	5265	5341	5417	5494	5570	5646	5722	5799	77
570	5875	5951	6027	6103	6180	6256	6332	6408	6484	6560	77
71	6636	6712	6788	6864	6940	7016	7092	7168	7244	7320	76
72	7396	7472	7548	7624	7700	7775	7851	7927	8003	8079	76
73	8155	8230	8306	8382	8458	8533	8609	8685	8761	8836	76
74	8912	8988	9063	9139	9214	9290	9366	9441	9517	9592	76
75	75 9668	9743	9819	9894	9970	*0045	*0121	*0196	*0272	*0347	76
76	76 0422	0498	0573	0649	0724	0799	0875	0950	1025	1101	76
77	1176	1251	1326	1402	1477	1552	1627	1702	1778	1853	76
78	1928	2003	2078	2153	2228	2303	2378	2453	2529	2604	76
79	2679	2754	2829	2904	2978	3053	3128	3203	3278	3353	75
580	3428	3503	3578	3653	3727	3802	3877	3952	4027	4101	75
81	4176	4251	4326	4400	4475	4550	4624	4699	4774	4848	75
82	4923	4998	5072	5147	5221	5296	5370	5445	5520	5594	75
83	5669	5743	5818	5892	5966	6041	6115	6190	6264	6338	75
84	6413	6487	6562	6636	6710	6785	6859	6933	7007	7082	75
85	7156	7230	7304	7379	7453	7527	7601	7675	7749	7823	75
86	7898	7972	8046	8120	8194	8268	8342	8416	8490	8564	74
87	8638	8712	8786	8860	8934	9008	9082	9156	9230	9303	74
88	76 9377	9451	9525	9599	9673	9746	9820	9894	9968	*0042	74
89	77 0115	0189	0263	0336	0410	0484	0557	0631	0705	0778	74
590	0852	0926	0999	1073	1146	1220	1293	1367	1440	1514	74
91	1587	1661	1734	1808	1881	1955	2028	2102	2175	2248	74
92	2322	2395	2468	2542	2615	2688	2762	2835	2908	2981	74
93	3055	3128	3201	3274	3348	3421	3494	3567	3640	3713	74
94	3786	3860	3933	4006	4079	4152	4225	4298	4371	4444	74
95	4517	4590	4663	4736	4809	4882	4955	5028	5100	5173	73
96	5246	5319	5392	5465	5538	5610	5683	5756	5829	5902	73
97	5974	6047	6120	6193	6265	6338	6411	6483	6556	6629	73
98	6701	6774	6846	6919	6992	7064	7137	7209	7282	7354	73
99	7427	7499	7572	7644	7717	7789	7862	7934	8006	8079	73
N	0	1	2	3	4	5	6	7	8	9	D

Table 2 **Logarithms of Numbers 6,000–6,499**
Six-Place Mantissas

N	0	1	2	3	4	5	6	7	8	9	D
600	77 8151	8224	8296	8368	8441	8513	8585	8658	8730	8802	73
01	8874	8947	9019	9091	9163	9236	9308	9380	9452	9524	73
02	77 9596	9669	9741	9813	9885	9957	*0029	*0101	*0173	*0245	73
03	78 0317	0389	0461	0533	0605	0677	0749	0821	0893	0965	72
04	1037	1109	1181	1253	1324	1396	1468	1540	1612	1684	72
05	1755	1827	1899	1971	2042	2114	2186	2258	2329	2401	72
06	2473	2544	2616	2688	2759	2831	2902	2974	3046	3117	72
07	3189	3260	3332	3403	3475	3546	3618	3689	3761	3832	72
08	3904	3975	4046	4118	4189	4261	4332	4403	4475	4546	72
09	4617	4689	4760	4831	4902	4974	5045	5116	5187	5259	72
610	5330	5401	5472	5543	5615	5686	5757	5828	5899	5970	72
11	6041	6112	6183	6254	6325	6396	6467	6538	6609	6680	71
12	6751	6822	6893	6964	7035	7106	7177	7248	7319	7390	71
13	7460	7531	7602	7673	7744	7815	7885	7956	8027	8098	71
14	8168	8239	8310	8381	8451	8522	8593	8663	8734	8804	71
15	8875	8946	9016	9087	9157	9228	9299	9369	9440	9510	71
16	78 9581	9651	9722	9792	9863	9933	*0004	*0074	*0144	*0215	71
17	79 0285	0356	0426	0496	0567	0637	0707	0778	0848	0918	71
18	0988	1059	1129	1199	1269	1340	1410	1480	1550	1620	71
19	1691	1761	1831	1901	1971	2041	2111	2181	2252	2322	71
620	2392	2462	2532	2602	2672	2742	2812	2882	2952	3022	70
21	3092	3162	3231	3301	3371	3441	3511	3581	3651	3721	70
22	3790	3860	3930	4000	4070	4139	4209	4279	4349	4418	70
23	4488	4558	4627	4697	4767	4836	4906	4976	5045	5115	70
24	5185	5254	5324	5393	5463	5532	5602	5672	5741	5811	70
25	5880	5949	6019	6088	6158	6227	6297	6366	6436	6505	70
26	6574	6644	6713	6782	6852	6921	6990	7060	7129	7198	70
27	7268	7337	7406	7475	7545	7614	7683	7752	7821	7890	70
28	7960	8029	8098	8167	8236	8305	8374	8443	8513	8582	70
29	8651	8720	8789	8858	8927	8996	9065	9134	9203	9272	69
630	79 9341	9409	9478	9547	9616	9685	9754	9823	9892	9961	69
31	80 0029	0098	0167	0236	0305	0373	0442	0511	0580	0648	69
32	0717	0786	0854	0923	0992	1061	1129	1198	1266	1335	69
33	1404	1472	1541	1609	1678	1747	1815	1884	1952	2021	69
34	2089	2158	2226	2295	2363	2432	2500	2568	2637	2705	69
35	2774	2842	2910	2979	3047	3116	3184	3252	3321	3389	69
36	3457	3525	3594	3662	3730	3798	3867	3935	4003	4071	69
37	4139	4208	4276	4344	4412	4480	4548	4616	4685	4753	69
38	4821	4889	4957	5025	5093	5161	5229	5297	5365	5433	68
39	5501	5569	5637	5705	5773	5841	5908	5976	6044	6112	68
640	6180	6248	6316	6384	6451	6519	6587	6655	6723	6790	68
41	6858	6926	6994	7061	7129	7197	7264	7332	7400	7467	68
42	7535	7603	7670	7738	7806	7873	7941	8008	8076	8143	68
43	8211	8279	8346	8414	8481	8549	8616	8684	8751	8818	68
44	8886	8953	9021	9088	9156	9223	9290	9358	9425	9492	68
45	80 9560	9627	9694	9762	9829	9896	9964	*0031	*0098	*0165	68
46	81 0233	0300	0367	0434	0501	0569	0636	0703	0770	0837	68
47	0904	0971	1039	1106	1173	1240	1307	1374	1441	1508	68
48	1575	1642	1709	1776	1843	1910	1977	2044	2111	2178	67
49	2245	2312	2379	2445	2512	2579	2646	2713	2780	2847	67
N	0	1	2	3	4	5	6	7	8	9	D

Table 2 **Logarithms of Numbers 6,500–6,999**
Six-Place Mantissas

N	0	1	2	3	4	5	6	7	8	9	D
650	81 2913	2980	3047	3114	3181	3247	3314	3381	3448	3514	67
51	3581	3648	3714	3781	3848	3914	3981	4048	4114	4181	67
52	4248	4314	4381	4447	4514	4581	4647	4714	4780	4847	67
53	4913	4980	5046	5113	5179	5246	5312	5378	5445	5511	67
54	5578	5644	5711	5777	5843	5910	5976	6042	6109	6175	67
55	6241	6308	6374	6440	6506	6573	6639	6705	6771	6838	67
56	6904	6970	7036	7102	7169	7235	7301	7367	7433	7499	67
57	7565	7631	7698	7764	7830	7896	7962	8028	8094	8160	67
58	8226	8292	8358	8424	8490	8556	8622	8688	8754	8820	66
59	8885	8951	9017	9083	9149	9215	9281	9346	9412	9478	66
660	81 9544	9610	9676	9741	9807	9873	9939	*0004	*0070	*0136	66
61	82 0201	0267	0333	0399	0464	0530	0595	0661	0727	0792	66
62	0858	0924	0989	1055	1120	1186	1251	1317	1382	1448	66
63	1514	1579	1645	1710	1775	1841	1906	1972	2037	2103	66
64	2168	2233	2299	2364	2430	2495	2560	2626	2691	2756	66
65	2822	2887	2952	3018	3083	3148	3213	3279	3344	3409	66
66	3474	3539	3605	3670	3735	3800	3865	3930	3996	4061	66
67	4126	4191	4256	4321	4386	4451	4516	4581	4646	4711	65
68	4776	4841	4906	4971	5036	5101	5166	5231	5296	5361	65
69	5426	5491	5556	5621	5686	5751	5815	5880	5945	6010	65
670	6075	6140	6204	6269	6334	6399	6464	6528	6593	6658	65
71	6723	6787	6852	6917	6981	7046	7111	7175	7240	7305	65
72	7369	7434	7499	7563	7628	7692	7757	7821	7886	7951	65
73	8015	8080	8144	8209	8273	8338	8402	8467	8531	8595	65
74	8660	8724	8789	8853	8918	8982	9046	9111	9175	9239	65
75	9304	9368	9432	9497	9561	9625	9690	9754	9818	9882	65
76	82 9947	*0011	*0075	*0139	*0204	*0268	*0332	*0396	*0460	*0525	65
77	83 0589	0653	0717	0781	0845	0909	0973	1037	1102	1166	65
78	1230	1294	1358	1422	1486	1550	1614	1678	1742	1806	64
79	1870	1934	1998	2062	2126	2189	2253	2317	2381	2445	64
680	2509	2573	2637	2700	2764	2828	2892	2956	3020	3083	64
81	3147	3211	3275	3338	3402	3466	3530	3593	3657	3721	64
82	3784	3848	3912	3975	4039	4103	4166	4230	4294	4357	64
83	4421	4484	4548	4611	4675	4739	4802	4866	4929	4993	64
84	5056	5120	5183	5247	5310	5373	5437	5500	5564	5627	64
85	5691	5754	5817	5881	5944	6007	6071	6134	6197	6261	64
86	6324	6387	6451	6514	6577	6641	6704	6767	6830	6894	64
87	6957	7020	7083	7146	7210	7273	7336	7399	7462	7525	64
88	7588	7652	7715	7778	7841	7904	7967	8030	8093	8156	64
89	8219	8282	8345	8408	8471	8534	8597	8660	8723	8786	63
690	8849	8912	8975	9038	9101	9164	9227	9289	9352	9415	63
91	83 9478	9541	9604	9667	9729	9792	9855	9918	9981	*0043	63
92	84 0106	0169	0232	0294	0357	0420	0482	0545	0608	0671	63
93	0733	0796	0859	0921	0984	1046	1109	1172	1234	1297	63
94	1359	1422	1485	1547	1610	1672	1735	1797	1860	1922	63
95	1985	2047	2110	2172	2235	2297	2360	2422	2484	2547	63
96	2609	2672	2734	2796	2859	2921	2983	3046	3108	3170	63
97	3233	3295	3357	3420	3482	3544	3606	3669	3731	3793	63
98	3855	3918	3980	4042	4104	4166	4229	4291	4353	4415	63
99	4477	4539	4601	4664	4726	4788	4850	4912	4974	5036	63
N	0	1	2	3	4	5	6	7	8	9	D

Table 2 **Logarithms of Numbers 7,000–7,499**
Six-Place Mantissas

N	0	1	2	3	4	5	6	7	8	9	D
700	84 5098	5160	5222	5284	5346	5408	5470	5532	5594	5656	62
01	5718	5780	5842	5904	5966	6028	6090	6151	6213	6275	62
02	6337	6399	6461	6523	6585	6646	6708	6770	6832	6894	62
03	6955	7017	7079	7141	7202	7264	7326	7388	7449	7511	62
04	7573	7634	7696	7758	7819	7881	7943	8004	8066	8128	62
05	8189	8251	8312	8374	8435	8497	8559	8620	8682	8743	62
06	8805	8866	8928	8989	9051	9112	9174	9235	9297	9358	62
07	84 9419	9481	9542	9604	9665	9726	9788	9849	9911	9972	62
08	85 0033	0095	0156	0217	0279	0340	0401	0462	0524	0585	62
09	0646	0707	0769	0830	0891	0952	1014	1075	1136	1197	62
710	1258	1320	1381	1442	1503	1564	1625	1686	1747	1809	62
11	1870	1931	1992	2053	2114	2175	2236	2297	2358	2419	61
12	2480	2541	2602	2663	2724	2785	2846	2907	2968	3029	61
13	3090	3150	3211	3272	3333	3394	3455	3516	3577	3637	61
14	3698	3759	3820	3881	3941	4002	4063	4124	4185	4245	61
15	4306	4367	4428	4488	4549	4610	4670	4731	4792	4852	61
16	4913	4974	5034	5095	5156	5216	5277	5337	5398	5459	61
17	5519	5580	5640	5701	5761	5822	5882	5943	6003	6064	61
18	6124	6185	6245	6306	6366	6427	6487	6548	6608	6668	61
19	6729	6789	6850	6910	6970	7031	7091	7152	7212	7272	61
720	7332	7393	7453	7513	7574	7634	7694	7755	7815	7875	61
21	7935	7995	8056	8116	8176	8236	8297	8357	8417	8477	61
22	8537	8597	8657	8718	8778	8838	8898	8958	9018	9078	61
23	9138	9198	9258	9318	9379	9439	9499	9559	9619	9679	61
24	85 9739	9799	9859	9918	9978	*0038	*0098	*0158	*0218	*0278	60
25	86 0338	0398	0458	0518	0578	0637	0697	0757	0817	0877	60
26	0937	0996	1056	1116	1176	1236	1295	1355	1415	1475	60
27	1534	1594	1654	1714	1773	1833	1893	1952	2012	2072	60
28	2131	2191	2251	2310	2370	2430	2489	2549	2608	2668	60
29	2728	2787	2847	2906	2966	3025	3085	3144	3204	3263	60
730	3323	3382	3442	3501	3561	3620	3680	3739	3799	3858	60
31	3917	3977	4036	4096	4155	4214	4274	4333	4392	4452	60
32	4511	4570	4630	4689	4748	4808	4867	4926	4985	5045	60
33	5104	5163	5222	5282	5341	5400	5459	5519	5578	5637	60
34	5696	5755	5814	5874	5933	5992	6051	6110	6169	6228	60
35	6287	6346	6405	6465	6524	6583	6642	6701	6760	6819	60
36	6878	6937	6996	7055	7114	7173	7232	7291	7350	7409	59
37	7467	7526	7585	7644	7703	7762	7821	7880	7939	7998	59
38	8056	8115	8174	8233	8292	8350	8409	8468	8527	8586	59
39	8644	8703	8762	8821	8879	8938	8997	9056	9114	9173	59
740	9232	9290	9349	9408	9466	9525	9584	9642	9701	9760	59
41	86 9818	9877	9935	9994	*0053	*0111	*0170	*0228	*0287	*0345	59
42	87 0404	0462	0521	0579	0638	0696	0755	0813	0872	0930	59
43	0989	1047	1106	1164	1223	1281	1339	1398	1456	1515	59
44	1573	1631	1690	1748	1806	1865	1923	1981	2040	2098	59
45	2156	2215	2273	2331	2389	2448	2506	2564	2622	2681	59
46	2739	2797	2855	2913	2972	3030	3088	3146	3204	3262	59
47	3321	3379	3437	3495	3553	3611	3669	3727	3785	3844	59
48	3902	3960	4018	4076	4134	4192	4250	4308	4366	4424	58
49	4482	4540	4598	4656	4714	4772	4830	4888	4945	5003	58
N	0	1	2	3	4	5	6	7	8	9	D

Table 2 **Logarithms of Numbers 7,500–7,999**
Six-Place Mantissas

N	0	1	2	3	4	5	6	7	8	9	D
750	87 5061	5119	5177	5235	5293	5351	5409	5466	5524	5582	58
51	5640	5698	5756	5813	5871	5929	5987	6045	6102	6160	58
52	6218	6276	6333	6391	6449	6507	6564	6622	6680	6737	58
53	6795	6853	6910	6968	7026	7083	7141	7199	7256	7314	58
54	7371	7429	7487	7544	7602	7659	7717	7774	7832	7889	58
55	7947	8004	8062	8119	8177	8234	8292	8349	8407	8464	58
56	8522	8579	8637	8694	8752	8809	8866	8924	8981	9039	58
57	9096	9153	9211	9268	9325	9383	9440	9497	9555	9612	58
58	87 9669	9726	9784	9841	9898	9956	*0013	*0070	*0127	*0185	58
59	88 0242	0299	0356	0413	0471	0528	0585	0642	0699	0756	58
760	0814	0871	0928	0985	1042	1099	1156	1213	1271	1328	58
61	1385	1442	1499	1556	1613	1670	1727	1784	1841	1898	57
62	1955	2012	2069	2126	2183	2240	2297	2354	2411	2468	57
63	2525	2581	2638	2695	2752	2809	2866	2923	2980	3037	57
64	3093	3150	3207	3264	3321	3377	3434	3491	3548	3605	57
65	3661	3718	3775	3832	3888	3945	4002	4059	4115	4172	57
66	4229	4285	4342	4399	4455	4512	4569	4625	4682	4739	57
67	4795	4852	4909	4965	5022	5078	5135	5192	5248	5305	57
68	5361	5418	5474	5531	5587	5644	5700	5757	5813	5870	57
69	5926	5983	6039	6096	6152	6209	6265	6321	6378	6434	57
770	6491	6547	6604	6660	6716	6773	6829	6885	6942	6998	57
71	7054	7111	7167	7223	7280	7336	7392	7449	7505	7561	57
72	7617	7674	7730	7786	7842	7898	7955	8011	8067	8123	57
73	8179	8236	8292	8348	8404	8460	8516	8573	8629	8685	57
74	8741	8797	8853	8909	8965	9021	9077	9134	9190	9246	57
75	9302	9358	9414	9470	9526	9582	9638	9694	9750	9806	56
76	88 9862	9918	9974	*0030	*0086	*0141	*0197	*0253	*0309	*0365	56
77	89 0421	0477	0533	0589	0645	0700	0756	0812	0868	0924	56
78	0980	1035	1091	1147	1203	1259	1314	1370	1426	1482	56
79	1537	1593	1649	1705	1760	1816	1872	1928	1983	2039	56
780	2095	2150	2206	2262	2317	2373	2429	2484	2540	2595	56
81	2651	2707	2762	2818	2873	2929	2985	3040	3096	3151	56
82	3207	3262	3318	3373	3429	3484	3540	3595	3651	3706	56
83	3762	3817	3873	3928	3984	4039	4094	4150	4205	4261	56
84	4316	4371	4427	4482	4538	4593	4648	4704	4759	4814	56
85	4870	4925	4980	5036	5091	5146	5201	5257	5312	5367	56
86	5423	5478	5533	5588	5644	5699	5754	5809	5864	5920	56
87	5975	6030	6085	6140	6195	6251	6306	6361	6416	6471	56
88	6526	6581	6636	6692	6747	6802	6857	6912	6967	7022	56
89	7077	7132	7187	7242	7297	7352	7407	7462	7517	7572	55
790	7627	7682	7737	7792	7847	7902	7957	8012	8067	8122	55
91	8176	8231	8286	8341	8396	8451	8506	8561	8615	8670	55
92	8725	8780	8835	8890	8944	8999	9054	9109	9164	9218	55
93	9273	9328	9383	9437	9492	9547	9602	9656	9711	9766	55
94	89 9821	9875	9930	9985	*0039	*0094	*0149	*0203	*0258	*0312	55
95	90 0367	0422	0476	0531	0586	0640	0695	0749	0804	0859	55
96	0913	0968	1022	1077	1131	1186	1240	1295	1349	1404	55
97	1458	1513	1567	1622	1676	1731	1785	1840	1894	1948	55
98	2003	2057	2112	2166	2221	2275	2329	2384	2438	2492	55
99	2547	2601	2655	2710	2764	2818	2873	2927	2981	3036	55
N	0	1	2	3	4	5	6	7	8	9	D

Table 2 **Logarithms of Numbers 8,000–8,499**
Six-Place Mantissas

N	0	1	2	3	4	5	6	7	8	9	D
800	90 3090	3144	3199	3253	3307	3361	3416	3470	3524	3578	55
01	3633	3687	3741	3795	3849	3904	3958	4012	4066	4120	55
02	4174	4229	4283	4337	4391	4445	4499	4553	4607	4661	55
03	4716	4770	4824	4878	4932	4986	5040	5094	5148	5202	54
04	5256	5310	5364	5418	5472	5526	5580	5634	5688	5742	54
05	5796	5850	5904	5958	6012	6066	6119	6173	6227	6281	54
06	6335	6389	6443	6497	6551	6604	6658	6712	6766	6820	54
07	6874	6927	6981	7035	7089	7143	7196	7250	7304	7358	54
08	7411	7465	7519	7573	7626	7680	7734	7787	7841	7895	54
09	7949	8002	8056	8110	8163	8217	8270	8324	8378	8431	54
810	8485	8539	8592	8646	8699	8753	8807	8860	8914	8967	54
11	9021	9074	9128	9181	9235	9289	9342	9396	9449	9503	54
12	90 9556	9610	9663	9716	9770	9823	9877	9930	9984	*0037	54
13	91 0091	0144	0197	0251	0304	0358	0411	0464	0518	0571	54
14	0624	0678	0731	0784	0838	0891	0944	0998	1051	1104	54
15	1158	1211	1264	1317	1371	1424	1477	1530	1584	1637	54
16	1690	1743	1797	1850	1903	1956	2009	2063	2116	2169	54
17	2222	2275	2328	2381	2435	2488	2541	2594	2647	2700	54
18	2753	2806	2859	2913	2966	3019	3072	3125	3178	3231	54
19	3284	3337	3390	3443	3496	3549	3602	3655	3708	3761	53
820	3814	3867	3920	3973	4026	4079	4132	4184	4237	4290	53
21	4343	4396	4449	4502	4555	4608	4660	4713	4766	4819	53
22	4872	4925	4977	5030	5083	5136	5189	5241	5294	5347	53
23	5400	5453	5505	5558	5611	5664	5716	5769	5822	5875	53
24	5927	5980	6033	6085	6138	6191	6243	6296	6349	6401	53
25	6454	6507	6559	6612	6664	6717	6770	6822	6875	6927	53
26	6980	7033	7085	7138	7190	7243	7295	7348	7400	7453	53
27	7506	7558	7611	7663	7716	7768	7820	7873	7925	7978	53
28	8030	8083	8135	8188	8240	8293	8345	8397	8450	8502	53
29	8555	8607	8659	8712	8764	8816	8869	8921	8973	9026	53
830	9078	9130	9183	9235	9287	9340	9392	9444	9496	9549	53
31	91 9601	9653	9706	9758	9810	9862	9914	9967	*0019	*0071	53
32	92 0123	0176	0228	0280	0332	0384	0436	0489	0541	0593	53
33	0645	0697	0749	0801	0853	0906	0958	1010	1062	1114	53
34	1166	1218	1270	1322	1374	1426	1478	1530	1582	1634	52
35	1686	1738	1790	1842	1894	1946	1998	2050	2102	2154	52
36	2206	2258	2310	2362	2414	2466	2518	2570	2622	2674	52
37	2725	2777	2829	2881	2933	2985	3037	3089	3140	3192	52
38	3244	3296	3348	3399	3451	3503	3555	3607	3658	3710	52
39	3762	3814	3865	3917	3969	4021	4072	4124	4176	4228	52
840	4279	4331	4383	4434	4486	4538	4589	4641	4693	4744	52
41	4796	4848	4899	4951	5003	5054	5106	5157	5209	5261	52
42	5312	5364	5415	5467	5518	5570	5621	5673	5725	5776	52
43	5828	5879	5931	5982	6034	6085	6137	6188	6240	6291	52
44	6342	6394	6445	6497	6548	6600	6651	6702	6754	6805	52
45	6857	6908	6959	7011	7062	7114	7165	7216	7268	7319	52
46	7370	7422	7473	7524	7576	7627	7678	7730	7781	7832	52
47	7883	7935	7986	8037	8088	8140	8191	8242	8293	8345	52
48	8396	8447	8498	8549	8601	8652	8703	8754	8805	8857	52
49	8908	8959	9010	9061	9112	9163	9215	9266	9317	9368	52
N	0	1	2	3	4	5	6	7	8	9	D

Table 2 **Logarithms of Numbers 8,500–8,999**
Six-Place Mantissas

N	0	1	2	3	4	5	6	7	8	9	D
850	92 9419	9470	9521	9572	9623	9674	9725	9776	9827	9879	52
51	92 9930	9981	*0032	*0083	*0134	*0185	*0236	*0287	*0338	*0389	51
52	93 0440	0491	0542	0592	0643	0694	0745	0796	0847	0898	51
53	0949	1000	1051	1102	1153	1204	1254	1305	1356	1407	51
54	1458	1509	1560	1610	1661	1712	1763	1814	1865	1915	51
55	1966	2017	2068	2118	2169	2220	2271	2322	2372	2423	51
56	2474	2524	2575	2626	2677	2727	2778	2829	2879	2930	51
57	2981	3031	3082	3133	3183	3234	3285	3335	3386	3437	51
58	3487	3538	3589	3639	3690	3740	3791	3841	3892	3943	51
59	3993	4044	4094	4145	4195	4246	4296	4347	4397	4448	51
860	4498	4549	4599	4650	4700	4751	4801	4852	4902	4953	51
61	5003	5054	5104	5154	5205	5255	5306	5356	5406	5457	51
62	5507	5558	5608	5658	5709	5759	5809	5860	5910	5960	51
63	6011	6061	6111	6162	6212	6262	6313	6363	6413	6463	51
64	6514	6564	6614	6665	6715	6765	6815	6865	6916	6966	51
65	7016	7066	7117	7167	7217	7267	7317	7367	7418	7468	51
66	7518	7568	7618	7668	7718	7769	7819	7869	7919	7969	51
67	8019	8069	8119	8169	8219	8269	8320	8370	8420	8470	51
68	8520	8570	8620	8670	8720	8770	8820	8870	8920	8970	50
69	9020	9070	9120	9170	9220	9270	9320	9369	9419	9469	50
870	93 9519	9569	9619	9669	9719	9769	9819	9869	9918	9968	50
71	94 0018	0068	0118	0168	0218	0267	0317	0367	0417	0467	50
72	0516	0566	0616	0666	0716	0765	0815	0865	0915	0964	50
73	1014	1064	1114	1163	1213	1263	1313	1362	1412	1462	50
74	1511	1561	1611	1660	1710	1760	1809	1859	1909	1958	50
75	2008	2058	2107	2157	2207	2256	2306	2355	2405	2455	50
76	2504	2554	2603	2653	2702	2752	2801	2851	2901	2950	50
77	3000	3049	3099	3148	3198	3247	3297	3346	3396	3445	50
78	3495	3544	3593	3643	3692	3742	3791	3841	3890	3939	50
79	3989	4038	4088	4137	4186	4236	4285	4335	4384	4433	50
880	4483	4532	4581	4631	4680	4729	4779	4828	4877	4927	50
81	4976	5025	5074	5124	5173	5222	5272	5321	5370	5419	50
82	5469	5518	5567	5616	5665	5715	5764	5813	5862	5912	50
83	5961	6010	6059	6108	6157	6207	6256	6305	6354	6403	50
84	6452	6501	6551	6600	6649	6698	6747	6796	6845	6894	50
85	6943	6992	7041	7090	7140	7189	7238	7287	7336	7385	50
86	7434	7483	7532	7581	7630	7679	7728	7777	7826	7875	49
87	7924	7973	8022	8070	8119	8168	8217	8266	8315	8364	49
88	8413	8462	8511	8560	8609	8657	8706	8755	8804	8853	49
89	8902	8951	8999	9048	9097	9146	9195	9244	9292	9341	49
890	9390	9439	9488	9536	9585	9634	9683	9731	9780	9829	49
91	94 9878	9926	9975	*0024	*0073	*0121	*0170	*0219	*0267	*0316	49
92	95 0365	0414	0462	0511	0560	0608	0657	0706	0754	0803	49
93	0851	0900	0949	0997	1046	1095	1143	1192	1240	1289	49
94	1338	1386	1435	1483	1532	1580	1629	1677	1726	1775	49
95	1823	1872	1920	1969	2017	2066	2114	2163	2211	2260	49
96	2308	2356	2405	2453	2502	2550	2599	2647	2696	2744	49
97	2792	2841	2889	2938	2986	3034	3083	3131	3180	3228	49
98	3276	3325	3373	3421	3470	3518	3566	3615	3663	3711	49
99	3760	3808	3856	3905	3953	4001	4049	4098	4146	4194	49
N	0	1	2	3	4	5	6	7	8	9	D

Table 2 **Logarithms of Numbers 9,000–9,499**
Six-Place Mantissas

N	0	1	2	3	4	5	6	7	8	9	D
900	95 4243	4291	4339	4387	4435	4484	4532	4580	4628	4677	49
01	4725	4773	4821	4869	4918	4966	5014	5062	5110	5158	49
02	5207	5255	5303	5351	5399	5447	5495	5543	5592	5640	49
03	5688	5736	5784	5832	5880	5928	5976	6024	6072	6120	48
04	6168	6216	6265	6313	6361	6409	6457	6505	6553	6601	48
05	6649	6697	6745	6793	6840	6888	6936	6984	7032	7080	48
06	7128	7176	7224	7272	7320	7368	7416	7464	7512	7559	48
07	7607	7655	7703	7751	7799	7847	7894	7942	7990	8038	48
08	8086	8134	8181	8229	8277	8325	8373	8421	8468	8516	48
09	8564	8612	8659	8707	8755	8803	8850	8898	8946	8994	48
910	9041	9089	9137	9185	9232	9280	9328	9375	9423	9471	48
11	9518	9566	9614	9661	9709	9757	9804	9852	9900	9947	48
12	95 9995	*0042	*0090	*0138	*0185	*0233	*0280	*0328	*0376	*0423	48
13	96 0471	0518	0566	0613	0661	0709	0756	0804	0851	0899	48
14	0946	0994	1041	1089	1136	1184	1231	1279	1326	1374	48
15	1421	1469	1516	1563	1611	1658	1706	1753	1801	1848	48
16	1895	1943	1990	2038	2085	2132	2180	2227	2275	2322	48
17	2369	2417	2464	2511	2559	2606	2653	2701	2748	2795	48
18	2843	2890	2937	2985	3032	3079	3126	3174	3221	3268	48
19	3316	3363	3410	3457	3504	3552	3599	3646	3693	3741	48
920	3788	3835	3882	3929	3977	4024	4071	4118	4165	4212	48
21	4260	4307	4354	4401	4448	4495	4542	4590	4637	4684	48
22	4731	4778	4825	4872	4919	4966	5013	5061	5108	5155	48
23	5202	5249	5296	5343	5390	5437	5484	5531	5578	5625	47
24	5672	5719	5766	5813	5860	5907	5954	6001	6048	6095	47
25	6142	6189	6236	6283	6329	6376	6423	6470	6517	6564	47
26	6611	6658	6705	6752	6799	6845	6892	6939	6986	7033	47
27	7080	7127	7173	7220	7267	7314	7361	7408	7454	7501	47
28	7548	7595	7642	7688	7735	7782	7829	7875	7922	7969	47
29	8016	8062	8109	8156	8203	8249	8296	8343	8390	8436	47
930	8483	8530	8576	8623	8670	8716	8763	8810	8856	8903	47
31	8950	8996	9043	9090	9136	9183	9229	9276	9323	9369	47
32	9416	9463	9509	9556	9602	9649	9695	9742	9789	9835	47
33	96 9882	9928	9975	*0021	*0068	*0114	*0161	*0207	*0254	*0300	47
34	97 0347	0393	0440	0486	0533	0579	0626	0672	0719	0765	47
35	0812	0858	0904	0951	0997	1044	1090	1137	1183	1229	47
36	1276	1322	1369	1415	1461	1508	1554	1601	1647	1693	47
37	1740	1786	1832	1879	1925	1971	2018	2064	2110	2157	47
38	2203	2249	2295	2342	2388	2434	2481	2527	2573	2619	47
39	2666	2712	2758	2804	2851	2897	2943	2989	3035	3082	47
940	3128	3174	3220	3266	3313	3359	3405	3451	3497	3543	47
41	3590	3636	3682	3728	3774	3820	3866	3913	3959	4005	47
42	4051	4097	4143	4189	4235	4281	4327	4374	4420	4466	47
43	4512	4558	4604	4650	4696	4742	4788	4834	4880	4926	46
44	4972	5018	5064	5110	5156	5202	5248	5294	5340	5386	46
45	5432	5478	5524	5570	5616	5662	5707	5753	5799	5845	46
46	5891	5937	5983	6029	6075	6121	6167	6212	6258	6304	46
47	6350	6396	6442	6488	6533	6579	6625	6671	6717	6763	46
48	6808	6854	6900	6946	6992	7037	7083	7129	7175	7220	46
49	7266	7312	7358	7403	7449	7495	7541	7586	7632	7678	46
N	0	1	2	3	4	5	6	7	8	9	D

Table 2 Logarithms of Numbers 9,500–9,999
Six-Place Mantissas

N	0	1	2	3	4	5	6	7	8	9	D
950	97 7724	7769	7815	7861	7906	7952	7998	8043	8089	8135	46
51	8181	8226	8272	8317	8363	8409	8454	8500	8546	8591	46
52	8637	8683	8728	8774	8819	8865	8911	8956	9002	9047	46
53	9093	9138	9184	9230	9275	9321	9366	9412	9457	9503	46
54	97 9548	9594	9639	9685	9730	9776	9821	9867	9912	9958	46
55	98 0003	0049	0094	0140	0185	0231	0276	0322	0367	0412	46
56	0458	0503	0549	0594	0640	0685	0730	0776	0821	0867	46
57	0912	0957	1003	1048	1093	1139	1184	1229	1275	1320	46
58	1366	1411	1456	1501	1547	1592	1637	1683	1728	1773	46
59	1819	1864	1909	1954	2000	2045	2090	2135	2181	2226	46
960	2271	2316	2362	2407	2452	2497	2543	2588	2633	2678	46
61	2723	2769	2814	2859	2904	2949	2994	3040	3085	3130	46
62	3175	3220	3265	3310	3356	3401	3446	3491	3536	3581	46
63	3626	3671	3716	3762	3807	3852	3897	3942	3987	4032	46
64	4077	4122	4167	4212	4257	4302	4347	4392	4437	4482	45
65	4527	4572	4617	4662	4707	4752	4797	4842	4887	4932	45
66	4977	5022	5067	5112	5157	5202	5247	5292	5337	5382	45
67	5426	5471	5516	5561	5606	5651	5696	5741	5786	5830	45
68	5875	5920	5965	6010	6055	6100	6144	6189	6234	6279	45
69	6324	6369	6413	6458	6503	6548	6593	6637	6682	6727	45
970	6772	6817	6861	6906	6951	6996	7040	7085	7130	7175	45
71	7219	7264	7309	7353	7398	7443	7488	7532	7577	7622	45
72	7666	7711	7756	7800	7845	7890	7934	7979	8024	8068	45
73	8113	8157	8202	8247	8291	8336	8381	8425	8470	8514	45
74	8559	8604	8648	8693	8737	8782	8826	8871	8916	8960	45
75	9005	9049	9094	9138	9183	9227	9272	9316	9361	9405	45
76	9450	9494	9539	9583	9628	9672	9717	9761	9806	9850	45
77	98 9895	9939	9983	*0028	*0072	*0117	*0161	*0206	*0250	*0294	45
78	99 0339	0383	0428	0472	0516	0561	0605	0650	0694	0738	45
79	0783	0827	0871	0916	0960	1004	1049	1093	1137	1182	45
980	1226	1270	1315	1359	1403	1448	1492	1536	1580	1625	45
81	1669	1713	1758	1802	1846	1890	1935	1979	2023	2067	45
82	2111	2156	2200	2244	2288	2333	2377	2421	2465	2509	45
83	2554	2598	2642	2686	2730	2774	2819	2863	2907	2951	45
84	2995	3039	3083	3127	3172	3216	3260	3304	3348	3392	45
85	3436	3480	3524	3568	3613	3657	3701	3745	3789	3833	45
86	3877	3921	3965	4009	4053	4097	4141	4185	4229	4273	44
87	4317	4361	4405	4449	4493	4537	4581	4625	4669	4713	44
88	4757	4801	4845	4889	4933	4977	5021	5065	5108	5152	44
89	5196	5240	5284	5328	5372	5416	5460	5504	5547	5591	44
990	5635	5679	5723	5767	5811	5854	5898	5942	5986	6030	44
91	6074	6117	6161	6205	6249	6293	6337	6380	6424	6468	44
92	6512	6555	6599	6643	6687	6731	6774	6818	6862	6906	44
93	6949	6993	7037	7080	7124	7168	7212	7255	7299	7343	44
94	7386	7430	7474	7517	7561	7605	7648	7692	7736	7779	44
95	7823	7867	7910	7954	7998	8041	8085	8129	8172	8216	44
96	8259	8303	8347	8390	8434	8477	8521	8564	8608	8652	44
97	8695	8739	8782	8826	8869	8913	8956	9000	9043	9087	44
98	9131	9174	9218	9261	9305	9348	9392	9435	9479	9522	44
99	99 9565	9609	9652	9696	9739	9783	9826	9870	9913	9957	44
N	0	1	2	3	4	5	6	7	8	9	D

Table 2 Logarithms of Numbers 10,000–10,499
Seven-Place Mantissas

N	0	1	2	3	4	5	6	7	8	9	D#
1000	000 0000	0434	0869	1303	1737	2171	2605	3039	3473	3907	435
1001	4341	4775	5208	5642	6076	6510	6943	7377	7810	8244	434
1002	8677	9111	9544	9977	*0411	*0844	*1277	*1710	*2143	*2576	434
1003	001 3009	3442	3875	4308	4741	5174	5607	6039	6472	6905	433
1004	7337	7770	8202	8635	9067	9499	9932	*0364	*0796	*1228	433
1005	002 1661	2093	2525	2957	3389	3821	4253	4685	5116	5548	432
1006	5980	6411	6843	7275	7706	8138	8569	9001	9432	9863	432
1007	003 0295	0726	1157	1588	2019	2451	2882	3313	3744	4174	432
1008	4605	5036	5467	5898	6328	6759	7190	7620	8051	8481	431
1009	8912	9342	9772	*0203	*0633	*1063	*1493	*1924	*2354	*2784	431
1010	004 3214	3644	4074	4504	4933	5363	5793	6223	6652	7082	430
1011	7512	7941	8371	8800	9229	9659	*0088	*0517	*0947	*1376	430
1012	005 1805	2234	2663	3092	3521	3950	4379	4808	5237	5666	429
1013	6094	6523	6952	7380	7809	8238	8666	9094	9523	9951	429
1014	006 0380	0808	1236	1664	2092	2521	2949	3377	3805	4233	429
1015	4660	5088	5516	5944	6372	6799	7227	7655	8082	8510	428
1016	8937	9365	9792	*0219	*0647	*1074	*1501	*1928	*2355	*2782	428
1017	007 3210	3637	4064	4490	4917	5344	5771	6198	6624	7051	427
1018	7478	7904	8331	8757	9184	9610	*0037	*0463	*0889	*1316	427
1019	008 1742	2168	2594	3020	3446	3872	4298	4724	5150	5576	426
1020	6002	6427	6853	7279	7704	8130	8556	8981	9407	9832	426
1021	009 0257	0683	1108	1533	1959	2384	2809	3234	3659	4084	426
1022	4509	4934	5359	5784	6208	6633	7058	7483	7907	8332	425
1023	8756	9181	9605	*0030	*0454	*0878	*1303	*1727	*2151	*2575	425
1024	010 3000	3424	3848	4272	4696	5120	5544	5967	6391	6815	424
1025	7239	7662	8086	8510	8933	9357	9780	*0204	*0627	*1050	424
1026	011 1474	1897	2320	2743	3166	3590	4013	4436	4859	5282	424
1027	5704	6127	6550	6973	7396	7818	8241	8664	9086	9509	423
1028	9931	*0354	*0776	*1198	*1621	*2043	*2465	*2887	*3310	*3732	423
1029	012 4154	4576	4998	5420	5842	6264	6685	7107	7529	7951	422
1030	8372	8794	9215	9637	*0059	*0480	*0901	*1323	*1744	*2165	422
1031	013 2587	3008	3429	3850	4271	4692	5113	5534	5955	6376	421
1032	6797	7218	7639	8059	8480	8901	9321	9742	*0162	*0583	421
1033	014 1003	1424	1844	2264	2685	3105	3525	3945	4365	4785	421
1034	5205	5625	6045	6465	6885	7305	7725	8144	8564	8984	420
1035	9403	9823	*0243	*0662	*1082	*1501	*1920	*2340	*2759	*3178	420
1036	015 3598	4017	4436	4855	5274	5693	6112	6531	6950	7369	419
1037	7788	8206	8625	9044	9462	9881	*0300	*0718	*1137	*1555	419
1038	016 1974	2392	2810	3229	3647	4065	4483	4901	5319	5737	419
1039	6155	6573	6991	7409	7827	8245	8663	9080	9498	9916	418
1040	017 0333	0751	1168	1586	2003	2421	2838	3256	3673	4090	418
1041	4507	4924	5342	5759	6176	6593	7010	7427	7844	8260	418
1042	8677	9094	9511	9927	*0344	*0761	*1177	*1594	*2010	*2427	417
1043	018 2843	3259	3676	4092	4508	4925	5341	5757	6173	6589	417
1044	7005	7421	7837	8253	8669	9084	9500	9916	*0332	*0747	416
1045	019 1163	1578	1994	2410	2825	3240	3656	4071	4486	4902	416
1046	5317	5732	6147	6562	6977	7392	7807	8222	8637	9052	415
1047	9467	9882	*0296	*0711	*1126	*1540	*1955	*2369	*2784	*3198	415
1048	020 3613	4027	4442	4856	5270	5684	6099	6513	6927	7341	415
1049	7755	8169	8583	8997	9411	9824	*0238	*0652	*1066	*1479	414
N	0	1	2	3	4	5	6	7	8	9	D

*Prefix first three places on next line.
Example: The mantissa for number (N) 10024 is *001 0411*.

#The *highest difference* between adjacent mantissas on the *individual line*. It is also the *lowest difference* between adjacent mantissas on the *preceding line* in many cases.

Table 2 **Logarithms of Numbers 10,500–11,009**
Seven-Place Mantissas

N	0	1	2	3	4	5	6	7	8	9	D
1050	021 1893	2307	2720	3134	3547	3961	4374	4787	5201	5614	414
1051	6027	6440	6854	7267	7680	8093	8506	8919	9332	9745	414
1052	022 0157	0570	0983	1396	1808	2221	2634	3046	3459	3871	413
1053	4284	4696	5109	5521	5933	6345	6758	7170	7582	7994	413
1054	8406	8818	9230	9642	*0054	*0466	*0878	*1289	*1701	*2113	412
1055	023 2525	2936	3348	3759	4171	4582	4994	5405	5817	6228	412
1056	6639	7050	7462	7873	8284	8695	9106	9517	9928	*0339	412
1057	024 0750	1161	1572	1982	2393	2804	3214	3625	4036	4446	411
1058	4857	5267	5678	6088	6498	6909	7319	7729	8139	8549	411
1059	8960	9370	9780	*0190	*0600	*1010	*1419	*1829	*2239	*2649	410
1060	025 3059	3468	3878	4288	4697	5107	5516	5926	6335	6744	410
1061	7154	7563	7972	8382	8791	9200	9609	*0018	*0427	*0836	410
1062	026 1245	1654	2063	2472	2881	3289	3698	4107	4515	4924	409
1063	5333	5741	6150	6558	6967	7375	7783	8192	8600	9008	409
1064	9416	9824	*0233	*0641	*1049	*1457	*1865	*2273	*2680	*3088	409
1065	027 3496	3904	4312	4719	5127	5535	5942	6350	6757	7165	408
1066	7572	7979	8387	8794	9201	9609	*0016	*0423	*0830	*1237	408
1067	028 1644	2051	2458	2865	3272	3679	4086	4492	4899	5306	408
1068	5713	6119	6526	6932	7339	7745	8152	8558	8964	9371	407
1069	9777	*0183	*0590	*0996	*1402	*1808	*2214	*2620	*3026	*3432	407
1070	029 3838	4244	4649	5055	5461	5867	6272	6678	7084	7489	406
1071	7895	8300	8706	9111	9516	9922	*0327	*0732	*1138	*1543	406
1072	030 1948	2353	2758	3163	3568	3973	4378	4783	5188	5592	405
1073	5997	6402	6807	7211	7616	8020	8425	8830	9234	9638	405
1074	031 0043	0447	0851	1256	1660	2064	2468	2872	3277	3681	405
1075	4085	4489	4893	5296	5700	6104	6508	6912	7315	7719	404
1076	8123	8526	8930	9333	9737	*0140	*0544	*0947	*1350	*1754	404
1077	032 2157	2560	2963	3367	3770	4173	4576	4979	5382	5785	404
1078	6188	6590	6993	7396	7799	8201	8604	9007	9409	9812	403
1079	033 0214	0617	1019	1422	1824	2226	2629	3031	3433	3835	403
1080	4238	4640	5042	5444	5846	6248	6650	7052	7453	7855	402
1081	8257	8659	9060	9462	9864	*0265	*0667	*1068	*1470	*1871	402
1082	034 2273	2674	3075	3477	3878	4279	4680	5081	5482	5884	402
1083	6285	6686	7087	7487	7888	8289	8690	9091	9491	9892	401
1084	035 0293	0693	1094	1495	1895	2296	2696	3096	3497	3897	401
1085	4297	4698	5098	5498	5898	6298	6698	7098	7498	7898	401
1086	8298	8698	9098	9498	9898	*0297	*0697	*1097	*1496	*1896	400
1087	036 2295	2695	3094	3494	3893	4293	4692	5091	5491	5890	400
1088	6289	6688	7087	7486	7885	8284	8683	9082	9481	9880	399
1089	037 0279	0678	1076	1475	1874	2272	2671	3070	3468	3867	399
1090	4265	4663	5062	5460	5858	6257	6655	7053	7451	7849	399
1091	8248	8646	9044	9442	9839	*0237	*0635	*1033	*1431	*1829	398
1092	038 2226	2624	3022	3419	3817	4214	4612	5009	5407	5804	398
1093	6202	6599	6996	7393	7791	8188	8585	8982	9379	9776	398
1094	039 0173	0570	0967	1364	1761	2158	2554	2951	3348	3745	397
1095	4141	4538	4934	5331	5727	6124	6520	6917	7313	7709	397
1096	8106	8502	8898	9294	9690	*0086	*0482	*0878	*1274	*1670	396
1097	040 2066	2462	2858	3254	3650	4045	4441	4837	5232	5628	396
1098	6023	6419	6814	7210	7605	8001	8396	8791	9187	9582	396
1099	9977	*0372	*0767	*1162	*1557	*1952	*2347	*2742	*3137	*3532	395
1100	041 3927	4322	4716	5111	5506	5900	6295	6690	7084	7479	395
N	0	1	2	3	4	5	6	7	8	9	D

Table 3 **Natural Logarithms of Numbers 1.00 to 5.49**

	.00	.01	.02	.03	.04	.05	.06	.07	.08	.09
1.0	0.0000	0.0100	0.0198	0.0296	0.0392	0.0488	0.0583	0.0677	0.0770	0.0862
1.1	0.0953	0.1044	0.1133	0.1222	0.1310	0.1398	0.1484	0.1570	0.1655	0.1740
1.2	0.1823	0.1906	0.1989	0.2070	0.2151	0.2231	0.2311	0.2390	0.2469	0.2546
1.3	0.2624	0.2700	0.2776	0.2852	0.2927	0.3001	0.3075	0.3148	0.3221	0.3293
1.4	0.3365	0.3436	0.3507	0.3577	0.3646	0.3716	0.3784	0.3853	0.3920	0.3988
1.5	0.4055	0.4121	0.4187	0.4253	0.4318	0.4383	0.4447	0.4511	0.4574	0.4637
1.6	0.4700	0.4762	0.4824	0.4886	0.4947	0.5008	0.5068	0.5128	0.5188	0.5247
1.7	0.5306	0.5365	0.5423	0.5481	0.5539	0.5596	0.5653	0.5710	0.5766	0.5822
1.8	0.5878	0.5933	0.5988	0.6043	0.6098	0.6152	0.6206	0.6259	0.6313	0.6366
1.9	0.6419	0.6471	0.6523	0.6575	0.6627	0.6678	0.6729	0.6780	0.6831	0.6881
2.0	0.6931	0.6981	0.7031	0.7080	0.7130	0.7178	0.7227	0.7275	0.7324	0.7372
2.1	0.7419	0.7467	0.7514	0.7561	0.7608	0.7655	0.7701	0.7747	0.7793	0.7839
2.2	0.7885	0.7930	0.7975	0.8020	0.8065	0.8109	0.8154	0.8198	0.8242	0.8286
2.3	0.8329	0.8372	0.8416	0.8459	0.8502	0.8544	0.8587	0.8629	0.8671	0.8713
2.4	0.8755	0.8796	0.8838	0.8879	0.8920	0.8961	0.9002	0.9042	0.9083	0.9123
2.5	0.9163	0.9203	0.9243	0.9282	0.9322	0.9361	0.9400	0.9439	0.9478	0.9517
2.6	0.9555	0.9594	0.9632	0.9670	0.9708	0.9746	0.9783	0.9821	0.9858	0.9895
2.7	0.9933	0.9969	1.0006	1.0043	1.0080	1.0116	1.0152	1.0188	1.0225	1.0260
2.8	1.0296	1.0332	1.0367	1.0403	1.0438	1.0473	1.0508	1.0543	1.0578	1.0613
2.9	1.0647	1.0682	1.0716	1.0750	1.0784	1.0818	1.0852	1.0886	1.0919	1.0953
3.0	1.0986	1.1019	1.1053	1.1086	1.1119	1.1151	1.1184	1.1217	1.1249	1.1282
3.1	1.1314	1.1346	1.1378	1.1410	1.1442	1.1474	1.1506	1.1537	1.1569	1.1600
3.2	1.1632	1.1663	1.1694	1.1725	1.1756	1.1787	1.1817	1.1848	1.1878	1.1909
3.3	1.1939	1.1970	1.2000	1.2030	1.2060	1.2090	1.2119	1.2149	1.2179	1.2208
3.4	1.2238	1.2267	1.2296	1.2326	1.2355	1.2384	1.2413	1.2442	1.2470	1.2499
3.5	1.2528	1.2556	1.2585	1.2613	1.2641	1.2669	1.2698	1.2726	1.2754	1.2782
3.6	1.2809	1.2837	1.2865	1.2892	1.2920	1.2947	1.2975	1.3002	1.3029	1.3056
3.7	1.3083	1.3110	1.3137	1.3164	1.3191	1.3218	1.3244	1.3271	1.3297	1.3324
3.8	1.3350	1.3376	1.3403	1.3429	1.3455	1.3481	1.3507	1.3533	1.3558	1.3584
3.9	1.3610	1.3635	1.3661	1.3686	1.3712	1.3737	1.3762	1.3788	1.3813	1.3838
4.0	1.3863	1.3888	1.3913	1.3938	1.3962	1.3987	1.4012	1.4036	1.4061	1.4085
4.1	1.4110	1.4134	1.4159	1.4183	1.4207	1.4231	1.4255	1.4279	1.4303	1.4327
4.2	1.4351	1.4376	1.4398	1.4422	1.4446	1.4469	1.4493	1.4516	1.4540	1.4563
4.3	1.4586	1.4609	1.4633	1.4656	1.4679	1.4702	1.4725	1.4748	1.4770	1.4793
4.4	1.4816	1.4839	1.4861	1.4884	1.4907	1.4929	1.4952	1.4974	1.4996	1.5019
4.5	1.5041	1.5063	1.5085	1.5107	1.5129	1.5151	1.5173	1.5195	1.5217	1.5239
4.6	1.5261	1.5282	1.5304	1.5326	1.5347	1.5369	1.5390	1.5412	1.5433	1.5454
4.7	1.5476	1.5497	1.5518	1.5539	1.5560	1.5581	1.5602	1.5623	1.5644	1.5665
4.8	1.5686	1.5707	1.5728	1.5748	1.5769	1.5790	1.5810	1.5831	1.5851	1.5872
4.9	1.5892	1.5913	1.5933	1.5953	1.5974	1.5994	1.6014	1.6034	1.6054	1.6074
5.0	1.6094	1.6114	1.6134	1.6154	1.6174	1.6194	1.6214	1.6233	1.6253	1.6273
5.1	1.6292	1.6312	1.6332	1.6351	1.6371	1.6390	1.6409	1.6429	1.6448	1.6467
5.2	1.6487	1.6506	1.6525	1.6544	1.6563	1.6582	1.6601	1.6620	1.6639	1.6658
5.3	1.6677	1.6696	1.6715	1.6734	1.6752	1.6771	1.6790	1.6808	1.6827	1.6845
5.4	1.6864	1.6882	1.6901	1.6919	1.6938	1.6956	1.6974	1.6993	1.7011	1.7029

$ln\ N = \log_e N = 2.302826\ (\log_{10} N)$

Table 3 **Natural Logarithms of Numbers 5.50 to 9.99**

	.00	.01	.02	.03	.04	.05	.06	.07	.08	.09
5.5	1.7047	1.7066	1.7084	1.7102	1.7120	1.7138	1.7156	1.7174	1.7192	1.7210
5.6	1.7228	1.7246	1.7263	1.7281	1.7299	1.7317	1.7334	1.7352	1.7370	1.7387
5.7	1.7405	1.7422	1.7440	1.7457	1.7475	1.7492	1.7509	1.7527	1.7544	1.7561
5.8	1.7579	1.7596	1.7613	1.7630	1.7647	1.7664	1.7682	1.7699	1.7716	1.7733
5.9	1.7750	1.7766	1.7783	1.7800	1.7817	1.7834	1.7851	1.7867	1.7884	1.7901
6.0	1.7918	1.7934	1.7951	1.7967	1.7984	1.8001	1.8017	1.8034	1.8050	1.8066
6.1	1.8083	1.8099	1.8116	1.8132	1.8148	1.8165	1.8181	1.8197	1.8213	1.8229
6.2	1.8245	1.8262	1.8278	1.8294	1.8310	1.8326	1.8342	1.8358	1.8374	1.8390
6.3	1.8406	1.8421	1.8437	1.8453	1.8469	1.8485	1.8500	1.8516	1.8532	1.8547
6.4	1.8563	1.8579	1.8594	1.8610	1.8625	1.8641	1.8656	1.8672	1.8687	1.8703
6.5	1.8718	1.8733	1.8749	1.8764	1.8779	1.8795	1.8810	1.8825	1.8840	1.8856
6.6	1.8871	1.8886	1.8901	1.8916	1.8931	1.8946	1.8961	1.8976	1.8991	1.9006
6.7	1.9021	1.9036	1.9051	1.9066	1.9081	1.9095	1.9110	1.9125	1.9140	1.9155
6.8	1.9169	1.9184	1.9199	1.9213	1.9228	1.9242	1.9257	1.9272	1.9286	1.9301
6.9	1.9315	1.9330	1.9344	1.9359	1.9373	1.9387	1.9402	1.9416	1.9430	1.9445
7.0	1.9459	1.9473	1.9488	1.9502	1.9516	1.9530	1.9544	1.9559	1.9573	1.9587
7.1	1.9601	1.9615	1.9629	1.9643	1.9657	1.9671	1.9685	1.9699	1.9713	1.9727
7.2	1.9741	1.9755	1.9769	1.9782	1.9796	1.9810	1.9824	1.9838	1.9851	1.9865
7.3	1.9879	1.9892	1.9906	1.9920	1.9933	1.9947	1.9961	1.9974	1.9988	2.0001
7.4	2.0015	2.0028	2.0042	2.0055	2.0069	2.0082	2.0096	2.0109	2.0122	2.0136
7.5	2.0149	2.0162	2.0176	2.0189	2.0202	2.0215	2.0229	2.0242	2.0255	2.0268
7.6	2.0282	2.0295	2.0308	2.0321	2.0334	2.0347	2.0360	2.0373	2.0386	2.0399
7.7	2.0412	2.0425	2.0438	2.0451	2.0464	2.0477	2.0490	2.0503	2.0516	2.0528
7.8	2.0541	2.0554	2.0567	2.0580	2.0592	2.0605	2.0618	2.0631	2.0643	2.0656
7.9	2.0669	2.0681	2.0694	2.0707	2.0719	2.0732	2.0744	2.0757	2.0769	2.0782
8.0	2.0794	2.0807	2.0819	2.0832	2.0844	2.0857	2.0869	2.0882	2.0894	2.0906
8.1	2.0919	2.0931	2.0943	2.0956	2.0968	2.0980	2.0992	2.1005	2.1017	2.1029
8.2	2.1041	2.1054	2.1066	2.1078	2.1090	2.1102	2.1114	2.1126	2.1138	2.1150
8.3	2.1163	2.1175	2.1187	2.1199	2.1211	2.1223	2.1235	2.1247	2.1258	2.1270
8.4	2.1282	2.1294	2.1306	2.1318	2.1330	2.1342	2.1353	2.1365	2.1377	2.1389
8.5	2.1401	2.1412	2.1424	2.1436	2.1448	2.1459	2.1471	2.1483	2.1494	2.1506
8.6	2.1518	2.1529	2.1541	2.1552	2.1564	2.1576	2.1587	2.1599	2.1610	2.1622
8.7	2.1633	2.1645	2.1656	2.1668	2.1679	2.1691	2.1702	2.1713	2.1725	2.1736
8.8	2.1748	2.1759	2.1770	2.1782	2.1793	2.1804	2.1815	2.1827	2.1838	2.1849
8.9	2.1861	2.1872	2.1883	2.1894	2.1905	2.1917	2.1928	2.1939	2.1950	2.1961
9.0	2.1972	2.1983	2.1994	2.2006	2.2017	2.2028	2.2039	2.2050	2.2061	2.2072
9.1	2.2083	2.2094	2.2105	2.2116	2.2127	2.2138	2.2148	2.2159	2.2170	2.2181
9.2	2.2192	2.2203	2.2214	2.2225	2.2235	2.2246	2.2257	2.2268	2.2279	2.2289
9.3	2.2300	2.2311	2.2322	2.2332	2.2343	2.2354	2.2364	2.2375	2.2386	2.2396
9.4	2.2407	2.2418	2.2428	2.2439	2.2450	2.2460	2.2471	2.2481	2.2492	2.2502
9.5	2.2513	2.2523	2.2534	2.2544	2.2555	2.2565	2.2576	2.2586	2.2597	2.2607
9.6	2.2618	2.2628	2.2638	2.2649	2.2659	2.2670	2.2680	2.2690	2.2701	2.2711
9.7	2.2721	2.2732	2.2742	2.2752	2.2762	2.2773	2.2783	2.2793	2.2803	2.2814
9.8	2.2824	2.2834	2.2844	2.2854	2.2865	2.2875	2.2885	2.2895	2.2905	2.2915
9.9	2.2925	2.2935	2.2946	2.2956	2.2966	2.2976	2.2986	2.2996	2.3006	2.3016

$\ln 10 = 2.3026$. Application: $\ln (N \cdot 10^m) = \ln N + m (\ln 10)$
$\ln (10^{-m}) = -m (\ln 10)$.

Table 4 Values of e^j and e^{-j} for Selected Values of j

j	e^j	e^{-j}	j	e^j	e^{-j}
.000	1.000 000 00	1.000 000 00	.21	1.233 678 06	.810 584 25
.001	1.001 000 50	.999 000 50	.22	1.246 076 73	.802 518 80
.002	1.002 002 00	.998 002 00	.23	1.258 600 01	.794 533 60
.003	1.003 004 50	.997 004 50	.24	1.271 249 15	.786 627 86
.004	1.004 008 01	.996 007 99	.25	1.284 025 42	.778 800 78
.005	1.005 012 52	.995 012 48	.26	1.296 930 09	.771 051 59
.006	1.006 018 04	.994 017 96	.27	1.309 964 45	.763 379 49
.007	1.007 024 56	.993 024 44	.28	1.323 129 81	.755 783 74
.008	1.008 032 09	.992 031 91	.29	1.336 427 49	.748 263 57
.009	1.009 040 62	.991 040 38	.30	1.349 858 81	.740 818 22
.01	1.010 050 17	.990 049 83	.31	1.363 425 11	.733 446 96
.02	1.020 201 34	.980 198 67	.32	1.377 127 76	.726 149 04
.03	1.030 454 53	.970 445 53	.33	1.390 968 13	.718 923 73
.04	1.040 810 77	.960 789 44	.34	1.404 947 59	.711 770 32
.05	1.051 271 10	.951 229 42	.35	1.419 067 55	.704 688 09
.06	1.061 836 55	.941 764 53	.36	1.433 329 41	.697 676 33
.07	1.072 508 18	.932 393 82	.37	1.447 734 61	.690 734 33
.08	1.083 287 07	.923 116 35	.38	1.462 284 59	.683 861 41
.09	1.094 174 28	.913 931 19	.39	1.476 980 79	.677 056 87
.10	1.105 170 92	.904 837 42	.40	1.491 824 70	.670 320 05
.11	1.116 278 07	.895 834 14	.41	1.506 817 79	.663 650 25
.12	1.127 496 85	.886 920 44	.42	1.521 961 56	.657 046 82
.13	1.138 828 38	.878 095 43	.43	1.537 257 52	.650 509 09
.14	1.150 273 80	.869 358 24	.44	1.552 707 22	.644 036 42
.15	1.161 834 24	.860 707 98	.45	1.568 312 19	.637 628 15
.16	1.173 510 87	.852 143 79	.46	1.584 073 99	.631 283 65
.17	1.185 304 85	.843 664 82	.47	1.599 994 19	.625 002 27
.18	1.197 217 36	.835 270 21	.48	1.616 074 40	.618 783 39
.19	1.209 249 60	.826 959 13	.49	1.632 316 22	.612 626 39
.20	1.221 402 76	.818 730 75	.50	1.648 721 27	.606 530 66

Table 4 Values of e^j and e^{-j} for Selected Values of j

j	e^j	e^{-j}	j	e^j	e^{-j}
.51	1.665 291 19	.600 495 58	.81	2.247 907 99	.444 858 07
.52	1.682 027 65	.594 520 55	.82	2.270 499 84	.440 431 65
.53	1.698 932 31	.588 604 97	.83	2.293 318 74	.436 049 29
.54	1.716 006 86	.582 748 25	.84	2.316 366 98	.431 710 52
.55	1.733 253 02	.576 949 81	.85	2.339 646 85	.427 414 93
.56	1.750 672 50	.571 209 06	.86	2.363 160 69	.423 162 08
.57	1.768 267 05	.565 525 44	.87	2.386 910 85	.418 951 55
.58	1.786 038 43	.559 898 37	.88	2.410 889 71	.414 782 91
.59	1.803 988 42	.554 327 28	.89	2.435 129 65	.410 655 75
.60	1.822 118 80	.548 811 64	.90	2.459 603 11	.406 569 66
.61	1.840 431 40	.543 350 87	.91	2.484 322 53	.402 524 22
.62	1.858 928 04	.537 944 44	.92	2.509 290 39	.398 519 04
.63	1.877 610 58	.532 591 80	.93	2.534 509 18	.394 553 71
.64	1.896 480 88	.527 292 42	.94	2.559 981 42	.390 627 84
.65	1.915 540 83	.522 045 78	.95	2.585 709 66	.386 741 02
.66	1.934 792 33	.516 851 33	.96	2.611 696 47	.382 892 89
.67	1.954 237 32	.511 708 58	.97	2.637 944 46	.379 083 04
.68	1.973 877 73	.506 616 99	.98	2.664 456 24	.375 311 10
.69	1.993 715 53	.501 576 07	.99	2.691 234 47	.371 576 69
.70	2.013 752 71	.496 585 30	1.00	2.718 281 83	.367 879 44
.71	2.033 991 26	.491 644 20	2.00	7.389 056 10	.135 335 28
.72	2.054 433 21	.486 752 26	3.00	20.085 536 92	.049 787 07
.73	2.075 080 61	.481 908 99	4.00	54.598 150 04	.018 315 64
.74	2.095 935 51	.477 113 92	5.00	148.413 159 11	.006 737 95
.75	2.117 000 02	.472 366 55	6.00	403.428 793 53	.002 478 75
.76	2.138 276 22	.467 666 43	7.00	1,096.633 158 54	.000 911 88
.77	2.159 766 25	.463 013 07	8.00	2,980.957 987 40	.000 335 46
.78	2.181 472 27	.458 406 01	9.00	8,103.083 928 66	.000 123 41
.79	2.203 396 43	.453 844 80	10.00	22,026.465 798 08	.000 045 40
.80	2.225 540 93	.449 328 96			

Table 5 **Binomial Probability Distribution**
 Values of Probability $P(X;n,P) = {}_nC_X \cdot P^X \cdot Q^{n-X}$

						P					
X	.01	.02	.03	.04	.05	.10	.15	.20	.30	.40	.50
						$n = 1$					
0	.9900	.9800	.9700	.9600	.9500	.9000	.8500	.8000	.7000	.6000	.5000
1	.0100	.0200	.0300	.0400	.0500	.1000	.1500	.2000	.3000	.4000	.5000
						$n = 2$					
0	.9801	.9604	.9409	.9216	.9025	.8100	.7225	.6400	.4900	.3600	.2500
1	.0198	.0392	.0582	.0768	.0950	.1800	.2550	.3200	.4200	.4800	.5000
2	.0001	.0004	.0009	.0016	.0025	.0100	.0225	.0400	.0900	.1600	.2500
						$n = 3$					
0	.9704	.9412	.9127	.8847	.8574	.7290	.6141	.5120	.3430	.2160	.1250
1	.0294	.0576	.0847	.1106	.1354	.2430	.3251	.3840	.4410	.4320	.3750
2	.0003	.0012	.0026	.0046	.0071	.0270	.0574	.0960	.1890	.2880	.3750
3	.0000	.0000	.0000	.0001	.0001	.0010	.0034	.0080	.0270	.0640	.1250
						$n = 4$					
0	.9606	.9224	.8853	.8493	.8145	.6561	.5220	.4096	.2401	.1296	.0625
1	.0388	.0753	.1095	.1416	.1715	.2916	.3685	.4096	.4116	.3456	.2500
2	.0006	.0023	.0051	.0088	.0135	.0486	.0975	.1536	.2646	.3456	.3750
3	.0000	.0000	.0001	.0002	.0005	.0036	.0115	.0256	.0756	.1536	.2500
4	.0000	.0000	.0000	.0000	.0000	.0001	.0005	.0016	.0081	.0256	.0625
						$n = 5$					
0	.9510	.9039	.8587	.8154	.7738	.5905	.4437	.3277	.1681	.0778	.0312
1	.0480	.0922	.1328	.1699	.2036	.3280	.3915	.4096	.3602	.2592	.1562
2	.0010	.0038	.0082	.0142	.0214	.0729	.1382	.2048	.3087	.3456	.3125
3	.0000	.0001	.0003	.0006	.0011	.0081	.0244	.0512	.1323	.2304	.3125
4	.0000	.0000	.0000	.0000	.0000	.0004	.0022	.0064	.0284	.0768	.1562
5	.0000	.0000	.0000	.0000	.0000	.0000	.0001	.0003	.0024	.0102	.0312
						$n = 6$					
0	.9415	.8858	.8330	.7828	.7351	.5314	.3771	.2621	.1176	.0467	.0156
1	.0571	.1085	.1546	.1957	.2321	.3543	.3993	.3932	.3025	.1866	.0938
2	.0014	.0055	.0120	.0204	.0305	.0984	.1762	.2458	.3241	.3110	.2344
3	.0000	.0002	.0005	.0011	.0021	.0146	.0415	.0819	.1852	.2765	.3125
4	.0000	.0000	.0000	.0000	.0001	.0012	.0055	.0154	.0595	.1382	.2344
5	.0000	.0000	.0000	.0000	.0000	.0001	.0004	.0015	.0102	.0369	.0938
6	.0000	.0000	.0000	.0000	.0000	.0000	.0000	.0001	.0007	.0041	.0156
						$n = 7$					
0	.9321	.8681	.8080	.7514	.6983	.4783	.3206	.2097	.0824	.0280	.0078
1	.0659	.1240	.1749	.2192	.2573	.3720	.3960	.3670	.2471	.1306	.0547
2	.0020	.0076	.0162	.0274	.0406	.1240	.2097	.2753	.3177	.2613	.1641
3	.0000	.0003	.0008	.0019	.0036	.0230	.0617	.1147	.2269	.2903	.2734
4	.0000	.0000	.0000	.0001	.0002	.0026	.0109	.0287	.0972	.1935	.2734

Table 5 **Binomial Probability Distribution**

X	.01	.02	.03	.04	.05	.10	.15	.20	.30	.40	.50
						P					
						n = 7					
5	.0000	.0000	.0000	.0000	.0009	.0002	.0012	.0043	.0250	.0774	.1641
6	.0000	.0000	.0000	.0000	.0000	.0000	.0001	.0004	.0036	.0172	.0547
7	.0000	.0000	.0000	.0000	.0000	.0000	.0000	.0000	.0002	.0016	.0078
						n = 8					
0	.9227	.8508	.7837	.7214	.6634	.4305	.2725	.1678	.0576	.0168	.0039
1	.0746	.1389	.1939	.2405	.2793	.3826	.3847	.3355	.1977	.0896	.0312
2	.0026	.0099	.0210	.0351	.0515	.1488	.2376	.2936	.2965	.2090	.1094
3	.0001	.0004	.0013	.0029	.0054	.0331	.0839	.1468	.2541	.2787	.2188
4	.0000	.0000	.0001	.0002	.0004	.0046	.0185	.0459	.1361	.2322	.2734
5	.0000	.0000	.0000	.0000	.0000	.0004	.0026	.0092	.0467	.1239	.2188
6	.0000	.0000	.0000	.0000	.0000	.0000	.0002	.0011	.0100	.0413	.1094
7	.0000	.0000	.0000	.0000	.0000	.0000	.0000	.0001	.0012	.0079	.0312
8	.0000	.0000	.0000	.0000	.0000	.0000	.0000	.0000	.0001	.0007	.0039
						n = 9					
0	.9135	.8337	.7602	.6925	.6302	.3874	.2316	.1342	.0404	.0101	.0020
1	.0830	.1531	.2116	.2597	.2985	.3874	.3679	.3020	.1556	.0605	.0176
2	.0034	.0125	.0262	.0433	.0629	.1722	.2597	.3020	.2668	.1612	.0703
3	.0001	.0006	.0019	.0042	.0077	.0446	.1069	.1762	.2668	.2508	.1641
4	.0000	.0000	.0001	.0003	.0006	.0074	.0283	.0661	.1715	.2508	.2461
5	.0000	.0000	.0000	.0000	.0000	.0008	.0050	.0165	.0735	.1672	.2461
6	.0000	.0000	.0000	.0000	.0000	.0001	.0006	.0028	.0210	.0743	.1641
7	.0000	.0000	.0000	.0000	.0000	.0000	.0000	.0003	.0039	.0212	.0703
8	.0000	.0000	.0000	.0000	.0000	.0000	.0000	.0000	.0004	.0035	.0176
9	.0000	.0000	.0000	.0000	.0000	.0000	.0000	.0000	.0000	.0003	.0020
						n = 10					
0	.9044	.8171	.7374	.6648	.5987	.3487	.1969	.1074	.0282	.0060	.0010
1	.0914	.1667	.2281	.2770	.3151	.3874	.3474	.2684	.1211	.0403	.0098
2	.0042	.0153	.0317	.0519	.0746	.1937	.2759	.3020	.2335	.1209	.0439
3	.0001	.0008	.0026	.0058	.0105	.0574	.1298	.2013	.2668	.2150	.1172
4	.0000	.0000	.0001	.0004	.0010	.0112	.0401	.0881	.2001	.2508	.2051
5	.0000	.0000	.0000	.0000	.0001	.0015	.0085	.0264	.1029	.2007	.2461
6	.0000	.0000	.0000	.0000	.0000	.0001	.0012	.0055	.0368	.1115	.2051
7	.0000	.0000	.0000	.0000	.0000	.0000	.0001	.0008	.0090	.0425	.1172
8	.0000	.0000	.0000	.0000	.0000	.0000	.0000	.0001	.0014	.0106	.0439
9	.0000	.0000	.0000	.0000	.0000	.0000	.0000	.0000	.0001	.0016	.0098
10	.0000	.0000	.0000	.0000	.0000	.0000	.0000	.0000	.0000	.0001	.0010
						n = 11					
0	.8953	.8007	.7153	.6382	.5688	.3138	.1673	.0859	.0198	.0036	.0005
1	.0995	.1798	.2433	.2925	.3293	.3835	.3248	.2362	.0932	.0266	.0054
2	.0050	.0183	.0376	.0609	.0867	.2131	.2866	.2953	.1998	.0887	.0269
3	.0002	.0011	.0035	.0076	.0137	.0710	.1517	.2215	.2568	.1774	.0806
4	.0000	.0000	.0002	.0006	.0014	.0158	.0536	.1107	.2201	.2365	.1611

Table 5 Binomial Probability Distribution

X	.01	.02	.03	.04	.05	.10	.15	.20	.30	.40	.50
						$n = 11$					
5	.0000	.0000	.0000	.0000	.0001	.0025	.0132	.0388	.1321	.2207	.2256
6	.0000	.0000	.0000	.0000	.0000	.0003	.0023	.0097	.0566	.1471	.2256
7	.0000	.0000	.0000	.0000	.0000	.0000	.0003	.0017	.0173	.0701	.1611
8	.0000	.0000	.0000	.0000	.0000	.0000	.0000	.0002	.0037	.0234	.0806
9	.0000	.0000	.0000	.0000	.0000	.0000	.0000	.0000	.0005	.0052	.0269
10	.0000	.0000	.0000	.0000	.0000	.0000	.0000	.0000	.0000	.0007	.0054
11	.0000	.0000	.0000	.0000	.0000	.0000	.0000	.0000	.0000	.0000	.0005
						$n = 12$					
0	.8864	.7847	.6938	.6127	.5404	.2824	.1422	.0687	.0138	.0022	.0002
1	.1074	.1922	.2575	.3064	.3413	.3766	.3012	.2062	.0712	.0174	.0029
2	.0060	.0216	.0438	.0702	.0988	.2301	.2924	.2835	.1678	.0639	.0161
3	.0002	.0015	.0045	.0098	.0173	.0852	.1720	.2362	.2397	.1419	.0537
4	.0000	.0001	.0003	.0009	.0021	.0213	.0683	.1329	.2311	.2128	.1208
5	.0000	.0000	.0000	.0001	.0002	.0038	.0193	.0532	.1585	.2270	.1934
6	.0000	.0000	.0000	.0000	.0000	.0005	.0040	.0155	.0792	.1766	.2256
7	.0000	.0000	.0000	.0000	.0000	.0000	.0006	.0033	.0291	.1009	.1934
8	.0000	.0000	.0000	.0000	.0000	.0000	.0000	.0005	.0078	.0420	.1208
9	.0000	.0000	.0000	.0000	.0000	.0000	.0000	.0001	.0015	.0125	.0537
10	.0000	.0000	.0000	.0000	.0000	.0000	.0000	.0000	.0002	.0025	.0161
11	.0000	.0000	.0000	.0000	.0000	.0000	.0000	.0000	.0000	.0003	.0029
12	.0000	.0000	.0000	.0000	.0000	.0000	.0000	.0000	.0000	.0000	.0002
						$n = 13$					
0	.8775	.7690	.6730	.5882	.5133	.2542	.1209	.0550	.0097	.0013	.0001
1	.1152	.2040	.2706	.3186	.3512	.3672	.2774	.1787	.0540	.0113	.0016
2	.0070	.0250	.0502	.0797	.1109	.2448	.2937	.2680	.1388	.0453	.0095
3	.0003	.0019	.0057	.0122	.0214	.0997	.1900	.2457	.2181	.1107	.0349
4	.0000	.0001	.0004	.0013	.0028	.0277	.0838	.1535	.2337	.1845	.0873
5	.0000	.0000	.0000	.0001	.0003	.0055	.0266	.0691	.1803	.2214	.1571
6	.0000	.0000	.0000	.0000	.0000	.0008	.0063	.0230	.1030	.1968	.2095
7	.0000	.0000	.0000	.0000	.0000	.0001	.0011	.0058	.0442	.1312	.2095
8	.0000	.0000	.0000	.0000	.0000	.0000	.0001	.0011	.0142	.0656	.1571
9	.0000	.0000	.0000	.0000	.0000	.0000	.0000	.0001	.0034	.0243	.0873
10	.0000	.0000	.0000	.0000	.0000	.0000	.0000	.0000	.0006	.0065	.0349
11	.0000	.0000	.0000	.0000	.0000	.0000	.0000	.0000	.0001	.0012	.0095
12	.0000	.0000	.0000	.0000	.0000	.0000	.0000	.0000	.0000	.0001	.0016
13	.0000	.0000	.0000	.0000	.0000	.0000	.0000	.0000	.0000	.0000	.0001
						$n = 14$					
0	.8687	.7536	.6528	.5647	.4877	.2288	.1028	.0440	.0068	.0008	.0001
1	.1229	.2153	.2827	.3294	.3593	.3559	.2539	.1539	.0407	.0073	.0009
2	.0081	.0286	.0568	.0892	.1229	.2570	.2912	.2501	.1134	.0317	.0056
3	.0003	.0023	.0070	.0149	.0259	.1142	.2056	.2501	.1943	.0845	.0222
4	.0000	.0001	.0006	.0017	.0037	.0349	.0998	.1720	.2290	.1549	.0611

Table 5 **Binomial Probability Distribution**

						P					
X	.01	.02	.03	.04	.05	.10	.15	.20	.30	.40	.50
n = 15											
5	.0000	.0000	.0000	.0001	.0004	.0078	.0352	.0860	.1963	.2066	.1222
6	.0000	.0000	.0000	.0000	.0000	.0013	.0093	.0322	.1262	.2066	.1833
7	.0000	.0000	.0000	.0000	.0000	.0002	.0019	.0092	.0618	.1574	.2095
8	.0000	.0000	.0000	.0000	.0000	.0000	.0003	.0020	.0232	.0918	.1833
9	.0000	.0000	.0000	.0000	.0000	.0000	.0000	.0003	.0066	.0408	.1222
10	.0000	.0000	.0000	.0000	.0000	.0000	.0000	.0000	.0014	.0136	.0611
11	.0000	.0000	.0000	.0000	.0000	.0000	.0000	.0000	.0002	.0033	.0222
12	.0000	.0000	.0000	.0000	.0000	.0000	.0000	.0000	.0000	.0005	.0056
13	.0000	.0000	.0000	.0000	.0000	.0000	.0000	.0000	.0000	.0001	.0009
14	.0000	.0000	.0000	.0000	.0000	.0000	.0000	.0000	.0000	.0000	.0001
n = 15											
0	.8601	.7386	.6333	.5421	.4633	.2059	.0874	.0352	.0047	.0005	.0000
1	.1303	.2261	.2938	.3388	.3658	.3432	.2312	.1319	.0305	.0047	.0005
2	.0092	.0323	.0636	.0988	.1348	.2669	.2856	.2309	.0916	.0219	.0032
3	.0004	.0029	.0085	.0178	.0307	.1285	.2184	.2501	.1700	.0634	.0139
4	.0000	.0002	.0008	.0022	.0049	.0428	.1156	.1876	.2186	.1268	.0417
5	.0000	.0000	.0001	.0002	.0006	.0105	.0449	.1032	.2061	.1859	.0916
6	.0000	.0000	.0000	.0000	.0000	.0019	.0132	.0430	.1472	.2066	.1527
7	.0000	.0000	.0000	.0000	.0000	.0003	.0030	.0138	.0811	.1771	.1964
8	.0000	.0000	.0000	.0000	.0000	.0000	.0005	.0035	.0348	.1181	.1964
9	.0000	.0000	.0000	.0000	.0000	.0000	.0001	.0007	.0116	.0612	.1527
10	.0000	.0000	.0000	.0000	.0000	.0000	.0000	.0001	.0030	.0245	.0916
11	.0000	.0000	.0000	.0000	.0000	.0000	.0000	.0000	.0006	.0074	.0417
12	.0000	.0000	.0000	.0000	.0000	.0000	.0000	.0000	.0001	.0016	.0139
13	.0000	.0000	.0000	.0000	.0000	.0000	.0000	.0000	.0000	.0003	.0032
14	.0000	.0000	.0000	.0000	.0000	.0000	.0000	.0000	.0000	.0000	.0005
n = 16											
0	.8515	.7238	.6143	.5204	.4401	.1853	.0743	.0281	.0033	.0003	.0000
1	.1376	.2363	.3040	.3469	.3706	.3294	.2097	.1126	.0228	.0030	.0002
2	.0104	.0362	.0705	.1084	.1463	.2745	.2775	.2111	.0732	.0150	.0018
3	.0005	.0034	.0102	.0211	.0359	.1423	.2285	.2463	.1465	.0468	.0085
4	.0000	.0002	.0010	.0029	.0061	.0514	.1311	.2001	.2040	.1014	.0278
5	.0000	.0000	.0001	.0003	.0008	.0137	.0555	.1201	.2099	.1623	.0667
6	.0000	.0000	.0000	.0000	.0001	.0028	.0180	.0550	.1649	.1983	.1222
7	.0000	.0000	.0000	.0000	.0000	.0004	.0045	.0197	.1010	.1889	.1746
8	.0000	.0000	.0000	.0000	.0000	.0001	.0009	.0055	.0487	.1417	.1964
9	.0000	.0000	.0000	.0000	.0000	.0000	.0001	.0012	.0185	.0840	.1746
10	.0000	.0000	.0000	.0000	.0000	.0000	.0000	.0002	.0056	.0392	.1222
11	.0000	.0000	.0000	.0000	.0000	.0000	.0000	.0000	.0013	.0142	.0667
12	.0000	.0000	.0000	.0000	.0000	.0000	.0000	.0000	.0002	.0040	.0278
13	.0000	.0000	.0000	.0000	.0000	.0000	.0000	.0000	.0000	.0008	.0085
14	.0000	.0000	.0000	.0000	.0000	.0000	.0000	.0000	.0000	.0001	.0018
15	.0000	.0000	.0000	.0000	.0000	.0000	.0000	.0000	.0000	.0000	.0002

Table 5 **Binomial Probability Distribution**

X	.01	.02	.03	.04	.05	.10	.15	.20	.30	.40	.50
							P				
						$n = 17$					
0	.8429	.7093	.5958	.4996	.4181	.1668	.0631	.0225	.0023	.0002	.0000
1	.1447	.2461	.3133	.3539	.3741	.3150	.1893	.0957	.0169	.0019	.0001
2	.0117	.0402	.0775	.1180	.1575	.2800	.2673	.1914	.0581	.0102	.0010
3	.0006	.0041	.0120	.0246	.0415	.1556	.2359	.2393	.1245	.0341	.0052
4	.0000	.0003	.0013	.0036	.0076	.0605	.1457	.2093	.1868	.0796	.0182
5	.0000	.0000	.0001	.0004	.0010	.0175	.0668	.1361	.2081	.1379	.0472
6	.0000	.0000	.0000	.0000	.0001	.0039	.0236	.0680	.1784	.1839	.0944
7	.0000	.0000	.0000	.0000	.0000	.0007	.0065	.0267	.1201	.1927	.1484
8	.0000	.0000	.0000	.0000	.0000	.0001	.0014	.0084	.0644	.1606	.1855
9	.0000	.0000	.0000	.0000	.0000	.0000	.0003	.0021	.0276	.1070	.1855
10	.0000	.0000	.0000	.0000	.0000	.0000	.0000	.0004	.0095	.0571	.1484
11	.0000	.0000	.0000	.0000	.0000	.0000	.0000	.0001	.0026	.0242	.0944
12	.0000	.0000	.0000	.0000	.0000	.0000	.0000	.0000	.0006	.0081	.0472
13	.0000	.0000	.0000	.0000	.0000	.0000	.0000	.0000	.0001	.0021	.0182
14	.0000	.0000	.0000	.0000	.0000	.0000	.0000	.0000	.0000	.0004	.0052
15	.0000	.0000	.0000	.0000	.0000	.0000	.0000	.0000	.0000	.0001	.0010
16	.0000	.0000	.0000	.0000	.0000	.0000	.0000	.0000	.0000	.0000	.0001
						$n = 18$					
0	.8345	.6951	.5780	.4796	.3972	.1501	.0536	.0180	.0016	.0001	.0000
1	.1517	.2554	.3217	.3597	.3763	.3002	.1704	.0811	.0126	.0012	.0001
2	.0130	.0443	.0846	.1274	.1683	.2835	.2556	.1723	.0458	.0069	.0006
3	.0007	.0048	.0140	.0283	.0473	.1680	.2406	.2297	.1046	.0246	.0031
4	.0000	.0004	.0016	.0044	.0093	.0700	.1592	.2153	.1681	.0614	.0117
5	.0000	.0000	.0001	.0005	.0014	.0218	.0787	.1507	.2017	.1146	.0327
6	.0000	.0000	.0000	.0000	.0002	.0052	.0301	.0816	.1873	.1655	.0708
7	.0000	.0000	.0000	.0000	.0000	.0010	.0091	.0350	.1376	.1892	.1214
8	.0000	.0000	.0000	.0000	.0000	.0002	.0022	.0120	.0811	.1734	.1669
9	.0000	.0000	.0000	.0000	.0000	.0000	.0004	.0033	.0386	.1284	.1855
10	.0000	.0000	.0000	.0000	.0000	.0000	.0001	.0008	.0149	.0771	.1669
11	.0000	.0000	.0000	.0000	.0000	.0000	.0000	.0001	.0046	.0374	.1214
12	.0000	.0000	.0000	.0000	.0000	.0000	.0000	.0000	.0012	.0145	.0708
13	.0000	.0000	.0000	.0000	.0000	.0000	.0000	.0000	.0002	.0045	.0327
14	.0000	.0000	.0000	.0000	.0000	.0000	.0000	.0000	.0000	.0011	.0117
15	.0000	.0000	.0000	.0000	.0000	.0000	.0000	.0000	.0000	.0002	.0031
16	.0000	.0000	.0000	.0000	.0000	.0000	.0000	.0000	.0000	.0000	.0006
17	.0000	.0000	.0000	.0000	.0000	.0000	.0000	.0000	.0000	.0000	.0001
						$n = 19$					
0	.8262	.6812	.5606	.4604	.3774	.1351	.0456	.0144	.0011	.0001	.0000
1	.1586	.2642	.3294	.3645	.3774	.2852	.1529	.0685	.0093	.0008	.0000
2	.0144	.0485	.0917	.1367	.1787	.2852	.2428	.1540	.0358	.0046	.0003
3	.0008	.0056	.0161	.0323	.0533	.1796	.2428	.2182	.0869	.0175	.0018
4	.0000	.0005	.0020	.0054	.0112	.0798	.1714	.2182	.1491	.0467	.0074

Table 5 **Binomial Probability Distribution**

						P					
X	.01	.02	.03	.04	.05	.10	.15	.20	.30	.40	.50

						n = 19					
5	.0000	.0000	.0002	.0007	.0018	.0266	.0907	.1636	.1916	.0933	.0222
6	.0000	.0000	.0000	.0001	.0002	.0069	.0374	.0955	.1916	.1451	.0518
7	.0000	.0000	.0000	.0000	.0000	.0014	.0122	.0443	.1525	.1797	.0961
8	.0000	.0000	.0000	.0000	.0000	.0002	.0032	.0166	.0981	.1797	.1442
9	.0000	.0000	.0000	.0000	.0000	.0000	.0007	.0051	.0514	.1464	.1762
10	.0000	.0000	.0000	.0000	.0000	.0000	.0001	.0013	.0220	.0976	.1762
11	.0000	.0000	.0000	.0000	.0000	.0000	.0000	.0003	.0077	.0532	.1442
12	.0000	.0000	.0000	.0000	.0000	.0000	.0000	.0000	.0022	.0237	.0961
13	.0000	.0000	.0000	.0000	.0000	.0000	.0000	.0000	.0005	.0085	.0518
14	.0000	.0000	.0000	.0000	.0000	.0000	.0000	.0000	.0001	.0024	.0222
15	.0000	.0000	.0000	.0000	.0000	.0000	.0000	.0000	.0000	.0005	.0074
16	.0000	.0000	.0000	.0000	.0000	.0000	.0000	.0000	.0000	.0001	.0018
17	.0000	.0000	.0000	.0000	.0000	.0000	.0000	.0000	.0000	.0000	.0003

						n = 20					
0	.8179	.6676	.5438	.4420	.3585	.1216	.0388	.0115	.0008	.0000	.0000
1	.1652	.2725	.3364	.3683	.3774	.2702	.1368	.0576	.0068	.0005	.0000
2	.0159	.0528	.0988	.1458	.1887	.2852	.2293	.1369	.0278	.0031	.0002
3	.0010	.0065	.0183	.0364	.0596	.1901	.2428	.2054	.0716	.0123	.0011
4	.0000	.0006	.0024	.0065	.0133	.0898	.1821	.2182	.1304	.0350	.0046
5	.0000	.0000	.0002	.0009	.0022	.0319	.1028	.1746	.1789	.0746	.0148
6	.0000	.0000	.0000	.0001	.0003	.0089	.0454	.1091	.1916	.1244	.0370
7	.0000	.0000	.0000	.0000	.0000	.0020	.0160	.0545	.1643	.1659	.0739
8	.0000	.0000	.0000	.0000	.0000	.0004	.0046	.0222	.1144	.1797	.1201
9	.0000	.0000	.0000	.0000	.0000	.0001	.0011	.0074	.0654	.1597	.1602
10	.0000	.0000	.0000	.0000	.0000	.0000	.0002	.0020	.0308	.1171	.1762
11	.0000	.0000	.0000	.0000	.0000	.0000	.0000	.0005	.0120	.0710	.1602
12	.0000	.0000	.0000	.0000	.0000	.0000	.0000	.0001	.0039	.0355	.1201
13	.0000	.0000	.0000	.0000	.0000	.0000	.0000	.0000	.0010	.0146	.0739
14	.0000	.0000	.0000	.0000	.0000	.0000	.0000	.0000	.0002	.0049	.0370
15	.0000	.0000	.0000	.0000	.0000	.0000	.0000	.0000	.0000	.0013	.0148
16	.0000	.0000	.0000	.0000	.0000	.0000	.0000	.0000	.0000	.0003	.0046
17	.0000	.0000	.0000	.0000	.0000	.0000	.0000	.0000	.0000	.0000	.0011
18	.0000	.0000	.0000	.0000	.0000	.0000	.0000	.0000	.0000	.0000	.0002

						n = 25					
0	.7778	.6035	.4670	.3604	.2774	.0718	.0172	.0038	.0001	.0000	.0000
1	.1964	.3079	.3611	.3754	.3650	.1994	.0759	.0236	.0014	.0000	.0000
2	.0238	.0754	.1340	.1877	.2305	.2659	.1607	.0708	.0074	.0004	.0000
3	.0018	.0118	.0318	.0600	.0930	.2265	.2174	.1358	.0243	.0019	.0001
4	.0001	.0013	.0054	.0137	.0269	.1384	.2110	.1867	.0572	.0071	.0004
5	.0000	.0001	.0007	.0024	.0060	.0646	.1564	.1960	.1030	.0199	.0016
6	.0000	.0000	.0001	.0003	.0010	.0239	.0920	.1633	.1472	.0442	.0053
7	.0000	.0000	.0000	.0000	.0001	.0072	.0441	.1108	.1712	.0800	.0143
8	.0000	.0000	.0000	.0000	.0000	.0018	.0175	.0623	.1651	.1200	.0322
9	.0000	.0000	.0000	.0000	.0000	.0004	.0058	.0294	.1336	.1511	.0609

Table 5 **Binomial Probability Distribution**

						P					
X	.01	.02	.03	.04	.05	.10	.15	.20	.30	.40	.50
						$n = 25$					
10	.0000	.0000	.0000	.0000	.0000	.0001	.0016	.0118	.0916	.1612	.0974
11	.0000	.0000	.0000	.0000	.0000	.0000	.0004	.0040	.0536	.1465	.1328
12	.0000	.0000	.0000	.0000	.0000	.0000	.0001	.0012	.0268	.1140	.1550
13	.0000	.0000	.0000	.0000	.0000	.0000	.0000	.0003	.0115	.0760	.1550
14	.0000	.0000	.0000	.0000	.0000	.0000	.0000	.0001	.0042	.0434	.1328
15	.0000	.0000	.0000	.0000	.0000	.0000	.0000	.0000	.0013	.0212	.0974
16	.0000	.0000	.0000	.0000	.0000	.0000	.0000	.0000	.0004	.0088	.0609
17	.0000	.0000	.0000	.0000	.0000	.0000	.0000	.0000	.0001	.0031	.0322
18	.0000	.0000	.0000	.0000	.0000	.0000	.0000	.0000	.0000	.0009	.0143
19	.0000	.0000	.0000	.0000	.0000	.0000	.0000	.0000	.0000	.0002	.0053
20	.0000	.0000	.0000	.0000	.0000	.0000	.0000	.0000	.0000	.0000	.0016
21	.0000	.0000	.0000	.0000	.0000	.0000	.0000	.0000	.0000	.0000	.0004
22	.0000	.0000	.0000	.0000	.0000	.0000	.0000	.0000	.0000	.0000	.0001
						$n = 30$					
0	.7397	.5455	.4010	.2939	.2146	.0424	.0076	.0012	.0000	.0000	.0000
1	.2242	.3340	.3721	.3673	.3389	.1413	.0404	.0093	.0003	.0000	.0000
2	.0328	.0988	.1669	.2219	.2586	.2277	.1034	.0337	.0018	.0000	.0000
3	.0031	.0188	.0482	.0863	.1270	.2361	.1703	.0785	.0072	.0003	.0000
4	.0002	.0026	.0101	.0243	.0451	.1771	.2028	.1325	.0208	.0012	.0000
5	.0000	.0003	.0016	.0053	.0124	.1023	.1861	.1723	.0464	.0041	.0001
6	.0000	.0000	.0002	.0009	.0027	.0474	.1368	.1795	.0829	.0115	.0006
7	.0000	.0000	.0000	.0001	.0005	.0180	.0828	.1538	.1219	.0263	.0019
8	.0000	.0000	.0000	.0000	.0001	.0058	.0420	.1106	.1501	.0505	.0055
9	.0000	.0000	.0000	.0000	.0000	.0016	.0181	.0676	.1573	.0823	.0133
10	.0000	.0000	.0000	.0000	.0000	.0004	.0067	.0355	.1416	.1152	.0280
11	.0000	.0000	.0000	.0000	.0000	.0001	.0022	.0161	.1103	.1396	.0509
12	.0000	.0000	.0000	.0000	.0000	.0000	.0006	.0064	.0749	.1474	.0806
13	.0000	.0000	.0000	.0000	.0000	.0000	.0001	.0022	.0444	.1360	.1115
14	.0000	.0000	.0000	.0000	.0000	.0000	.0000	.0007	.0231	.1101	.1354
15	.0000	.0000	.0000	.0000	.0000	.0000	.0000	.0002	.0106	.0783	.1445
16	.0000	.0000	.0000	.0000	.0000	.0000	.0000	.0000	.0042	.0489	.1354
17	.0000	.0000	.0000	.0000	.0000	.0000	.0000	.0000	.0015	.0269	.1115
18	.0000	.0000	.0000	.0000	.0000	.0000	.0000	.0000	.0005	.0129	.0806
19	.0000	.0000	.0000	.0000	.0000	.0000	.0000	.0000	.0001	.0054	.0509
20	.0000	.0000	.0000	.0000	.0000	.0000	.0000	.0000	.0000	.0020	.0280
21	.0000	.0000	.0000	.0000	.0000	.0000	.0000	.0000	.0000	.0006	.0133
22	.0000	.0000	.0000	.0000	.0000	.0000	.0000	.0000	.0000	.0002	.0055
23	.0000	.0000	.0000	.0000	.0000	.0000	.0000	.0000	.0000	.0000	.0019
24	.0000	.0000	.0000	.0000	.0000	.0000	.0000	.0000	.0000	.0000	.0006
25	.0000	.0000	.0000	.0000	.0000	.0000	.0000	.0000	.0000	.0000	.0001

Table 6 **Areas Under the Normal Curve**
Values of $A(z)$ Between Ordinate at Mean (Y_o) and Ordinate at z

$z\left(=\frac{x}{o}\right)$	.00	.01	.02	.03	.04	.05	.06	.07	.08	.09
0.0	.00000	.00399	.00798	.01197	.01595	.01994	.02392	.02790	.03188	.03586
0.1	.03983	.04380	.04776	.05172	.05567	.05962	.06356	.06749	.07142	.07535
0.2	.07926	.08317	.08706	.09095	.09483	.09871	.10257	.10642	.11026	.11409
0.3	.11791	.12172	.12552	.12930	.13307	.13683	.14058	.14431	.14803	.15173
0.4	.15542	.15910	.16276	.16640	.17003	.17364	.17724	.18082	.18439	.18793
0.5	.19146	.19497	.19847	.20194	.20540	.20884	.21226	.21566	.21904	.22240
0.6	.22575	.22907	.23237	.23565	.23891	.24215	.24537	.24857	.25175	.25490
0.7	.25804	.26115	.26424	.26730	.27035	.27337	.27637	.27935	.28230	.28524
0.8	.28814	.29103	.29389	.29673	.29955	.30234	.30511	.30785	.31057	.31327
0.9	.31594	.31859	.32121	.32381	.32639	.32894	.33147	.33398	.33646	.33891
1.0	.34134	.34375	.34614	.34850	.35083	.35314	.35543	.35769	.35993	.36214
1.1	.36433	.36650	.36864	.37076	.37286	.37493	.37698	.37900	.38100	.38298
1.2	.38493	.38686	.38877	.39065	.39251	.39435	.39617	.39796	.39973	.40147
1.3	.40320	.40490	.40658	.40824	.40988	.41149	.41309	.41466	.41621	.41774
1.4	.41924	.42073	.42220	.42364	.42507	.42647	.42786	.42922	.43056	.43189
1.5	.43319	.43448	.43574	.43699	.43822	.43943	.44062	.44179	.44295	.44408
1.6	.44520	.44630	.44738	.44845	.44950	.45053	.45154	.45254	.45352	.45449
1.7	.45543	.45637	.45728	.45818	.45907	.45994	.46080	.46164	.46246	.46327
1.8	.46407	.46485	.46562	.46638	.46712	.46784	.46856	.46926	.46995	.47062
1.9	.47128	.47193	.47257	.47320	.47381	.47441	.47500	.47558	.47615	.47670
2.0	.47725	.47778	.47831	.47882	.47932	.47982	.48030	.48077	.48124	.48169
2.1	.48214	.48257	.48300	.48341	.48382	.48422	.48461	.48500	.48537	.48574
2.2	.48610	.48645	.48679	.48713	.48745	.48778	.48809	.48840	.48870	.48899
2.3	.48928	.48956	.48983	.49010	.49036	.49061	.49086	.49111	.49134	.49158
2.4	.49180	.49202	.49224	.49245	.49266	.49286	.49305	.49324	.49343	.49361
2.5	.49379	.49396	.49413	.49430	.49446	.49461	.49477	.49492	.49506	.49520
2.6	.49534	.49547	.49560	.49573	.49585	.49598	.49609	.49621	.49632	.49643
2.7	.49653	.49664	.49674	.49683	.49693	.49702	.49711	.49720	.49728	.49736
2.8	.49744	.49752	.49760	.49767	.49774	.49781	.49788	.49795	.49801	.49807
2.9	.49813	.49819	.49825	.49831	.49386	.49841	.49846	.49851	.49856	.49861
3.0	.49865	.49869	.49874	.49878	.49882	.49886	.49889	.49893	.49897	.49900
3.1	.49903	.49906	.49910	.49913	.49916	.49918	.49921	.49924	.49926	.49929
3.2	.49931	.49934	.49936	.49938	.49940	.49942	.49944	.49946	.49948	.49950
3.3	.49952	.49953	.49955	.49957	.49958	.49960	.49961	.49962	.49964	.49965
3.4	.49966	.49968	.49969	.49970	.49971	.49972	.49973	.49974	.49975	.49976
3.5	.49977	.49978	.49978	.49979	.49980	.49981	.49981	.49982	.49983	.49983
3.6	.49984	.49985	.49985	.49986	.49986	.49987	.49987	.49988	.49988	.49989
3.7	.49989	.49990	.49990	.49990	.49991	.49991	.49992	.49992	.49992	.49992
3.8	.49993	.49993	.49993	.49994	.49994	.49994	.49994	.49995	.49995	.49995
3.9	.49995	.49995	.49996	.49996	.49996	.49996	.49996	.49996	.49997	.49997
4.0	.49997									

Answers to Odd-Numbered Problems

Exercise 1–1 (p. 5)

1. (a) Finite   (c) Empty   (e) Finite
    (b) Infinite   (d) Finite

3.

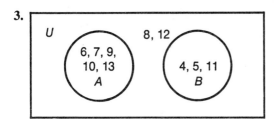

5. (a) $A = \{a, b, c, d, e, f, g, h\}$
      $A = \{x | x$ is a letter and is one of the first eight letters of the alphabet$\}$
   (b) $A = \{3, 4, 5, 6, 7, 8, 9\}$
      $A = \{x | x$ is an integer and $3 \leq x \leq 9\}$
   (c) $A = \{$Carson, Estes, Frenton, Nixton$\}$
      $A = \{x | x$ is a partner of Tex Oil Company$\}$
   (d) $A = \{$Begg, Griffin, Smith, Meyers, Taby$\}$
      $A = \{x | x$ is a professor at King College$\}$
7. (a) $b \in S$ (c) $Q = \phi$ (e) $H \subset M$
   (b) $h \notin S$ (d) $P \subseteq G$

Exercise 1–2 (p. 9)

1. (a) $A \cup B = \{$Anderson, Bell, Darr, Griffin, Kello$\}$
   (b) $A \cap B = \{$Darr, Kello$\}$

3.

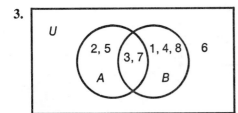

(a) $A \cup B = \{1, 2, 3, 4, 5, 7, 8\}$
(b) $A \cap B = \{3, 7\}$
(c) $A' \cup B = \{1, 3, 4, 6, 7, 8\}$
(d) $A' \cap B' = \{6\}$

497

5.

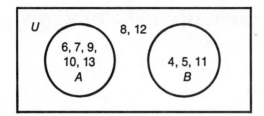

(a) $A \cup B = \{4, 5, 6, 7, 9, 10, 11, 13\}$
(b) $A \cap B = \phi$
(c) $A \cup B' = \{6, 7, 8, 9, 10, 12, 13\} = B'$
(d) $A \cap B' = \{6, 7, 9, 10, 13\} = A$

7.

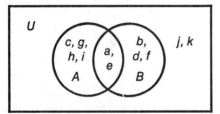

(a) $A \cup B = \{a, b, c, d, e, f, g, h, i\}$ (c) $A' = \{b, d, f, j, k\}$
(b) $A \cap B = \{a, e\}$ (d) $A \cup A' = U$

Exercise 1–3 (p. 16)

1.

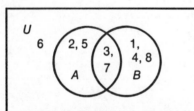

(a) $A - B = \{2, 5\}$
(b) $B - A = \{1, 4, 8\}$
(c) $A \cap B' = \{2, 5\}$
(d) $B \cap A' = \{1, 4, 8\}$

3.

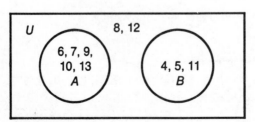

(a) $A - B = A$
(b) $A \cap B' = A$

5.

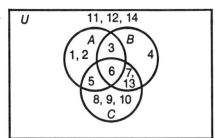

U   11, 12, 14

A   B

1, 2   3   4

6   7, 13

5

8, 9, 10

C

(a) $A \cup B = \{1, 2, 3, 4, 5, 6, 7, 13\}$
(b) $B \cap C = \{6, 7, 13\}$
(c) $A \cap B = \{3, 6\}$
(d) $A - C = \{1, 2, 3\}$
(e) $A \cap B \cap C = \{6\}$
(f) $(A \cup B) \cap C = \{5, 6, 7, 13\}$
(g) $A' \cap C \cap B' = \{8, 9, 10\}$
(h) $A \cap B \cap C' = \{3\}$

7. (a) $(P \cap Q') \cup R$ (//////)

(b) $(P - Q) - R$ (\\\\\\\)

(c) $(Q - R) \cap P'$ (≡≡≡)

(d) $(R - Q) - P$ (||||||||)

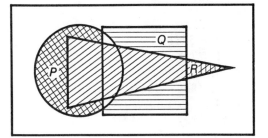

Q

P

R

9. (a) $C \cap (A \cup B) = (C \cap A) \cup (C \cap B) = \{5, 6\} \cup \{6, 7, 13\}$
$= \{5, 6, 7, 13\}$
(b) $B \cup (A \cap C) = (B \cup A) \cap (B \cup C) = \{1, 2, 3, 4, 5, 6, 7, 13\} \cap$
$\{3, 4, 5, 6, 7, 8, 9, 10, 13\}$
$= \{3, 4, 5, 6, 7, 13\}$
(c) $A \cap (B \cup C) = (A \cap B) \cup (A \cap C) = \{3, 6\} \cup \{5, 6\} = \{3, 5, 6\}$
(d) $C \cup (A \cap B) = (C \cup A) \cap (C \cup B) = \{1, 2, 3, 5, 6, 7, 8, 9, 10, 13\} \cap$
$\{3, 4, 5, 6, 7, 8, 9, 10, 13\}$
$= \{3, 5, 6, 7, 8, 9, 10, 13\}$

11. (a) $B \cup (B \cup A) = (B \cup B) \cup A = B \cup A$
(b) $A \cup (A' \cap B) = (A \cup A') \cap (A \cup B) = U \cap (A \cup B) = A \cup B$

13. Let $P = $ Set $(A \cup B)$.

$(A \cup B) \cap (A' \cup C) = P \cap (A' \cup C)$
$= (P \cap A') \cup (P \cap C)$
$= [(A \cup B) \cap A'] \cup [(A \cup B) \cap C]$
$= [A' \cap (A \cup B)] \cup [C \cap (A \cup B)]$
$= [(A' \cap A) \cup (A' \cap B)] \cup [(C \cap A) \cup (C \cap B)]$
$= \phi \cup (A' \cap B) \cup (C \cap A) \cup (C \cap B)$

Check:

$(A \cup B) \cap (A' \cup C) = \{1, 2, 3, 4, 5, 6\} \cap \{3, 4, 5, 6, 7, 8, 9\}$
$= \{3, 4, 5, 6\}$

$(A' \cap A) \cup (A' \cap B) \cup (C \cap A) \cup (C \cap B)$
$= \phi \cup \{5, 6\} \cup \{3, 4\} \cup \{4, 6\} = \{3, 4, 5, 6\}$

Exercise 2–1 (p. 24)

1. -8
3. -20.984
5. -24.766
7. 1.537
9. 7.44

11. 2.546
13. -4.848
15. 5.306
17. 7.56
19. -29.52

21. -576
23. 97.60
25. -7
27. 7.1
29. -4

31. 2
33. 36
35. -1.97
37. 20.536
39. -38.63

Exercise 2–2 (p. 31)

1. $39x$
3. $-38ab$
5. $27g$
7. $-20x + 26y$
9. $13xy - 11ab$
11. $7a^2b + 11cd^2 - 10abc$
13. $13xy$
15. $-33ab$
17. $9e$
19. $2x - 4y$
21. $12fg + 58hi$
23. $-4xy^2 - 9x^2y^3 - 7xy$
25. $3^6 = 729$
27. a^5
29. c^3
31. $56^2 = 3,136$
33. $5^6 = 15,625$
35. $6^{6/5}$
37. $-20x^3y^2z^3$

39. $-12x^2 + 15xy - 18x$
41. $6x^3 + 15x^2 - 8x - 20$
43. 5
45. 1

47. $\dfrac{y}{x^3}$

49. $\left(\dfrac{1}{2}\right)^2 = \dfrac{1}{4}$

51. $\dfrac{-9yz^2}{w^2}$

53. $\dfrac{8b}{5a} - 3ac^5 + 4b^2$

55. $4x - 5$

57. $(2x - 1) + \dfrac{-19x + 42}{5x^2 - 5x + 7}$

Exercise 2–3 (p. 36)

1. $7\,(a + b)$
3. $5\,(x + 2)$
5. $6\,(3x + y - 2z)$
7. $(a + 3b)(2x - y)$
9. $(c + x)(d + y)$
11. $(2x + y)(2x - y)$

13. $(5b + 2m)(5b - 2m)$
15. $(2x + y)^2$
17. $(5b - 2m)^2$
19. $(2x + 3)(4x + 5)$
21. $(5x - 2)(3x + 4)$
23. $(2x + 9)(5x - 8)$

Exercise 2–4 (p. 42)

1. $\dfrac{11}{8} = 1\dfrac{3}{8}$

3. $\dfrac{71a}{70}$

5. $9\dfrac{23}{42}x$

7. $\dfrac{13x + 8y}{x^2 - y^2}$

9. $\dfrac{1}{18}$

11. $27\dfrac{49}{60}x$

13. $\dfrac{-x}{21}$

15. $\dfrac{b}{a+b}$

17. $\dfrac{-5(3x-4)-(-20x+15)}{(x+1)(3x-4)} = \dfrac{5x+5}{(x+1)(3x-4)}$

$$= \dfrac{5}{3x-4}$$

19. $\dfrac{48}{161}$

21. $\dfrac{4x^4y}{9}$

23. $\dfrac{5}{9}$

25. $\dfrac{2}{3}$

27. $\dfrac{2(2x+5)}{(x-3)(2x+5)} \cdot \dfrac{x-3}{2(x+2)} = \dfrac{1}{x+2}$

29. $\dfrac{20y^4}{9x^3}$

31. $\dfrac{221}{2x}$

33. $\dfrac{x+1}{3x-2}$

35. $\dfrac{xy^2+3}{3+x}$

Exercise 3–1 (p. 47)

1. $x=3$

3. $x=-2$

5. $x=-3$

7. $x=4$

9. $x=\dfrac{2}{7}$

11. $x=\dfrac{4-3a}{2}$

13. $x=\dfrac{6b-5}{7}$

15. $x=\dfrac{6-8a}{-19}$

17. Let $x=$ Peter's money.

$$
\begin{aligned}
x + (x - 131.50) &= 181.00 \\
2x &= 312.50 \\
x &= \$156.25
\end{aligned}
$$

19. Let $x=$ the number.

$$
\begin{aligned}
x + 8x &= 63 \\
x &= 7
\end{aligned}
$$

21. Let $x =$ Steve's age today.

$$x - 10 = 2(32 - x - 10)$$
$$x - 10 = 44 - 2x$$
$$3x = 54$$
$$x = 18$$

23. Let $x =$ number of pounds of the $1.20 a pound grade to be used in the mixture.

$$1.20x + 1.80(30 - x) = 1.40(30)$$
$$1.20x - 1.80x + 54.00 = 42.00$$
$$-.60x = -12.00$$
$$x = 20$$
$$30 - 20 = 10$$

Answer: 20 pounds – $1.20 grade
          10 pounds – $1.80 grade

25. Let $x =$ hours needed for the second car to catch up with the first car.

$$55x = 45x + 45\left(\frac{40}{60}\right)$$
$$10x = 30$$
$$x = 3 \text{ hours}$$

Cars have traveled: $55(3) = 165$ miles

27. Let $x =$ number of quarters. Then,
    $2x =$ number of dimes, and
$3(2x) = 6x =$ number of half-dollars.

$$x(.25) + 2x(.10) + 6x(.50) = 34.50$$
$$.25x + .20x + 3.00x = 34.50$$
$$3.45x = 34.50$$
$$x = 10 \text{ quarters}$$
$$2x = 20 \text{ dimes}$$
$$6x = 60 \text{ half-dollars}$$

29. Let $x =$ Peter's share of profit. Then, $(7/10)x =$ John's share, and $(3/5)(7/10)x = (21/50)x =$ Larry's share.

$$x + \frac{7}{10}x + \frac{21}{50}x = 10,600$$

$$\frac{50x + 35x + 21x}{50} = 10,600$$

$$\frac{106x}{50} = 10,600$$

$$x = \$5,000 \text{ (Peter's share)}$$

$$\frac{7}{10}x = \$3,500 \text{ (John's share)}$$

$$\frac{21}{50}x = \$2,100 \text{ (Larry's share)}$$

Exercise 3–2 (p. 51)

1. $x = 3, y = 2$
3. $x = 2, y = -1$
5. $x = -1, y = -3$
7. $x = 3, y = -2$
9. $x = 1, y = 5$

11. $x = 3, y = 1$
13. $x = 4, y = -1$
15. $x = 2, y = 3$
17. $x = 1, y = 2, z = 3$
19. $x = -1, y = 3, z = 2$

21. Let $x =$ price per pound of beef,
    $y =$ price per can of coffee.

$$2x + 7y = 18.81 \qquad\qquad x = \$2.58 \text{ per pound}$$
$$1x + 5y = 12.33 \qquad\qquad y = \$1.95 \text{ per can}$$

23. Let $x =$ the larger number,
    $y =$ the smaller number.

$$x + y = 235 \qquad\qquad x = 150$$
$$\frac{1}{3}x + \frac{1}{17}y = 55 \qquad\qquad y = \ \ 85$$

25. Let $x =$ number of tickets sold to adults,
    $y =$ number of tickets sold to children.

$$x + y = 242 \qquad\qquad x = 180 \text{ tickets}$$
$$5.40x + 2.00y = 1,096.00 \qquad\qquad y = \ \ 62 \text{ tickets}$$

27. Let $x =$ Peter's money,
    $y =$ Nancy's money.

$$x + y = 181.00 \qquad\qquad x = \$156.25$$
$$x - y = 131.50 \qquad\qquad y = \$ \ \ 24.75$$

29. Let $x =$ Steve's age today,
    $y =$ Alan's age today.

$$x + y = 32 \qquad\qquad x = 18 \text{ years}$$
$$x - 10 = 2(y - 10) \qquad\qquad y = 14 \text{ years}$$

31. Let $x =$ number of pounds at \$1.20 per pound,
    $y =$ number of pounds at \$1.80 per pound.

$$x + y = 30 \qquad\qquad x = 20 \text{ pounds}$$
$$1.20x + 1.80y = 1.40(30) \qquad\qquad y = 10 \text{ pounds}$$

Exercise 3–3 (p. 57)

1. $x = 9(x - 40)$
   $8x = 360$
   $x = 45$

3. $\dfrac{x + 6x}{8} = 14$
   $7x = 112$
   $x = 16$

5. $(x + 1)(x + 3) = (x + 9)(x - 2)$
   $x^2 + 4x + 3 = x^2 + 7x - 18$
   $-3x = -21$
   $x = 7$

7. $(x + a)(x - 3) = (x + 2a)(x - 1)$
   $x^2 + ax - 3x - 3a = x^2 + 2ax - x - 2a$
   $- ax - 2x = a$
   $x = \dfrac{a}{-a - 2}$

9. $15x = 4(20)$
   $x = \dfrac{80}{15} = 5\frac{1}{3}$

11. $8x = 12(\frac{1}{3})$
    $x = \frac{1}{2}$

13. $\dfrac{\$140}{40} = \dfrac{x}{25}$
    $40x = 25(140)$
    $x = \$87.50$

15. $\dfrac{375}{1\frac{1}{2}} = \dfrac{x}{4\frac{1}{2}}$
    $1.5x = 4.5(375)$
    $x = 1,125$ miles

17. $\dfrac{630}{35} = \dfrac{x}{42}$
    $35x = 42(630)$
    $x = 756$ toys

19. $\$350(18\%) = \$63$

21. $\dfrac{145}{580} = .25$ or 25%

23. $\dfrac{84}{24\%} = 350$

25. $\$50(1 + 35\%) = \$67.50$

27. $\$465 - \$465(x) = \$279$
    $1 - x = 279/465$
    $x = .40$ or 40%

29. $\dfrac{\$110}{1 - 12\%} = \$125$

Exercise 3–4 (p. 61)

1. $x = 3, y = 2$
3. $x = 2, y = -1$
5. $x = -1, y = -3$

7. $x = 3, y = -2$
9. $x = 1, y = 5$
11. $x = 3, y = 1$

1.

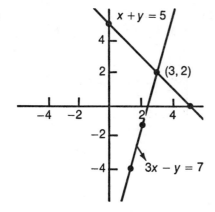

7.

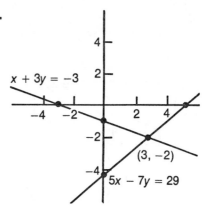

3.

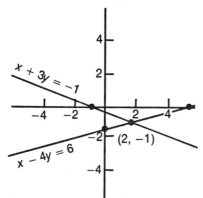

9.

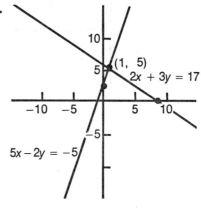

5.

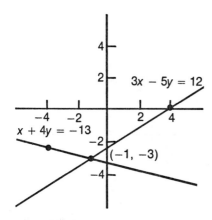

11.

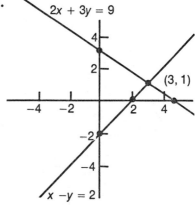

Exercise 3–5 (p. 64)

1. $f(2) = 10(2) = 20$

3. $f(4) = 7 + 3(4) = 19$

5. $f(-1) = \dfrac{1}{2(-1) + 3} = 1$

7. $g(-3) = 5(-3) + 14 = -1$

9. $f(y) = 4(4x + 5) + 5$
$= 16x + 25$

11. $f(x + 5) - f(x)$
$= (2(x + 5) - 3) - (2x - 3)$
$= 10$

13. $4x - 12 = 0$
$y = 4x - 12$
$y = 0, x = 3$

x	y = 4x − 12
0	−12
3	0
5	8

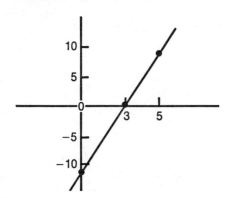

17. $x + 3 = 0$
$y = x + 3$
$y = 0, x = -3$

x	y = x + 3
3	6
0	3
−3	0

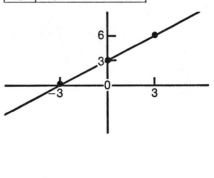

15. $3x + 6 = 0$
$y = 3x + 6$
$y = 0, x = -2$

x	y = 3x + 6
2	12
0	6
−2	0

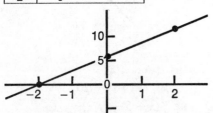

19. $9x - 36 = 0$
$y = 9x - 36$
$y = 0, x = 4$

x	y = 9x − 36
4	0
2	−18
0	−36

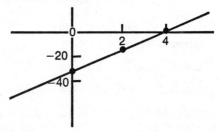

21. $7x - 2 = 0$
            $y = 7x - 2$
            $y = 0, x = {}^2/_7$

x	$y = 7x - 2$
1	5
${}^2/_7$	0
0	-2

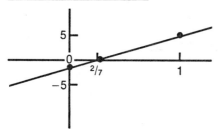

Exercise 4–1 (p. 70)

1. $\sqrt[4]{a}$

3. $\sqrt[5]{c^2}$

5. $\sqrt[3]{43}$

7. $\sqrt[8]{(26x)^3}$

9. $e^{2/3}$

11. $g^{5/6}$

13. $(3x)^{1/2}$

15. $(34xy^3)^{1/5}$

17. $\sqrt{25(3)} = \sqrt{75}$

19. $\sqrt{2x^2y}$

21. $\sqrt{405}$

23. $\sqrt[5]{a^{10}b^5(4ab)} = \sqrt[5]{4a^{11}b^6}$

25. $6\sqrt{3}$

27. $5\sqrt{2} + 7\sqrt{2} - 3\sqrt{2} = 9\sqrt{2}$

29. $10\sqrt{5a} + 3\sqrt{5b}$

31. $15\sqrt{8a^2b} = 30\sqrt{2a^2b}$

33. $2(3^{1/2})(4^{1/3}) = 2(3^{3/6})(4^{2/6}) = 2(3^3 \cdot 4^2)^{1/6} = 2(\sqrt[6]{432})$

35. $\dfrac{5}{3} = 1\dfrac{2}{3}$

37. $\dfrac{\sqrt{15}}{\sqrt{7}} \cdot \dfrac{\sqrt{7}}{\sqrt{7}} = \dfrac{\sqrt{105}}{7}$

39. $\dfrac{3^{1/2}}{2^{1/3}} = \dfrac{3^{3/6}}{2^{2/6}} \cdot \dfrac{2^{4/6}}{2^{4/6}} = \dfrac{(3^3 \cdot 2^4)^{1/6}}{2} = \dfrac{\sqrt[6]{432}}{2}$

Exercise 4–2 (p. 74)

1. 25

3. 132

5. 23.5

7. 14.26

9. $x + 4 = 9$
        $x = 5$

11. $x - 6 = x^2 - 16x + 64$
        $x^2 - 17x + 70 = 0$
        $(x - 7)(x - 10) = 0$

$x - 7 = 0, x = 7$
$x - 10 = 0, x = 10$
10 is the correct root.

13. $\sqrt{x + 1} = 5 - \sqrt{x - 4}$
            $x + 1 = 25 - 10\sqrt{x - 4} + (x - 4)$
        $10\sqrt{x - 4} = 20$
            $x - 4 = 4$
            $x = 8$

15. $\sqrt{2x-3}=1+\sqrt{x-2}$
$\quad 2x-3=1+2\sqrt{x-2}+(x-2)$
$\quad 2\sqrt{x-2}=x-2$
$\quad 4(x-2)=x^2-4x+4$
$\quad x^2-8x+12=0$
$\quad (x-6)(x-2)=0$

$\quad x-6=0,\ x=6$
$\quad x-2=0,\ x=2$

6 and 2 are the correct roots.

17. $(3-x)+2\sqrt{3-x}\cdot\sqrt{x+2}+(x+2)=2x+5$
$\quad 2\sqrt{(3-x)(x+2)}=2x$
$\quad\quad 6+x-x^2=x^2$
$\quad\quad 2x^2-x-6=0$
$\quad (x-2)(2x+3)=0$

$\quad x-2=0,\ x=2$
$\quad 2x+3=0,\ x=-\frac{3}{2}$
$\quad$ 2 is the correct root.

Exercise 4–3 (p. 80)

1. $3(x+4)(x-4)=0$
$\quad x+4=0,\ x=-4$
$\quad x-4=0,\ x=4$

3. $5(x+2)(x-2)=0$
$\quad x+2=0,\ x=-2$
$\quad x-2=0,\ x=2$

5. $(x+2)(2x+3)=0$
$\quad x+2=0,\ x=-2$
$\quad 2x+3=0,\ x=-\frac{3}{2}$

7. $(2x-1)(x+3)=0$
$\quad 2x-1=0,\ x=\frac{1}{2}$
$\quad x+3=0,\ x=-3$

9. $(x+2)(-3x+5)=0$
$\quad x+2=0,\ x=-2$
$\quad -3x+5=0,\ x=\frac{5}{3}$

11.

$a=2, b=1, c=-6$

$$x=\frac{-1\pm\sqrt{1^2-4(2)\,(-6)}}{2(2)}$$

$$=\frac{-1\pm 7}{4}$$

$$x=\frac{-1+7}{4}=3/2$$

$$x=\frac{-1-7}{4}=-2$$

$y=2x^2+x-6$

x	−3	−2	−1	0	1	2	3
y	9	0	−5	−6	−3	4	15

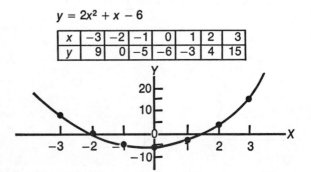

13. $a = 4, b = 8, c = 3$

$$x = \frac{-8 \pm \sqrt{8^2 - 4(4)(3)}}{2(4)}$$

$$= \frac{-8 \pm 4}{8}$$

$$x = \frac{-8 + 4}{8} = -\frac{1}{2}$$

$$x = \frac{-8 - 4}{8} = -1\frac{1}{2}$$

$y = 4x^2 + 8x + 3$

x	-3	-2	-1	0	1	2	3
y	15	3	-1	3	15	35	63

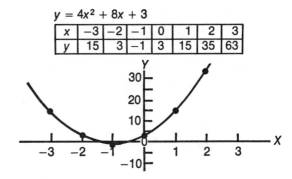

15. $a = 1, b = -6, c = 9$

$$x = \frac{-(-6) \pm \sqrt{(-6)^2 - 4(1)(9)}}{2(1)}$$

$$= \frac{6 \pm 0}{2} = 3$$

$y = x^2 - 6x + 9$

x	-1	0	1	2	3	4	5
y	16	9	4	1	0	1	4

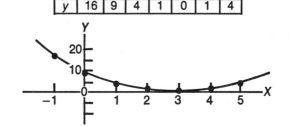

17. $a = 1, b = -2, c = 10$

$$x = \frac{-(-2) \pm \sqrt{(-2)^2 - 4(1)(10)}}{2(1)}$$

$$= \frac{2 \pm 6i}{2}$$

$$x = \frac{2 + 6i}{2} = 1 + 3i$$

$$x = \frac{2 - 6i}{2} = 1 - 3i$$

$y = x^2 - 2x + 10$

x	-2	-1	0	1	2	3	4
y	18	13	10	9	10	13	18

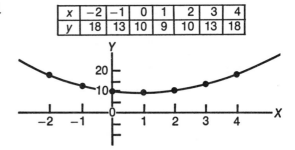

19. $a = -2, b = 3, c = 2$

$$x = \frac{-3 \pm \sqrt{3^2 - 4(-2)(2)}}{2(-2)}$$

$$= \frac{-3 \pm 5}{-4}$$

$$x = \frac{-3 + 5}{-4} = -\frac{1}{2}$$

$$x = \frac{-3 - 5}{-4} = 2$$

$y = -2x^2 + 3x + 2$

x	-3	-2	-1	0	1	2	3
y	-25	-12	-3	2	3	0	-7

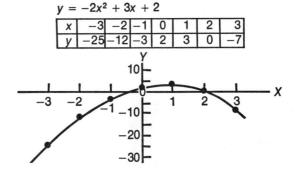

Exercise 5–1 (p. 85)

1. 2 −3   5 |6_
        12 54
   ────────────
   2   9 59

   Quotient: $2x + 9$

   Remainder: 59

3. 4 −3   0   7 |−1
       −4   7 −7
   ──────────────────
   4 −7   7   0

   Quotient: $4x^2 − 7x + 7$

   Remainder: 0

5. 5   14 −3   9 |−4
      −20 24 −84
   ────────────────────
   5 − 6 21 −75

   Quotient: $5x^2 − 6x + 21$

   Remainder: −75

7. 2   0 −5   3 −108 |−3
      −6 18 −39   108
   ──────────────────────────
   2 −6 13 −36   0

   Quotient: $2x^3 − 6x^2 + 13x − 36$

   Remainder: 0

9. 1 0 −14   0 −7   10 |4_
      4 16   8 32 100
   ────────────────────────────
   1 4   2   8 25 110

   Quotient:
   $x^4 + 4x^3 + 2x^2 + 8x + 25$

   Remainder: 110

11. 2 1 2   −2k   −1 |1_ $4 − 2k = 0$
       2 3   5   5 − 2k $k = 2$
   ──────────────────────────
   2 3 5 5 − 2k 4 − 2k

13. (a) $f(6) = 2(6)^2 − 3(6) + 5 = 59$
    (b) $f(−6) = 2(−6)^2 − 3(−6) + 5 = 95$

15. (a) $f(−1) = 4(−1)^3 − 3(−1)^2 + 7 = 0$
    (b) $f(1) = 4(1)^3 − 3(1)^2 + 7 = 8$

17. (a) $f(−4) = 5(−4)^3 + 14(−4)^2 − 3(−4) + 9 = −75$
    (b) $f(4) = 5(4)^3 + 14(4)^2 − 3(4) + 9 = 541$

19. $f(6) = 59 \neq 0$. (See solution to problem 13(a).)
    Thus, $(x − 6)$ is not a factor of the dividend.

21. $f(−1) = 0$. (See solution to problem 15(a).)
    Thus, $(x + 1)$ is a factor of the dividend.

23. $f(−2) = 3(−2)^4 − 7(−2)^2 + 8(−2) − 4 = 0$
    Thus, $(x + 2)$ is a factor.

Exercise 5–2 (p. 92)

1. $(x − 1)(x − 2)(x + 3) = 0$
   $x^3 − 7x + 6 = 0$

3. $(x − 1)(x + 2)(x − 3)(x + 1) = 0$
   $x^4 − x^3 − 7x^2 + x + 6 = 0$

5. $(x − 1)(x^2 − x − 6) = 0$
    $(x − 1)(x + 2)(x − 3) = 0$
    $x = 1, −2, 3$

   1 −2 −5   6 |1_
         1 −1 −6
   ──────────────────
   1 −1 −6   0

7.
$$(x - 1)(x^2 - 2) = 0$$
$$(x - 1)(x + \sqrt{2})(x - \sqrt{2}) = 0$$
$$x = 1, -\sqrt{2}, \sqrt{2}$$

```
1  -1  -2   2 |1
       1   0  -2
   _____
   1   0  -2   0
```

9. $(x - 1)(2x^2 + 7x + 5) = 0$
$(x - 1)(x + 1)(2x + 5) = 0$
$$x = 1, -1, -\tfrac{5}{2}$$

```
2   5  -2  -5 |1
        2   7   5
    _____
    2   7   5   0
```

11. $(x + 1)(3x^3 - 5x^2 - 6x + 8) = 0$
$(x + 1)(x - 1)(3x^2 - 2x - 8) = 0$
$(x + 1)(x - 1)(x - 2)(3x + 4) = 0$

$$x = -1, 1, 2, -\tfrac{4}{3}$$

```
3  -2  -11   2   8 |-1
       -3    5   6  -8
   _____
3  -5   -6    8   0 |1
        3    -2  -8
   _____
3  -2   -8    0
```

13. $f(x) = x^3 - 2x^2 + x - 3$

$f(1) = 1 - 2 + 1 - 3 = -3$ $\qquad$ $f(3) = 27 - 18 + 3 - 3 = 9$
$f(-1) = -1 - 2 - 1 - 3 = -7$ $\qquad$ $f(-3) = -27 - 18 - 3 - 3 = -51$

Thus, there are no rational roots.

x	3	2	1	0	−1	−2	−3
y	9	−1	−3	−3	−7	−21	−51

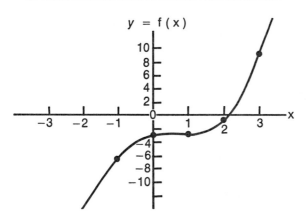

$y = f(x)$

$f(2.1) = 9.261 - 8.820 + 2.100 - 3.000 = -.459$
$f(2.2) = 10.648 - 9.680 + 2.200 - 3.000 = +.168$

x	$f(x)$	
2.2	+0.168	(1)
?	0	(2)
2.1	−0.459	(3)

$(2) - (3) \quad \dfrac{? - 2.1}{0.1} = \dfrac{0.459}{0.627}$
$(1) - (3)$

$$? = 2.1 + 0.1 \left(\frac{0.459}{0.627}\right) = 2.1 + 0.073 = 2.173, \text{ or round to } 2.17.$$

Exercise 6–1 (p. 104)

1. Two-component row vector. $a^t = \begin{bmatrix} 3 \\ 7 \end{bmatrix}$

3. Two-component column vector. $c^t = [4, 9]$

5. Three-component column vector. $e^t = [3, -4, 9]$

7. $[3, 7] + [-2, 5] = [1, 12]$

9. $\begin{bmatrix} 2 \\ 1 \\ 6 \end{bmatrix} - \begin{bmatrix} 3 \\ -4 \\ 9 \end{bmatrix} = \begin{bmatrix} -1 \\ 5 \\ -3 \end{bmatrix}$

11. $\begin{bmatrix} 3 \\ -4 \\ 9 \end{bmatrix} - \begin{bmatrix} 2 \\ 1 \\ 6 \end{bmatrix} = \begin{bmatrix} 1 \\ -5 \\ 3 \end{bmatrix}$

13. $4[3, 7] = [12, 28]$

15. $5\begin{bmatrix} 4 \\ 9 \end{bmatrix} = \begin{bmatrix} 20 \\ 45 \end{bmatrix}$

17. $6\begin{bmatrix} 2 \\ 1 \\ 6 \end{bmatrix} = \begin{bmatrix} 12 \\ 6 \\ 36 \end{bmatrix}$

19. $[3, 7]\begin{bmatrix} 4 \\ 9 \end{bmatrix} = 12 + 63 = 75$

21. $[1, 5, 7]\begin{bmatrix} 2 \\ 1 \\ 6 \end{bmatrix} = 2 + 5 + 42 = 49$

23. Length of $g = \sqrt{2^2 + 3^2} = \sqrt{13}$
$= 3.61$

25. Length of $i = \sqrt{3^2 + (-2)^2}$
$= \sqrt{13} = 3.61$

Graph for 23 and 25:

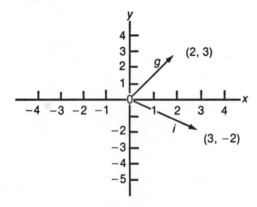

27.

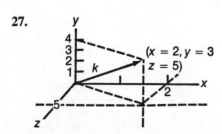

29.

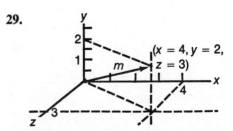

31.

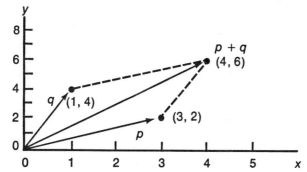

$$p + q = [3, 2] + [1, 4]$$
$$= [4, 6]$$

$$2p = 2[3, 2]$$
$$= [6, 4]$$

33.

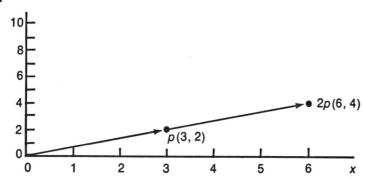

35. $1x + 1y = 8$
  $2x - 1y = 7$

   $3x = 15$
    $x = 5$
    $y = 3$

Independent

37. $2x + 3y = 23$
  $3x - 2y = 28$

   $x = 10$
   $y = 1$

Independent

39. $3x + 6y = 15$ (1)
  $1x + 2y = 5$ (2)

 $(2) \cdot 3: 3x + 6y = 15$ (3)

 $(1) = (3)$: Dependent
      equations

41. $1x + 0y = 5$
  $0x + 1y = 7$

   $x = 5$
   $y = 7$

Independent

43. $3\begin{bmatrix} 2 \\ 5 \end{bmatrix} + (-2)\begin{bmatrix} 4 \\ 1 \end{bmatrix} = \begin{bmatrix} -2 \\ 13 \end{bmatrix} = s$

Exercise 6-2 (p. 112)

1. 2×2 square matrix $\qquad\qquad A^t = \begin{bmatrix} 4 & 3 \\ -8 & 2 \end{bmatrix}$

3. 3×2 matrix $\qquad\qquad\qquad C^t = \begin{bmatrix} 2 & 4 & 1 \\ 3 & -6 & 8 \end{bmatrix}$

5. 2×3 matrix $\qquad\qquad\qquad E^t = \begin{bmatrix} 3 & 8 \\ 2 & 9 \\ 6 & -1 \end{bmatrix}$

7. 3×3 square matrix $\qquad\qquad G^t = \begin{bmatrix} 1 & -2 & 3 \\ 4 & 6 & -7 \\ 0 & -5 & 9 \end{bmatrix}$

9. $\begin{bmatrix} 4 & -8 \\ 3 & 2 \end{bmatrix} + \begin{bmatrix} 1 & 5 \\ 7 & 6 \end{bmatrix} = \begin{bmatrix} 5 & -3 \\ 10 & 8 \end{bmatrix}$

11. $\begin{bmatrix} 2 & 3 \\ 4 & -6 \\ 1 & 8 \end{bmatrix} - \begin{bmatrix} 7 & -1 \\ 2 & 5 \\ 4 & 9 \end{bmatrix} = \begin{bmatrix} -5 & 4 \\ 2 & -11 \\ -3 & -1 \end{bmatrix}$

13. $\begin{bmatrix} 4 & -8 \\ 3 & 2 \end{bmatrix} - \begin{bmatrix} 4 & -8 \\ 3 & 2 \end{bmatrix} = \begin{bmatrix} 0 & 0 \\ 0 & 0 \end{bmatrix}$

15. $2\begin{bmatrix} 4 & -8 \\ 3 & 2 \end{bmatrix} = \begin{bmatrix} 8 & -16 \\ 6 & 4 \end{bmatrix}$

17. $\frac{1}{5}\begin{bmatrix} 2 & 3 \\ 4 & -6 \\ 1 & 8 \end{bmatrix} = \begin{bmatrix} \frac{2}{5} & \frac{3}{5} \\ \frac{4}{5} & -\frac{6}{5} \\ \frac{1}{5} & \frac{8}{5} \end{bmatrix}$

19. $\begin{bmatrix} 4 & -8 \\ 3 & 2 \end{bmatrix}\begin{bmatrix} 1 & 5 \\ 7 & 6 \end{bmatrix} = \begin{bmatrix} 4-56 & 20-48 \\ 3+14 & 15+12 \end{bmatrix} = \begin{bmatrix} -52 & -28 \\ 17 & 27 \end{bmatrix}$

21. $\begin{bmatrix} 4 & -8 \\ 3 & 2 \end{bmatrix}\begin{bmatrix} 3 & 2 & 6 \\ 8 & 9 & -1 \end{bmatrix} = \begin{bmatrix} 12-64 & 8-72 & 24+8 \\ 9+16 & 6+18 & 18-2 \end{bmatrix} = \begin{bmatrix} -52 & -64 & 32 \\ 25 & 24 & 16 \end{bmatrix}$

23. $\begin{bmatrix} 4 & -8 \\ 3 & 2 \end{bmatrix}\begin{bmatrix} 1 & 4 & -9 \\ 6 & 2 & 3 \end{bmatrix} = \begin{bmatrix} 4-48 & 16-16 & -36-24 \\ 3+12 & 12+4 & -27+6 \end{bmatrix} = \begin{bmatrix} -44 & 0 & -60 \\ 15 & 16 & -21 \end{bmatrix}$

25. $\begin{bmatrix} 3 & 2 & 6 \\ 8 & 9 & -1 \end{bmatrix}\begin{bmatrix} 2 & 3 \\ 4 & -6 \\ 1 & 8 \end{bmatrix} = \begin{bmatrix} 6+8+6 & 9-12+48 \\ 16+36-1 & 24-54-8 \end{bmatrix} = \begin{bmatrix} 20 & 45 \\ 51 & -38 \end{bmatrix}$

27. $\begin{bmatrix} 2 & 3 \\ 4 & -6 \\ 1 & 8 \end{bmatrix}\begin{bmatrix} 4 & -8 \\ 3 & 2 \end{bmatrix} = \begin{bmatrix} 8+9 & -16+6 \\ 16-18 & -32-12 \\ 4+24 & -8+16 \end{bmatrix} = \begin{bmatrix} 17 & -10 \\ -2 & -44 \\ 28 & 8 \end{bmatrix}$

29. $\begin{bmatrix} 1 & 4 & -9 \\ 6 & 2 & 3 \end{bmatrix}\begin{bmatrix} 2 & 3 \\ 4 & -6 \\ 1 & 8 \end{bmatrix} = \begin{bmatrix} 2+16-9 & 3-24-72 \\ 12+8+3 & 18-12+24 \end{bmatrix} = \begin{bmatrix} 9 & -93 \\ 23 & 30 \end{bmatrix}$

31. $\begin{bmatrix} 2 & 3 \\ 4 & -6 \\ 1 & 8 \end{bmatrix}\begin{bmatrix} 3 & 2 & 6 \\ 8 & 9 & -1 \end{bmatrix} = \begin{bmatrix} 6+24 & 4+27 & 12-3 \\ 12-48 & 8-54 & 24+6 \\ 3+64 & 2+72 & 6-8 \end{bmatrix} = \begin{bmatrix} 30 & 31 & 9 \\ -36 & -46 & 30 \\ 67 & 74 & -2 \end{bmatrix}$

33. $\begin{bmatrix} 2 & 3 \\ 4 & -6 \\ 1 & 8 \end{bmatrix}\begin{bmatrix} 1 & 4 & -9 \\ 6 & 2 & 3 \end{bmatrix} = \begin{bmatrix} 2+18 & 8+6 & -18+9 \\ 4-36 & 16-12 & -36-18 \\ 1+48 & 4+16 & -9+24 \end{bmatrix} = \begin{bmatrix} 20 & 14 & -9 \\ -32 & 4 & -54 \\ 49 & 20 & 15 \end{bmatrix}$

35. $\begin{bmatrix} 1 & 4 & 0 \\ -2 & 6 & -5 \\ 3 & -7 & 9 \end{bmatrix} + \begin{bmatrix} 0 & 0 & 0 \\ 0 & 0 & 0 \\ 0 & 0 & 0 \end{bmatrix} = \begin{bmatrix} 1 & 4 & 0 \\ -2 & 6 & -5 \\ 3 & -7 & 9 \end{bmatrix}$

37. $\begin{bmatrix} 1 & 4 & 0 \\ -2 & 6 & -5 \\ 3 & -7 & 9 \end{bmatrix}\begin{bmatrix} 2 & 3 \\ 4 & -6 \\ 1 & 8 \end{bmatrix} = \begin{bmatrix} 2+16+0 & 3-24+0 \\ -4+24-5 & -6-36-40 \\ 6-28+9 & 9+42+72 \end{bmatrix} = \begin{bmatrix} 18 & -21 \\ 15 & -82 \\ -13 & 123 \end{bmatrix}$

39. $\begin{bmatrix} 1 & 4 & -9 \\ 6 & 2 & 3 \end{bmatrix}\begin{bmatrix} 1 & 4 & 0 \\ -2 & 6 & -5 \\ 3 & -7 & 9 \end{bmatrix} = \begin{bmatrix} 1-8-27 & 4+24+63 & 0-20-81 \\ 6-4+9 & 24+12-21 & 0-10+27 \end{bmatrix}$

$$= \begin{bmatrix} -34 & 91 & -101 \\ 11 & 15 & 17 \end{bmatrix}$$

41. $\begin{bmatrix} 5 & -1 \\ 2 & 3 \end{bmatrix}\begin{bmatrix} 1 & 0 \\ 0 & 1 \end{bmatrix} = \begin{bmatrix} 5+0 & 0-1 \\ 2+0 & 0+3 \end{bmatrix} = \begin{bmatrix} 5 & -1 \\ 2 & 3 \end{bmatrix}$

$\begin{bmatrix} 1 & 0 \\ 0 & 1 \end{bmatrix}\begin{bmatrix} 5 & -1 \\ 2 & 3 \end{bmatrix} = \begin{bmatrix} 5+0 & -1+0 \\ 0+2 & 0+3 \end{bmatrix} = \begin{bmatrix} 5 & -1 \\ 2 & 3 \end{bmatrix}$

43. $\begin{bmatrix} 1 & 3 & 8 \\ 2 & -6 & 7 \\ 5 & 4 & 9 \end{bmatrix}\begin{bmatrix} 1 & 0 & 0 \\ 0 & 1 & 0 \\ 0 & 0 & 1 \end{bmatrix} = \begin{bmatrix} 1+0+0 & 0+3+0 & 0+0+8 \\ 2+0+0 & 0-6+0 & 0+0+7 \\ 5+0+0 & 0+4+0 & 0+0+9 \end{bmatrix} = \begin{bmatrix} 1 & 3 & 8 \\ 2 & -6 & 7 \\ 5 & 4 & 9 \end{bmatrix}$

$\begin{bmatrix} 1 & 0 & 0 \\ 0 & 1 & 0 \\ 0 & 0 & 1 \end{bmatrix}\begin{bmatrix} 1 & 3 & 8 \\ 2 & -6 & 7 \\ 5 & 4 & 9 \end{bmatrix} = \begin{bmatrix} 1+0+0 & 3+0+0 & 8+0+0 \\ 0+2+0 & 0-6+0 & 0+7+0 \\ 0+0+5 & 0+0+4 & 0+0+9 \end{bmatrix} = \begin{bmatrix} 1 & 3 & 8 \\ 2 & -6 & 7 \\ 5 & 4 & 9 \end{bmatrix}$

45. $X = \begin{bmatrix} 7 & -1 \\ 2 & 6 \end{bmatrix} - \begin{bmatrix} 2 & -1 \\ 3 & 4 \end{bmatrix} = \begin{bmatrix} 5 & 0 \\ -1 & 2 \end{bmatrix}$

47. $X = \begin{bmatrix} 3 & 7 \\ 2 & 5 \end{bmatrix} + \begin{bmatrix} -5 & -2 \\ 1 & 0 \end{bmatrix} - \begin{bmatrix} 5 & 2 \\ -1 & 4 \end{bmatrix} = \begin{bmatrix} -7 & 3 \\ 4 & 1 \end{bmatrix}$

49. $c\left(\begin{bmatrix} 4 & -8 \\ 3 & 2 \end{bmatrix} + \begin{bmatrix} 1 & 5 \\ 7 & 6 \end{bmatrix}\right) = c\begin{bmatrix} 5 & -3 \\ 10 & 8 \end{bmatrix} = \begin{bmatrix} 5c & -3c \\ 10c & 8c \end{bmatrix}$

$c\begin{bmatrix} 4 & -8 \\ 3 & 2 \end{bmatrix} + c\begin{bmatrix} 1 & 5 \\ 7 & 6 \end{bmatrix} = \begin{bmatrix} 4c & -8c \\ 3c & 2c \end{bmatrix} + \begin{bmatrix} 1c & 5c \\ 7c & 6c \end{bmatrix} = \begin{bmatrix} 5c & -3c \\ 10c & 8c \end{bmatrix}$

Exercise 7–1 (p. 121)

1. (a) $\begin{vmatrix} 2 & 7 \\ 3 & 4 \end{vmatrix} = 8 - 21 = -13$

  (b) $2|4| - 3|7| = -13$

3. (a) $\begin{vmatrix} 3 & -7 \\ 4 & 5 \end{vmatrix} = 15 + 28 = 43$

  (b) $3|5| - 4|-7| = 43$

5. (a) $\begin{vmatrix} -1 & 6 \\ 12 & 0 \end{vmatrix} = 0 - 72 = -72$

  (b) $-1|0| - 12|6| = -72$

7. (a) $\begin{vmatrix} 2 & -5 & 1 \\ 4 & 8 & 6 \\ 9 & 3 & -7 \end{vmatrix} = -112 - 270 + 12 - 72 - 36 - 140 = -618$

  (b) $2 \begin{vmatrix} 8 & 6 \\ 3 & -7 \end{vmatrix} - 4 \begin{vmatrix} -5 & 1 \\ 3 & -7 \end{vmatrix} + 9 \begin{vmatrix} -5 & 1 \\ 8 & 6 \end{vmatrix} = 2(-56 - 18)$
  $- 4(35 - 3) + 9(-30 - 8) = -148 - 128 - 342 = -618$

9. (a) $\begin{vmatrix} 1 & 8 & 0 \\ -2 & 4 & -3 \\ 7 & -6 & 5 \end{vmatrix} = 20 - 168 + 0 - 0 - 18 + 80 = -86$

  (b) $1 \begin{vmatrix} 4 & -3 \\ -6 & 5 \end{vmatrix} - (-2) \begin{vmatrix} 8 & 0 \\ -6 & 5 \end{vmatrix} + 7 \begin{vmatrix} 8 & 0 \\ 4 & -3 \end{vmatrix} = (20 - 18)$
  $+ 2(40 + 0) + 7(-24 - 0) = 2 + 80 - 168 = -86$

11. (a) $\begin{vmatrix} 5 & 4 & -2 \\ 6 & 0 & 1 \\ -7 & 3 & -8 \end{vmatrix} = 0 - 28 - 36 - 0 - 15 + 192 = 113$

  (b) $5 \begin{vmatrix} 0 & 1 \\ 3 & -8 \end{vmatrix} - 6 \begin{vmatrix} 4 & -2 \\ 3 & -8 \end{vmatrix} + (-7) \begin{vmatrix} 4 & -2 \\ 0 & 1 \end{vmatrix} = 5(-3) - 6(-32 + 6)$
  $- 7(4) = -15 + 156 - 28 = 113$

13. $\begin{vmatrix} 2 & 7 \\ 3 & 4 \end{vmatrix} = 2|4| - 7|3| = -13$

15. $\begin{vmatrix} 3 & -7 \\ 4 & 5 \end{vmatrix} = -4|-7| + 5|3| = 43$

17. $\begin{vmatrix} 2 & -5 & 1 \\ 4 & 8 & 6 \\ 9 & 3 & -7 \end{vmatrix} = -(-5) \begin{vmatrix} 4 & 6 \\ 9 & -7 \end{vmatrix} + 8 \begin{vmatrix} 2 & 1 \\ 9 & -7 \end{vmatrix} - 3 \begin{vmatrix} 2 & 1 \\ 4 & 6 \end{vmatrix}$

  $= 5(-28 - 54) + 8(-14 - 9) - 3(12 - 4)$
  $= -410 - 184 - 24 = -618$

19. $\begin{vmatrix} 1 & 8 & 0 \\ -2 & 4 & -3 \\ 7 & -6 & 5 \end{vmatrix} = -(-2)\begin{vmatrix} 8 & 0 \\ -6 & 5 \end{vmatrix} + 4\begin{vmatrix} 1 & 0 \\ 7 & 5 \end{vmatrix} - (-3)\begin{vmatrix} 1 & 8 \\ 7 & -6 \end{vmatrix}$

$\qquad = 2(40) + 4(5) + 3(-6 - 56)$

$\qquad = 80 + 20 - 186 = -86$

21. $\begin{vmatrix} 0 & 2 & 0 & 0 \\ -1 & 5 & 7 & 3 \\ 3 & 1 & 2 & 4 \\ 1 & 3 & 0 & 0 \end{vmatrix} = -2\begin{vmatrix} -1 & 7 & 3 \\ 3 & 2 & 4 \\ 1 & 0 & 0 \end{vmatrix} = -2(1)\begin{vmatrix} 7 & 3 \\ 2 & 4 \end{vmatrix} = -2(22) = -44$

23. $\begin{vmatrix} 3 & 0 & 1 & -1 \\ 4 & 1 & 3 & 2 \\ 0 & 0 & 2 & 0 \\ -1 & 1 & 0 & 1 \end{vmatrix} = 2\begin{vmatrix} 3 & 0 & -1 \\ 4 & 1 & 2 \\ -1 & 1 & 1 \end{vmatrix} = 2(3 - 4 - 1 - 6) = -16$

Exercise 7-2 (p. 126)

1. $\begin{vmatrix} 0 & 1 \\ 0 & 2 \end{vmatrix} = 0 - 0 = 0$ Property (1)

3. $\begin{vmatrix} 2 & -3 \\ 5 & 1 \end{vmatrix} = 2 + 15 = 17$ Property (3)

$\quad -\begin{vmatrix} -3 & 2 \\ 1 & 5 \end{vmatrix} = -(-15 - 2) = 17$

5. $\begin{vmatrix} 5 & 9 \\ 2 & 6 \end{vmatrix} = 30 - 18 = 12$

$\quad \begin{vmatrix} 5 & 1 \\ 2 & 7 \end{vmatrix} + \begin{vmatrix} 5 & 8 \\ 2 & -1 \end{vmatrix} = (35 - 2) + (-5 - 16) = 33 - 21 = 12$ Property (4)

7. $\begin{vmatrix} 1 & 4 & 1 \\ -2 & 0 & -2 \\ 3 & 7 & 3 \end{vmatrix} = 0 - 24 - 14 - 0 + 14 + 24 = 0$ Property (2)

9. $\begin{vmatrix} 1 & 2 & 3 \\ 4 & 0 & 7 \\ 2 & 1 & 5 \end{vmatrix} = 0 + 28 + 12 - 0 - 7 - 40 = -7$

$\quad -\begin{vmatrix} 4 & 0 & 7 \\ 1 & 2 & 3 \\ 2 & 1 & 5 \end{vmatrix} = -(40 + 0 + 7 - 28 - 12 - 0) = -7$ Property (3)

11. $2\begin{vmatrix} 1 & 3 & 5 \\ 2 & 0 & -6 \\ -1 & 4 & 1 \end{vmatrix} = 2(0 + 18 + 40 - 0 + 24 - 6) = 2(76) = 152$

$\quad -\begin{vmatrix} 1 & 3 & -10 \\ 2 & 0 & 12 \\ -1 & 4 & -2 \end{vmatrix} = -(0 - 36 - 80 - 0 - 48 + 12) = -(-152) = 152$

$\qquad\qquad\qquad\qquad\qquad\qquad\qquad$ Property (5)

13. $\begin{vmatrix} 1 & 4 & 5 & -2 \\ 3 & 0 & 1 & 0 \\ 1 & 4 & 5 & -2 \\ 2 & 1 & -3 & 1 \end{vmatrix} = -3 \begin{vmatrix} 4 & 5 & -2 \\ 4 & 5 & -2 \\ 1 & -3 & 1 \end{vmatrix} -1 \begin{vmatrix} 1 & 4 & -2 \\ 1 & 4 & -2 \\ 2 & 1 & 1 \end{vmatrix}$

$$= -3(0) - 1(0) = 0.$$ 　　　　　　　　Property (2)

15. $\begin{vmatrix} 2 & 3 & 1 & 4 \\ 0 & 1 & 5 & -1 \\ 1 & -1 & 2 & 3 \\ 0 & 2 & 1 & -2 \end{vmatrix} = 2 \begin{vmatrix} 1 & 5 & -1 \\ -1 & 2 & 3 \\ 2 & 1 & -2 \end{vmatrix} +1 \begin{vmatrix} 3 & 1 & 4 \\ 1 & 5 & -1 \\ 2 & 1 & -2 \end{vmatrix}$

$$= 2(-4 + 30 + 1 + 4 - 3 - 10) + $$
$$1(-30 - 2 + 4 - 40 + 3 + 2$$
$$= 2(18) + (-63) = -27$$

$\begin{vmatrix} 0 & 5 & -3 & -2 \\ 0 & 1 & 5 & -1 \\ 1 & -1 & 2 & 3 \\ 0 & 2 & 1 & -2 \end{vmatrix} = 1 \begin{vmatrix} 5 & -3 & -2 \\ 1 & 5 & -1 \\ 2 & 1 & -2 \end{vmatrix} = 1(-50 + 6 - 2 + 20 + 5 - 6)$

$$= -27$$ 　　　　　　　　Property (6)

17.　$\begin{array}{cc} 2 & 3 \\ (-) \underline{4k} & \underline{6k} \\ 0 & x \end{array}$ 　$\begin{array}{l} 2 - 4k = 0 \\ k = \frac{2}{4} = \frac{1}{2} \\ x = 3 - 6k = 3 - 6(\frac{1}{2}) = 0 \end{array}$

19.　$\begin{array}{ccc} 2 & 1 & 3 \\ (-) \underline{1k} & \underline{-3k} & \underline{4k} \\ 0 & x_1 & x_2 \end{array}$ 　$\begin{array}{l} 2 - 1k = 0 \\ k = 2 \\ x_1 = 1 - (-3k) = 1 + 3k \\ \quad = 1 + 3(2) = 7 \\ x_2 = 3 - 4k = 3 - 4(2) = -5 \end{array}$

21.　$\begin{array}{cccc} -4 & 1 & 2 & 1 \\ (-) \underline{2k} & \underline{-1k} & \underline{-2k} & \underline{4k} \\ 0 & x_1 & x_2 & x_3 \end{array}$ 　$\begin{array}{l} -4 - 2k = 0 \\ k = -2 \\ x_1 = 1 - (-1k) = 1 - (-(-2)) = -1 \\ x_2 = 2 - (-2k) = 2 - (-2(-2)) = -2 \\ x_3 = 1 - 4k = 1 - 4(-2) = 9 \end{array}$

23. $\begin{vmatrix} 1 & -3 & 4 \\ 2 & 1 & 3 \\ 0 & 5 & 1 \end{vmatrix} = \begin{vmatrix} 1 & -3 & 4 \\ 0 & 7 & -5 \\ 0 & 5 & 1 \end{vmatrix} = 1 \begin{vmatrix} 7 & -5 \\ 5 & 1 \end{vmatrix} = 7 - (-25) = 32$

　　　　　　　　　　　　　　Also see problem 19 solution.

25. $\begin{vmatrix} -4 & 1 & 2 & 1 \\ 0 & 2 & 1 & 3 \\ 2 & -1 & -2 & 4 \\ 0 & 3 & 2 & 1 \end{vmatrix} = \begin{vmatrix} 0 & -1 & -2 & 9 \\ 0 & 2 & 1 & 3 \\ 2 & -1 & -2 & 4 \\ 0 & 3 & 2 & 1 \end{vmatrix} = 2 \begin{vmatrix} -1 & -2 & 9 \\ 2 & 1 & 3 \\ 3 & 2 & 1 \end{vmatrix}$

$$= 2(-1 - 18 + 36 - 27 + 6 + 4) = 2(0) = 0$$

　　　　　　　　　　　　　　Also see problem 21 solution.

Exercise 7–3 (p. 138)

1. $\begin{vmatrix} 7 & 5 \\ 7 & 5 \end{vmatrix} = 0$

No inverse, since $|A| = 0$.

3. (a) $\begin{bmatrix} 6 & -10 \\ 2 & -5 \end{bmatrix} \begin{bmatrix} b_{11} & b_{12} \\ b_{21} & b_{22} \end{bmatrix} = \begin{bmatrix} 1 & 0 \\ 0 & 1 \end{bmatrix}$

$6b_{11} - 10b_{21} = 1 \qquad\qquad 6b_{12} - 10b_{22} = 0$
$2b_{11} - 5b_{21} = 0 \qquad\qquad 2b_{12} - 5b_{22} = 1$

$b_{11} = \frac{1}{2} \qquad\qquad\qquad b_{12} = -1$
$b_{21} = \frac{1}{5} \qquad\qquad\qquad b_{22} = -\frac{6}{10} = -\frac{3}{5}$

$A^{-1} = \begin{bmatrix} \frac{1}{2} & -1 \\ \frac{1}{5} & -\frac{3}{5} \end{bmatrix}$

(b) $|A| = \begin{vmatrix} 6 & -10 \\ 2 & -5 \end{vmatrix} = -30 + 20 = -10$

$A^{-1} = \frac{1}{-10} \begin{bmatrix} -5 & 10 \\ -2 & 6 \end{bmatrix} = \begin{bmatrix} \frac{1}{2} & -1 \\ \frac{1}{5} & -\frac{3}{5} \end{bmatrix}$

(c) $\begin{bmatrix} 6 & -10 & | & 1 & 0 \\ 2 & -5 & | & 0 & 1 \end{bmatrix} \qquad \begin{matrix} (1) \\ (2) \end{matrix}$

$\begin{matrix} (1) \div 6 \\ (2) - (1)' \times 2 \end{matrix} \begin{bmatrix} 1 & -\frac{5}{3} & | & \frac{1}{6} & 0 \\ 0 & -\frac{5}{3} & | & -\frac{1}{3} & 1 \end{bmatrix} \begin{matrix} (1)' \\ (2)' \end{matrix}$

$\begin{matrix} (1)' - (2)'' \times (-\frac{5}{3}) \\ (2)' \div (-\frac{5}{3}) \end{matrix} \begin{bmatrix} 1 & 0 & | & \frac{1}{2} & -1 \\ 0 & 1 & | & \frac{1}{5} & -\frac{3}{5} \end{bmatrix} \begin{matrix} (1)'' \\ (2)'' \end{matrix}$

5. (a) $\begin{bmatrix} 4 & -5 \\ 2 & 7 \end{bmatrix} \begin{bmatrix} b_{11} & b_{12} \\ b_{21} & b_{22} \end{bmatrix} = \begin{bmatrix} 1 & 0 \\ 0 & 1 \end{bmatrix}$

$4b_{11} - 5b_{21} = 1 \qquad\qquad 4b_{12} - 5b_{22} = 0$
$2b_{11} + 7b_{21} = 0 \qquad\qquad 2b_{12} + 7b_{22} = 1$
$b_{11} = \frac{7}{38} \qquad\qquad\qquad b_{12} = \frac{5}{38}$
$b_{21} = -\frac{1}{19} \qquad\qquad\quad b_{22} = \frac{2}{19}$

$A^{-1} = \begin{bmatrix} \frac{7}{38} & \frac{5}{38} \\ -\frac{1}{19} & \frac{2}{19} \end{bmatrix}$

(b) $|A| = \begin{vmatrix} 4 & -5 \\ 2 & 7 \end{vmatrix} = 28 + 10 = 38$

$A^{-1} = \frac{1}{38} \begin{bmatrix} 7 & 5 \\ -2 & 4 \end{bmatrix} = \begin{bmatrix} \frac{7}{38} & \frac{5}{38} \\ -\frac{1}{19} & \frac{2}{19} \end{bmatrix}$

(c) $\begin{bmatrix} 4 & -5 & | & 1 & 0 \\ 2 & 7 & | & 0 & 1 \end{bmatrix}$ (1)
                                       (2)

(1) $\div$ 4 $\begin{bmatrix} 1 & -\frac{5}{4} & | & \frac{1}{4} & 0 \\ 0 & \frac{19}{2} & | & -\frac{1}{2} & 1 \end{bmatrix}$ (1)'
(2) $-$ (1)' $\times$ 2   (2)'

(1)' $-$ (2)" $\times$ $(-\frac{5}{4})$ $\begin{bmatrix} 1 & 0 & | & \frac{7}{38} & \frac{5}{38} \\ 0 & 1 & | & -\frac{1}{19} & \frac{2}{19} \end{bmatrix}$ (1)"
(2)' $\div$ $(\frac{19}{2})$ (2)"

7. (a) $|A| = -618$ (Problem 7, Exercise 7–1)

$$B = \begin{bmatrix} \begin{vmatrix} 8 & 6 \\ 3 & -7 \end{vmatrix} & -\begin{vmatrix} 4 & 6 \\ 9 & -7 \end{vmatrix} & \begin{vmatrix} 4 & 8 \\ 9 & 3 \end{vmatrix} \\ -\begin{vmatrix} -5 & 1 \\ 3 & -7 \end{vmatrix} & \begin{vmatrix} 2 & 1 \\ 9 & -7 \end{vmatrix} & -\begin{vmatrix} 2 & -5 \\ 9 & 3 \end{vmatrix} \\ \begin{vmatrix} -5 & 1 \\ 8 & 6 \end{vmatrix} & -\begin{vmatrix} 2 & 1 \\ 4 & 6 \end{vmatrix} & \begin{vmatrix} 2 & -5 \\ 4 & 8 \end{vmatrix} \end{bmatrix} = \begin{bmatrix} -74 & 82 & -60 \\ -32 & -23 & -51 \\ -38 & -8 & 36 \end{bmatrix}$$

$$A^{-1} = \frac{1}{-618} \begin{bmatrix} -74 & -32 & -38 \\ 82 & -23 & -8 \\ -60 & -51 & 36 \end{bmatrix} = \frac{1}{618} \begin{bmatrix} 74 & 32 & 38 \\ -82 & 23 & 8 \\ 60 & 51 & -36 \end{bmatrix}$$

(b) $\begin{bmatrix} 2 & -5 & 1 & | & 1 & 0 & 0 \\ 4 & 8 & 6 & | & 0 & 1 & 0 \\ 9 & 3 & -7 & | & 0 & 0 & 1 \end{bmatrix}$ (1)
                                                          (2)
                                                          (3)

(1) $\div$ 2 $\begin{bmatrix} 1 & -\frac{5}{2} & \frac{1}{2} & | & \frac{1}{2} & 0 & 0 \\ 0 & 18 & 4 & | & -2 & 1 & 0 \\ 0 & \frac{51}{2} & -\frac{23}{2} & | & -\frac{9}{2} & 0 & 1 \end{bmatrix}$ (1)'
(2) $-$ (1)' $\times$ 4   (2)'
(3) $-$ (1)' $\times$ 9   (3)'

(1)' $-$ (2)" $\times$ $(-\frac{5}{2})$ $\begin{bmatrix} 1 & 0 & \frac{19}{18} & | & \frac{2}{9} & \frac{5}{36} & 0 \\ 0 & 1 & \frac{2}{9} & | & -\frac{1}{9} & \frac{1}{18} & 0 \\ 0 & 0 & -\frac{103}{6} & | & -\frac{5}{3} & -\frac{17}{12} & 1 \end{bmatrix}$ (1)"
(2)' $\div$ 18   (2)"
(3)' $-$ (2)" $\times$ $(\frac{51}{2})$ (3)"

(1)" $-$ (3)"' $\times$ $(\frac{19}{18})$ $\begin{bmatrix} 1 & 0 & 0 & | & \frac{74}{618} & \frac{32}{618} & \frac{38}{618} \\ 0 & 1 & 0 & | & -\frac{82}{618} & \frac{23}{618} & \frac{8}{618} \\ 0 & 0 & 1 & | & \frac{60}{618} & \frac{51}{618} & -\frac{36}{618} \end{bmatrix}$ (1)"'
(2)" $-$ (3)"' $\times$ $(\frac{2}{9})$ (2)"'
(3)" $\div$ $(-\frac{103}{6})$ (3)"'

9. (a) $|A| = -86$ (Problem 9, Exercise 7–1)

$$B = \begin{bmatrix} \begin{vmatrix} 4 & -3 \\ -6 & 5 \end{vmatrix} & -\begin{vmatrix} -2 & -3 \\ 7 & 5 \end{vmatrix} & \begin{vmatrix} -2 & 4 \\ 7 & -6 \end{vmatrix} \\ -\begin{vmatrix} 8 & 0 \\ -6 & 5 \end{vmatrix} & \begin{vmatrix} 1 & 0 \\ 7 & 5 \end{vmatrix} & -\begin{vmatrix} 1 & 8 \\ 7 & -6 \end{vmatrix} \\ \begin{vmatrix} 8 & 0 \\ 4 & -3 \end{vmatrix} & -\begin{vmatrix} 1 & 0 \\ -2 & -3 \end{vmatrix} & \begin{vmatrix} 1 & 8 \\ -2 & 4 \end{vmatrix} \end{bmatrix} = \begin{bmatrix} 2 & -11 & -16 \\ -40 & 5 & 62 \\ -24 & 3 & 20 \end{bmatrix}$$

$$A^{-1} = \frac{1}{-86} \begin{bmatrix} 2 & -40 & -24 \\ -11 & 5 & 3 \\ -16 & 62 & 20 \end{bmatrix} = \frac{1}{86} \begin{bmatrix} -2 & 40 & 24 \\ 11 & -5 & -3 \\ 16 & -62 & -20 \end{bmatrix}$$

(b) $\left[\begin{array}{rrr|rrr} 1 & 8 & 0 & 1 & 0 & 0 \\ -2 & 4 & -3 & 0 & 1 & 0 \\ 7 & -6 & 5 & 0 & 0 & 1 \end{array}\right]$ (1) (2) (3)

$\begin{array}{l} (2)-(1)\times(-2) \\ (3)-(1)\times 7 \end{array}$ $\left[\begin{array}{rrr|rrr} 1 & 8 & 0 & 1 & 0 & 0 \\ 0 & 20 & -3 & 2 & 1 & 0 \\ 0 & -62 & 5 & -7 & 0 & 1 \end{array}\right]$ (1) (2)' (3)'

$\begin{array}{l} (1)-(2)''\times 8 \\ (2)'\div 20 \\ (3)'-(2)''\times(-62) \end{array}$ $\left[\begin{array}{rrr|rrr} 1 & 0 & \frac{6}{5} & \frac{1}{5} & -\frac{2}{5} & 0 \\ 0 & 1 & -\frac{3}{20} & \frac{1}{10} & \frac{1}{20} & 0 \\ 0 & 0 & -\frac{43}{10} & -\frac{4}{5} & \frac{31}{10} & 1 \end{array}\right]$ (1)' (2)'' (3)''

$\begin{array}{l} (1)'-(3)'''\times(\frac{6}{5}) \\ (2)''-(3)'''\times(-\frac{3}{20}) \\ (3)''\div(-\frac{43}{10}) \end{array}$ $\left[\begin{array}{rrr|rrr} 1 & 0 & 0 & -\frac{2}{86} & \frac{40}{86} & \frac{24}{86} \\ 0 & 1 & 0 & \frac{11}{86} & -\frac{5}{86} & -\frac{3}{86} \\ 0 & 0 & 1 & \frac{16}{86} & -\frac{62}{86} & -\frac{20}{86} \end{array}\right]$ (1)'' (2)''' (3)'''

11. (a) $|A| = 113$ (Problem 11, Exercise 7-1)

$$B = \left[\begin{array}{ccc} \begin{vmatrix}0 & 1\\3 & -8\end{vmatrix} & -\begin{vmatrix}6 & 1\\-7 & -8\end{vmatrix} & \begin{vmatrix}6 & 0\\-7 & 3\end{vmatrix} \\[2mm] -\begin{vmatrix}4 & -2\\3 & -8\end{vmatrix} & \begin{vmatrix}5 & -2\\-7 & -8\end{vmatrix} & -\begin{vmatrix}5 & 4\\-7 & 3\end{vmatrix} \\[2mm] \begin{vmatrix}4 & -2\\0 & 1\end{vmatrix} & -\begin{vmatrix}5 & -2\\6 & 1\end{vmatrix} & \begin{vmatrix}5 & 4\\6 & 0\end{vmatrix} \end{array}\right] = \left[\begin{array}{rrr} -3 & 41 & 18 \\ 26 & -54 & -43 \\ 4 & -17 & -24 \end{array}\right]$$

$$A^{-1} = \frac{1}{113}\left[\begin{array}{rrr} -3 & 26 & 4 \\ 41 & -54 & -17 \\ 18 & -43 & -24 \end{array}\right]$$

(b) $\left[\begin{array}{rrr|rrr} 5 & 4 & -2 & 1 & 0 & 0 \\ 6 & 0 & 1 & 0 & 1 & 0 \\ -7 & 3 & -8 & 0 & 0 & 1 \end{array}\right]$ (1) (2) (3)

$\begin{array}{l} (1)\div 5 \\ (2)-(1)'\times 6 \\ (3)-(1)'\times(-7) \end{array}$ $\left[\begin{array}{rrr|rrr} 1 & \frac{4}{5} & -\frac{2}{5} & \frac{1}{5} & 0 & 0 \\ 0 & -\frac{24}{5} & \frac{17}{5} & -\frac{6}{5} & 1 & 0 \\ 0 & \frac{43}{5} & -\frac{54}{5} & \frac{7}{5} & 0 & 1 \end{array}\right]$ (1)' (2)' (3)'

$\begin{array}{l} (1)'-(2)''\times(\frac{4}{5}) \\ (2)'\div(-\frac{24}{5}) \\ (3)'-(2)''\times(\frac{43}{5}) \end{array}$ $\left[\begin{array}{rrr|rrr} 1 & 0 & \frac{1}{6} & 0 & \frac{1}{6} & 0 \\ 0 & 1 & -\frac{17}{24} & \frac{6}{24} & -\frac{5}{24} & 0 \\ 0 & 0 & -\frac{113}{24} & -\frac{3}{4} & \frac{43}{24} & 1 \end{array}\right]$ (1)'' (2)'' (3)''

$\begin{array}{l} (1)''-(3)'''\times(\frac{1}{6}) \\ (2)''-(3)'''\times(-\frac{17}{24}) \\ (3)''\div(-\frac{113}{24}) \end{array}$ $\left[\begin{array}{rrr|rrr} 1 & 0 & 0 & -\frac{3}{113} & \frac{26}{113} & \frac{4}{113} \\ 0 & 1 & 0 & \frac{41}{113} & -\frac{54}{113} & -\frac{17}{113} \\ 0 & 0 & 1 & \frac{18}{113} & -\frac{43}{113} & -\frac{24}{113} \end{array}\right]$ (1)''' (2)''' (3)'''

Exercise 7-4 (p. 146)

1. (a) $A = \begin{bmatrix} 3 & 1 \\ 4 & -1 \end{bmatrix}$ $\quad |A| = \begin{vmatrix} 3 & 1 \\ 4 & -1 \end{vmatrix} = -3 - 4 = -7$

$\begin{bmatrix} x \\ y \end{bmatrix} = \frac{1}{-7}\begin{bmatrix} -1 & -1 \\ -4 & 3 \end{bmatrix} \cdot \begin{bmatrix} 11 \\ 3 \end{bmatrix} = \begin{bmatrix} 2 \\ 5 \end{bmatrix}$ $\qquad \begin{array}{l} x = 2 \\ y = 5 \end{array}$

(b) $x = \dfrac{1}{-7} \begin{vmatrix} 11 & 1 \\ 3 & -1 \end{vmatrix} = \dfrac{-14}{-7} = 2$

$y = \dfrac{1}{-7} \begin{vmatrix} 3 & 11 \\ 4 & 3 \end{vmatrix} = \dfrac{-35}{-7} = 5$

(c) $\begin{bmatrix} 3 & 1 & | & 11 \\ 4 & -1 & | & 3 \end{bmatrix}$ (1)
                                         (2)

$(1) \div 3$ $\begin{bmatrix} 1 & \frac{1}{3} & | & \frac{11}{3} \\ 0 & -\frac{7}{3} & | & -\frac{35}{3} \end{bmatrix}$ (1)′
$(2) - (1)′ \times 4$ (2)′

$(1)′ - (2)″ \times (\frac{1}{3})$ $\begin{bmatrix} 1 & 0 & | & 2 \\ 0 & 1 & | & 5 \end{bmatrix}$ (1)″
$(2)′ \div (-\frac{7}{3})$ (2)″

3. (a) $A = \begin{bmatrix} 1 & 3 \\ 5 & -2 \end{bmatrix}$ $|A| = \begin{vmatrix} 1 & 3 \\ 5 & -2 \end{vmatrix} = -2 - 15 = -17$

$\begin{bmatrix} x \\ y \end{bmatrix} = \dfrac{1}{-17} \begin{bmatrix} -2 & -3 \\ -5 & 1 \end{bmatrix} \begin{bmatrix} 18 \\ -29 \end{bmatrix} = \begin{bmatrix} -3 \\ 7 \end{bmatrix}$ $\begin{matrix} x = -3 \\ y = 7 \end{matrix}$

(b) $x = \dfrac{1}{-17} \begin{vmatrix} 18 & 3 \\ -29 & -2 \end{vmatrix} = \dfrac{-51}{17} = -3$

$y = \dfrac{1}{-17} \begin{vmatrix} 1 & 18 \\ 5 & -29 \end{vmatrix} = \dfrac{-119}{-17} = 7$

(c) $\begin{bmatrix} 1 & 3 & | & 18 \\ 5 & -2 & | & -29 \end{bmatrix}$ (1)
                                          (2)

(1) $\begin{bmatrix} 1 & 3 & | & 18 \\ 0 & -17 & | & -119 \end{bmatrix}$ (1)
$(2) - (1) \times 5$ (2)′

$(1) - (2)″ \times 3$ $\begin{bmatrix} 1 & 0 & | & -3 \\ 0 & 1 & | & 7 \end{bmatrix}$ (1)′
$(2)′ \div (-17)$ (2)″

5. (a) $A = \begin{bmatrix} 6 & -2 \\ 1 & 5 \end{bmatrix}$ $|A| = \begin{vmatrix} 6 & -2 \\ 1 & 5 \end{vmatrix} = 30 + 2 = 32$

$\begin{bmatrix} x \\ y \end{bmatrix} = \dfrac{1}{32} \begin{bmatrix} 5 & 2 \\ -1 & 6 \end{bmatrix} \begin{bmatrix} 46 \\ -3 \end{bmatrix} = \begin{bmatrix} 7 \\ -2 \end{bmatrix}$ $\begin{matrix} x = 7 \\ y = -2 \end{matrix}$

(b) $x = \dfrac{1}{32} \begin{vmatrix} 46 & -2 \\ -3 & 5 \end{vmatrix} = \dfrac{224}{32} = 7$

$y = \dfrac{1}{32} \begin{vmatrix} 6 & 46 \\ 1 & -3 \end{vmatrix} = \dfrac{-64}{32} = -2$

(c) $\begin{bmatrix} 6 & -2 & | & 46 \\ 1 & 5 & | & -3 \end{bmatrix}$ (1)
                                        (2)

$(1) \div 6$ $\begin{bmatrix} 1 & -\frac{1}{3} & | & \frac{23}{3} \\ 0 & \frac{16}{3} & | & -\frac{32}{3} \end{bmatrix}$ (1)′
$(2) - (1)′$ (2)′

$(1)′ - (2)″ \times (-\frac{1}{3})$ $\begin{bmatrix} 1 & 0 & | & 7 \\ 0 & 1 & | & -2 \end{bmatrix}$ (1)″
$(2)′ \div (\frac{16}{3})$ (2)″

7. (a) $|A| = \begin{vmatrix} 2 & 1 & 1 \\ 1 & 3 & 2 \\ 4 & -5 & -6 \end{vmatrix} = -36 + 8 - 5 - 12 + 20 + 6 = -19$

$$x = \frac{1}{-19} \begin{vmatrix} 3 & 1 & 1 \\ 1 & 3 & 2 \\ -4 & -5 & -6 \end{vmatrix} = \frac{1}{-19}(-54 - 8 - 5 + 12 + 30 + 6)$$

$$= \frac{-19}{-19} = 1$$

$$y = \frac{1}{-19} \begin{vmatrix} 2 & 3 & 1 \\ 1 & 1 & 2 \\ 4 & -4 & -6 \end{vmatrix} = \frac{1}{-19}(-12 + 24 - 4 - 4 + 16 + 18)$$

$$= \frac{38}{-19} = -2$$

$$z = \frac{1}{-19} \begin{vmatrix} 2 & 1 & 3 \\ 1 & 3 & 1 \\ 4 & -5 & -4 \end{vmatrix} = \frac{1}{-19}(-24 + 4 - 15 - 36 + 10 + 4)$$

$$= \frac{-57}{-19} = 3$$

(b) $\begin{bmatrix} 2 & 1 & 1 & | & 3 \\ 1 & 3 & 2 & | & 1 \\ 4 & -5 & -6 & | & -4 \end{bmatrix}$ (1) / (2) / (3)

$\begin{array}{l}(1) \div 2 \\ (2) - (1)' \\ (3) - (1)' \times 4\end{array}$ $\begin{bmatrix} 1 & \frac{1}{2} & \frac{1}{2} & | & \frac{3}{2} \\ 0 & \frac{5}{2} & \frac{3}{2} & | & -\frac{1}{2} \\ 0 & -7 & -8 & | & -10 \end{bmatrix}$ $\begin{array}{l}(1)' \\ (2)' \\ (3)'\end{array}$

$\begin{array}{l}(1)' - (2)'' \times (\frac{1}{2}) \\ (2)' \div (\frac{5}{2}) \\ (3)' - (2)'' \times (-7)\end{array}$ $\begin{bmatrix} 1 & 0 & \frac{1}{5} & | & \frac{8}{5} \\ 0 & 1 & \frac{3}{5} & | & -\frac{1}{5} \\ 0 & 0 & -\frac{19}{5} & | & -\frac{57}{5} \end{bmatrix}$ $\begin{array}{l}(1)'' \\ (2)'' \\ (3)''\end{array}$

$\begin{array}{l}(1)'' - (3)''' \times (\frac{1}{5}) \\ (2)'' - (3)''' \times (\frac{3}{5}) \\ (3)'' \div (-\frac{19}{5})\end{array}$ $\begin{bmatrix} 1 & 0 & 0 & | & 1 \\ 0 & 1 & 0 & | & -2 \\ 0 & 0 & 1 & | & 3 \end{bmatrix}$ $\begin{array}{l}(1)''' \\ (2)''' \\ (3)'''\end{array}$

9. (a) $|A| = \begin{vmatrix} 2 & 3 & 1 \\ 1 & -2 & -1 \\ 3 & 0 & 2 \end{vmatrix} = -8 - 9 + 6 - 6 = -17$

$$x = \frac{1}{-17} \begin{vmatrix} 3 & 3 & 1 \\ 7 & -2 & -1 \\ 14 & 0 & 2 \end{vmatrix} = \frac{1}{-17}(-12 - 42 + 28 - 42) = \frac{-68}{-17} = 4$$

$$y = \frac{1}{-17} \begin{vmatrix} 2 & 3 & 1 \\ 1 & 7 & -1 \\ 3 & 14 & 2 \end{vmatrix} = \frac{1}{-17}(28 - 9 + 14 - 21 + 28 - 6)$$

$$= \frac{34}{-17} = -2$$

$$z = \frac{1}{-17} \begin{vmatrix} 2 & 3 & 3 \\ 1 & -2 & 7 \\ 3 & 0 & 14 \end{vmatrix} = \frac{1}{-17}(-56 + 63 + 18 - 42) = \frac{-17}{-17} = 1$$

(b)
$$\left[\begin{array}{ccc|c} 2 & 3 & 1 & 3 \\ 1 & -2 & -1 & 7 \\ 3 & 0 & 2 & 14 \end{array} \right] \quad \begin{array}{l} (1) \\ (2) \\ (3) \end{array}$$

$$\begin{array}{l} (1) \div 2 \\ (2) - (1)' \\ (3) - (1)' \times 3 \end{array} \left[\begin{array}{ccc|c} 1 & \frac{3}{2} & \frac{1}{2} & \frac{3}{2} \\ 0 & -\frac{7}{2} & -\frac{3}{2} & \frac{11}{2} \\ 0 & -\frac{9}{2} & \frac{1}{2} & \frac{19}{2} \end{array} \right] \quad \begin{array}{l} (1)' \\ (2)' \\ (3)' \end{array}$$

$$\begin{array}{l} (1)' - (2)'' \times (\frac{3}{2}) \\ (2)' \div (-\frac{7}{2}) \\ (3)' - (2)'' \times (-\frac{9}{2}) \end{array} \left[\begin{array}{ccc|c} 1 & 0 & -\frac{1}{7} & \frac{27}{7} \\ 0 & 1 & \frac{3}{7} & -\frac{11}{7} \\ 0 & 0 & \frac{17}{7} & \frac{17}{7} \end{array} \right] \quad \begin{array}{l} (1)'' \\ (2)'' \\ (3)'' \end{array}$$

$$\begin{array}{l} (1)'' - (3)''' \times (-\frac{1}{7}) \\ (2)'' - (3)''' \times (\frac{3}{7}) \\ (3)'' \div (\frac{17}{7}) \end{array} \left[\begin{array}{ccc|c} 1 & 0 & 0 & 4 \\ 0 & 1 & 0 & -2 \\ 0 & 0 & 1 & 1 \end{array} \right] \quad \begin{array}{l} (1)''' \\ (2)''' \\ (3)''' \end{array}$$

11. (a) $|A| = \begin{vmatrix} 2 & 1 & -4 \\ 3 & -2 & 2 \\ 1 & 1 & 1 \end{vmatrix} = -4 + 2 - 12 - 8 - 4 - 3 = -29$

$$x = \frac{1}{-29} \begin{vmatrix} 12 & 1 & -4 \\ 3 & -2 & 2 \\ 4 & 1 & 1 \end{vmatrix} = \frac{1}{-29}(-24 + 8 - 12 - 32 - 24 - 3)$$

$$= \frac{-87}{-29} = 3$$

$$y = \frac{1}{-29} \begin{vmatrix} 2 & 12 & -4 \\ 3 & 3 & 2 \\ 1 & 4 & 1 \end{vmatrix} = \frac{1}{-29}(6 + 24 - 48 + 12 - 16 - 36)$$

$$= \frac{-58}{-29} = 2$$

$$z = \frac{1}{-29} \begin{vmatrix} 2 & 1 & 12 \\ 3 & -2 & 3 \\ 1 & 1 & 4 \end{vmatrix} = \frac{1}{-29}(-16 + 3 + 36 + 24 - 6 - 12)$$

$$= \frac{29}{-29} = -1$$

(b)
$$\left[\begin{array}{ccc|c} 2 & 1 & -4 & 12 \\ 3 & -2 & 2 & 3 \\ 1 & 1 & 1 & 4 \end{array} \right] \quad \begin{array}{l} (1) \\ (2) \\ (3) \end{array}$$

$$\begin{array}{l} (1) \div 2 \\ (2) - (1)' \times 3 \\ (3) - (1)' \end{array} \left[\begin{array}{ccc|c} 1 & \frac{1}{2} & -2 & 6 \\ 0 & -\frac{7}{2} & 8 & -15 \\ 0 & \frac{1}{2} & 3 & -2 \end{array} \right] \quad \begin{array}{l} (1)' \\ (2)' \\ (3)' \end{array}$$

$(1)' - (2)'' \times (\frac{1}{2})$
$(2)' \div (-\frac{7}{2})$
$(3)' - (2)'' \times (\frac{1}{2})$
$$\begin{bmatrix} 1 & 0 & -\frac{6}{7} & \frac{27}{7} \\ 0 & 1 & -\frac{16}{7} & \frac{30}{7} \\ 0 & 0 & \frac{29}{7} & -\frac{29}{7} \end{bmatrix}$$
$(1)''$
$(2)''$
$(3)''$

$(1)'' - (3)'''\times(-\frac{6}{7})$
$(2)'' - (3)'''\times(-\frac{16}{7})$
$(3)'' \div (\frac{29}{7})$
$$\begin{bmatrix} 1 & 0 & 0 & 3 \\ 0 & 1 & 0 & 2 \\ 0 & 0 & 1 & -1 \end{bmatrix}$$
$(1)'''$
$(2)'''$
$(3)'''$

13. (a)

$$|A| = \begin{vmatrix} 2 & 3 & 0 \\ 3 & 0 & -2 \\ 0 & 4 & -5 \end{vmatrix} = 16 + 45 = 61$$

$$x = \frac{1}{61}\begin{vmatrix} 5 & 3 & 0 \\ -8 & 0 & -2 \\ 7 & 4 & -5 \end{vmatrix} = \frac{1}{61}(-42 + 40 - 120) = \frac{-122}{61} = -2$$

$$y = \frac{1}{61}\begin{vmatrix} 2 & 5 & 0 \\ 3 & -8 & -2 \\ 0 & 7 & -5 \end{vmatrix} = \frac{1}{61}(80 + 28 + 75) = \frac{183}{61} = 3$$

$$z = \frac{1}{61}\begin{vmatrix} 2 & 3 & 5 \\ 3 & 0 & -8 \\ 0 & 4 & 7 \end{vmatrix} = \frac{1}{61}(60 + 64 - 63) = \frac{61}{61} = 1$$

(b)
$$\begin{bmatrix} 2 & 3 & 0 & 5 \\ 3 & 0 & -2 & -8 \\ 0 & 4 & -5 & 7 \end{bmatrix}$$
(1)
(2)
(3)

$(1) \div 2$
$(2) - (1)' \times 3$
$$\begin{bmatrix} 1 & \frac{3}{2} & 0 & \frac{5}{2} \\ 0 & -\frac{9}{2} & -2 & -\frac{31}{2} \\ 0 & 4 & -5 & 7 \end{bmatrix}$$
$(1)'$
$(2)'$
(3)

$(1)' - (2)'' \times (\frac{3}{2})$
$(2)' \div (-\frac{9}{2})$
$(3) - (2)'' \times 4$
$$\begin{bmatrix} 1 & 0 & -\frac{2}{3} & -\frac{8}{3} \\ 0 & 1 & \frac{4}{9} & \frac{31}{9} \\ 0 & 0 & -\frac{61}{9} & -\frac{61}{9} \end{bmatrix}$$
$(1)''$
$(2)''$
$(3)'$

$(1)'' - (3)'' \times (-\frac{2}{3})$
$(2)'' - (3)'' \times (\frac{4}{9})$
$(3)' \div (-\frac{61}{9})$
$$\begin{bmatrix} 1 & 0 & 0 & -2 \\ 0 & 1 & 0 & 3 \\ 0 & 0 & 1 & 1 \end{bmatrix}$$
$(1)'''$
$(2)'''$
$(3)''$

Exercise 8–1 (p. 160)

1. (a) **(b)**

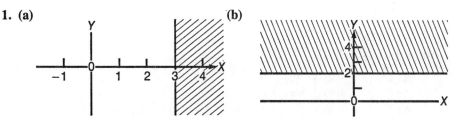

(c)

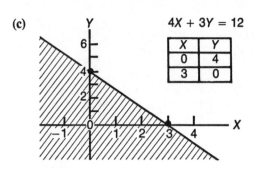

$4X + 3Y = 12$

X	Y
0	4
3	0

(d)

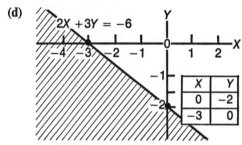

$2X + 3Y = -6$

X	Y
0	-2
-3	0

3.

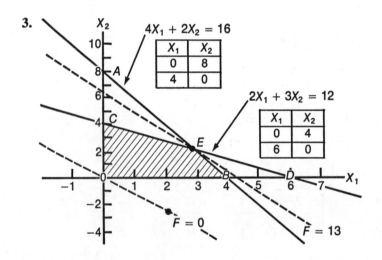

$4X_1 + 2X_2 = 16$

X_1	X_2
0	8
4	0

$2X_1 + 3X_2 = 12$

X_1	X_2
0	4
6	0

F lines:

$F = 3X_1 + 2X_2$

$$X_2 = \frac{F}{2} - \frac{3X_1}{2}$$

At O: $F = 0$

X_1	X_2
0	0
2	-3

At E: $F = 13$

X_1	X_2
0	13/2
3	2

Point O: $X_1 = 0$, $X_2 = 0$
Point C: $X_1 = 0$, $X_2 = 4$
Point E: $X_1 = 3$, $X_2 = 2$
Point B: $X_1 = 4$, $X_2 = 0$

(a) $F = 3X_1 + 2X_2$

At O: $F = 3(0) + 2(0) = 0$
At C: $F = 3(0) + 2(4) = 8$
At E: $F = 3(3) + 2(2) = 13$ E is the optimum point.
At B: $F = 3(4) + 2(0) = 12$

(b) $F = 2X_1 + 3X_2$

At O: $F = 2(0) + 3(0) = 0$
At C: $F = 2(0) + 3(4) = 12$ Any point on the CE line
At E: $F = 2(3) + 3(2) = 12$ is an optimum point.
At B: $F = 2(4) + 3(0) = 8$

(c) $F = 5X_1 + 2X_2$

At O: $F = 5(0) + 2(0) = 0$
At C: $F = 5(0) + 2(4) = 8$
At E: $F = 5(3) + 2(2) = 19$
At B: $F = 5(4) + 2(0) = 20$ B is the optimum point.

(d) $F = 3X_1 + 3X_2$

At O: $F = 3(0) + 3(0) = 0$
At C: $F = 3(0) + 3(4) = 12$
At E: $F = 3(3) + 3(2) = 15$ E is the optimum point.
At B: $F = 3(4) + 3(0) = 12$

5.

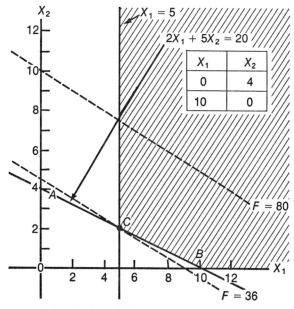

F lines:

$$F = 4X_1 + 8X_2$$

$$X_2 = \frac{F}{8} - \frac{X_1}{2}$$

$F = 80$ $X_2 = 10 - \tfrac{1}{2}X_1$

X_1	X_2
0	10
10	5

At C: $F = 36$
        $X_2 = 4\tfrac{1}{2} - \tfrac{1}{2}X_1$

X_1	X_2
0	4½
9	0

$2X_1 + 5X_2 = 20$

X_1	X_2
0	4
10	0

Point C: $X_1 = 5, X_2 = 2$
Point B: $X_1 = 10, X_2 = 0$

(a) $F = 4X_1 + 8X_2$

At C: $F = 4(5) + 8(2) = 36$ C is the optimum point.
At B: $F = 4(10) + 8(0) = 40$

(b) $F = 2X_1 + 7X_2$

At C: $F = 2(5) + 7(2) = 24$

At B: $F = 2(10) + 7(0) = 20$ B is the optimum point.

7. Let $X_1 =$ number of doors,

$X_2 =$ number of windows.

$1X_1 + 3X_2 \leq 9$ (hours available on machine I)

$2X_1 + 1X_2 \leq 8$ (hours available on machine II)

$X_1 \geq 0$

$X_2 \geq 0$

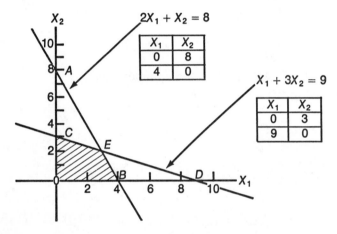

Point C: $X_1 = 0, X_2 = 3$

Point E: $X_1 = 3, X_2 = 2$

Point B: $X_1 = 4, X_2 = 0$ Maximize profit, F

(a) $F = \$10X_1 + \$20X_2$ At C: $F = 10(0) + 20(3) = \$60$

At E: $F = 10(3) + 20(2) = \$70$

At B: $F = 10(4) + 20(0) = \$40$

E is the optimum point: Produce 3 doors and 2 windows to make $70 profit.

(b) $F = \$10X_1 + \$35X_2$ At C: $F = 10(0) + 35(3) = \$105$

At E: $F = 10(3) + 35(2) = \$100$

At B: $F = 10(4) + 35(0) = \$40$

C is the optimum point: Produce 3 windows to make $105 profit.

9. Let $X_1 =$ amount invested in stock A in thousands of dollars,

$X_2 =$ amount invested in stock B in thousands of dollars.

$6\%X_1 + 2\%X_2 \geq \$300$

$1X_1 + 2X_2 \geq \$10,000$

$X_1 \geq 0$

$X_2 \geq 0$

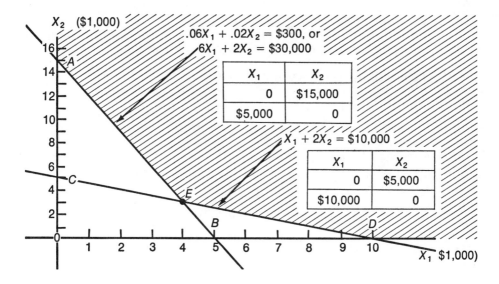

Point A: $X_1 = 0$, $X_2 = 15$
Point E: $X_1 = 4$, $X_2 = 3$
Point D: $X_1 = 10$, $X_2 = 0$

Minimize total amount invested, F. $F = X_1 + X_2$

At A: $F = 0 + 15 = 15$, or $15,000
At E: $F = 4 + 3 = 7$, or $7,000
At D: $F = 10 + 0 = 10$, or $10,000

E is the optimum point: Buy $4,000 stock A and $3,000 stock B to meet
                          his wishes.

Check: Dividend each year—$4,000(6%) + $3,000(2%) = $300
         Increase on investment in one year—
         1($4,000) + 2($3,000) = $10,000

Exercise 9–1 (p. 173)

1. $2X_1 + 3X_2 + S_3 = 12$ (1)
$4X_1 + 2X_2 + S_4 = 16$ (2)

Finding the initial solution:

$\qquad$ Let $X_1 = 0$ and $X_2 = 0$.
$\qquad$ From (1), $\quad S_3 = 12 - 2X_1 - 3X_2$ (3)
$\qquad\qquad\qquad\quad = 12$
$\qquad$ From (2), $\quad S_4 = 16 - 4X_1 - 2X_2$ (4)
$\qquad\qquad\qquad\quad = 16$
$\qquad\quad F = 3X_1 + 2X_2 = 3(0) + 2(0) = 0$

Initial solution: $X_1 = 0$, $X_2 = 0$, $S_3 = 12$, $S_4 = 16$, $F = 0$

Finding the second solution:

X_1 is more profitable since its coefficient is 3 in F equation.

If S_3 is replaced by X_1, $S_3 = 0$ and $X_2 = 0$.

$$\text{From (3), } X_1 = \frac{12 - 3X_2 - S_3}{2} = 6 - \frac{3X_2}{2} - \frac{S_3}{2}$$

$$= 6$$

If S_4 is replaced by X_1, $S_4 = 0$ and $X_2 = 0$.

$$\text{From (4), } X_1 = \frac{16 - 2X_2 - S_4}{4} = 4 - \frac{X_2}{2} - \frac{S_4}{4} \tag{5}$$

$$= 4 \text{ (smaller positive)}$$

$$\text{From (3), } S_3 = 12 - 2\left(4 - \frac{X_2}{2} - \frac{S_4}{4}\right) - 3X_2$$

$$= 4 - 2X_2 + \frac{S_4}{2} \tag{6}$$

$$= 4$$

$$F = 3X_1 + 2X_2 = 3\left(4 - \frac{X_2}{2} - \frac{S_4}{4}\right) + 2X_2 = 12 + \frac{X_2}{2} - \frac{3S_4}{4}$$

$$= 12 + \frac{0}{2} - \frac{3(0)}{4} = 12$$

Second solution: $X_1 = 4$, $X_2 = 0$, $S_3 = 4$, $S_4 = 0$, $F = 12$

Finding the third solution:

X_2 is a profitable one since its coefficient is $+\frac{1}{2}$ in the F equation.

If X_1 is replaced by X_2, $X_1 = 0$ and $S_4 = 0$.

$$\text{From (5), } X_2 = 2\left(4 - X_1 - \frac{S_4}{4}\right) = 8 - 2X_1 - \frac{S_4}{2}$$

$$= 8$$

If S_3 is replaced by X_2, $S_3 = 0$ and $S_4 = 0$.

$$\text{From (6), } X_2 = \frac{4 - S_3 + S_4/2}{2} = 2 - \frac{S_3}{2} + \frac{S_4}{4} \tag{7}$$

$$= 2 \text{ (smaller positive)}$$

$$\text{From (5), } X_1 = 4 - \frac{1}{2}\left(2 - \frac{S_3}{2} + \frac{S_4}{4}\right) - \frac{S_4}{4}$$

$$= 3 + \frac{1}{4}S_3 - \frac{3}{8}S_4 \tag{8}$$

$$= 3$$

$$F = 12 + \frac{X_2}{2} - \frac{3S_4}{4} = 12 + \frac{1}{2}\left(2 - \frac{S_3}{2} + \frac{S_4}{4}\right) - \frac{3S_4}{4}$$

$$= 13 - \frac{1}{4}S_3 - \frac{5}{8}S_4$$

$$= 13$$

Third solution: $X_1 = 3$, $X_2 = 2$, $S_3 = 0$, $S_4 = 0$, $F = 13$

The third solution is optimal since both the coefficients of S_3 and S_4 are negative in the F solution.

3. $1X_1 + 3X_2 + S_3 = 9$ (1) $X_1 =$ number of doors
   $2X_1 + 1X_2 + S_4 = 8$ (2) $X_2 =$ number of windows

Finding the initial solution:

Let $X_1 = 0$ and $X_2 = 0$.

From (1), $S_3 = 9 - X_1 - 3X_2$ (3)
                    $= 9$

From (2), $S_4 = 8 - 2X_1 - X_2$ (4)
                    $= 8$

$$F = 10X_1 + 20X_2 = 10(0) + 20(0) = 0$$

Initial solution: $X_1 = 0$, $X_2 = 0$, $S_3 = 9$, $S_4 = 8$, $F = 0$

Finding the second solution:

X_2 is more profitable since its coefficient is $+20$ in the F equation.

If S_4 is replaced by X_2, $S_4 = 0$ and $X_1 = 0$.

From (4), $X_2 = 8 - 2X_1 - S_4$
                      $= 8$

If S_3 is replaced by X_2, $S_3 = 0$ and $X_1 = 0$.

From (3), $X_2 = \dfrac{9 - X_1 - S_3}{3} = 3 - \dfrac{X_1}{3} - \dfrac{S_3}{3}$ (5)
                  $= 3$ (smaller positive)

From (4), $S_4 = 8 - 2X_1 - \left(3 - \dfrac{X_1}{3} - \dfrac{S_3}{3}\right)$

$$= 5 - \frac{5}{3}X_1 + \frac{S_3}{3} \qquad (6)$$

$$= 5$$

$$F = 10X_1 + 20X_2 = 10X_1 + 20\left(3 - \frac{X_1}{3} - \frac{S_3}{3}\right)$$

$$= 60 + \frac{10}{3}X_1 - \frac{20}{3}S_3 = 60$$

Second solution: $X_1 = 0$, $X_2 = 3$, $S_3 = 0$, $S_4 = 5$, $F = 60$

Finding the third solution:

X_1 is a profitable one since its coefficient is $+\frac{10}{3}$ in the F equation.

If X_2 is replaced by X_1, $X_2 = 0$ and $S_3 = 0$.

$$\text{From (5), } X_1 = 3\left(3 - X_2 - \frac{S_3}{3}\right) = 9 - 3X_2 - S_3$$
$$= 9$$

If S_4 is replaced by X_1, $S_4 = 0$ and $S_3 = 0$.

$$\text{From (6), } X_1 = \frac{3}{5}\left(5 + \frac{S_3}{3} - S_4\right) = 3 + \frac{1}{5}S_3 - \frac{3}{5}S_4 \qquad (7)$$
$$= 3 \text{ (smaller positive)}$$

$$\text{From (5), } X_2 = 3 - \frac{1}{3}\left(3 + \frac{1}{5}S_3 - \frac{3}{5}S_4\right) - \frac{S_3}{3}$$
$$= 2 - \frac{2}{5}S_3 + \frac{1}{5}S_4 \qquad (8)$$
$$= 2$$

$$F = 60 + \frac{10}{3}X_1 - \frac{20}{3}S_3 = 60 + \frac{10}{3}\left(3 + \frac{1}{5}S_3 - \frac{3}{5}S_4\right) - \frac{20}{3}S_3$$
$$= 70 - 6S_3 - 2S_4 = 70$$

Third solution: $X_1 = 3$, $X_2 = 2$, $S_3 = 0$, $S_4 = 0$, $F = 70$

The third solution is optimal since both the coefficients of S_3 and S_4 are negative in the F solution.

Exercise 9–2 (p. 188)

1. $2X_1 + 3X_2 \leq 12$ converted to $2X_1 + 3X_2 + 1S_3 + 0S_4 = 12$
   $4X_1 + 2X_2 \leq 16$ $4X_1 + 2X_2 + 0S_3 + 1S_4 = 16$

$$F = 3X_1 + 2X_2 + 0S_3 + 0S_4$$

Initial Simplex Table

C_j	3	2	0	0	Feasible Solution		$\dfrac{V}{X_1}$
C_i	X_1	X_2	S_3	S_4	Basic Variable	V	
0	2	3	1	0	S_3	12	$\frac{12}{2} = 6$
0	④	2	0	1	S_4	16	$\frac{16}{4} = 4$ smaller positive
Z_j	0	0	0	0	$F = 0$		
$C_j - Z_j$	3	2	0	0			

↑

Second Simplex Table

						V	$\dfrac{V}{X_2}$
0	0	(2)	1	$-\frac{1}{2}$	S_3	4	$\frac{4}{2}=2$ smaller positive
3	1	$\frac{1}{2}$	0	$\frac{1}{4}$	X_1	4	$\frac{4}{\left(\frac{1}{2}\right)}=8$
Z_j	3	$\frac{3}{2}$	0	$\frac{3}{4}$	$F=12$		
C_j-Z_j	0	$\frac{1}{2}$	0	$-\frac{3}{4}$			

Third Simplex Table (Optimal)

						V
2	0	1	$\frac{1}{2}$	$-\frac{1}{4}$	X_2	2
3	1	0	$-\frac{1}{4}$	$\frac{3}{8}$	X_1	3
Z_j	3	2	$\frac{1}{4}$	$\frac{5}{8}$	$F=13$	
C_j-Z_j	0	0	$-\frac{1}{4}$	$-\frac{5}{8}$		

Optimum solution: $X_1 = 3,\ X_2 = 2;\ F = 13$

3. $\begin{aligned} 2X_1 + 6X_2 &\le 12 \\ X_1 + 4X_2 &\le 7 \end{aligned}\Biggr\}$ converted to $\begin{cases} 2X_1 + 6X_2 + 1S_3 + 0S_4 = 12 \\ 1X_1 + 4X_2 + 0S_3 + 1S_4 = 7 \end{cases}$

Maximize $F = 3X_1 + 5X_2 + 0S_3 + 0S_4$

Initial Simplex Table

C_j	3	5	0	0	Feasible Solution		$\dfrac{V}{X_2}$
C_i	X_1	X_2	S_3	S_4	Basic Variable	V	
0	2	6	1	0	S_3	12	$\frac{12}{6}=2$
0	1	(4)	0	1	S_4	7	$\frac{7}{4}$ smaller positive
Z_j	0	0	0	0	$F=0$		
C_j-Z_j	3	5	0	0			

Second Simplex Table | | | | | | V | $\dfrac{V}{X_1}$

						V	$\dfrac{V}{X_1}$
0	$\textcircled{\tfrac{1}{2}}$	0	1	$-\tfrac{3}{2}$	S_3	$\tfrac{3}{2}$	$\dfrac{\tfrac{3}{2}}{\tfrac{1}{2}} = 3$ smaller positive
5	$\tfrac{1}{4}$	1	0	$\tfrac{1}{4}$	X_2	$\tfrac{7}{4}$	$\dfrac{\tfrac{7}{4}}{\tfrac{1}{4}} = 7$
Z_j	$\tfrac{5}{4}$	5	0	$\tfrac{5}{4}$	$F = \tfrac{35}{4}$		
$C_j - Z_j$	$\tfrac{7}{4}$	0	0	$-\tfrac{5}{4}$			

↑

Third Simplex Table | | | | | | V | $\dfrac{V}{S_4}$

						V	$\dfrac{V}{S_4}$
3	1	0	2	-3	X_1	3	$\dfrac{3}{-3} = -1$
5	0	1	$-\tfrac{1}{2}$	$\textcircled{1}$	X_2	1	$\tfrac{1}{1} = 1$ positive
Z_j	3	5	$\tfrac{7}{2}$	-4	$F = 14$		
$C_j - Z_j$	0	0	$-\tfrac{7}{2}$	4			

↑

Fourth Simplex Table (Optimal) | | | | | | V |

						V
3	1	3	$\tfrac{1}{2}$	0	X_1	6
0	0	1	$-\tfrac{1}{2}$	1	S_4	1
Z_j	3	9	$\tfrac{3}{2}$	0	$F = 18$	
$C_j - Z_j$	0	-4	$-\tfrac{3}{2}$	0		

Optimum solution: $X_1 = 6$, $X_2 = 0$, $S_3 = 0$, $S_4 = 1$; $F = 18$

5. $\left.\begin{array}{l} X_1 \le 5 \\ X_2 \le 10 \end{array}\right\}$ converted to $\left\{\begin{array}{l} 1X_1 + 0X_2 + 1S_3 + 0S_4 = 5 \\ 0X_1 + 1X_2 + 0S_3 + 1S_4 = 10 \end{array}\right.$

    $X_2 \le 14$ (redundant)

           Maximize $F = 2X_1 + 7X_2 + 0S_3 + 0S_4$

Initial Simplex Table

C_j	2	7	0	0	Feasible Solution		$\dfrac{V}{X_2}$
C_i	X_1	X_2	S_3	S_4	Basic Variable	V	
0	1	0	1	0	S_3	5	$\frac{5}{0}$ undefined
0	0	①	0	1	S_4	10	$\frac{10}{1} = 10$ positive
Z_j	0	0	0	0	$F = 0$		
$C_j - Z_j$	2	7	0	0			

↑

Second Simplex Table

							V	$\dfrac{V}{X_1}$
0	①	0	1	0	S_3		5	$\frac{5}{1} = 5$ positive
7	0	1	0	1	X_2		10	$\frac{10}{0}$ undefined
Z_j	0	7	0	7	$F = 70$			
$C_j - Z_j$	2	0	0	-7				

↑

Third Simplex Table (Optimal)

							V
2	1	0	1	0	X_1		5
7	0	1	0	1	X_2		10
Z_j	2	7	2	7	$F = 80$		
$C_j - Z_j$	0	0	-2	-7			

Optimum solution: $X_1 = 5$, $X_2 = 10$, $S_3 = 0$, $S_4 = 0$; $F = 80$

7. $X_1 =$ number of doors $X_2 =$ number of windows

$$\left. \begin{array}{l} X_1 + 3X_2 \leq 9 \\ 2X_1 + X_2 \leq 8 \end{array} \right\} \text{ converted to } \left\{ \begin{array}{l} 1X_1 + 3X_2 + 1S_3 + 0S_4 = 9 \\ 2X_1 + 1X_2 + 0S_3 + 1S_4 = 8 \end{array} \right.$$

Maximize $F = 10X_1 + 20X_2 + 0S_3 + 0S_4$

Initial Simplex Table

C_j	10	20	0	0	Feasible Solution		$\dfrac{V}{X_2}$
C_i	X_1	X_2	S_3	S_4	Basic Variable	V	
0	1	③	1	0	S_3	9	$\frac{9}{3}=3$ smaller positive
0	2	1	0	1	S_4	8	$\frac{8}{1}=8$
Z_j	0	0	0	0	$F=0$		
C_j-Z_j	10	20	0	0			

↑

Second Simplex Table | | | | | | V | $\dfrac{V}{X_1}$ |

20	$\frac{1}{3}$	1	$\frac{1}{3}$	0	X_2	3	$\dfrac{3}{(\frac{1}{3})}=9$
0	⑤⁄₃	0	$-\frac{1}{3}$	1	S_4	5	$\dfrac{5}{(\frac{5}{3})}=3$ smaller positive
Z_j	$\frac{20}{3}$	20	$\frac{20}{3}$	0	$F=60$		
C_j-Z_j	$\frac{10}{3}$	0	$-\frac{20}{3}$	0			

↑

Third Simplex Table (Optimal) | | | | | | V |

20	0	1	$\frac{2}{5}$	$-\frac{1}{5}$	X_2	2
10	1	0	$-\frac{1}{5}$	$\frac{3}{5}$	X_1	3
Z_j	10	20	6	2	$F=70$	
C_j-Z_j	0	0	-6	-2		

Optimum solution: $X_1=3$, $X_2=2$, $S_3=0$, $S_4=0$; $F=70$

9. $X_1=$ amount of industrial loan in \$ millions
$X_2=$ amount of residential loan in \$ millions

$$\left.\begin{array}{r} X_1+X_2\le 10 \\ X_1\le\ 6\ (=.6(10)) \\ X_2\le\ 8\ (=.8(10)) \end{array}\right\} \begin{array}{c}\text{converted}\\ \text{to}\end{array} \left\{\begin{array}{l} 1X_1+1X_2+1S_3+0S_4+0S_5=10 \\ 1X_1+0X_2+0S_3+1S_4+0S_5=\ 6 \\ 0X_1+1X_2+0S_3+0S_4+1S_5=\ 8 \end{array}\right.$$

Maximize $F=15\%X_1+10\%X_2+0S_3+0S_4+0S_5$

Initial Simplex Table

C_j	.15	.10	0	0	0	Feasible Solution		$\dfrac{V}{X_1}$
C_i	X_1	X_2	S_3	S_4	S_5	Basic Variable	V	
0	1	1	1	0	0	S_3	10	$\frac{10}{1}=10$
0	①	0	0	1	0	S_4	6	$\frac{6}{1}=6$ smaller positive
0	0	1	0	0	1	S_5	8	$\frac{8}{0}$ undefined
Z_j	0	0	0	0	0	$F=0$		
C_j-Z_j	.15	.10	0	0	0			

↑

Second Simplex Table V $\dfrac{V}{X_2}$

							V	$\dfrac{V}{X_2}$
0	0	①	1	−1	0	S_3	4	$\frac{4}{1}=4$ smaller positive
.15	1	0	0	1	0	X_1	6	$\frac{6}{0}$ undefined
0	0	1	0	0	1	S_5	8	$\frac{8}{1}=8$
Z_j	.15	0	0	.15	0	$F=.90$		
C_j-Z_j	0	.10	0	−.15	0			

↑

Third Simplex Table (Optimal) V

							V
.10	0	1	1	−1	0	X_2	4
.15	1	0	0	1	0	X_1	6
0	0	0	−1	1	1	S_5	4
Z_j	.15	.10	.10	.05	0	$F=1.30$	
C_j-Z_j	0	0	−.10	−.05	0		

Optimum solution: $X_1=6$, $X_2=4$, $S_3=4$, $S_4=0$, $S_5=4$
$F=1.30$ ($1,300,000)

Exercise 10-1 (p. 208)

1. $\left.\begin{array}{r}2X_1+5X_2\geq 20\\ X_1\geq5\end{array}\right\}$ converted to $\left\{\begin{array}{l}2X_1+5X_2-1S_3+0S_4+1A_5+0A_6=20\\ 1X_1+0X_2+0S_3-1S_4+0A_5+1A_6=5\end{array}\right.$

Minimize $F=4X_1+8X_2+0S_3+0S_4+MA_5+MA_6$

Initial Simplex Table

C_j	4	8	0	0	M	M	Feasible Solution		$\dfrac{V}{X_2}$
C_i	X_1	X_2	S_3	S_4	A_5	A_6	Basic Variable	V	
M	2	⑤	-1	0	1	0	A_5	20	$\frac{20}{5}=4$ positive
M	1	0	0	-1	0	1	A_6	5	$\frac{5}{0}$ undefined
Z_j	$3M$	$5M$	$-M$	$-M$	M	M	$F = 25M$		
$C_j - Z_j$	$4-3M$	$8-5M$	M	M	0	0			

↑

Second Simplex Table

								V	$\dfrac{V}{X_1}$
8	$\frac{2}{5}$	1	$-\frac{1}{5}$	0	$\frac{1}{5}$	0	X_2	4	$\frac{4}{\frac{2}{5}}=10$
M	①	0	0	-1	0	1	A_6	5	$\frac{5}{1}=5$ smaller
Z_j	$\frac{16}{5}+M$	8	$-\frac{8}{5}$	$-M$	$\frac{8}{5}$	M	$F = 32 + 5M$		
$C_j - Z_j$	$\frac{4}{5}-M$	0	$\frac{8}{5}$	M	$M-\frac{8}{5}$	0			

↑

Third Simplex Table (Optimal)

								V
8	0	1	$-\frac{1}{5}$	$\frac{2}{5}$	$\frac{1}{5}$	$-\frac{2}{5}$	X_2	2
4	1	0	0	-1	0	1	X_1	5
Z_j	4	8	$-\frac{8}{5}$	$-\frac{4}{5}$	$\frac{8}{5}$	$\frac{4}{5}$	$F = 36$	
$C_j - Z_j$	0	0	$\frac{8}{5}$	$\frac{4}{5}$	$M-\frac{8}{5}$	$M-\frac{4}{5}$		

Optimum solution: $X_1 = 5$, $X_2 = 2$, $S_3 = S_4 = A_5 = A_6 = 0$
$$F = 36$$

3. $\left.\begin{array}{l} X_1 \geq 5 \\ X_2 \geq 10 \end{array}\right\}$ converted to $\left\{\begin{array}{l} 1X_1 + 0X_2 - 1S_3 + 0S_4 + 1A_5 + 0A_6 = 5 \\ 0X_1 + 1X_2 + 0S_3 - 1S_4 + 0A_5 + 1A_6 = 10 \end{array}\right.$

Minimize $F = 2X_1 + 7X_2 + 0S_3 + 0S_4 + MA_5 + MA_7$

Initial Simplex Table

C_j	2	7	0	0	M	M	Feasible Solution		$\dfrac{V}{X_1}$
C_i	X_1	X_2	S_3	S_4	A_5	A_6	Basic Variable	V	
M	①	0	−1	0	1	0	A_5	5	$\frac{5}{1}=5$ positive
M	0	1	0	−1	0	1	A_6	10	$\frac{10}{0}$ undefined
Z_j	M	M	−M	−M	M	M	$F=15M$		
C_j-Z_j	$2-M$	$7-M$	M	M	0	0			

↑

Second Simplex Table | | V | $\dfrac{V}{X_2}$

2	1	0	−1	0	1	0	X_1	5	$\frac{5}{0}$ undefined
M	0	①	0	−1	0	1	A_6	10	$\frac{10}{1}=10$ positive
Z_j	2	M	−2	−M	2	M	$F=10+10M$		
C_j-Z_j	0	$7-M$	2	M	$M-2$	0			

↑

Third Simplex Table | | V

2	1	0	−1	0	1	0	X_1	5
7	0	1	0	−1	0	1	X_2	10
Z_j	2	7	−2	−7	2	7	$F=80$	
C_j-Z_j	0	0	2	7	$M-2$	$M-7$		

Optimum solution: $X_1 = 5$, $X_2 = 10$, $S_3 = S_4 = A_5 = A_6 = 0$
$F = 80$

No. The answers to X_1, X_2, and F are the same for both this problem and Problem 5 of Exercise 9-2. A graphical presentation for the two problems may help students to understand the reason of the similarity. The two answers are represented by the same point on the graph.

5. $\left.\begin{array}{r}4X_1 + 2X_2 \geq 100 \\ X_2 \geq 20 \\ \text{Minimize:} \\ F = 8X_1 + 5X_2\end{array}\right\}$ converted to $\left\{\begin{array}{l}4X_1 + 2X_2 - 1S_3 + 0S_4 + 1A_5 + 0A_6 = 100 \\ 0X_1 + 1X_2 + 0S_3 - 1S_4 + 0A_5 + 1A_6 = 20 \\ \text{Maximize } G = -F: \\ G = -8X_1 - 5X_2 - 0S_3 - 0S_4 - MA_5 - MA_6\end{array}\right.$

Initial Simplex Table

C_j	-8	-5	0	0	$-M$	$-M$	Feasible Solution		$\dfrac{V}{X_1}$
C_i	X_1	X_2	S_3	S_4	A_5	A_6	Basic Variable	V	
$-M$	④	2	-1	0	1	0	A_5	100	$\frac{100}{4}=25$ positive
$-M$	0	1	0	-1	0	1	A_6	20	$\frac{20}{0}$ undefined
Z_j	$-4M$	$-3M$	M	M	$-M$	$-M$	$G=-120M$		
C_j-Z_j	$4M-8$	$3M-5$	$-M$	$-M$	0	0			

↑

Second Simplex Table V $\dfrac{V}{X_2}$

	-8	1	$\frac{1}{2}$	$-\frac{1}{4}$	0	$\frac{1}{4}$	0	X_1	25	$\frac{25}{\frac{1}{2}}=50$
	$-M$	0	①	0	-1	0	1	A_6	20	$\frac{20}{1}=20$ smaller
Z_j	-8	$-M-4$	2	M	-2	$-M$		$G=-200-20M$		
C_j-Z_j	0	$M-1$	-2	$-M$	$2-M$	0				

↑

Third Simplex Table (Optimal) V

	-8	1	0	$-\frac{1}{4}$	$\frac{1}{2}$	$\frac{1}{4}$	$-\frac{1}{2}$	X_1	15
	-5	0	1	0	-1	0	1	X_2	20
Z_j	-8	-5	2	1	-2	-1	$G=-220$		
C_j-Z_j	0	0	-2	-1	$2-M$	$1-M$			

Optimum Solution: $X_1=15$, $X_2=20$, $S_3=S_4=A_5=A_6=0$
$$F=-G=-(-220)=220 \text{ (minimized value)}$$

No. The answers to X_1, X_2, and F are the same for both this problem and Problem 4 of Exercise 9–2. The two answers can be represented by the same point in a graph.

7. $\left.\begin{array}{l} 4X_1+2X_2=100 \\ X_1\geq 12 \\ X_2\leq 20 \end{array}\right\}$ $\begin{array}{c}\text{converted}\\\text{to}\end{array}$ $\left\{\begin{array}{l} 4X_1+2X_2+0S_3+0S_4+1A_5+0A_6=100 \\ 1X_1+0X_2-1S_3+0S_4+0A_5+1A_6=12 \\ 0X_1+1X_2+0S_3+1S_4+0A_5+0A_6=20 \end{array}\right.$

$$\text{Maximize} \quad F=8X_1+5X_2+0S_3+0S_4-MA_5-MA_6$$

Initial Simplex Table

C_j	8	5	0	0	$-M$	$-M$	Feasible Solution		$\dfrac{V}{X_1}$
C_i	X_1	X_2	S_3	S_4	A_5	A_6	Basic Variable	V	
$-M$	4	2	0	0	1	0	A_5	100	$\frac{100}{4}=25$
$-M$	①	0	-1	0	0	1	A_6	12	$\frac{12}{1}=12$ smaller
0	0	1	0	1	0	0	S_4	20	$\frac{20}{0}$ undefined
Z_j	$-5M$	$-2M$	M	0	$-M$	$-M$	$F=-112M$		
C_j-Z_j	$8+5M$	$5+2M$	$-M$	0	0	0			

↑

Second Simplex Table V $\dfrac{V}{S_3}$

$-M$	0	2	④	0	1	-4	A_5	52	$\frac{52}{4}=13$ positive
8	1	0	-1	0	0	1	X_1	12	$\frac{12}{-1}=-12$
0	0	1	0	1	0	0	S_4	20	$\frac{20}{0}$ undefined
Z_j	8	$-2M$	$-4M-8$	0	$-M$	$4M+8$	$F=96-52M$		
C_j-Z_j	0	$5+2M$	$4M+8$	0	0	$-5M-8$			

↑

Third Simplex Table V $\dfrac{V}{X_2}$

0	0	$\frac{1}{2}$	1	0	$\frac{1}{4}$	-1	S_3	13	$\frac{13}{\frac{1}{2}}=26$
8	1	$\frac{1}{2}$	0	0	$\frac{1}{4}$	0	X_1	25	$\frac{25}{\frac{1}{2}}=50$
0	0	①	0	1	0	0	S_4	20	$\frac{20}{1}=20$ smallest
Z_j	8	4	0	0	2	0	$F=200$		
C_j-Z_j	0	1	0	0	$-M-2$	$-M$			

↑

Fourth Simplex Table (Optimal) V

0	0	0	1	$-\frac{1}{2}$	$\frac{1}{4}$	-1	S_3	3
8	1	0	0	$-\frac{1}{2}$	$\frac{1}{4}$	0	X_1	15
5	0	1	0	1	0	0	X_2	20
Z_j	8	5	0	1	2	0	$F=220$	
C_j-Z_j	0	0	0	-1	$-M-2$	$-M$		

Optimum solution: $X_1 = 15$, $X_2 = 20$, $S_3 = 3$, $S_4 = A_5 = A_6 = 0$; $F = 220$.

Yes. The answers to X_1, X_2, and F are different for the two problems. Also, see the graph in Figure 8-7. The answer to Example 3 ($X_1 = 12$, $X_2 = 26$) is represented by point B, and the answer to our problem ($X_1 = 15$, $X_2 = 20$) is represented by point E in the figure.

9. X_1 = amount invested in stock A in thousands of dollars,
X_2 = amount invested in stock B in thousands of dollars.

$$\left.\begin{array}{l} 6\%X_1 + 2\%X_2 \geq \$300 \\ X_1 + 2X_2 \geq \$10,000 \\ \text{Minimize } F \text{ (amount)} \\ F = X_1 + X_2 \end{array}\right\} \begin{array}{l} \text{converted} \\ \text{to} \end{array} \left\{\begin{array}{l} 6X_1 + 2X_2 - 1S_3 + 0S_4 + 1A_5 + 0A_6 = 30* \\ 1X_1 + 2X_2 + 0S_3 - 1S_4 + 0A_5 + 1A_6 = 10* \\ \text{Minimize} \qquad\qquad (\text{* in } \$1,000\text{'s}) \\ F = 1X_1 + 1X_2 + 0S_3 + 0S_4 + MA_5 + MA_6 \end{array}\right.$$

Initial Simplex Table

C_j	1	1	0	0	M	M	Feasible Solution		$\dfrac{V}{X_1}$
C_i	X_1	X_2	S_3	S_4	A_5	A_6	Basic Variable	V	
M	⑥	2	-1	0	1	0	A_5	30	$\frac{30}{6} = 5$ smaller
M	1	2	0	-1	0	1	A_6	10	$\frac{10}{1} = 10$
Z_j	$7M$	$4M$	$-M$	$-M$	M	M	$F = 40M$		
$C_j - Z_j$	$1-7M$	$1-4M$	M	M	0	0			

$\uparrow$

Second Simplex Table, V, $\dfrac{V}{X_2}$

1	1	$\frac{1}{3}$	$-\frac{1}{6}$	0	$\frac{1}{6}$	0	X_1	5	$\frac{5}{(\frac{1}{3})} = 15$
M	0	$⑤\frac{5}{3}$	$\frac{1}{6}$	-1	$-\frac{1}{6}$	1	A_6	5	$\frac{5}{(\frac{5}{3})} = 3$ smaller
Z_j	1	$\frac{1}{3}+\frac{5M}{3}$	$\frac{M}{6}-\frac{1}{6}$	$-M$	$\frac{1}{6}-\frac{M}{6}$	M		$F = 5 + 5M$	
C_j-Z_j	0	$\frac{2}{3}-\frac{5M}{3}$	$\frac{1}{6}-\frac{M}{6}$	M	$\frac{7M}{6}-\frac{1}{6}$	0			

$\uparrow$

Third Simplex Table (Optimal), V

1	1	0	$-\frac{1}{8}$	$\frac{1}{8}$	$\frac{1}{8}$	$-\frac{1}{8}$	X_1	4
1	0	1	$\frac{1}{10}$	$-\frac{3}{5}$	$-\frac{1}{10}$	$\frac{3}{5}$	X_2	3
Z_j	1	1	$-\frac{1}{10}$	$-\frac{2}{5}$	$\frac{1}{10}$	$\frac{2}{5}$	$F = 7$	
C_j-Z_j	0	0	$\frac{1}{10}$	$\frac{2}{5}$	$M-\frac{1}{10}$	$M-\frac{2}{5}$		

Optimum solution: $X_1 = 4$, $X_2 = 3$, $S_3 = S_4 = A_5 = A_6 = 0$; $F = 7$
(all in 1,000's.)

Answer: Invest \$4,000 in stock A and \$3,000 in stock B. The minimum amount of investment is \$7,000.

11. Pivot X_1 column in Table 10–11:

Second Simplex Table

X_1	X_2	S_3	S_4	A_5	A_6	Basic Variable	V
1	$\frac{1}{2}$	$-\frac{1}{4}$	0	$\frac{1}{4}$	0	X_1	25
0	1	0	-1	0	1	A_6	20
0	$1 - M$	2	M	$M - 2$	0	$F - (200 + 20M)$	

$$
\begin{array}{lcccccc}
\text{Old } F \text{ row} & 8-4M & 5-3M & M & M & 0 & 0 \\
(-)\ (8-4M)\cdot X_1 \text{ row} & 1(8-4M) & \tfrac{1}{2}(8-4M) & -\tfrac{1}{4}(8-4M) & 0 & \tfrac{1}{4}(8-4M) & 0 \\
\text{New } F \text{ row} & 0 & 1-M & 2 & M & M-2 & 0
\end{array}
$$

$$F - 120M$$
$$\underline{25(8-4M)}$$
$$F - 200 - 20M$$

The second solution: $X_1 = 25$, $X_2 = S_3 = S_4 = A_5 = 0$, $A_6 = 20$

The F equation is:

$$0X_1 + (1 - M)X_2 + 2S_3 + MS_4 + (M - 2)A_5 + 0A_6 = F - 200 - 20M$$
$$0 = F - 200 - 20M$$
$$F = 200 + 20M$$

Exercise 11–1 (p. 216)

1. (a) $_4P_4 = 4 \cdot 3 \cdot 2 \cdot 1 = 24$

(b) $_7P_3 = 7 \cdot 6 \cdot 5 = 210$

(c) $_3C_3 = \dfrac{3 \cdot 2 \cdot 1}{3 \cdot 2 \cdot 1} = 1$

(d) $_6C_2 = \dfrac{6 \cdot 5}{2 \cdot 1} = 15$

(e) $_6C_4 = \dfrac{6 \cdot 5 \cdot 4 \cdot 3}{4 \cdot 3 \cdot 2 \cdot 1} = 15$

(f) $_{100}C_{97} = {}_{100}C_3 = \dfrac{100 \cdot 99 \cdot 98}{3 \cdot 2 \cdot 1}$
   $= 161,700$

3. (a) $_5P_5 = 5 \cdot 4 \cdot 3 \cdot 2 \cdot 1 = 120$

(b) $_5P_4 = 5 \cdot 4 \cdot 3 \cdot 2 = 120$

5. (a) $10 \cdot 10 \cdot 10 = 1,000$

(b) $10 \cdot 9 \cdot 8 = 720$

7. $_6C_0 + {}_6C_1 + {}_6C_2 + {}_6C_3 + {}_6C_4 + {}_6C_5 + {}_6C_6 = 1 + 6 + 15 + 20 + 15 + 6 + 1$
   $= 64$, or
$$2^6 = 64$$

9. (a) $_{20}P_3 = 20 \cdot 19 \cdot 18 = 6,840$

(b) $_{20}C_3 = \dfrac{20 \cdot 19 \cdot 18}{3 \cdot 2 \cdot 1} = 1,140$

11. $_5C_2 + {_5C_3} + {_5C_4} + {_5C_5} = 10 + 10 + 5 + 1 = 26$

Exercise 11–2 (p. 223)

1. (a) $\dfrac{9}{25}$ (b) $\dfrac{16}{25}$ (c) $\dfrac{16}{25}$ (d) 1   (e) 0

7. (a) $\dfrac{2}{5} \cdot \dfrac{3}{5} = \dfrac{6}{25}$

3. (a) $\dfrac{9,647,694}{9,664,994} = .99821$, or round to .998

(b) $\dfrac{3}{5} \cdot \dfrac{3}{5} = \dfrac{9}{25}$

(b) $1 - .998 = .002$

9. $\dfrac{2}{6} \cdot \dfrac{4}{6} = \dfrac{8}{36} = \dfrac{2}{9}$

5. $\dfrac{45}{100} + \dfrac{30}{100} - \dfrac{10}{100} = \dfrac{65}{100} = \dfrac{13}{20}$

11. (a) $P(\text{Mary on the committee}) = \dfrac{{_7C_2}}{{_8C_3}} = \dfrac{\dfrac{7 \cdot 6}{2}}{\dfrac{8 \cdot 7 \cdot 6}{3 \cdot 2 \cdot 1}} = \dfrac{21}{56}$

$P(\text{Mary not on the committee}) = 1 - \dfrac{21}{56} = \dfrac{35}{56}$

(b) $P(\text{Mary on the committee}) = \dfrac{21}{56}$

$P(\text{Nancy on the committee}) = \dfrac{21}{56}$

$P(\text{Mary and Nancy both on the committee}) = \dfrac{{_6C_1}}{{_8C_3}} = \dfrac{6}{56}$

$P(\text{Mary or Nancy or both on the committee}) = \dfrac{21}{56} + \dfrac{21}{56} - \dfrac{6}{56} = \dfrac{36}{56}$

$= \dfrac{9}{14}$

Or:

P(Mary and two girls, including Nancy; that is, Mary or both on the committee) $= \dfrac{21}{56}$

P(Nancy and two girls, excluding Mary, on the committee) $= \dfrac{{_6C_2}}{{_8C_3}} = \dfrac{15}{56}$

P(Mary or Nancy or both on the committee) $= \dfrac{21}{56} + \dfrac{15}{56} = \dfrac{36}{56} = \dfrac{9}{14}$

Exercise 11–3 (p. 233)

1. $(a - b)^3 = {}_3C_0a^3 + {}_3C_1a^2(-b) + {}_3C_2a(-b)^2 + {}_3C_3(-b)^3$
$= a^3 - 3a^2b + 3ab^2 - b^3$

3. $(3a + 2b)^5 = {}_5C_0(3a)^5 + {}_5C_1(3a)^4(2b) + {}_5C_2(3a)^3(2b)^2 + {}_5C_3(3a)^2(2b)^3$
$+ {}_5C_4(3a)(2b)^4 + {}_5C_5(2b)^5$
$= 243a^5 + 810a^4b + 1{,}080a^3b^2 + 720a^2b^3 + 240ab^4 + 32b^5$

5. $(.2 + .8)^4 = {}_4C_0(.2)^4 + {}_4C_1(.2)^3(.8) + {}_4C_2(.2)^2(.8)^2 + {}_4C_3(.2)(.8)^3 + {}_4C_4(.8)^4$
$= .0016 + .0256 + .1536 + .4096 + .4096 = 1$

7. $(Q + P)^5 = {}_5C_0Q^5 + {}_5C_1Q^4P + {}_5C_2Q^3P^2 + {}_5C_3Q^2P^3 + {}_5C_4QP^4 + {}_5C_5P^5$
$= Q^5 + 5Q^4P + 10Q^3P^2 + 10Q^2P^3 + 5QP^4 + P^5$

9. (a) $P\left(0; 3, \dfrac{3}{5} = .6\right) = {}_3C_0(.6)^0(.4)^3 = \quad .064$

$\begin{aligned}
P(1; 3, .6) &= {}_3C_1(.6)^1(.4)^2 = \quad .288 \\
P(2; 3, .6) &= {}_3C_2(.6)^2(.4)^1 = \quad .432 \\
P(3; 3, .6) &= {}_3C_3(.6)^3(.4)^0 = \quad \underline{.216} \\
&\qquad\qquad\text{Total} \quad 1.000
\end{aligned}$

(b)   <u>Discrete Form</u>   <u>Continuous Form</u>

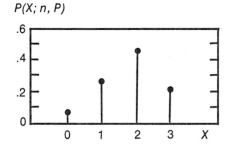 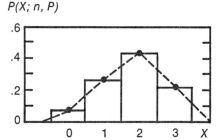

(c) and (d)(1)

X	$P(X; n, P)$ or f	fX	$X - \mu$ $\mu = 1.8$	$f(X - \mu)$	$f(X - \mu)^2$
0	.064	0	−1.8	−.1152	.20736
1	.288	.288	− .8	−.2304	.18432
2	.432	.864	.2	.0864	.01728
3	.216	.648	1.2	.2592	.31104
Total	1.000	1.800		.0000	.72000

$$\mu = \frac{1.800}{1.000} = 1.8$$

$$\sigma = \sqrt{\frac{.72000}{1.000}} = \sqrt{.72} = .848528, \text{ or round to } 0.85$$

(c)(2) $\mu = nP = 3(.6) = 1.8$

(d)(2) $\sigma = \sqrt{nPQ} = \sqrt{3(.6)(.4)} = \sqrt{.72}$

11. $P(4; 8, .6) = {}_8C_4(.6)^4(.4)^4 = 70(.1296)(.0256) = .23224320$, or round to .23

$$\left(\text{or } \frac{18,144}{78,125} \right)$$

$\mu = nP = 8(.6) = 4.8$

$\sigma = \sqrt{nPQ} = \sqrt{8(.6)(.4)} = \sqrt{1.92} = 1.38564$, or round to 1.39

Exercise 11–4 (p. 241)

1. (a) $z = \dfrac{70 - 60}{10} = 1$ $\qquad A(1) = .34134$

$10,000(.5 - .34134) = 1,586.6$, or 1,587 sales tickets

(b) $z = \dfrac{90 - 60}{10} = 3$ $\qquad A(3) = .49865$

$10,000(.5 - .49865) = 13.5$, or 14 tickets

(c) $z = \dfrac{60 - 60}{10} = 0$ $\qquad A(0) = 0$

$10,000(.5) = 5,000$ tickets

(d) $z = \dfrac{40 - 60}{10} = -2$ $\qquad A(-2) = .47725$

$10,000(.5 - .47725) = 227.5$, or 228 tickets

3. (a) $z = \dfrac{70 - 60}{10} = 1$ $\qquad A(1) = .34134$

$10,000(.34134) = 3,413.4$ or 3,413 sales

(b) $z = \dfrac{80 - 60}{10} = 2$ $\qquad A(2) = .47725$

$10,000(.47725) = 4,772.5$ or 4,773 sales

(c) $z = \dfrac{90 - 60}{10} = 3$ $\qquad A(3) = .49865$

$10,000(.49865) = 4,986.5$ or 4,987 sales

(d) $z = \dfrac{50 - 60}{10} = -1$ $\qquad A(-1) = .34134$

$10,000(.34134) = 3,413.4$ or 3,413 sales

(e) $z = \dfrac{40 - 60}{10} = -2$ $\qquad A(-2) = .47725$

$10,000(.47725) = 4,772.5$ or 4,773 sales

(f) $z = \dfrac{30 - 60}{10} = -3$ $\qquad A(-3) = .49865$

$10,000(.49865) = 4,986.5$ or 4,987 sales

5. (a) $A(z) = 50\% - 10\% = 40\% = .4$ $z = 1.28$

$1.28 = \dfrac{X - 60}{10}$ $X = 1.28(10) + 60 = \$72.8$

(b) $A(z) = 70\% - 50\% = 20\% = .2$ $z = -.52$

$-.52 = \dfrac{X - 60}{10}$ $X = (-.52)(10) + 60 = \$54.8$

(c) $A(z) = 50\% - 10\% = 40\% = .4$ $z = -1.28$

$-1.28 = \dfrac{X - 60}{10}$ $X = (-1.28)(10) + 60 = \$47.2$

(d) $A(z) = 80\% - 50\% = 30\% = .3$ $z = .84$

$.84 = \dfrac{X - 60}{10}$ $X = .84(10) + 60 = \$68.4$

Exercise 12–1 (p. 256)

1. (a)

Mark / Color	X	Y	Row Total
White	$P(\text{white and } X)$ $= \frac{3}{40}$	$P(\text{white and } Y)$ $= \frac{7}{40}$	$P(\text{white})$ $= \frac{10}{40}$
Yellow	$P(\text{yellow and } X)$ $= \frac{12}{40}$	$P(\text{yellow and } Y)$ $= \frac{18}{40}$	$P(\text{yellow})$ $= \frac{30}{40}$
Column Total	$P(X)$ $= \frac{15}{40}$	$P(Y)$ $= \frac{25}{40}$	$\frac{40}{40} = 1$

(b) $P(\text{white}|X) = \dfrac{P(\text{white and } X)}{P(X)} = \dfrac{\frac{3}{40}}{\frac{15}{40}} = \dfrac{3}{15} = \dfrac{1}{5}$

$P(\text{yellow}|X) = \dfrac{P(\text{yellow and } X)}{P(X)} = \dfrac{\frac{12}{40}}{\frac{15}{40}} = \dfrac{12}{15} = \dfrac{4}{5}$

3. Let $A_1 =$ the event of selecting an accounting employee,
$A_2 =$ the event of selecting a purchasing employee,
$A_3 =$ the event of selecting a marketing employee,
$B \ =$ the event of selecting a college educated employee.

(1) Event A_i	(2) $P(A_i)$	(3) $P(B\|A_i)$	(4) $P(A_i \text{ and } B)$ (2) × (3)	(5) $P(A_i\|B)$ (4) ÷ $P(B)$
A_1	$\dfrac{12+8}{60}=\dfrac{1}{3}$	$\dfrac{12}{8+12}=\dfrac{12}{20}$	$\dfrac{4}{20}$	$\dfrac{4}{20} \div \dfrac{7}{20}=\dfrac{4}{7}$
A_2	$\dfrac{3+17}{60}=\dfrac{1}{3}$	$\dfrac{3}{3+17}=\dfrac{3}{20}$	$\dfrac{1}{20}$	$\dfrac{1}{20} \div \dfrac{7}{20}=\dfrac{1}{7}$
A_3	$\dfrac{6+14}{60}=\dfrac{1}{3}$	$\dfrac{6}{6+14}=\dfrac{6}{20}$	$\dfrac{2}{20}$	$\dfrac{2}{20} \div \dfrac{7}{20}=\dfrac{2}{7}$
Total	$\frac{3}{3}=1$		$P(B)=\frac{7}{20}$	$\frac{7}{7}=1$

Answer: The employee is most likely from the Accounting
Department, 4/7.

5.

(1) Event A_i	(2) $P(A_i)$	(3) $P(B\|A_i)$ (B = Event of drawing spoiled apples)	(4) $P(A_i \text{ and } B)$ (2) × (3)	(5) $P(A_i\|B)$ (4) ÷ $P(B)$
A_1 (Good Shipment)	.9	$P(3; 30, .02) = .0188$	.01692	.5712
A_2 (Poor Shipment)	.1	$P(3; 30, .05) = .1270$	.01270	.4288
Total	1.0		$P(B) = .02962$	1.0000

7.

(1) Event A_i	(2) $P(A_i)$	(3) $P(B\|A_i)$ (B = Event of drawing nondefective)	(4) $P(A_i \text{ and } B)$ (2) × (3)	(5) $P(A_i\|B)$ (4) ÷ $P(B)$
A_1 (Plant 1)	.4	$\frac{9}{10}=.90$	.36	.3871
A_2 (Plant 2)	.6	$\frac{19}{20}=.95$	.57	.6129
Total	1.0		$P(B) = .93$	1.0000

Check:

Plant	Production (Bulbs)	Defective (Bulbs)	Nondefective (Bulbs)
1	$10,000 \times 40\% = 4,000$	$4,000 \times \frac{1}{10} = 400$	3,600
2	$10,000 \times 60\% = 6,000$	$6,000 \times \frac{1}{20} = 300$	5,700
Total	10,000	700	9,300

$P(\text{Plant 1: defective}) = \frac{400}{700} = \frac{4}{7} = .5714$
$P(\text{Plant 2: defective}) = \frac{300}{700} = \frac{3}{7} = .4286$

$P(\text{Plant 1: nondefective}) = \frac{3,600}{9,300} = \frac{12}{31} = .3871$
$P(\text{Plant 2: nondefective}) = \frac{5,700}{9,300} = \frac{19}{31} = .6129$

9.

(1) Event A_i	(2) $P(A_i)$	(3) $P(B\|A_i)$ (B = Event of drawing 4 defective bulbs from 25 bulbs in A_i)	(4) $P(A_i \text{ and } B)$ (2) × (3)	(5) $P(A_i\|B)$ (4) ÷ P(B)
A_1 (Plant 1)	.30	$P(4; 25, .01) = .0001$	.00003	.0004
A_2 (Plant 2)	.20	$P(4; 25, .05) = .0269$	.00538	.0677
A_3 (Plant 3)	.40	$P(4; 25, .10) = .1384$	.05536	.6969
A_4 (Plant 4)	.10	$P(4; 25, .20) = .1867$	.01867	.2350
Total	1.00		.07944	1.0000

Exercise 12–2 (p. 263)

1.

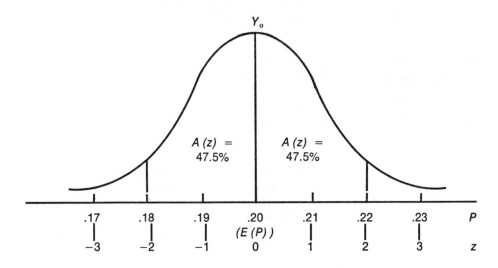

Prior Probability Distribution

$$1.96 = \frac{.22 - .20}{\sigma},$$

$\sigma = .0102$, or round to .01.

Computation of Prior Probabilities

P(Event)		P − E(P) = (1) − .20	$z = \frac{(3)}{.01}$	Area Between Y_o and z	Prior Probability P(p)
Interval limits	Midpoint				
(1)	(2)	(3)	(4)	(5)	(6)
.185		−.015	−1.5	.43319	
	.19				.24173
.195		−.005	−.5	.19146−┐	
	.20			+——.38292	
.205		.005	.5	.19146−┘	
	.21				.24173
.215		.015	1.5	.43319	
	.22				.06060
.225		.025	2.5	.49379	
	.23				.00598
.235		.035	3.5	.49977	

3.

Computation of Posterior Probability and Expected Value

(1) P (Event)		(2) Prior Proba- bility	(3) Conditional Probability	(4) Joint Probability (2) × (3)	(5) Posterior Probability (4) ÷ .23513	(6) Posterior Expected P (1) × (5)
Interval limits	Mid- point					
.185–.195	.19	.24173	.00135	.00033	.00140	.00027
.195–.205	.20	.38292	.15731	.06024	.25620	.05124
.205–.215	.21	.24173	.68268	.16502	.70182	.14738
.215–.225	.22	.06060	.15731	.00953	.04053	.00892
.225–.235	.23	.00598	.00135	.00001	.00004	.00001
Total		1.00000		.23513	1.00000	.20782

Exercise 13–1 (p. 273)

1. (a)

A Payoff Table

Possible Daily Sales (Events)	Possible Number of Dozens of Doughnuts Stocked (Act)		
	10	11	12
	Conditional Profit by Each Act		
10 (Dozens)	(10 × 1.6) −(10 × 1.0) = 6.00	(10 × 1.6) −(11 × 1.0) = 5.00	(10 × 1.6) −(12 × 1.0) = 4.00
11	6.00	(11 × 1.6) −(11 × 1.0) = 6.60	(11 × 1.6) −(12 × 1.0) = 5.60
12	6.00	6.60	(12 × 1.6) −(12 × 1.0) = 7.20

(b) and (c)

(1) Daily Sales (Event)	(2) Number of Days	(3) Probability of Each Event	(4) Expected Daily Sales
		(2) ÷ 100	(1) × (3)
10 (Dozens)	20	.20	2.0
11	30	.30	3.3
12	50	.50	6.0
Total	100	1.00	11.3

(d)

Computing Expected Profits

		Act (Doughnuts Stocked)					
		10 (Dozens)		11		12	
		Computing Expected Profit					
		Profit		Profit		Profit	
Event (Dozens)	Probability of Event	Condi-tional	Expected	Condi-tional	Expected	Condi-tional	Expected
10	.2	$6.00	$1.20	$5.00	$1.00	$4.00	$0.80
11	.3	6.00	1.80	6.60	1.98	5.60	1.68
12	.5	6.00	3.00	6.60	3.30	7.20	3.60
Total	1.0		$6.00		$6.28		$6.08
					Optimum Act: 11		

(e)

Expected Profit Under Certainty

(1) Event (Dozens)	(2) Probability of Event	(3) Conditional Profit under Certainty	(4) Expected Profit under Certainty
			(2) × (3)
10	.2	$6.00	$1.20
11	.3	6.60	1.98
12	.5	7.20	3.60
Total	1.0		$6.78

(f) Expected value of perfect information
$6.78 − $6.28 = $0.50

3. (a)

A Payoff Table (Also see problem 1(a) table)

	Act (Dozens Stocked)		
	10	11	12
Events (Dozens)	Conditional Profit by Each Act		
10	$6	$5 + .80 = \$5.80$	$4 + 2(.80) = \$5.60$
11	6	6.60	$5.6 + .80 =\ 6.40$
12	6	6.60	7.20

(b) and (c): Same as in problem 1.

(d)

Computing Expected Profits

		Act (Doughnuts Stocked)					
		10 (Dozens)		11		12	
		Computing Expected Profit					
		Profit		Profit		Profit	
Event (Dozens)	Probability of Event	Condi-tional	Expected	Condi-tional	Expected	Condi-tional	Expected
10	.2	$6.00	$1.20	$5.80	$1.16	$5.60	$1.12
11	.3	6.00	1.80	6.60	1.98	6.40	1.92
12	.5	6.00	3.00	6.60	3.30	7.20	3.60
Total	1.0		6.00		6.44		6.64

Optimum Act: 12

(e) Same as in problem 1.
(f) Expected value of perfect information
    $\$6.78 - \$6.64 = \$0.14$

5. (a)

A Payoff Table

	Dozens of Doughnuts Stocked (Act)				
Event	10	20	30	40	50
(Dozens)	Conditional Profit by Each Act				
10	$6	−$ 4	−$14	−$24	−$34
20	6	$12	$ 2	−$ 8	−$18
30	6	12	18	$ 8	−$ 2
40	6	12	18	$24	$14
50	6	12	18	24	30

(b) and (c)

(1) Event (Dozens)	(2) Number of Days	(3) Probability of Each Event	(4) Expected Daily Sales
		(2) ÷ 200	(1) × (3)
10	60	.30	3.0
20	50	.25	5.0
30	40	.20	6.0
40	30	.15	6.0
50	20	.10	5.0
Total	200	1.00	25.0

(d)

Computing Expected Profits

		Dozens of Doughnuts Stocked (Act)									
		10		20		30		40		50	
	Proba-	Computing Expected Profit									
	bility	Profit		Profit		Profit		Profit		Profit	
Event	of										
(Dozens)	Event	Con.	Exp.	Con.	Exp.	Con.	Exp.	Con.	Exp.	Con.	Exp.
10	.30	$6	$1.80	−$ 4	−1.20	−$14	−4.20	−$24	−7.20	−$34	−10.20
20	.25	6	1.50	12	3.00	2	.50	− 8	−2.00	− 18	− 4.50
30	.20	6	1.20	12	2.40	18	3.60	8	1.60	− 2	− .40
40	.15	6	.90	12	1.80	18	2.70	24	3.60	14	2.10
50	.10	6	.60	12	1.20	18	1.80	24	2.40	30	3.00
Total	1.00		6.00		$7.20		4.40		−1.60		−10.00

Optimum
Act: 20

(e)

Expected Profit Under Certainty

(1) Event (Dozens)	(2) Probability of Event	(3) Conditional Profit under Certainty	(4) Expected Profit under Certainty
			(2) × (3)
10	.30	$ 6	$ 1.80
20	.25	$12	3.00
30	.20	18	3.60
40	.15	24	3.60
50	.10	30	3.00
Total	1.00		$15.00

(f) Expected value of perfect information
$$15.00 - 7.20 = \$7.80$$

7. (a)

A Payoff Table (Also see problem 5(a) table.)

Event (Dozens)	Dozens of Doughnuts Stocked (Act)				
	10	20	30	40	50
	Conditional Profit by Each Act				
10	$6	−4 + .80(10) = $4	−14 + .80(20) = $2	−24 + .80(30) = $0	−34 + .80(40) = −$2
20	6	12	2 + .80(10) = 10	− 8 + .80(20) = 8	−18 + .80(30) = $6
30	6	12	18	8 + .80(10) = 16	− 2 + .80(20) = 14
40	6	12	18	24	14 + .80(10) = 22
50	6	12	18	24	30

(b) and (c): Same as in problem 5.

(d)

Computing Expected Profits

Event (Dozens)	Probability of Event	Dozens of Doughnuts Stocked (Act) 10 Profit Con.	Exp.	20 Profit Con.	Exp.	30 Profit Con.	Exp.	40 Profit Con.	Exp.	50 Profit Con.	Exp.
10	.30	$6	$1.80	$ 4	$1.20	$ 2	$ 0.60	$ 0	$ 0	−$ 2	−$ .60
20	.25	6	1.50	12	3.00	10	2.50	8	2.00	6	1.50
30	.20	6	1.20	12	2.40	18	3.60	16	3.20	14	2.80
40	.15	6	.90	12	1.80	18	2.70	24	3.60	22	3.30
50	.10	6	.60	12	1.20	18	1.80	24	2.40	30	3.00
Total	1.00		6.00		9.60		11.20		11.20		10.00
						Optimum act: 30 or 40					

(e) Same as in problem 5(e).
(f) Expected value of perfect information
$15.00 - 11.20 = \$3.80$

Exercise 13–2 (p. 282)

1. $U(\$700) = .5 \times U(\$100) + .5 \times U(\$4{,}000)$
$= .5(30) + .5(80)$
$= 15 + 40$
$= 55$

3. $U(\$1{,}000) = .3 \times U(\$100) + .7 \times U(\$4{,}000)$
$= .3(30) + .7(80)$
$= 9 + 56$
$= 65$

5. $U(\$4{,}000) = .3 \times U(\$100) + .7 \times U(\$9{,}000)$
$80 = .3(30) + .7 \times U(\$9{,}000)$

$$U(\$9{,}000) = \frac{80 - 9}{.7} = \frac{71}{\frac{7}{10}} = 101\frac{3}{7} = 101.43$$

7. $U(\$100) = .4 \times U(-\$2{,}500) + .6 \times U(\$4{,}000)$
$30 = .4 \times U(-\$2{,}500) + .6(80)$

$$U(-\$2{,}500) = \frac{30 - 48}{.4} = -45$$

9. $U(\$1,200) = .7 \times U(\$100) + .3 \times U(\$4,000)$
$= .7(30) + .3(80)$
$= 21 + 24$
$= 45$

No. His opinion is not being consistent. The result that $U(\$700) = 55$ and $U(\$1,200) = 45$ is not logical, since $\$700 < \$1,200$ but $55 > 45$. The probability of gaining $4,000 in this problem should be higher than .5 (problem 1) instead of only .3.

11. Let P be the probability of giving $1,000 and $(1 - P)$ be the probability of giving $6,000.

$U(\$4,000) = P \times U(\$1,000) + (1 - P) \times U(\$6,000)$
$90 = P(50) + (1 - P)(100)$
$= 50P + 100 - 100P$
$50P = 10$
$P = .2$
$1 - P = 1 - .2 = .8$

13. (a) Maximum possible profit $= \$0 -$ Do not buy insurance.
(b) Minimum possible loss $= \$0 -$ Do not buy insurance.
(c) Average profit if buy $= \frac{1}{2}(-100) + \frac{1}{2}(-100) = -\$100$. Average profit if not buy $= \frac{1}{2}(0) + \frac{1}{2}(-50,000) = -\$25,000$. Buy insurance since $(-\$100)$ is higher.
(d) No fire – event occurring at .999. Highest profit $= \$0 -$ Do not buy insurance.

Exercise 14–1 (p. 297)

1. (a) $I = 1,000 \times 5\% \times 3 = \$150$ $S = 1,000 + 150 = \$1,150$
(b) $I = 1,000 \times 5\% \times \frac{6}{12} = \$25$ $S = 1,000 + 25 = \$1,025$

3. (a) $I = 4,000 \times 12\% \times \frac{60}{360} = \$80$
(b) $S = 4,000 + 80 = \$4,080$

5. (a) $S = 1,000(1 + 5\%)^3 = 1,000(1.157625) = \$1,157.63$
(b) $I = 1,157.63 - 1,000 = \$157.63$

7. (a) $S = 1,000(1 + 4\%)^8 = 1,000(1.368569) = \$1,368.57$
$I = 1,368.57 - 1,000 = \$368.57$
(b) $S = 1,000(1 + 2\%)^{16} = 1,000(1.372786) = \$1,372.79$
$I = 1,372.79 - 1,000 = \$372.79$

9. $S = 5,000(1 + 5\%)^{12} = 5,000(1.795856) = \$8,979.28$

11. $P = 1,500(1 + 7\%)^{-8} = 1,500(0.582009) = \$873.01$

13. $S = 1,000(1 + 5\%)^6 = 1,000(1.340096) = \$1,340.10$
$P = 1,340.10(1 + 2\%)^{-6} = 1,340.10(0.837971) = \$1,122.96$ (Proceeds)
$1,340.10 - 1,122.96 = \$217.14$ (Compound discount)

15. (a) $f = (1 + \frac{1}{2}\%)^{12} - 1 = 1.061678 - 1 = .061678$, or 6.17%

(b) $f = (1 + 1\frac{1}{2}\%)^4 - 1 = 1.061364 - 1 = .061364$, or 6.14%

(c) $f = (1 + 3\%)^2 - 1 = 1.060900 - 1 = .060900$, or 6.09%

(d) $f = 6\%$

(e) $f = e^{.06} - 1 = 1.06183655 - 1 = .06183655$, or 6.18%

17. (a) $S = 1,000(e^{.05(1)}) = 1,000(1.05127110) = \$1,051.27$

$I = 1,051.27 - 1,000 = \$51.27$

(b) $S = 1,000(e^{.05(4)}) = 1,000(e^{.2}) = 1,000(1.22140276)$

$= \$1,221.40$

$I = 1,221.40 - 1,000 = \$221.40$

19. (a) $P = 1,000(e^{-.06(5)}) = 1,000(e^{-.3}) = 1,000(.74081822)$

$= \$740.82$

(b) Compound discount $= 1,000.00 - 740.82 = \$259.18$

Exercise 14–2 (p. 316)

1. $S_n = 1,000 \ s_{\overline{24)}1\%} = 1,000(26.973465) = \$26,973.47$

3. (a) $S_n = 200 \ s_{\overline{20)}1\frac{1}{2}\%} = 200(23.123667) = \$4,624.73$

(b) $I = 4,624.73 - 20(200) = \$624.73$

5. (a) $S = 4,624.73(1 + 1\frac{1}{2}\%)^{12} = 4,624.73(1.195618) = \$5,529.41$

(b) $I = 5,529.41 - 20(200) = \$1,529.41$

7. $A_n = 150 \ a_{\overline{36)}\frac{1}{2}\%} = 150(32.871016) = \$4,930.65$

9. $A_n = 400 \ a_{\overline{6)}4\%} = 400(5.242137) = \$2,096.85$

11. $S_n = 100 \ s_{\overline{36)}1\%} = 100(43.076878) = \$4,307.69$

$A_n = 100 \ a_{\overline{36)}1\%} = 100(30.107505) = \$3,010.75$

13. $A_n = 80,000 - 20,000 = \$60,000$

$60,000 = R \ a_{\overline{48)}\frac{1}{2}\%}$ $R = \dfrac{60,000}{42.580318} = \$1,409.10$

15. There are 36 (= 48 − 12) remaining payments of $1,409.10 each at $\frac{1}{2}\%$ interest rate per month.

Unpaid balance $= A_n = 1,409.10 \ a_{\overline{36)}\frac{1}{2}\%} = 1,409.10(32.871016)$

$= \$46,318.55$

17. $S_n(\text{due}) = 1,000(s_{\overline{24 + 1)}1\%} - 1) = 1,000(28.243200 - 1)$

$= \$27,243.20$

19. (a) $S_n(\text{due}) = 200(s_{\overline{20 + 1)}1\frac{1}{2}\%} - 1) = 200(24.470522 - 1)$

$= \$4,694.10$

(b) $I = 4,694.10 - 20(200) = \$694.10$

21. $A_n(\text{due}) = 150(a_{\overline{36-1}|\frac{1}{2}\%} + 1) = 150(32.035371 + 1) = \$4,955.31$

23. $A_n(\text{due}) = 400(a_{\overline{6-1}|4\%} + 1) = 400(4.451822 + 1) = \$2,180.73$

25. $S_n(\text{due}) = 10,000 = R(s_{\overline{12+1}|2\%} - 1) = R(14.680332 - 1)$

$$R = \frac{10,000}{13.680332} = \$730.98$$

27. $S_n(\text{defer}) = 1,000\ s_{\overline{24}|1\%} = 1,000(26.973465) = \$26,973.47$

$A_n(\text{defer}) = 1,000(a_{\overline{11+24}|1\%} - a_{\overline{11}|1\%})$
$= 1,000(29.408580 - 10.367628) = \$19,040.95$

29. $A_n(\text{defer}) = 80,000 - 20,000 = R(a_{\overline{11+36}|\frac{1}{2}\%} - a_{\overline{11}|\frac{1}{2}\%})$
$$60,000 = R(41.793219 - 10.677027)$$
$$R = \frac{60,000}{31.116192} = \$1,928.26$$

Exercise 15–1 (p. 322)

1. $L = 2 + (10 - 1)2 = 20$ $\qquad S_n = \dfrac{10}{2}(2 + 20) = 110$

3. $L = \dfrac{1}{5} + (8 - 1)\dfrac{1}{5} = \dfrac{1}{5} + \dfrac{7}{5} = \dfrac{8}{5}$

$S_n = \dfrac{8}{2}\left(\dfrac{1}{5} + \dfrac{8}{5}\right) = 4\left(\dfrac{9}{5}\right) = \dfrac{36}{5}$

5. $L = .1 + (7 - 1).3 = 1.9$ $\qquad S_n = \dfrac{7}{2}(.1 + 1.9) = 7$

7. $L = 2a + (6 - 1)c = 2a + 5c$

$S_n = \dfrac{6}{2}(2a + (2a + 5c)) = 12a + 15c$

9. $L = 3 + (6 - 1)2 = 13$ $\qquad S_n = \dfrac{6}{2}(3 + 13) = 48$

11. $L = 4 + (n - 1).5 = 3.5 + .5n$

$S_n = 62.5 = \dfrac{n}{2}(4 + (3.5 + .5n))$

$125 = n(7.5 + .5n) = 7.5n + .5n^2$
$n^2 + 15n - 250 = 0$
$(n - 10)(n + 25) = 0$

$n = 10$ ($n = -25$ is not logical.)
$L = 3.5 + .5(10) = 8.5$

13. $L = \tfrac{1}{4} + (9-1)d = \tfrac{1}{4} + 8d$

$$S_n = 20\tfrac{1}{4} = \frac{9}{2}(\tfrac{1}{4} + (\tfrac{1}{4} + 8d)) = \frac{9}{4} + 36d$$

$$d = \frac{18}{36} = \tfrac{1}{2}$$

$$L = \tfrac{1}{4} + 8(\tfrac{1}{2}) = 4\tfrac{1}{4}$$

15. $L = 7\dfrac{2}{3} = a + (12 - 1)\left(\dfrac{2}{3}\right)$

$$a = 7\frac{2}{3} - \frac{22}{3} = \frac{1}{3}$$

$$S_n = \frac{12}{2}\left(\frac{1}{3} + 7\frac{2}{3}\right) = 6(8) = 48$$

17. $a = 2,\ L = 17,\ n = 4 + 2 = 6$

$17 = 2 + (6 - 1)d$
$5d = 17 - 2$
$d = 3$

The four means are: 5, 8, 11, and 14.

19. $a = 12,\ L = 8.4,\ n = 5 + 2 = 7$

$8.4 = 12 + (7 - 1)d$
$6d = 8.4 - 12 = -3.6$
$d = -.6$

The five means are: 11.4, 10.8, 10.2, 9.6, and 9.

21. $a = \$10,000,\ d = \$2,000,\ n = 20$
$L = 10,000 + (20 - 1)2,000 = \$48,000$

$$S_n = \frac{20}{2}(10,000 + 48,000) = \$580,000 \text{ (total sales for 20 months)}$$

Exercise 15–2 (p. 327)

1. $L = 1 \cdot 2^5 = 32$ $S_n = \dfrac{1(1 - 2^6)}{1 - 2} = 63$

3. $L = 1 \cdot (-2)^6 = 64$ $S_n = \dfrac{1(1 - (-2)^7)}{1 - (-2)} = 43$

5. $L = 1 \cdot (.5)^5 = .03125$ $S_n = \dfrac{1(1 - .5^6)}{1 - .5} = 1.96875$

7. $L = 2(\tfrac{1}{2})^{n-1}$, which approaches 0 since $n \to \infty$ and $(\tfrac{1}{2})^{n-1} \to 0$.

$$\lim_{n \to \infty} S_n = \frac{a}{1 - r} = \frac{2}{1 - \tfrac{1}{2}} = 4$$

9. $L = 64 = 1(2)^{n-1}$ $S_n = \dfrac{1(1-2^7)}{1-2} = 127$
   $2^6 = 2^{n-1}$
   $6 = n - 1$
   $n = 7$

11. $L = -32 = 1 \cdot r^{6-1}$ $S_n = \dfrac{1(1-(-2)^6)}{1-(-2)} = -21$
    $(-2)^5 = r^5$
    $r = -2$

13. $L = .0625 = 1 \cdot r^{n-1}$ $r^{n-1} = .0625$

    $$S_n = \dfrac{1(1-r^n)}{1-r} = 1.9375$$

    $1 - r^n = 1.9375(1-r)$
    $1 - 1.9375 = -1.9375r + r^n = r(r^{n-1} - 1.9375)$
    $-.9375 = r(.0625 - 1.9375)$

    $$r = \dfrac{-.9375}{-1.8750} = .5$$

    $.5^{n-1} = .0625 = .5^4$
    $n - 1 = 4$
    $n = 5$

15. $S_n = -42 = \dfrac{a(1-(-2)^6)}{1-(-2)} = \dfrac{a(-63)}{3} = -21a$

    $a = \dfrac{-42}{-21} = 2$

    $L = 2 \cdot (-2)^{6-1} = -64$

17. $a = 2, L = 32; n = 3 + 2 = 5$
    $32 = 2 \cdot r^{5-1}$
    $16 = r^4$
    $r = \pm 2$

    (1) When $r = 2$, the three means are 4, 8, and 16.
    (2) When $r = -2$, the three means are -4, 8, and -16.

19. In $0.0454545 \ldots$, $a = .045$, $r = .01$, $n \to \infty$, and

    $\underset{n \to \infty}{\text{limit}} \; S_n = \dfrac{.045}{1 - .01} = \dfrac{.045}{.990} = \dfrac{1}{22}$

    $0.3454545 \ldots = 0.3 + (0.0454545 \ldots) = \dfrac{3}{10} + \dfrac{1}{22} = \dfrac{19}{55}$

21. $a = \$100, r = 100\% + 10\% = 1.1, n = 4$

    $S_n = \dfrac{100(1-1.1^4)}{1-1.1} = \dfrac{100(1-1.4641)}{-.1} = \$464.10$

23. This is a G. P.: $1(2) = 2$ (the couple)

$$1 \cdot 4(2) = 8 \text{ (sons and their wives)}$$
$$1 \cdot 4 \cdot 4(2) = 32 \text{ (grandsons and their wives)}$$
$$1 \cdot 4 \cdot 4 \cdot 4(2) = 128 \text{ (great-grandsons and their wives)}$$

$a = 2,\ r = 4,\ n = 4$

$$S_n = \frac{2(1 - 4^4)}{1 - 4} = \frac{2(1 - 256)}{-3} = 170 \text{ (persons in the family)}$$

Exercise 15–3 (p. 331)

1. $\dfrac{2(1)}{3(1) - 1} = 1 \qquad \dfrac{2(2)}{3(2) - 1} = \dfrac{4}{5} \qquad \dfrac{2(n + 1)}{3(n + 1) - 1} = \dfrac{2n + 2}{3n + 2}$

3. $\dfrac{1 + 4}{1^2 - 3} = \dfrac{5}{-2} \qquad \dfrac{2 + 4}{2^2 - 3} = 6 \qquad \dfrac{(n + 1) + 4}{(n + 1)^2 - 3} = \dfrac{n + 5}{n^2 + 2n - 2}$

5. $(-1)^1 \left(\dfrac{2(1)^2}{2(1) - 1} \right) = -2 \qquad\qquad (-1)^2 \left(\dfrac{2(2)^2}{2(2) - 1} \right) = \dfrac{8}{3}$

$(-1)^{n+1} \left(\dfrac{2(n + 1)^2}{2(n + 1) - 1} \right) = (-1)^{n+1} \left(\dfrac{2n^2 + 4n + 2}{2n + 1} \right)$

7. $L = 3 + (n - 1)2 = 2n + 1$

$S_n = 3 + 5 + 7 + 9 + 11 = 35,\text{ or } S_n = \dfrac{5}{2}(3 + 11) = 35$

9. $L = 5 + (n - 1)2 = 2n + 3$

$S_n = 5 + 7 + 9 + 11 + 13 = 45,\text{ or } S_n = \dfrac{5}{2}(5 + 13) = 45$

11. $(-1)^{n+1} (n + 1)$ The 5th term: $(-1)^{5+1} (5 + 1) = 6$

$S_n = 2 - 3 + 4 - 5 + 6 = 4$

13. $\dfrac{3n - 2}{2n + 2}$ The 5th term: $\dfrac{3(5) - 2}{2(5) + 2} = \dfrac{13}{12}$

$S_n = \dfrac{1}{4} + \dfrac{4}{6} + \dfrac{7}{8} + \dfrac{10}{10} + \dfrac{13}{12} = \dfrac{6 + 16 + 21 + 24 + 26}{24} = \dfrac{93}{24} = 3\dfrac{7}{8}$

15. $(-1)^{n+1} \left(\dfrac{1}{n^3} \right)$ The 5th term: $(-1)^{5+1} \left(\dfrac{1}{5^3} \right) = \dfrac{1}{125}$

$S_n = 1 - \dfrac{1}{8} + \dfrac{1}{27} - \dfrac{1}{64} + \dfrac{1}{125} = \dfrac{216{,}000 - 27{,}000 + 8{,}000 - 3{,}375 + 1{,}728}{216{,}000}$

$= \dfrac{195{,}353}{216{,}000}$

Exercise 15–4 (p. 337)

1. $t_n = \dfrac{3n}{n+1} = \dfrac{\dfrac{3n}{n}}{\dfrac{n+1}{n}} = \dfrac{3}{1+\dfrac{1}{n}}$ $\qquad \underset{n\to\infty}{\text{limit}}\; t_n = 3$

3. $t_n = \dfrac{n}{n(n+1)} = \dfrac{1}{n+1}$ $\qquad \underset{n\to\infty}{\text{limit}}\; t_n = 0$

5. $t_n = \dfrac{n^2}{3n+1} = \dfrac{\dfrac{n^2}{n^2}}{\dfrac{3n+1}{n^2}} = \dfrac{1}{\dfrac{3}{n}+\dfrac{1}{n^2}}$ $\qquad \underset{n\to\infty}{\text{limit}}\; t_n = \infty$

7. $t_n = \dfrac{(n+1)^2}{3n(n+2)} = \dfrac{n^2+2n+1}{3n^2+6n} = \dfrac{1+\dfrac{2}{n}+\dfrac{1}{n^2}}{3+\dfrac{6}{n}}$ $\qquad \underset{n\to\infty}{\text{limit}}\; t_n = \dfrac{1}{3}$

9. The terms form an $A.\ P.$: $a = 1,\ d = 2,\ L = 1 + (n-1)2 = 2n - 1$

$S_n = \dfrac{n}{2}(1 + (2n-1)) = n^2$ $\qquad \underset{n\to\infty}{\text{limit}}\; S_n = \infty$ (Divergence)

11. The terms form an $A.\ P.$: $a = 2,\ d = 2,\ L = 2 + (n-1)2 = 2n.$

$S_n = \dfrac{n}{2}(2 + 2n) = n + n^2$ $\qquad \underset{n\to\infty}{\text{limit}}\; S_n = \infty$ (Divergence)

13. The terms form a $G.\ P.$: $a = 1,\ r = .5,\ L = 1(.5)^{n-1}.$

$S_n = \dfrac{1(1-.5^n)}{1-.5} = 2 - \dfrac{.5^n}{.5}$ $\qquad \underset{n\to\infty}{\text{Limit}}\; S_n = 2$ (Convergence)

15. The terms form a $G.\ P.$: $a = 2,\ r = \frac{2}{3},\ L = 2(\frac{2}{3})^{n-1}.$

$S_n = \dfrac{2\left(1 - \left(\frac{2}{3}\right)^n\right)}{1 - \dfrac{2}{3}} = 6\left(1 - \left(\frac{2}{3}\right)^n\right)$ $\qquad \underset{n\to\infty}{\text{Limit}}\; S_n = 6$ (Convergence)

17. The terms form an $A.\ P.$: $a = \frac{2}{5},\ d = \frac{1}{5},\ L = \frac{2}{5} + (n-1)(\frac{1}{5}) = \frac{n+1}{5}.$

$S_n = \dfrac{n}{2}\left(\dfrac{2}{5} + \dfrac{n+1}{5}\right) = \dfrac{3n+n^2}{10}$ $\qquad \underset{n\to\infty}{\text{Limit}}\; S_n = \infty$ (Divergence)

19. The terms form a $G.\ P.$: $a = 1,\ r = 2,\ L = 1 \cdot 2^{n-1}.$

$S_n = \dfrac{1(1-2^n)}{1-2} = 2^n - 1$ $\qquad \underset{n\to\infty}{\text{Limit}}\; S_n = \infty$ (Divergence)

Exercise 16–1 (p. 346)

1. $\log_5 25 = 2$
3. $\log_a c = 4$
5. $\log 7 = 0.845098$
7. $\log 68 = 1.832509$
9. $5^3 = 125$
11. $a^x = M$
13. $10^{1.093422} = 12.4$
15. $10^{8.507721-10} = 0.03219$

17. 2.416641
19. $-2 + .331630$
21. 2.669596
23. 0.0171168
25. 186
27. 3.871
29. 0.02882
31. 0.1452

33.

	N	M	
	68250	834103	(1)
	68247	x	(2)
	68240	834039	(3)
(2) − (3)	$\dfrac{7}{10}$	$= \dfrac{x - 834039}{64}$	
(1) − (3)			

$$\log 0.068247 = -2 + .8340838$$

$x = 834039 \quad + 64(\tfrac{7}{10})$
$\quad = 834039$
$\quad\quad + 44.8$
$\quad\quad 834083.8$

35.

	N	M	
	57130	756864	(1)
	57124	x	(2)
	57120	756788	(3)
(2) − (3)	$\dfrac{4}{10}$	$= \dfrac{x - 756788}{76}$	
(1) − (3)			

$$\log 0.57124 = -1 + .7568184$$

$x = 756788 \quad + 76(\tfrac{4}{10})$
$\quad = 756788$
$\quad\quad + 30.4$
$\quad\quad 756818.4$

37.

	N	M	
	28210	450403	(1)
	x	450327	(2)
	28200	450249	(3)
(2) − (3)	$\dfrac{x - 28200}{10}$	$= \dfrac{78}{154}$	
(1) − (3)			

$x = 28200 \quad + 10(\tfrac{78}{154})$
$\quad = 28200$
$\quad\quad + 5.06$
$\quad\quad 28205.06 \quad N = 282.05$

39.

	N	M	
	182500	261263	(1)
	x	261135	(2)
	182400	261025	(3)
(2) − (3)	$\dfrac{x - 182400}{100}$	$= \dfrac{110}{238}$	
(1) − (3)			

$x = 182400 \quad + 100(\tfrac{110}{238})$
$\quad = 182400$
$\quad\quad + 46.218$
$\quad\quad 182446.218$
$N = 0.182446$

Exercise 16–2 (p. 350)

1. $\log 3.45 \quad = 0.537819$
   $(+) \log 0.246 \ = \underline{9.390935 - 10}$
   $\quad\quad\quad\quad\quad 9.928754 - 10$

$3.45 \times 0.246 = 0.8487$

3. log   367   = 2.564666
(+) log 1,258   = 3.099681
        5.664347 $367 \times 1,258 = 461,700$

5. log 0.5832 =   9.765818 − 10
(+) log 43.12   =   1.634679
        11.400497 − 10 $0.5832 \times 43.12 = 25.15$

7. log 35.72 = 1.552911
(+) log 1.0253 = 0.0108510
        1.5637620
                    $-35.72 \times 1.0253 = -36.62$

9. log 5.62   = 0.749736
(−) log 1.682   = 0.225826
        0.523910 $5.62 \div 1.682 = 3.341$

11. log 32.5   = 11.511883 − 10
(−) log 0.1562 =   9.193681 − 10
        2.318202 $32.5 \div 0.1562 = 208.1$

13. log 5.924   = 10.772615 − 10
(−) log 6,349   =   3.802705
        6.969910 − 10 $5.924 \div 6,349 = 0.0009331$

15. log 14.64   = 11.165541 − 10
(−) log 526.8   =   2.721646
        8.443895 − 10 $14.64 \div 526.8 = 0.02779$

17. log 4 = 0.602060
  $\times$ 2
  1.204120 $4^2 = 16.00$

19. log 34.2 = 1.534026
  $\times$ 3
  4.602078 $34.2^3 = 40,000$

21. log 1.045 = 0.019116
  $\times$ 20
  0.382320 $1.045^{20} = 2.412$

23. log 524.3 =   2.719580
  $\times$ 10 $524.3^{10} = 1.57 \times 10^{27}$, or
  27.195800   = 1,570 and 24
                              zeros

25. log 128 = 2.107210
  $\times$ $\frac{1}{7}$
  0.301030 $\sqrt[7]{128} = 2.000$

27. log 5,628 = 3.750354
  $\times$ $\frac{1}{10}$
  0.3750354 $\sqrt[10]{5,628} = 2.372$

29. $\log 20.25 = 1.306425$

$$\frac{\times \qquad \frac{1}{2}}{0.6532125}$$

$\sqrt{20.25} = 4.500$

31. $\log 623.4 = 2.794767$

$$\frac{\times \qquad \frac{1}{3}}{0.931589}$$

$\sqrt[3]{623.4} = 8.543$

Exercise 16–3 (p. 359)

1. (a) $\log_e 52 = 2.302826(\log 52) = 2.302826(1.716003) = 3.951656$
   (b) $\log_e 52 = \ln (5.2 \times 10) = \ln 5.2 + \ln 10 = 1.6487 + 2.3026$
$$= 3.9513$$

3. (a) $\log_e 186 = 2.302826(\log 186) = 2.302826(2.269513) = 5.226294$
   (b) $\log_e 186 = \ln (1.86 \times 10^2) = \ln 1.86 + 2(\ln 10)$
$$= 0.6206 + 2(2.3026) = 5.2258$$

5. $\log_6 36 = \dfrac{\log 36}{\log 6} = \dfrac{1.556303}{0.778151} = 2$

7. $\log_8 512 = \dfrac{\log 512}{\log 8} = \dfrac{2.709270}{0.903090} = 3$

9. $\log_5 100 = \dfrac{\log 100}{\log 5} = \dfrac{2.000000}{0.698970} = 2.861353$

11. $\qquad\qquad 2^{x+1} = 256$

$$(x + 1) \log 2 = \log 256$$
$$x + 1 = \frac{\log 256}{\log 2} = \frac{2.408240}{0.301030} = 8$$
$$x = 8 - 1 = 7$$

13. $\qquad (2 + x)^{10} = 2{,}500$

$$10 \log (2 + x) = \log 2{,}500$$
$$\log (2 + x) = \frac{\log 2{,}500}{10} = \frac{3.397940}{10} = 0.339794$$

$$2 + x = 2.187$$
$$x = 0.187$$

15. $\qquad (3 - x)^{-6} = 0.04621$

$$(-6) \log (3 - x) = \log 0.04621 = -2 + .664736 = -1.335264$$
$$\log (3 - x) = \frac{-1.335264}{-6} = 0.222544$$

$$3 - x = 1.669$$
$$x = 1.331$$

17. $\log (x + 5) = 4$

    $x + 5 = 10^4 = 10,000$

         $x = 9,995$

19. $\log_2 (x^2 - 1) = 5 + \log_2 (x + 1)$

    $\log_2 \left(\dfrac{x^2 - 1}{x + 1}\right) = 5$

        $x - 1 = 2^5 = 32$

           $x = 33$

21. $y = \log_2 x$ $x = 2^y$

y	0	1	2	3	4	5	6
x	1	2	4	8	16	32	64

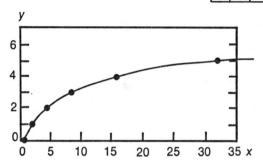

23.

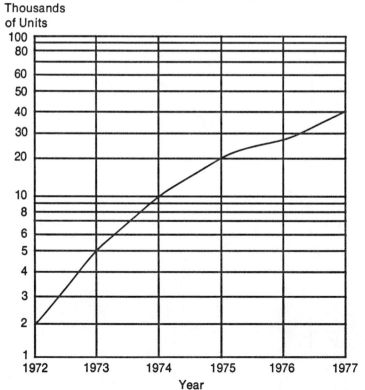

Begler Television Company
Production, 1972–1977

Thousands
of Units

Year

25. $y = 3(2^x)$

x	-1	0	1	2	3	4	5	6	7	8
y	1.5	3	6	12	24	48	96	192	384	768

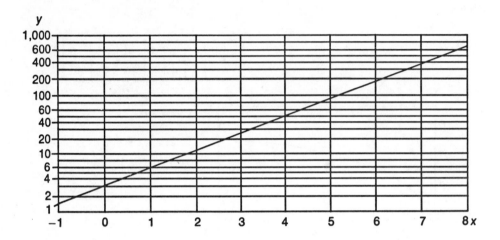

Exercise 17–1 (p. 367)

1, 3, and 5(a)

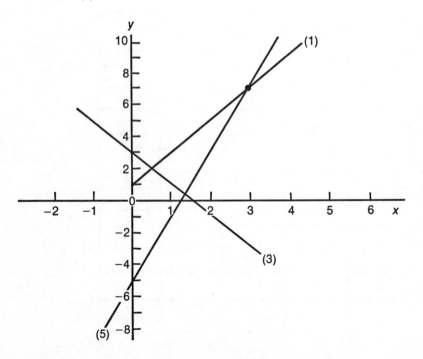

1. (b) $b = \dfrac{5-3}{2-1} = 2$

$3 = 2(1) + a$

$a = 3 - 2 = 1$

(c) $y_s = 2x + 1$

3. (b) $b = \dfrac{1-(-1)}{1-2} = \dfrac{2}{-1} = -2$

$-1 = -2(2) + a$

$a = -1 + 4 = 3$

(c) $y_s = -2x + 3$

5. (b) $b = \dfrac{7-(-1)}{3-1} = \dfrac{8}{2} = 4$

$-1 = 4(1) + a$

$a = -1 - 4 = -5$

(c) $y_s = 4x - 5$

7. $b = \dfrac{((x_1 + \Delta x)^2 + 3) - (x_1^2 + 3)}{\Delta x} = \dfrac{x_1^2 + 2x_1\Delta x + (\Delta x)^2 + 3 - x_1^2 - 3}{\Delta x}$

$\qquad = 2x_1 + \Delta x = 2x_1 + 0 = 2x_1 \qquad$ (Since $\Delta x \to 0$)

9. $b = \dfrac{((x_1 + \Delta x)^3 - 5) - (x_1^3 - 5)}{\Delta x}$

$\qquad = \dfrac{x_1^3 + 3x_1^2\Delta x + 3x_1(\Delta x)^2 + (\Delta x)^3 - 5 - x_1^3 + 5}{\Delta x}$

$\qquad = 3x_1^2 + 3x_1\Delta x + (\Delta x)^2 = 3x_1^2 + 0 + 0 = 3x_1^2$

11. $b = \dfrac{3(x_1 + \Delta x)^2 + 4 - (3x_1^2 + 4)}{\Delta x}$

$\qquad = \dfrac{3x_1^2 + 6x_1\Delta x + 3(\Delta x)^2 + 4 - 3x_1^2 - 4}{\Delta x}$

$\qquad = 6x_1 + 3\Delta x = 6x_1 + 0 = 6x_1$

13. $b = \dfrac{3(x_1 + \Delta x)^2 - 5(x_1 + \Delta x) - (3x_1^2 - 5x_1)}{\Delta x}$

$\qquad = \dfrac{3x_1^2 + 6x_1\Delta x + 3(\Delta x)^2 - 5x_1 - 5\Delta x - 3x_1^2 + 5x_1}{\Delta x}$

$\qquad = 6x_1 + 3\Delta x - 5 = 6x_1 + 0 - 5 = 6x_1 - 5$

15.

$b = \dfrac{\dfrac{3}{x_1 + \Delta x + 1} - \dfrac{3}{x_1 + 1}}{\Delta x} = \dfrac{3(x_1 + 1) - 3(x_1 + \Delta x + 1)}{\Delta x(x_1 + 1)(x_1 + \Delta x + 1)}$

$\qquad = \dfrac{3x_1 + 3 - 3x_1 - 3\Delta x - 3}{\Delta x(x_1 + 1)(x_1 + \Delta x + 1)} = \dfrac{-3\Delta x}{\Delta x(x_1 + 1)(x_1 + \Delta x + 1)}$

$\qquad = \dfrac{-3}{(x_1 + 1)(x_1 + \Delta x + 1)} = \dfrac{-3}{(x_1 + 1)^2}$

17.

(a)
$y = x^2 + 3$

x	y
0	3
1	4
2	7
3	12
−1	4
−2	7
−3	12

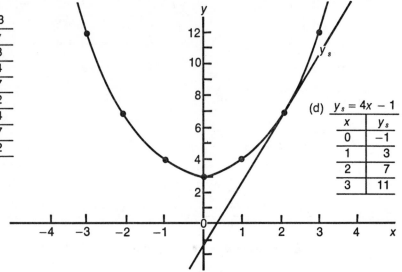

(d) $y_s = 4x - 1$

x	y_s
0	−1
1	3
2	7
3	11

(b) $b = 2x_1 = 2(2) = 4$

(c) $x = 2$
$y = x^2 + 3 = 2^2 + 3 = 7$

$7 = 4(2) + a$
$a = 7 - 8 = -1$

$y_s = 4x - 1$

19.

(a)
$y = x^3 - 5$

x	y
0	−5
1	−4
2	3
3	22
−1	−6
−2	−13
−3	−32

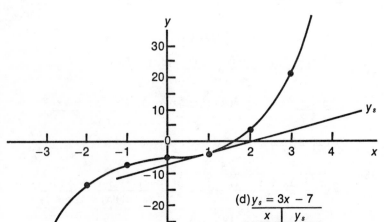

(d) $y_s = 3x - 7$

x	y_s
0	−7
1	−4
2	−1
3	2

(b) $\quad b = 3x_1^2 = 3(1^2) = 3$

(c) $\quad x = 1$

$$y = x^3 - 5 = 1^3 - 5 = -4$$

$$-4 = 3(1) + a$$

$$a = -4 - 3 = -7$$

$$y_s = 3x - 7$$

Exercise 17-2 (p. 377)

1. $y' = 0$
3. $y' = 2x$
5. $y' = 12x^{11}$
7. $y' = \dfrac{8}{7}(x^{1/7})$
9. $y' = 6x^2$
11. $y' = 12x^2$
13. $y' = 3x^{-1/2} = \dfrac{3}{x^{1/2}}$

15. $y' = 2x$
17. $y' = 4$
19. $y' = 3x^2$
21. $y' = 2x^9 - 2$
23. $y' = 6x - 4$
25. $y' = 16x + 5$
27. $y' = 12x^3 + 6x^2 - 30x^5$

29. $y' = 2x^5(3x^2 + 4) + (x^3 + 4x)(10x^4)$
$\qquad = 6x^7 + 8x^5 + 10x^7 + 40x^5 = 16x^7 + 48x^5$

31. $y' = (x - 1)(6x^2) + (2x^3 + 5)(1) = 8x^3 - 6x^2 + 5$

33. $y' = \dfrac{(x + 1)(0) - 3(1)}{(x + 1)^2} = \dfrac{-3}{(x + 1)^2}$

35. $y' = \dfrac{3x(4x) - (2x^2 - 5)(3)}{9x^2} = \dfrac{12x^2 - 6x^2 - 15}{9x^2} = \dfrac{2x^2 - 5}{3x^2}$

37. $y' = \dfrac{1}{3}(x^{-2/3}) = \dfrac{1}{3x^{2/3}}$

At point $(8,2)$, $x = 8$, $y = 2$

Slope $y' = \dfrac{1}{3(8^{2/3})} = \dfrac{1}{3(\sqrt[3]{8^2})} = \dfrac{1}{3(4)} = \dfrac{1}{12} = b$ at the point

$$y_s = bx + a$$

$$2 = \dfrac{1}{12}(8) + a$$

$$a = 2 - \dfrac{2}{3} = 1\dfrac{1}{3}$$

$$y_s = \dfrac{1x}{12} + 1\dfrac{1}{3} \quad \text{(Equation of slope)}$$

Exercise 18–1 (p. 386)

1. $y' = 2(x^2 + 3)(2x) = 4x^3 + 12x$

3. $y' = \dfrac{3}{2}(3x^4 - 5)^{1/2}(12x^3) = 18x^3(3x^4 - 5)^{1/2}$

5. $y' = 6(5x^2 - 4x + 7)^5(10x - 4)$

7. $y' = -5(6x^5 - 3x^4 + 2x - 3)^{-6}(30x^4 - 12x^3 + 2)$

9. $y' = 20\left(\dfrac{x+3}{x-2}\right)^{19}\left(\dfrac{(x-2)-(x+3)}{(x-2)^2}\right) = \dfrac{-100(x+3)^{19}}{(x-2)^{21}}$

11. $y' = \dfrac{dy}{du} \cdot \dfrac{du}{dv} \cdot \dfrac{dv}{dx} = 3(3v^2)(2x) = 3(3(x^2)^2)(2x) = 18x^5$

13. $y' = 5^x(\ln 5)$

15. $y' = 15^{2x}(\ln 15)2 = 2(15^{2x})(\ln 15)$

17. $y' = 2\,e^x$

19. $y' = 3\,e^{3x}$

21. $y' = \dfrac{1}{x}\log_2 e$

23. $y' = \dfrac{3}{3x}\log_{10} e = \dfrac{1}{x}\log e$

25. $y' = \dfrac{4x - 3}{2x^2 - 3x}\log e$

27. $y' = \dfrac{2}{2x} = \dfrac{1}{x}$

29. $y' = \dfrac{12x^2 + 6x}{4x^3 + 3x^2 - 8}$

31. $y' = \dfrac{\frac{1}{2}(x^2 + 1)^{-1/2}(2x)}{(x^2 + 1)^{1/2}} = \dfrac{x}{x^2 + 1}$

33. $y' = (2x)^x\left(x\left(\dfrac{2}{2x}\right) + (1)(\ln 2x)\right)$

$\quad = (2x)^x(1 + \ln 2x)$

35. $y' = (x^2)^{3x+1}\left((3x + 1)\left(\dfrac{2x}{x^2}\right) + 3(\ln x^2)\right)$

$\quad = (x^2)^{3x+1}\left(\dfrac{6x + 2}{x} + 3(\ln x^2)\right)$

Exercise 18–2 (p. 396)

1. $\dfrac{dy}{dx} = 3 \qquad\qquad \dfrac{dx}{dy} = \dfrac{1}{3}$

3. $\dfrac{dy}{dx} = 10x - 7 \qquad \dfrac{dx}{dy} = \dfrac{1}{10x - 7}$

5. $\dfrac{dy}{dx} = 15x^2 + 8x - 3 \qquad \dfrac{dx}{dy} = \dfrac{1}{15x^2 + 8x - 3}$

7. $\dfrac{dy}{dx} = \dfrac{(x - 1) - (x + 2)}{(x - 1)^2} = \dfrac{-3}{(x - 1)^2} \qquad \dfrac{dx}{dy} = \dfrac{(x - 1)^2}{-3}$

9. $\dfrac{dx}{dy} = 1 + y + y^3 \qquad \dfrac{dy}{dx} = \dfrac{1}{1 + y + y^3}$

11. $\dfrac{dy}{dx} - 3 = 0$

$\quad \dfrac{dy}{dx} = 3$

13. $\dfrac{dy}{dx} + 7 = 5(2x) + 0$

$\quad \dfrac{dy}{dx} = 10x - 7$

15. $2y \dfrac{dy}{dx} + 4 = 6x$

$$\dfrac{dy}{dx} = \dfrac{6x - 4}{2y} = \dfrac{3x - 2}{y}$$

17. $3x^2 + 4y^3 \left(\dfrac{dy}{dx}\right) = 0 - 3\dfrac{dy}{dx}$

$$\dfrac{dy}{dx}(4y^3 + 3) = -3x^2$$

$$\dfrac{dy}{dx} = \dfrac{-3x^2}{4y^3 + 3}$$

19. $6x^2 - 5\left(2y\dfrac{dy}{dx}\right) - \left(x^2\dfrac{dy}{dx} + y(2x)\right) = 0$

$$6x^2 - 10y\dfrac{dy}{dx} - x^2\dfrac{dy}{dx} - 2xy = 0$$

$$\dfrac{dy}{dx}(10y + x^2) = 6x^2 - 2xy$$

$$\dfrac{dy}{dx} = \dfrac{6x^2 - 2xy}{(10y + x^2)}$$

21. (a)

Original function
$y = x^2 - 2$

x	y
0	-2
1	-1
2	2
3	7
-1	-1
-2	2
-3	7

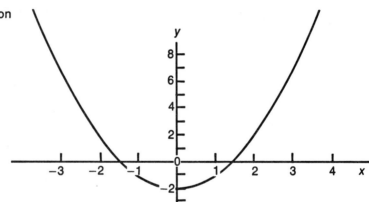

(b)

First derivative
$y' = 2x$

x	y'
0	0
1	2
2	4
-1	-2
-2	-4

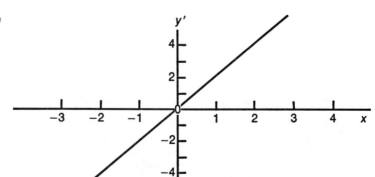

(c)

Second derivative
$y'' = 2$
for all
x values

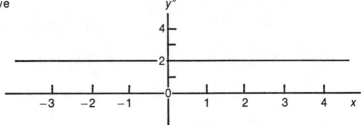

(d) Third derivative

$$y''' = 0$$

23. $y' = 6x$
$y'' = 6$
$y''' = 0$

25. $y' = 12x^2 + 4x - 1$
$y'' = 24x + 4$
$y''' = 24$
$y'''' = 0$

27. $y = (x + 1)(2x - 3)$
$= 2x^2 - x - 3$

$y' = 4x - 1$
$y'' = 4$
$y''' = 0$

29. $y = \dfrac{x - 5}{x + 1}$

$y' = \dfrac{(x + 1)(1) - (x - 5)(1)}{(x + 1)^2}$

$= \dfrac{6}{(x + 1)^2} = 6(x + 1)^{-2}$

$y'' = 6(-2)(x + 1)^{-3}$
$= -12(x + 1)^{-3}$

$y''' = -12(-3)(x + 1)^{-4}$
$= \dfrac{36}{(x + 1)^4}$

Exercise 19–1 (p. 409)

1. (a)

$y = -x^2 + 2x + 3$

x	y
0	3
1	4
2	3
3	0
4	-5
-1	0
-2	-5
-3	-12

(b) $y' = -2x + 2 = 0$

$x = 1$ (Critical point at)

$y'' = -2$

$y = f(x) = f(1) = -(1)^2 + 2(1) + 3 = 4$ (maximum)

(c) Check: $x_1 = 1 - .1 = .9$

$x_2 = 1 + .1 = 1.1$

First way: $f(.9) = -(.9)^2 + 2(.9) + 3 = 3.99$

$f(1.1) = -(1.1)^2 + 2(1.1) + 3 = 3.99$

Second way: $f'(.9) = -2(.9) + 2 = +0.2$

$f'(1.1) = -2(1.1) + 2 = -0.2$

3. (a)

$y = 3x^2 + 12x - 10$

x	y
0	−10
1	5
2	26
−1	−19
−2	−22
−3	−19
−4	−10
−5	5
−6	26

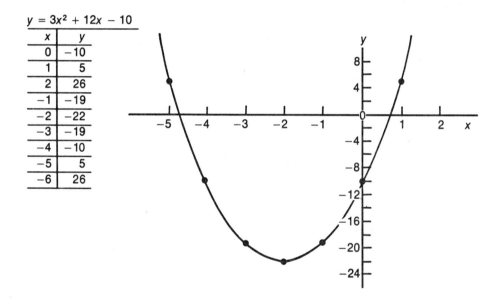

(b) $y' = 6x + 12 = 0$

$x = -2$ (Critical point at)

$y'' = 6$

$y = f(x) = f(-2) = 3(-2)^2 + 12(-2) - 10 = -22$ (minimum)

(c) Check: $x_1 = -2 - .1 = -2.1$

$x_2 = -2 + .1 = -1.9$

First way: $f(-2.1) = 3(-2.1)^2 + 12(-2.1) - 10 = -21.97$

$f(-1.9) = 3(-1.9)^2 + 12(-1.9) - 10 = -21.97$

Second way: $f'(-2.1) = 6(-2.1) + 12 = -0.6$

$f'(-1.9) = 6(-1.9) + 12 = +0.6$

5. $y = 4x - x^2 + 5 = -x^2 + 4x + 5$

$y' = -2x + 4 = 0$

$\qquad x = 2$

$y'' = -2$

$\quad y = f(x) = f(2) = -(2)^2 + 4(2) + 5 = 9 \text{ (maximum)}$

7. $y = 3x^2 - 4x + 5$

$y' = 6x - 4 = 0$

$\qquad x = \frac{2}{3}$

$y'' = 6$

$\quad y = f(x) = f(\frac{2}{3}) = 3(\frac{2}{3})^2 - 4(\frac{2}{3}) + 5 = 3\frac{2}{3} \text{ (minimum)}$

9. $y = (\frac{1}{3})x^3 + x^2 - 3x - 4$

$y' = x^2 + 2x - 3 = 0$

$\quad (x + 3)(x - 1) = 0$

$\quad x + 3 = 0, x = -3$

$\quad x - 1 = 0, x = 1$

$y'' = 2x + 2 \quad$ At $x = -3$: $y'' = 2(-3) + 2 = -4 \text{ (maximum)}$

$\qquad\qquad\qquad$ At $x = 1$: $y'' = 2(1) + 2 = +4 \text{ (minimum)}$

$y = f(-3) = (\frac{1}{3})(-3)^3 + (-3)^2 - 3(-3) - 4 = 5 \text{ (maximum)}$

$y = f(1) = (\frac{1}{3})(1)^3 + 1^2 - 3(1) - 4 = -5\frac{2}{3} \text{ (minimum)}$

11. $y = 2x^3 - 7x^2 + 4x - 5$

$y' = 6x^2 - 14x + 4 = 0$

$\quad 3x^2 - 7x + 2 = 0$

$\quad (x - 2)(3x - 1) = 0$

$\quad x - 2 = 0, x = 2$

$\quad 3x - 1 = 0, x = \frac{1}{3}$

$y'' = 12x - 14 \quad$ At $x = 2$: $y'' = 12(2) - 14 = +10 \text{ (minimum)}$

$\qquad\qquad\qquad$ At $x = \frac{1}{3}$: $y'' = 12(\frac{1}{3}) - 14 = -10 \text{ (maximum)}$

$y = f(2) = 2(2)^3 - 7(2)^2 + 4(2) - 5 = -9 \text{ (minimum)}$

$y = f(\frac{1}{3}) = 2(\frac{1}{3})^3 - 7(\frac{1}{3})^2 + 4(\frac{1}{3}) - 5 = -4\frac{19}{27} \text{ (maximum)}$

13. $y = x^3 - x^2 - 8x + 5$

$y' = 3x^2 - 2x - 8 = 0,$

$\quad (x - 2)(3x + 4) = 0$

$\quad x - 2 = 0, x = 2$

$\quad 3x + 4 = 0, x = -\frac{4}{3}$

$y'' = 6x - 2. \quad$ At $x = 2$: $y'' = 6(2) - 2 = +10. \text{ (minimum)}$

$\qquad\qquad\qquad$ At $x = -\frac{4}{3}$: $y'' = 6(-\frac{4}{3}) - 2 = -10. \text{ (maximum)}$

$y = f(2) = 2^3 - 2^2 - 8(2) + 5 = -7$ (minimum)

$y = f(-\frac{4}{3}) = (-\frac{4}{3})^3 - (-\frac{4}{3})^2 - 8(-\frac{4}{3}) + 5 = 11\frac{14}{27}$ (maximum)

In the interval (a): $-3 \le x \le 3$

$x = 2$, $y = -7$ is the local minimum.

$x = -\frac{4}{3}$, $y = 11\frac{14}{27}$ is the local maximum.

In the interval (b): $0 \le x \le 3$

$x = 2$, $y = -7$ is the local minimum.

$x = -\frac{4}{3}$ (which is smaller than 0), $y = 11\frac{14}{27}$ must be disregarded.

15. $y = 2x^3 + 1$

$y' = 6x^2 = 0$

$\qquad x = 0$

$y'' = 12x \quad$ At $x = 0$, $y'' = 12(0) = 0$.

$y = f(0) = 2(0)^3 + 1 = 1$ is neither maximum nor minimum at $x = 0$.

Thus, there are no local maxima and minima.

$y = f(-3) = 2(-3)^3 + 1 = -53$ (absolute minimum)

$y = f(1) = 2(1)^3 + 1 = 3$ (absolute maximum)

17. $y = \dfrac{1}{x} + 2 = x^{-1} + 2 \qquad y$ is discontinuous at $x = 0$. $\dfrac{1}{0}$ is undefined.

$y' = (-1)x^{-2} = \dfrac{-1}{x^2} = 0 \quad$ No value for x that can make $y' = 0$.

$\qquad\qquad\qquad\qquad\qquad$ Thus, there is no critical point.

$y'' = 2x^{-3} = \dfrac{2}{x^3} \quad$ At $x = 0, y'' = \dfrac{2}{0}$, which is undefined.

No local maxima and minima.

Exercise 19–2 (p. 421)

1. Let $p =$ price per gallon of paint,

$\qquad x =$ number of gallons sold.

Given: $\quad p = \$10, x = 40$

$\qquad\qquad p = \$7, x = 100$

Slope $\quad b = \dfrac{10 - 7}{40 - 100} = -.05$

$\qquad\qquad 10 = -.05(40) + a$

$\qquad\qquad a = 10 + 2 = 12$

$\qquad\qquad p = -.05x + 12$

Let $S = $ total sales.

$$S = px = (-.05x + 12)x$$
$$S = -.05x^2 + 12x$$

$$S' = -.05(2)x + 12$$
$$S' = -.10x + 12 = 0$$
$$x = 120, \text{ which is the maximum since}$$

$$S'' = -.10 \text{ (negative)}$$

(a) At $x = 120$ gallons,

$$S = (-.05(120) + 12)120 = (-6 + 12)120 = \$720.00.$$

(b) At $x = 120$ gallons,

$$p = -.05(120) + 12 = \$6 \text{ per gallon.}$$

3. Let $P = $ total profit $= $ total sales $- $ total cost.

$$P = (-.05x^2 + 12x) - 4x$$
$$P = -.05x^2 + 8x$$

$$P' = -.05(2)x + 8 = -.10x + 8 = 0$$
$$x = 80, \text{ which is the maximum since}$$

$$P'' = -.10 \text{ (negative)}$$

At $x = 80$ gallons:

(a) The total sales, $S = -.05(80)^2 + 12(80) = -320 + 960 = \$640.00.$
(b) Price per gallon, $p = -.05x + 12 = -.05(80) + 12 = \$8.$
(c) Profit, $P = -.05(80)^2 + 8(80) = -320 + 640 = \$320.00.$

5. Let $n = $ units of machine part used in each year $= 10,000$ units,
    $a = $ cost of placing an order $= \$40,$
    $b = $ cost of carrying one unit of inventory $= \$5,$
    $Q = $ quantity of ordering each time,
    $\dfrac{Q}{2} = $ average inventory size during a year,
    $I = $ the total inventory cost during a year.

$$Q = \sqrt{\frac{2na}{b}} = \sqrt{\frac{2(10,000)(40)}{5}} = \sqrt{160,000} = 400 \text{ units}$$
$$\text{(most economic order size)}$$

$$I = \frac{na}{Q} + \frac{bQ}{2} = \frac{10,000(40)}{400} + \frac{5(400)}{2} = 1,000 + 1,000 = \$2,000$$
$$\text{(minimum cost)}$$

7. Let $w = $ the width of the parking zone,
    $x = $ the length of the parking zone,
    $y = $ the size of the parking lot.

Then, $wx = 20,000$, $w = \dfrac{20,000}{x}$, and

$$y = (w + 2 + 2)(x + 2 + 40) = (w + 4)(x + 42)$$
$$= wx + 4x + 42w + 168$$
$$= 20,000 + 4x + 42\left(\dfrac{20,000}{x}\right) + 168$$
$$= 20,168 + 4x + 840,000x^{-1}$$
$$y' = 4 + (-1)(840,000)x^{-2} = 0$$
$$\dfrac{1}{x^2} = \dfrac{-4}{-840,000} = \dfrac{1}{210,000}$$
$$x = \sqrt{210,000} = 458.26 \text{ feet}$$

$$w = \dfrac{20,000}{458.26} = 43.64 \text{ feet}$$

Parking lot size:

$$\text{Width} = 43.64 + 2 + 2 = 47.64 \text{ feet}$$
$$\text{Length} = 458.26 + 2 + 40 = 500.26 \text{ feet}$$

The parking lot size is the minimum since

$$y'' = (-1)(-2)(840,000)x^{-3} = 1,680,000x^{-3}$$

At $x = 458.26$,

$$y'' = 1,680,000(458.26)^{-3} = \dfrac{1,680,000}{(458.26)^3}, \text{ which is positive.}$$

9. $y = 4x^3 + (xz)^2 - z^3$

$$\dfrac{\partial y}{\partial x} = 12x^2 + 2(xz)(z) - 0 = 12x^2 + 2xz^2$$

$$\dfrac{\partial y}{\partial z} = 0 + 2(xz)(x) - 3z^2 = 2x^2z - 3z^2$$

11. $y = (2x^3z + z^2)^4$

$$\dfrac{\partial y}{\partial x} = 4(2x^3z + z^2)^3(6zx^2) = 24x^2z(2x^3 + z^2)^3$$

$$\dfrac{\partial y}{\partial z} = 4(2x^3z + z^2)^3(2x^3 + 2z)$$

13. $z = 3xy + x + 2y$

$$\dfrac{\partial z}{\partial x} = 3y + 1$$

$$\dfrac{\partial z}{\partial y} = 3x + 2$$

15. $z = 2x^3y^4 + 3x^2y^3 - 4xy^2 + 5$

$$\frac{\partial z}{\partial x} = 6x^2y^4 + 6xy^3 - 4y^2$$

$$\frac{\partial z}{\partial y} = 8x^3y^3 + 9x^2y^2 - 8xy$$

17. (a) and (c)

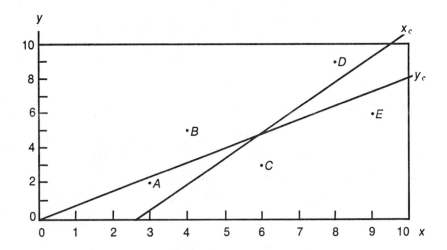

(b)

(1)	(2)	(3)	(4)	(5)	(6)
Worker	Hourly Wage x	Weekly Spending y	xy	x^2	y^2
A	\$ 3	\$ 2	6	9	4
B	4	5	20	16	25
C	6	3	18	36	9
D	8	9	72	64	81
E	9	6	54	81	36
Total Σ	30	25	170	206	155

Normal equations:

$$\begin{cases} (1) \quad \Sigma y = b\Sigma x + na \\ (2) \; \Sigma(xy) = b\Sigma x^2 + a\Sigma x \end{cases} \qquad \begin{cases} (1) \; 25 = b(30) + a(5) \\ (2) \; 170 = b(206) + a(30) \end{cases}$$

Solve the two normal equations: $b = \dfrac{10}{13}, \qquad a = \dfrac{5}{13}$

Linear equation: $y_c = \dfrac{10}{13}x + \dfrac{5}{13} = 0.77x + 0.38$

(c) $y_c = 0.77x + 0.38$

x	y_c
0	0.38
1	1.15
10	8.08

The straight line based on the equation is shown in the graph.

(d) $x = \$10$

$\quad y_c = 0.77(10) + 0.38 = \$8.08$ (Estimated weekly spending on recreation)

Exercise 20–1 (p. 433)

1. $8x + C$

3. $-6x + C$

5. $\dfrac{x^4}{4} + C$

7. $-x^{-1} + C = -\dfrac{1}{x} + C$

9. $3\left(\dfrac{x^3}{3}\right) + C = x^3 + C$

11. $-21\left(\dfrac{x^{-3}}{-3}\right) + C = 7x^{-3} + C$

13. $\dfrac{x^3}{3} + 5x + C$

15. $\dfrac{x^4}{4} - 2x^2 + C$

17. $20\left(\dfrac{x^4}{4}\right) + 12\left(\dfrac{x^3}{3}\right) - 3x + C = 5x^4 + 4x^3 - 3x + C$

19. $8x\Big|_1^3 = 8(3) - 8(1) = 24 - 8 = 16$

21. $\dfrac{x^2}{2}\Big|_0^4 = \dfrac{16}{2} - 0 = 8$

23. $\dfrac{x^4}{4}\Big|_{-1}^2 = \dfrac{16}{4} - \dfrac{1}{4} = \dfrac{15}{4} = 3\dfrac{3}{4}$

25. $x^3\Big|_2^3 = 27 - 8 = 19$

27. $\left(\dfrac{x^3}{3} + 5x\right)\Big|_1^4 = \left(\dfrac{64}{3} + 20\right) - \left(\dfrac{1}{3} + 5\right) = 36$

29. $(x^3 - x^2 + 5x)\Big|_1^2 = (8 - 4 + 10) - (1 - 1 + 5) = 14 - 5 = 9$

Exercise 20–2 (p. 440)

1. Let $u = x^2 + 1$

$$\frac{du}{dx} = 2x$$

$$du = 2x(dx)$$

$$x(dx) = \frac{du}{2}$$

$$\int (x^2 + 1)^4 x \, dx = \int u^4 \frac{du}{2} = \frac{1}{2} \int u^4 \, du = \frac{1}{2} \left(\frac{u^5}{5}\right) + C = \frac{(x^2 + 1)^5}{10} + C$$

3. Let $u = x^5 + 7$; $du = 5x^4 \, dx$

$$\int \frac{15x^4}{x^5 + 7} \, dx = 3 \int \frac{5x^4}{x^5 + 7} \, dx = 3 \int u^{-1} \, du = 3 \ln u + C$$
$$= 3 \ln (x^5 + 7) + C$$

5. Let $u = x^2$; $du = 2x \, dx$

$$\int 6xe^{x^2} \, dx = 3 \int e^{x^2} 2x \, dx = 3 \int e^u \, du = 3 e^u + C = 3 e^{x^2} + C$$

7. Let $u = 3x^2 + 5$; $du = 6x \, dx$.

$$\int 2x(3x^2 + 5) \, dx = \frac{1}{3} \int (3x^2 + 5) 6x \, dx = \frac{1}{3} \int u \, du$$

$$= \frac{1}{3} \left(\frac{u^2}{2}\right) + C = \frac{(3x^2 + 5)^2}{6} + C$$

9. $\qquad \int_1^5 2x(3x^2 + 5) \, dx = \frac{(3x^2 + 5)^2}{6}\Big|_1^5 = \frac{(3(5)^2 + 5)^2}{6} - \frac{(3(1)^2 + 5)^2}{6}$

$$= \frac{6{,}400}{6} - \frac{64}{6} = 1{,}056 \quad \text{(See problem 7.)}$$

11. Let $u = x$ and $v' = e^{ax}$. Then,

$$u' = 1 \text{ and } v = \int e^{ax} \, dx = \frac{e^{ax}}{a} + C$$

$$\int x e^{ax} \, dx = x\left(\frac{e^{ax}}{a}\right) - \int 1\left(\frac{e^{ax}}{a}\right) dx = \frac{x \, e^{ax}}{a} - \frac{1}{a}\left(\frac{e^{ax}}{a}\right)$$

$$= \frac{e^{ax}}{a}\left(x - \frac{1}{a}\right) + C$$

13. Let $u = x$ and $v' = e^{-x}$. Then,

$$u' = 1 \text{ and } v = \int e^{-x} \, dx = \frac{e^{-1x}}{-1} = -e^{-x}$$

$$\left(\int e^{ax} \, dx = \frac{e^{ax}}{a} + C \text{ Here } a = -1.\right)$$

$$\int xe^{-x}\,dx = x(-e^{-x}) - \int 1(-e^{-x})\,dx = -xe^{-x} + \int e^{-x}\,dx$$

$$= -xe^{-x} + \frac{e^{-x}}{-1} + C = -e^{-x}(x+1) + C$$

15. Let $u = \ln x$ and $v' = x^2$. Then,

$$u' = \frac{1}{x} \text{ and } v = \int x^2\,dx = \frac{x^3}{3}$$

$$\int (\ln x)x^2\,dx = (\ln x)\left(\frac{x^3}{3}\right) - \int \frac{1}{x}\left(\frac{x^3}{3}\right)\,dx = \frac{x^3 \ln x}{3} - \frac{1}{3}\int x^2\,dx$$

$$= \frac{x^3 \ln x}{3} - \frac{1}{3}\left(\frac{x^3}{3}\right) + C = \frac{x^3 \ln x}{3} - \frac{x^3}{9} + C$$

17. $\displaystyle\int_0^2 xe^{-x}\,dx = -e^{-x}(x+1)\Big|_0^2 = (-e^{-2}(2+1)) - (-e^{-0}(0+1))$

$$= -3e^{-2} - (-1) = 1 - 3e^{-2} \quad \text{(See problem 13.)}$$

19. $\displaystyle\int \frac{2x-3}{x^2-3x+5}\,dx = \ln(x^2-3x+5) + C$ (#6)

21. $\displaystyle\int e^x\,dx = e^x + C$ (#10)

23. $\displaystyle\int e^{5x}\,dx = \frac{e^{5x}}{5} + C$ (#8)

25. $\displaystyle\int \ln x\,dx = x(\ln x) - x + C$ (#14)

27. $\displaystyle\int \frac{1}{x \ln x}\,dx = \ln(\ln x) + C$ (#16)

29. $\displaystyle\int \frac{1}{16-x^2}\,dx = \frac{1}{8}\ln\frac{4+x}{4-x} + C.$ (#20: $x^2 < 16$)

Exercise 20-3 (p. 453)

1.

$y = 5$

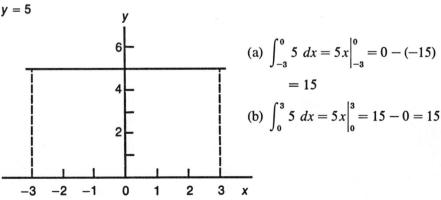

(a) $\displaystyle\int_{-3}^0 5\,dx = 5x\Big|_{-3}^0 = 0 - (-15)$

$$= 15$$

(b) $\displaystyle\int_0^3 5\,dx = 5x\Big|_0^3 = 15 - 0 = 15$

3. $y = x - 5$

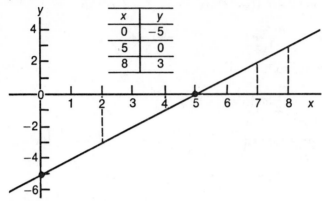

x	y
0	-5
5	0
8	3

(a) $\displaystyle\int_2^8 (x - 5)\, dx$

$$= \left(\frac{x^2}{2} - 5x\right)\Big|_2^8$$

$$= (32 - 40) - (2 - 10)$$

$$= 0$$

(b) $\displaystyle\int_4^7 (x - 5)\, dx$

$$= \left(\frac{x^2}{2} - 5x\right)\Big|_4^7$$

$$= (24.5 - 35) - (8 - 20)$$

$$= -10.5 + 12 = 1.5$$

5. $y = x^3 + x^2 - 2x$

x	y
0	0
½	-⅝
1	0
2	8
-1	2
-2	0

(a) $\displaystyle\int_{-2}^1 (x^3 + x^2 - 2x)\, dx$

$$= \left(\frac{x^4}{4} + \frac{x^3}{3} - x^2\right)\Big|_{-2}^1$$

$$= \left(\frac{1}{4} + \frac{1}{3} - 1\right) - \left(4 - \frac{8}{3} - 4\right)$$

$$= \frac{9}{4}$$

(b) $\displaystyle\int_{-1}^{1} (x^3 + x^2 - 2x)\, dx = \left(\frac{x^4}{4} + \frac{x^3}{3} - x^2\right)\Big|_{-1}^{1}$

$$= \left(\frac{1}{4} + \frac{1}{3} - 1\right) - \left(\frac{1}{4} - \frac{1}{3} - 1\right) = \frac{2}{3}$$

7. $y = x^2 + 5$

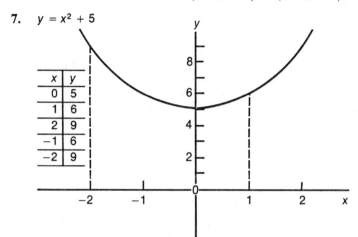

x	y
0	5
1	6
2	9
−1	6
−2	9

$\displaystyle\int_{-2}^{1} (x^2 + 5)\, dx$

$= \left(\dfrac{x^3}{3} + 5x\right)\Big|_{-2}^{1}$

$= \left(\dfrac{1}{3} + 5\right) - \left(\dfrac{-8}{3} - 10\right)$

$= 18$

9. $y = x^2 + 3$

x	y
0	3
1	4
2	7
−1	4
−2	7

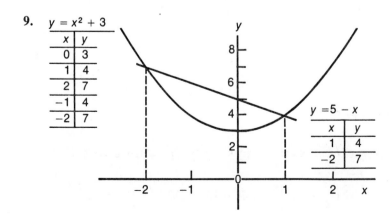

$y = 5 - x$

x	y
1	4
−2	7

$$x^2 + 3 = 5 - x$$

$$x^2 + x - 2 = 0$$
$$(x - 1)(x + 2) = 0$$

$$x - 1 = 0, x = 1$$
$$x + 2 = 0, x = -2$$

$$A_1 = \int_{-2}^{1} (5 - x)\, dx = \left(5x - \frac{x^2}{2}\right)\Big|_{-2}^{1} = \left(5 - \frac{1}{2}\right) - (-10 - 2) = 16\tfrac{1}{2}$$

$$A_2 = \int_{-2}^{1} (x^2 + 3)\, dx = \left(\frac{x^3}{3} + 3x\right)\Big|_{-2}^{1} = \left(\frac{1}{3} + 3\right) - \left(\frac{-8}{3} - 6\right) = 12$$

$$A = A_1 - A_2 = 16\tfrac{1}{2} - 12 = 4\tfrac{1}{2}$$

11.

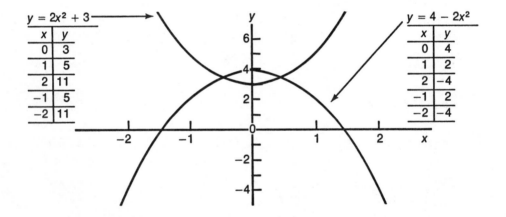

$y = 2x^2 + 3$	
x	y
0	3
1	5
2	11
-1	5
-2	11

$y = 4 - 2x^2$	
x	y
0	4
1	2
2	-4
-1	2
-2	-4

$$2x^2 + 3 = 4 - 2x^2$$

$$4x^2 = 1, x^2 = \frac{1}{4}$$

$$x = \pm\frac{1}{2}$$

$$A = \int_{-1/2}^{1/2} (4 - 2x^2) - (2x^2 + 3)\, dx$$

$$= \int_{-1/2}^{1/2} (1 - 4x^2)\, dx = \left(x - \frac{4x^3}{3}\right)\Big|_{-1/2}^{1/2}$$

$$= \left(\frac{1}{2} - \frac{1}{6}\right) - \left(-\frac{1}{2} + \frac{1}{6}\right) = \frac{2}{3}$$

13. $y = x - 5 = 0, x = 5$

(a) $\int_{2}^{5} (x - 5)\, dx = \left(\frac{x^2}{2} - 5x\right)\Big|_{2}^{5} = (12.5 - 25) - (2 - 10) = -4.5$

$\int_{5}^{8} (x - 5)\, dx = \left(\frac{x^2}{2} - 5x\right)\Big|_{5}^{8} = (32 - 40) - (12.5 - 25) = 4.5$

$|A| = |-4.5| + |4.5| = |9|$

(b) $\int_4^5 (x-5)\, dx = \left(\frac{x^2}{2} - 5x\right)\Big|_4^5 = (12.5 - 25) - (8 - 20) = -.5$

$\int_5^7 (x-5)\, dx = \left(\frac{x^2}{2} - 5x\right)\Big|_5^7 = (24.5 - 35) - (12.5 - 25) = 2$

$|A| = |-.5| + |2| = |2.5|$

15. $y = x^3 + x^2 - 2x = x(x+2)(x-1) = 0$

$\qquad x = 0,\ x = -2,\ x = 1$

y is positive when x is between -2 and 0.

$\int_{-2}^0 (x^3 + x^2 - 2x)\, dx = \left(\frac{x^4}{4} + \frac{x^3}{3} - x^2\right)\Big|_{-2}^0 = 0 - \left(4 - \frac{8}{3} - 4\right) = \frac{8}{3}$

y is negative when x is between 0 and 1.

$\int_0^1 (x^3 + x^2 - 2x)\, dx = \left(\frac{x^4}{4} + \frac{x^3}{3} - x^2\right)\Big|_0^1 = \left(\frac{1}{4} + \frac{1}{3} - 1\right) - 0 = -\frac{5}{12}$

$|A| = \left|\frac{8}{3}\right| + \left|-\frac{5}{12}\right| = \left|\frac{37}{12}\right|$

17. $F(x) = \text{Total revenue} = \int (-.1x + 12)\, dx = -.05x^2 + 12x + C$

$\qquad\qquad$ If $x = 0$, total revenue $= 0$. Thus, $C = 0$.

(a) $x = 100$ gallons

$\qquad$ Total revenue $= -.05(100)^2 + 12(100) = -500 + 1,200 = \$700$

(b) $x = 200$ gallons

$\qquad$ Total revenue $= -.05(200)^2 + 12(200) = -2,000 + 2,400 = \$400$

(c) $F'(x) = f(x) = -.1x + 12 = 0$

$\qquad x = 120$ gallons $-$ maximum since $F''(x) = -.1$ (negative)

(d) $x = 120$ gallons

$\qquad$ Total revenue $= -.05(120)^2 + 12(120) = -720 + 1,440 = \$720$

19. Let $x =$ number of sewing machines produced,

$\qquad f(x) = \$50$, the marginal cost.

$$\text{Total cost} = \int 50\, dx = \$50 + C$$

$$C = \$2,000, \text{ the fixed cost}$$

The total cost of producing 100 machines is:

$$\$50(100) + \$2,000 = \$7,000$$

The average cost per machine is:

$$\$7,000 \div 100 = \$70$$

21. $16 = 32 - x^2$

$x^2 = 16$

$x = \pm 4$

$$\int_0^4 ((32 - x^2) - 16)\, dx = \int_0^4 (16 - x^2)\, dx$$

$$= \left(16x - \frac{x^3}{3}\right)\Big|_0^4 = \left(64 - \frac{64}{3}\right) - 0 = 42\frac{2}{3} = \$42.67 \text{ (Consumers' surplus)}$$

23. $2x^2 = 16 - 2x^2$ When $x = 2$,

$\quad 4x^2 = 16$ $\quad y = 2x^2 = 2(2)^2 = 8$ (based on

$\quad\ \ x^2 = 4$ demand function),

$\quad\ \ x = \pm 2$ or $y = 16 - 2x^2 = 16 - 2(2)^2$

$\qquad\qquad\qquad\qquad\qquad = 8$ (based on supply function).

(a) The consumers' surplus:

$$\int_0^2 ((16 - 2x^2) - 8)\, dx = \int_0^2 (8 - 2x^2)\, dx$$

$$= \left(8x - \frac{2x^3}{3}\right)\Big|_0^2 = \left(16 - \frac{16}{3}\right) - 0 = 10\frac{2}{3} = \$10.67$$

(b) The producers' surplus:

$$\int_0^2 (8 - 2x^2)\, dx = 10\frac{2}{3} = \$10.67$$

25. (a) The exact method:

Here $R = \$100$ per year, $j = 8\% = .08$, $n = 10$ annual payments.

Combine formulas (14–6) and (14–8),

$$S_n = R \cdot \frac{e^{jn} - 1}{e^j - 1} = 100 \cdot \frac{e^{.08(10)} - 1}{e^{.08} - 1} = 100 \cdot \frac{e^{.8} - 1}{e^{.08} - 1}$$

$$= 100 \cdot \frac{2.22554093 - 1}{1.08328707 - 1} = 100(14.71466015) = \$1,471.47$$

(b) The approximate method:

$$\int_{-.5}^{9.5} 100(e^{.08x})\, dx = 100 \left(\frac{e^{.08x}}{.08}\right)\Big|_{-.5}^{9.5}$$

$$= 1{,}250(e^{.76} - e^{-.04})$$
$$= 1{,}250(2.13827622 - .96078944)$$
$$= 1{,}250(1.17748678)$$
$$= 1{,}471.85847500 \text{ or } \$1{,}471.86$$

Index

A

B

C